Recent Advances in Biotechnology

(Volume 8)

Recent Progress in Pharmaceutical Nanobiotechnology: A Medical Perspective

Edited by

Habibe Yılmaz

Department of Pharmaceutical Biotechnology
Faculty of Pharmacy
Trakya University
Edirne
Türkiye

Recent Advances in Biotechnology

(Volume 8)

Recent Progress in Pharmaceutical Nanobiotechnology: A Medical Perspective

Editor: Habibe Yilmaz

ISSN (Online): 2468-5372

ISSN (Print): 2468-5364

ISBN (Online): 978-981-5179-42-2

ISBN (Print): 978-981-5179-43-9

ISBN (Paperback): 978-981-5179-44-6

Published by Bentham Science Publishers Pte. Ltd. Singapore. All Rights Reserved.

First published in 2023.

need for a court order if at any point you breach any terms of this License Agreement. In no event will any delay or failure by Bentham Science Publishers in enforcing your compliance with this License Agreement constitute a waiver of any of its rights.

3. You acknowledge that you have read this License Agreement, and agree to be bound by its terms and conditions. To the extent that any other terms and conditions presented on any website of Bentham Science Publishers conflict with, or are inconsistent with, the terms and conditions set out in this License Agreement, you acknowledge that the terms and conditions set out in this License Agreement shall prevail.

Bentham Science Publishers Pte. Ltd.
80 Robinson Road #02-00
Singapore 068898
Singapore
Email: subscriptions@benthamscience.net

BENTHAM SCIENCE

CONTENTS

FOREWORD

In a world that has become global and where all kinds of technology have entered our lives rapidly with the industrial revolution, it has brought many diseases with it, even though the quality of life and duration of people have increased. The fact that we have started to use technological developments more effectively has enabled us to develop the awareness of coping with all new and old diseases that we encounter. In addition to repositioning conventional drugs, biotechnological drugs and new nano-drugs are being developed from the combination of nanotechnology with biology. It has become a necessity for every individual who wants to improve herself/himself in this field to follow the data that is revealed quickly and to follow the applications in life. Current practices and research aim to increase patient compliance and to restore patients' health as soon as possible without creating any financial and social burden.

In this context, this book focuses on recent advancements and applications of nanobiotechnology in medical application areas.

This edited book covers 14 high-quality chapters. The chapters of this book have a large variety of interesting and relevant subjects such as precision medicine, biomimetic design, exosomes, glycobiology, targeting strategies of nanomedicines, biocompatibility, *in vitro* and *in vivo* evaluation of nanomedicines as well as specific applications at different pathologies and technologies such as photodynamic therapy and biosensors. I highly recommend this book to students, academicians, researchers and industrial organizations who are interested in the new era of medicinal applications of nanobiotechnology.

Ercüment Karasulu
Department of Pharmaceutical Technology
Faculty of Pharmacy
Ege University
Bornova, 35100
İzmir
Türkiye

PREFACE

Biotechnology is a science that has many application areas such as medical, microbial, forensic, plant, agricultural, marine and food. In recent years, developments in the medical biotechnology field have accelerated. In particular, with the emergence of knowledge on the molecular and biochemical basis of diseases, biotechnological drugs have been preferred over conventional drugs. With the development of omics technologies, our increasing knowledge about diseases has revealed that diseases should be handled individually rather than in general. This, in turn, has given rise to precision medicine and the need for more specific tools for more effective treatment. Based on this information, the range of pharmaceutical products developed by medical biotechnology has expanded from recombinant proteins to nano drug delivery systems. Nanobiotechnological approaches allow many alternatives in terms of both treatment and diagnosis. There are many advantages going down to nano sizes, such as targeting, selecting, and penetrating a single cell, and providing more effective treatment with a lower drug dose.

This book aims to give the reader general information about some of the current therapeutic and diagnostic nanopharmaceutical products as well as their *in vitro* and *in vivo* applications. The reader will be able to grasp the wide possibilities and most current applications of nanobiotechnology.

Many different nanodrug systems, such as biomimetic systems, lipid and metallic nanoparticles, exosomes and glycoconjugates and photodynamic therapy, are included. In addition, information about the emergence of nanobiotechnology in personalized therapy, nanobiotechnological approaches in cancer stem cells and glioblastoma, and their use as biosensors will be conveyed. Emphasizing the importance of the biocompatibility and toxicological profile of these products, information about the relationship and modulation of bio corona, which is one of the most current issues of recent years, will also be included. Following targeting strategies, which is one of the most important advantages of nanodrugs, the book is concludes with two chapters on *in vitro* and *in vivo* studies of nanopharmaceuticals.

The application areas and tools of nanobiotechnology are so wide that it has not been possible to include them all in this book. However, we expect that readers will be informed about the most current and promising approaches to nanobiotechnology from a medical point of view.

Habibe Yılmaz
Department of Pharmaceutical Biotechnology
Faculty of Pharmacy
Trakya University
Edirne
Türkiye

List of Contributors

Abdullah Tuli — Çukurova University, Faculty of Medicine, Medical Biochemistry Department, Adana, Turkey

Aslı Sade Memişoğlu — Faculty of Education, Department of Science Education, Dokuz Eylul University, İzmir, Turkey

Ayça Erek — Department of Pharmaceutical Biotechnology, Faculty of Pharmacy, Trakya University, Edirne, Türkiye

Aysa Ostovaneh — Department of Biotechnology, Ege University, Graduate School of Natural and Applied Sciences, 35100, İzmir, Türkiye

Bakiye Göker Bağca — Department of Medical Biology, Aydin Adnan Menderes University, Aydin, Turkey

Buket Özel — Ege University, Faculty of Medicine, Department of Medical Biology, Izmir, Turkey

Ceren Sarı — Department of Medical Biology, Faculty of Medicine, Karadeniz Technical University, Trabzon, Turkey

Cigir Biray Avcı — Department of Medical Biology, Ege University, Izmir, Turkey

Elvan Bakar — Trakya University, Faculty of Pharmacy, Department of Basic Science, Edirne, Turkey

Emine Taşhan — Zoleant LLC, Zoeuticals, Regulatory Affairs and Quality, New York, USA

Fulya Oz Tuncay — Karadeniz Technical University, Faculty of Science, Department of Chemistry, Trabzon, Türkiye

Figen Celep Eyüpoğlu — Department of Medical Biology, Faculty of Medicine, Karadeniz Technical University, Trabzon, Turkey

Güliz Ak — Department of Biochemistry, Faculty of Science, Ege University, 35100, İzmir, Türkiye

Habibe Yılmaz — Department of Pharmaceutical Biotechnology, Faculty of Pharmacy, Trakya University, Edirne, Türkiye

Hande Balyapan — Department of Biochemistry, Faculty of Science, Ege University, 35100, İzmir, Türkiye

Nebiye Pelin Türker — Development Application and Research Center, Trakya University, Technology Research, 22030, Edirne, Türkiye

Nur Selvi Günel — Ege University, Faculty of Medicine, Department of Medical Biology, Izmir, Turkey

Özlem Çoban — Department of Pharmaceutical Technology, Faculty of Pharmacy, Karadeniz Technical University, Trabzon, Türkiye

Sezgi Kıpçak — Ege University, Faculty of Medicine, Department of Medical Biology, Izmir, Turkey

Sultan Eda Kuş — Faculty of Engineering, Department of Genetics and Bioengineering, Alanya Alaaddin Keykubat University, Antalya, Turkey

Şenay Hamarat Şanlıer	Ege University, Faculty of Science, Department of Biochemistry, İzmir, Turkey
Şeref Akay	Faculty of Engineering, Department of Genetics and Bioengineering, Alanya Alaaddin Keykubat University, Antalya, Turkey
Tuğba Karakayalı	Ege University, Faculty of Science, Department of Biochemistry, Izmir, Turkey
Ummuhan Cakmak	Karadeniz Technical University, Faculty of Science, Department of Chemistry, Trabzon, Türkiye
Ümmühan Fulden Aydın	Çukurova University, Faculty of Medicine, Medical Biochemistry Department, Adana, Turkey
Yakup Kolcuoğlu	Karadeniz Technical University, Faculty of Science, Department of Chemistry, Trabzon, Türkiye
Yeliz Yıldırım	Department of Chemistry, Faculty of Science, Ege University, İzmir, Türkiye Center of Drug, R&D, and Pharmacokinetic Aplications (ARGEFAR), Ege University, İzmir, Türkiye
Zehra Tavşan	Faculty of Education, Department of Science Education, Dokuz Eylul University, İzmir, Turkey
Zeynep Erim	Department of Biotechnology and Genetics, Trakya University, Institute of Natural and Applied Sciences, 22030, Edirne, Türkiye

.

CHAPTER 1

Bioinspired, Biomimetic Nanomedicines

Şenay Hamarat Şanlıer[1,*], Ayça Erek[2] and Habibe Yılmaz[2]

[1] *Ege University, Faculty of Science, Department of Biochemistry, İzmir, Turkey*

[2] *Department of Pharmaceutical Biotechnology, Faculty of Pharmacy, Trakya University, Edirne, Türkiye*

Abstract: Bio-inspired nanotechnology (biomimetic nanotechnology) is defined as the acquisition of nanomaterials or nanodevices and systems using the principles of biology during design or synthesis. Transferring a mechanism, an idea, or a formation from living systems to inanimate systems is an essential strategy. In this context, nanoparticles inspired by nature have many advantages, such as functionality, biocompatibility, low toxicity, diversity, and tolerability. It is known that biomimetic approaches have been used in materials science since ancient times. Today, it plays a crucial role in the development of drug delivery systems, imaging, and diagnostics in medical science. There is no doubt that interest and research in biomimetic approaches, which is an innovative approach and inspired by nature, will continue in the field of medicine and life sciences hereafter. Within the scope of this chapter, polymeric nanomedicines, monoclonal antibodies and related structures, cell and cell-membrane-derived biomimetic nanomedicines, bacteria-inspired nanomedicines, viral biomimetic nanomedicines, organelle-related nanomedicines, nanozymes, protein corona, and nanomedicine concepts and new developments will be elucidated.

Keywords: Bacteria-inspired, Bioinspired nanomedicine, Biomaterials, Biomimetic nanomedicine, Bionanotechnology, Cell membrane-derived, Cell, DNA, Monoclonal antibody, Nanobiotechnology, Nanomedicine, Nanoparticle, Nanotechnology, Nanozyme, Nature-inspired, Nature, Organelle-related nano-medicine, Protein corona, Viral, Virus-inspired.

INTRODUCTION

Before moving on to the topics that follow in the book chapter, terminology related to nanotechnology, biomimetics, and bioinspiration elaboration would be beneficial. In the last 10 years, standards related to nanotechnology or nanomaterials have been retrieved, published, or withdrawn. There are 223 stan-

* **Corresponding author Şenay Hamarat Şanlıer:** Ege University, Faculty of Science, Department of Biochemistry, İzmir, Turkey; Tel: +90 232 311 2323; E-mail: senay.sanlier@ege.edu.tr

Habibe Yılmaz (Ed.)

dards published or under development regarding nanotechnology on the International Organization for Standardization (ISO) website.

ISO/DIS 80004-1(en) defines the nanoscale as a size in the range of approximately 1 to 100 nm. Nanomaterials, on the other hand, are defined as materials with nanoscale external dimensions or nanoscale in their internal structure or surface structure. Nanoparticles are defined as "nanoobjects with all external dimensions in the nanoscale" [1]. ISO/TS 80004-5:2011(en) defines the relationship between nanomaterials and biology. Nanobiotechnology, bionanotechnology, and bioinspired nanotechnology are defined separately. According to the standard, nanobiotechnology is the application of nanoscience or nanotechnology to biology or biotechnology, while bionanotechnology is the application of biology to nanotechnology. Bio-inspired nanotechnology (in other words, biomimetic nanotechnology) is defined as the acquisition of nanomaterials, or nanodevices, and nanosystems by using the principles of biology during design or synthesis [2].

It has been stated in a previous review that biomimetic nanomedicines exhibited minimal interaction with the biological environment when applied at the beginning of the technology development period. Interaction with the biological environment is enhanced by modifying these relatively inert nanomedicines with targeting molecules or by designing them to respond to stimuli from the biological environment. Finally, in the review, it is stated that the third generation is obtained by coating these nanoparticles on cell membranes [3].

However, bio-inspired/biomimetic nanomedicine has moved further beyond this classification. Nanoparticles are now encapsulated in organelles or can be targeted to the organelle. In addition, they can be encapsulated into bacteria, yeast, and viruses. Virosomes, or virus-like particles (VLPs), are also among the bio-inspired structures, especially in vaccines, as part of preventive therapy [4].

In addition, as will be discussed under the following headings, many other nanomedicines, such as exosomes, nanozymes, monoclonal antibodies and derivatives, and reprogramming of dendritic cells are among the bioinspired/biomimetic nanomedicines. Lab-on-a-chip or organ-on-a-chip applications are also among biomimetic/bio-inspired nanotechnological designs. However, it will not be discussed as it is out of the scope of the chapter.

Liposomes

Liposomes were discovered in England in the 1960s by Dr. Alex D. Bangham *et al.* and published in 1964. Since its discovery, much research has been done on it, and it has taken its place in the market as a liposomal drug [5].

Liposomes are nanostructures of various lipid components in the form of a lipid bilayer, like a cell. Its core offers the advantage of encapsulating hydrophilic components, while the lipid bilayer layer offers the advantage of encapsulating hydrophobic components. At the same time, it has many other advantages, such as biocompatibility, self-forming capacity, good reproducibility, derivatization of the outer surface, and suitable physicochemical behavior. Since their discovery, apart from their conventional use, liposomes have been PEGylated, derivatized with targeting molecules, and even developed as a theranostic structure [6].

In this chapter, attention is drawn to other uses of liposomes other than their conventional synthesis and derivatization. Nowadays, liposomes are used as models to understand cell membranes, and even their qualities are taken a step further by gaining features that mimic cell membranes, which are prepared as hybrid membranes that fuse with liposomes. A group of researchers in Spain carried out a study to synthesize liposomes that mimic HeLa cell membranes to facilitate liposome cell recognition and thus deliver drugs to their targets. They proved that the presence of cholesterol in the liposome structure is important for such an interaction and that it is effective as SNARE (Soluble NSF Attachment Protein Receptor) proteins [7]. They used artificial liposomes as decoy targets for toxin neutralization against infectious diseases, thus preventing the devastating effects of infection. Researchers have shown that liposomes prepared using sphingomyelin, cholesterol, phosphatidylcholine, and phosphatidylserine protect monocytes against S. *pyogenes, S. pneumoniae,* and *S. aureus* [8]. Another group prepared reconstituted high-density lipoprotein (HDL) nanostructures containing a low concentration of ganglioside monosialotetrahexosylganglioside (GM1) lipoprotein. Thus, they were able to neutralize the cholera toxin. During their study, they determined that lipoprotein configuration is essential in receptor-toxin interaction [9].

On the other hand, fusogenic liposomes are used so that the liposomes can more effectively transport drugs to the cytoplasm or to the target. Fusogenic liposomes were designed with the knowledge that cells use membrane fusion for inter- and intracellular molecule transport. Among the important factors in the preparation of fusogenic liposomes are the use of the neutral lipid dioleoylphosphatidy-lethanolamine (DOPE) and the cationic lipid 1,2-dioleoyl-3-trimethylammon-iumpropane (DOTAP) at appropriate rates in liposome synthesis and the use of SNARE or proteins mimicking SNARE, which allows membranes to be positioned close to each other [10 - 12].

The Role of DNA in Nanomedicine Design

Deoxyribonucleic acid (DNA) is the nucleic acid that carries the hereditary information of all living things as well as some viruses. DNA is a biopolymer composed of nucleotide units. DNA is also used in the preparation of bio-inspired nanostructures due to its unique properties.

By using DNA, it is possible to obtain nanostructures of the desired size and geometry. Because it is a natural biopolymer, it is biocompatible, biodegradable, stable, and modifiable. Moreover, other features of DNA for designing a nanomedicine are its reproducibility, predictability, and scalability. Another advantage is that it can be self-assembled to the desired size and geometry by arranging the sequence. They can be modified with molecules such as proteins or signal ligands to be targeted, or they can be sensitized to the environment (such as pH) by chemical modifications. All these features make it possible to transport molecules such as drugs or siRNA within DNA nanostructures [13, 14].

One of the most striking examples of DNA nanostructures are DNA nanorobots. Researchers in China have designed a DNA nanorobot and developed an anticoagulant nanodevice. This nanorobot has the ability to sense the amount of thrombin in the blood while it is in circulation and produce a nucleic acid-based anticoagulant when the amount of thrombin rises above a certain level and releases it into the environment. They designed two main structures on their platform, all made of DNA. The first is the DNA origami structure, which acts as a cage and provides stability. The second, embedded in it, is the computing core where the molecular reaction cascade occurs. The computing core is also made up of three separate functional components: which are the Input Sensor, the Threshold Controller, and the Inhibitor Generator. The input module is the thrombin aptamer (TA-29)-DNA duplex in the ssDNA structure that reacts with thrombin. Threshold module is a DNA duplex structure that determines the threshold level. Up to a certain level, only the Input and the Threshold modules are in communication, and although the released TA-29 aptamer binds to thrombin, thrombin can maintain its normal function. When the threshold value is exceeded, the ssDNA released from the input module consumes the DNA duplex released from the threshold module and reacts with the output part of the generator module, causing ssDNA H release. In this process, the TA-15 aptamer released from the generator module has an anticoagulant effect by binding to the exocite I part, which inactivates the function of thrombin. This designed structure can regulate the anticoagulant level without requiring external dose adjustment by detecting the thrombin level in the circulation spontaneously. However, the anticoagulant effect mainly depends on the amount of DNA duplex in the threshold module. Therefore, since both the threshold value and the amount of

TA-15 to be released were dependent on the amount of DNA duplex in this module, the researchers stated that it should continue to be developed [15]. Another research team has developed a DNA nanorobot that acts as a cancer therapeutic. The DNA nanorobot causes the cancer cell to undergo necrosis by releasing the thrombin it carries into the blood vessels in the cancerous area. The aptamer in the nanorobot structure recognizes the nucleolin-1 protein overexpressed in tumor-associated endothelial cells. In addition, this interaction acts as a trigger to release the thrombin in its content [16]. In another study, a DNA nanorobot was developed that triggers the lysosomal degradation of the HER2 receptor in breast cancer, thereby driving the cell into apoptosis. After the anti-HER2 aptamer- tetrahedral framework nucleic acid (HApt-tFNA) nanorobot they designed specifically binds to HER2, it is taken into the cell by endocytosis as the HER2-HApt-tFNA complex and undergoes lysosomal degradation together with the HER2 receptor. In this process, the cell is dragged into apoptosis *via* the Akt pathway [17].

One of the major drawbacks of cancer immunotherapy that is difficulty to deliver the desired amount of antigens and adjuvants to where the immune response will be regulated. A sufficient amount of antigen and adjuvant is needed to be transported to the area where the immune response will occur. Therefore, DNA nanostructures are among the many carrier systems investigated. Researchers have prepared rectangular DNA origami using the M13 bacteriophage DNA strands. Three different peptide antigens were attached to the DNA chains, which extended from each rectangler with azide bonds and were assembled on the surface by DNA hybridization. According to the authors' findings, studies in mice induced a high and prolonged T cell response when the system containing pH-sensitive DNA sequences was fragmented in lysosomes inside antigen-presenting cells [18]. Another group developed a highly effective DNA/RNA nanovaccine in a more complex way. Two different components were used as adjuvants. Unmethylated cytosine-guanine oligonucleotides (CpG) are DNA molecules that activate APCS *via* Toll-like receptor 9 and are preferred as immunostimulators. On the other hand, since the STAT3 pathway must be suppressed to get an effective immune response, stat3 shRNA was the other adjuvant used, which would act with RNA interference. Microflowers were synthesized using a combination of rolling circle replication and rolling circle transcription techniques for the first time, to obtain hybrid DNA and RNA for the nanostructure. Since efficient transport to lymph nodes requires a reduction in size, PEG-grafted polypeptides were synthesized and integrated into the surface of the structure. PEG-grafted polypeptides used to reduce the size also allowed hydrophobic neoantigens to physically bind to the peptide through hydrophobic interactions. It has been determined that this nanovaccine, which is obtained in quite complex ways, produces an 8 times more effective T cell response compared to neoantigens [19].

Monoclonal Antibodies and Related Structures

Monoclonal antibodies (mAbs) are immunoglobulins from a single clonal pool specific to their antigen. Monoclonal antibodies are among the molecules that scientists have inspired from biology and one of the best mimicked. In this section, information will be given on the molecules included in the guidelines as well as on other structures associated with monoclonal antibodies.

As stated in the guide published by the European Medicines Agency (EMA) in 2016, technologies such as hybridoma, recombinant DNA (rDNA) technology, and phage display are used in the production of monoclonal antibodies. In addition to its antigen specificity, it also has an effector function mediated by immune system cells [20]. Contrary to the advantage of their high specificity, they have some disadvantages, such as complex and expensive production and low penetration into the target tissue due to their large molecular weight. Therefore, it has been considered to develop new biological molecules that can offer the advantages of monoclonal antibodies without exhibiting the disadvantages they have [21]. With these considerations and the developments in recombinant DNA technology, new biological molecules have been obtained that are inspired by monoclonal antibodies. Some of them have been included in the World Health Organization's (WHO) guidelines. They can be listed as follows: all mAb isotypes, antibody fragments, single-domain antibodies, bispecific or multispecific antibodies, Fc-fusion proteins, chemically modified mAbs, or related proteins [22]. Moreover, there are also monoclonal antibody-derived biomimetic nanostructures that have not yet entered the guidelines, such as diabody, affibody, nanobody, and synthetic single domain antibodies. Some of the structures related to monoclonal antibodies can be seen in Fig. (**1**).

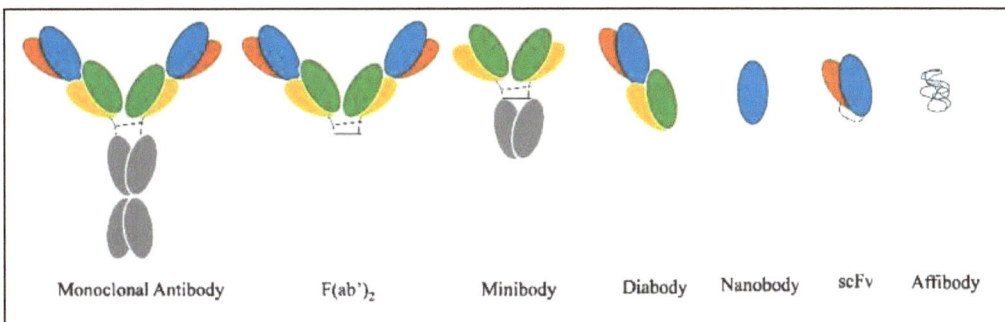

Fig. (1). Illustration of monoclonal antibody structure and its related structures.

Affibodies are non-immunoglobulin affinity proteins with a molecular weight of 6-6.5 kDa. It was discovered and patented by the Swedish company Affibody AB in 1998. Affibodies prepared as libraries are initially derived from the protein A produced by *Staphylococcus aureus* and are able to bind to the Fc region of antibodies. By changing the combination of several amino acids on one or two helix groups of the structure consisting of a three-helix bundle, molecular structures with high affinity can be obtained. These structures, which were used as molecular diagnostic tools at first, have shown that they can be used therapeutically as a result of later modification by genetic fusion to the protein or by binding of chemically toxic agents. Since their small molecular mass limits their circulation time, strategies to create disulfide bridges or increase circulation time by chemical modifications are also developed to increase their stability [23 - 25].

Nanobodies are the smallest units of nanometer-sized antibodies with a variable heavy chain structure with a single-unit, antigen-recognizing paratope region. Nanobodies used for therapeutic purposes are derived from naturally occurring heavy-chain structures in camelid species or immunoglobulin new antigen receptors in sharks [21]. Immune libraries are one of the most widely used production methods. Lymphocytes from immunized camelid species can be collected and produced recombinantly in bacteria or yeast after the sequence resulting in nanobody production is detected by a method such as phage display. In the use of naive libraries, instead of a targeting strategy as in immunization, the appropriate nanobody is screened in the pooled blood. In the use of synthetic libraries, it is aimed to produce the nanobody with the highest affinity and the lowest immunogenicity by preserving the conserved region and randomly changing the remaining CDR region in synthetic/semisynthetic ways [26].

Cell and Cell Membrane-Derived Biomimetic Nanomedicines

Many different approaches are being investigated to increase the efficacy and therapeutic index, regulate their elimination, and reduce the toxicity of nanomaterials developed for treatment or diagnosis. One of these approaches is to cover the synthesized nano-drug delivery systems with biological membranes.

For nano-drug delivery systems or therapeutic proteins to escape from the immune system and reach their target site, blood cells such as erythrocytes, monocytes, neutrophils, and platelets were the first cells to apply this concept [27, 28]. During the youth of this concept and process, within the scope of my dissertation study, urease/ALA dehydrogenase enzymes were co-encapsulated into erythrocytes to bring an alternative solution to renal failure patients that encountered hemodialysis many times. The enzymes were first covalently modified with polyethylene glycol and then encapsulated into erythrocytes by slow dialysis [29]. This concept was then taken a step further by our research team. Magnetic nanoparticles were synthesized in the presence of glucose and then encapsulated by extrusion into erythrocyte vesicles after the nano-drug carrier was coupled to doxorubicin by a hydrazone bond, making them sensitive to pH changes in the biological environment. The obtained results showed that the side effects of doxorubicin were reduced, and the targeting was successful [30]. An advantage of this concept is that it is an approach that can be preferred in personalized treatment since the patient's own blood components can be utilized. However, erythrocytes were not only used to cloak the nanoparticles. Researchers have developed red blood cell hitchhiking as a new strategy for targeting nanoparticles to desired organs. Nanomaterials with certain physicochemical properties were adsorbed on red blood cells *ex vivo* and then administered into the circulation. They showed that erythrocytes leave the nano-drug carriers in the lung capillaries, which are their first stop, and targeting to the lung increases by at least 30% [31, 32].

Blood cells are not the only cells used for cloaking nano-drug delivery systems. Tumor cells have also begun to be used in this manner. Due to the altered metabolic activities of cancer cells, the self-recognition elements they express on their cell surface also mean that they have good targeting components. In this approach, it is possible to encapsulate the nano-drug delivery system into the cancer cells of interest, and it is also possible to target the tissues in which the relevant cancer type prefers to metastasize. This improves access to organs and tissues that are difficult to reach and treat. For example, glioblastoma is a highly aggressive and difficult-to-treat tumor because therapeutics cannot cross the blood-brain barrier; thus, the chance of treatment success is very low. In the study, in which therapeutics were encapsulated into glioma cells by crossing the blood-brain barrier, it was stated that nanomedicines coated with glioma cells were promising and could offer personalized treatment [33]. Apart from the self-targeting coating strategy, the work of Gdowski *et al.* can be given as an example of targeting the tissue where cancer has a high potential to metastasize. After detecting the molecular component that is effective in metastasis to the bone as a result of screening performed on individuals with prostate cancer, prostate cancer cells were reprogrammed to express the relevant molecular component, and these reprogrammed cells were used for coating. Thus, they were able to target the nano-drug delivery systems they prepared for the bone tissue [34].

Cancer cells or hybrid cells obtained from the combination of cancer cells and healthy cells can be used for immunomodulation. Lin *et al.* engineered a biomimetic nanomedicine for the treatment of osteomyelitis by creating a hybrid vesicle derived from both macrophages and mammary carcinoma cells, which encapsulated MnOx. They demonstrated that the prepared biomimetic nanomedicine regresses bacterial infection, evoking the development of systemic antibacterial immunity and even long-term immune memory that prevents infection from recurring [35]. The strategy of developing vaccines to develop immunity against cancer cells for the treatment and prevention of cancer has been studied for a long time. 4T1 cells were used in the preparation of micrometer-sized vesicles (HMVs). Hyaluronic acid-coated dendritic polymer (HDDTs) NBs containing doxorubicin were added to 4T1 cells and incubated or sonicated. The resulting biomimetic constructs were then evaluated in terms of the immune response. They showed that the biomimetic nanomedicine they prepared triggered a systemic tumor-specific immune response and was superior to antigen-containing cancer vaccines [36].

Bacteria-Inspired Nanomedicines

Since the discovery that Coley's toxins regress cancer, many nanomedicines have been designed using the products of bacteria, whole bacteria, or fragments of membranes. Bacteria-inspired nanomedicine has been developed not only for cancer but also for the treatment of infection, diabetes mellitus, and inflammatory bowel disease. It has been shown that the chemotaxis and mobility of bacteria can actively target the drugs that they carry or that are provided by recombinant DNA technology to the target site. In addition, it has been determined that some bacteria can hijack antigen-presenting cells and direct cancer cells to immune system cells. In nanovaccines, components such as bacterial lipopolysaccharide (LPS) and exotoxin can be used as adjuvants. Still, there are safety concerns with the use of bacteria for therapeutic purposes [37].

In order to overcome the safety problems in bacterial therapy, the approach of eliminating the expression of LPS has been tried before, but it has been found that it reduces the bacterial colonization needed in clinical studies and cannot provide treatment. Therefore, instead of reducing the expression of LPS, researchers tried the approach of encapsulating bacteria with capsular polysaccharides, which ensures the survival and colonization of bacteria in the human body. Therefore, instead of reducing the expression of LPS, the researchers tried the approach of encapsulating bacteria with capsular polysaccharide, which ensures the survival and colonization of bacteria in the human body. With this approach, they achieved a 10-fold improvement in the application of dose-limiting bacteria [38].

In another study, the tumor microenvironment was rearranged using recombinant bacteria, and the success of cancer immunotherapy was increased. To increase the antitumor T cell response, researchers prepared a recombinant bacterium that converts ammonia in the cancer microenvironment to L-arginine, which is necessary for the T cell response [39].

Bacteria secrete outer membrane vesicles (OMVs) in the proteoliposome structures as part of their natural processes to communicate with their surroundings for survival. The inside of the OMVs carries the bacterial periplasm, and many bacterial biomolecules are located on the surface [40, 41]. The fact that the surfaces of OMVs are equipped with antigenic determinants to stimulate the immune system makes them a vaccine candidate. The meningococcal B vaccine, which has already been approved by the FDA in 2014, is in OMV format and is on the market [42]. To utilize OMVs as vaccines, it is crucial to produce and purify them effectively. If the OMVs produced during the period of high bacterial populations are not harvested before the death of the bacteria, they may become contaminated with residues. At this point, the importance of purifying of OMVs becomes prominent. Although classically, several-step centrifugation and precipitation are used in purification, density gradient separation and gel filtration techniques have also been used more recently. In addition to being used as a vaccine, OMVs has also been used as an adjuvant.

Inspired by the knowledge that OMVs also carry virulence factors of bacteria, the idea was born that these structures could be used as drug delivery systems. Although the encapsulation rate is not as high as expected, it can deliver therapeutics such as small molecules, antibiotics or chemotherapeutics [43, 44] and siRNA to the target site. In addition, by recombination with bacteria, targeting proteins such as affibody can be expressed on the cell surface. Thus, both targeting and therapeutic function can be achieved [45, 46].

Viral Biomimetic Nanomedicines

Another strategy used to deliver therapeutic moieties to their targets is the use of viruses or virus-related components. In fact, viruses have been in our lives for years as a preventive treatment in vaccine form. In addition to viruses, the vaccine form is also available on the market as virus-like particles. In recent years, their use as cancer immunotherapy and drug delivery systems has been investigated.

The use of viral nanovectors in cancer immunotherapy, whether in the form of whole virus or virus-like particles, constitutes one of the most current research topics. The researchers conjugated Au nanoparticles, an oligodeoxynucleotide adjuvant containing an unmethylated CpG motif, which is known to stimulate macrophages, natural killers, B cells, and dendritic cells but cannot be directly taken into the cell. This nanostructure was then encapsulated into the VLP structures obtained from the hepatitis B core protein and analyzed for the immune response. As a result, an increase in HBc-specific antibody response was obtained

[47]. Another approach has been to decorate the antigen on the VLP surface, which requires an immune response. The researchers developed an alternative solution because genetic fusion or chemical modification on the VLP surface is a lengthy procedure that can result in misfolding of the proteins of the VLP or antigen and reduce the protective response. Therefore, they synthesized a genetically encoded protein called SpyCatcher, which spontaneously covalently binds to its peptide partner, which they call SpyTag. They synthesized SpyCatcher expressing VLPs in *E. coli* and incubated them with SpyTag-conjugated malaria antigens. They determined that the prepared nanostructure improved the malarial antibody response after the first application and increased it in the second application [48].

In addition to infectious diseases, research on platforms for cancer immunotherapy or chemotherapy is also ongoing. For this purpose, mostly peptides or empty virus particles are preferred. However, Fusciello *et al.* have introduced a new approach. Ad5D24-CpG oncolytic virus was coated by extrusion with membranes of murine melanoma, lung, bladder, and ovarian cancer cells, respectively. As a result of the application in mice, it has been revealed that there is a significant slowdown in the progression of cancer for all cancer types and that the oncolytic virus is covered on the cancer membrane, increasing both the adjuvant and antigen-presenting capacity, resulting in a more positive response [49].

Plant viruses, known to be non-pathogenic to humans, are among the preferred platforms in this sense. However, the lifespan of plant viruses in the human body is limited to a few minutes. For this reason, there are studies on coating the surface with molecules such as albumin or formulating nanoparticles such as dendrimers and delivering them to their target before they are not recognized by the body's immune cells. In addition, plant viruses prefer cell lines that are sensitive to them, and therefore, care should be taken to use viruses to which the cell is susceptible in such a study [50]. Notably, brome mosaic virus (BMV) and cowpea chlorotic mottle virus (CCMV) are among the potential nanodrug carriers. In an exemplary study, the immunogenicity of PEGylated BMV and CCMV was investigated by first loading a fluorophore. After it was determined that BMV did not trigger macrophages, BMV VLPs were synthesized, and siAkt1 was encapsulated. The resulting VLP-based drug delivery system has been shown to inhibit tumor growth in mice. Thus, it was concluded that BMW, which does not stimulate the immune system, is a suitable nanocarrier [51].

Viruses and VLPs have been considered suitable constructs as drug targeting vehicles in recent years. The fact that the desired targeting molecule can already be expressed on their surface and that the desired drug can be loaded into it makes them an ideal carrier candidate. Shan *et al.* used HBc virus-like particles for this purpose. As a target for B16-F10 murine melanoma cancer, the lipophilic NS5A peptide, 6xHis tag, and tumor-targeting peptide (RGD) were expressed in the C terminal and major immunodominant loop regions of the Hepatitis B core protein using recombinant DNA technology. Doxorubicin-VLPs were produced by simultaneous encapsulation of doxorubicin while dialyzed in assembling buffer after the prepared VLP was separated into HBc-144 subunits by denaturant. The resulting targeted, DOX-encapsulated HBc VLPs have been shown to regress tumors by 90% and are less cardiotoxic than DOX [52]. Shapira *et al.*, on the other hand, developed a drug delivery system against cancer by toxin/antitoxin transport with an adenoviral system. Based on the knowledge that the *E. coli* MazF-MazE toxin-antitoxin system can eradicate RAS-mutated cancer cells, the related system was expressed in adenovirus under Ras and p53 regulation. This new platform has selective tumor regression and lack of toxicity, suggesting that it is an important advantage and a promising platform [53].

Organelle-Related Nanomedicines

Organelle-related nanomedicine should be discussed under a few headings, such as artificial organelles, organelle membrane-coated nano-drug delivery systems, and organelle targeting. It is possible to treat cancer at several compartmental levels in the cellular sense. It benefits from the knowledge that each organelle has different functions. In the cell, the lysosome is involved in autophagy, mitochondria in apoptosis, the nucleus in cell division and proliferation, and the endoplasmic reticulum (ER) in protein synthesis and transport. Biomolecules such as triphenyl phosphonium (TPP), cyclic guanidium, and mitochondrial targeting sequences (MTS) have been used in mitochondrial targeting, and many drugs such as coumarin, cisplatin, doxorubicin, and chlorambucil have been delivered [54]. It has been determined that the lysosome is a vehicle for providing the appropriate environment for the cell to perform its function as a cleaning center for intracellular waste, as well as being associated with diseases such as autoimmune diseases and cancer. Lysosome biogenesis is regulated by the transcription factor EB (TFEB), which also regulates autophagy, which plays a prominent role in immunity [55]. TFEB is also in charge of regulating the p-glycoprotein (p-gp) traffic which is responsible for drug resistance. Therefore, it is suggested that the lysosome is also associated with drug resistance and that lysosomal targeting may contribute to standard therapy performance [56]. In the targeting of lysosomes, molecules such as mannose-6-phosphate and β-glucocerebrosidase on the nanoparticle surface, as well as magnetic nanoparticles, were used [57]. The endoplasmic reticulum is involved in the synthesis and post-translational modification of proteins required for intra- and extracellular communication and traffic, as well as in the removal of unfolded/misfolded proteins associated with disease. Unfolded/misfolded proteins are directed to the cytosol by the ER and degraded by proteosomes. Otherwise, the remaining and accumulating of these faulty products in the ER lumen causes ER stress, which leads to many diseases such as diabetes, cancer, bipolar disorder, and renal diseases. With this insight, researchers realized that by targeting the ER, many diseases could be cured. Nanoparticle designs in which caveola-mediated intracellular uptake is triggered, such as proper choice of lipid composition of liposomes and targeting with small molecules such as sulfonamides, chloride, glibenclamide, or amphoteric ionic ligands, can be performed. In addition to small molecules, ER targeting was achieved with peptides such as KDEL, pardaxin from *Pardachirus marmoratus*, the ER-insertion signal sequence (Eriss), and poly(aspartic acid) (PAsp). The Lys-Lys-X-X pattern has been shown to be effective in targeting ER-targeting peptides design. As the X's herein can be any amino acid, it is obvious that the presence of Lys-Lys- is needed [58]. On the other hand, TAT and RGD peptides are frequently preferred for nucleus targeting. These peptides have been used for the modification of many nanomaterial surfaces, such as mesoporous silica nanoparticles, quantum dots, and liposomes [59]. However, one of the most important issues to be considered in intracellular and subcellular organelle targeting is that the nano-drug delivery system should be 100 nm and below.

On the other hand, over time, the idea of using functional artificial organelles as a treatment tool has emerged in the treatment of diseases related to organelle dysfunction. One of the two approaches used for this is the transfer of organelles from one organism to another, while the other is completely artificially creating the membrane and the catalytic function and integrating them into the cell. The catalytic properties of ruthenium, copper, palladium, and gold are used to obtain catalytic properties in artificially prepared organelles. The hydrophobic nature of the organometallic complexes prepared with these metals facilitates their uptake into the cell. However, the transport of enzymes into the cell requires more effort. For this reason, nanoreactors have been developed to localize both enzymes and their cofactors inside the cell. Polymer-based nanoreactors are the most commonly used for this purpose. In the preparation of the mentioned polymerosomes, triblock co-polymer poly(methyloxazoline)-poly(dimethylsiloxane)-poly(methyl-oxazoline) (PMOXA-b-PDMS-b-PMOXA), polystyrene-b-poly(iso-cyano-alanine(2-thiophen-3-yl-ethyl)amide) (PS-PIAT), polystyrene-b-poly(ethylene glycol) (PS-PEG), and poly(ethylene glycol)-b-poly(ε-caprolactone-g-trimethylene carbonate) (PEG) di- and tri-block copolymers such as -b-P(CL-g-TMC) are preferred. In addition, RGD-peptide-decorated liposomes can also be used for this purpose. Besides these structures, there are also nanocages prepared from ferritin and chaperones. Creating an artificial organelle inside the cell is another option. It has been discovered that elastin-like polypeptides containing the

VPGVG sequence spontaneously form an organelle or cell-like structure. On the other hand, it is possible to target the encapsulins, carboxysomes, and Pdu microcompartments already existing in nature with new functions and target them into the cell. Lumazine synthase, which is found in many plants and microorganisms such as encapsulins, has been used as a nanoreactor due to its ability to encapsulate riboflavin synthase as well as its catalytic properties. Cowpea chlorotic mottle virus (CCMV) capsids, Qβ, MS2, and P22 bacteriophages can also be used to form organelle-like structures [60]. One of the studies achieved redox sensitivity by encapsulating the horseradish peroxidase (HRP) enzyme with the modified outer membrane protein F (mombF) porins into polymerosomes. It has been shown that the artificial organelle is sensitive to the intracellular glutathione level and can control the ROS level [61]. In another study, catalase-encapsulated, semi-permeable, biodegradable poly(ethylene glycol)-block-poly(caprolactone-gradient-trimethylene carbonate) (PEG–PCLg-TMC) polymersomal nanoreactors functionalized with cell-penetrating peptide (CPP) were prepared. It has been shown that these nanoreactors can protect human complex-I-deficient primary fibroblasts from hydrogen peroxide damage [62].

Nanozymes

Nanozymes are artificial enzymes that mimic enzymes. They are nanomaterials obtained from cerium, ferrum, copper, manganese, molybdenum, platinum, cobalt, gold, iridium, and ruthenium complexes. The concept of nanozymes came into our lives for the first time in 2007 with the announcement that Fe_3O_4 exhibits enzyme-like activity [63, 64]. It has been shown that the catalase- (CAT), peroxidase- (POD), cytochrome c oxidase (COX)-, and superoxide dismutase-mimetic (SOD) activities of cerium nanozymes are affected by the reduction and oxidation states. Cerium compounds, which can be used in different morphologies, also increase the serum levels of tumor necrosis factor alpha (TNF-α) and lactate dehydrogenase (LDH) and thus show cytotoxicity. It has been shown that ferrum compounds can regulate ROS levels through catalase and peroxidase activities. While nanozymes synthesized with copper also showed activity like those synthesized with ferrum, it was also determined that when synthesized in the presence of phenylalanine, they exhibited multienzyme activity such as GPx, POD, and superoxide dismutase (SOD). Nanozyme structures synthesized from manganese, molybdenum, gold, iridium, and platinum also show more catalase and peroxidase activities. It is obvious that nanozymes obtained from metallic components exert their effects by scavenging or revealing reactive oxygen species, and their morphology also contributes to cytotoxicity [63]. However, it has been discovered that carbon-based nanomaterials other than metals also have enzyme-mimicking properties. Carbon-based nanostructures exhibiting nanozyme activity are nanomaterials such as carbon nanotubes, fullerenes, graphene oxides, and carbon nitrite. These structures also exhibit catalase, peroxidase, superoxide dismutase, and hydrolase activities [64]. The catalytic activity of nanozymes can also be increased by synthesizing them as bimetallic alloy nanocages, and they are generally preferred as a photodynamic therapy tool [65, 66].

Although these structures are stable, biocompatibility problems may be encountered. In addition, their small size causes them to be rapidly eliminated from circulation by the renal route and to leave the system without showing the desired effect. Strategies such as coating with polymers or encapsulation into cellular membranes can be used to increase the biocompatibility of these metallic nanostructures with nanozyme activity [67 - 70].

Protein Corona and Nanomedicine

In recent years, special attention has been drawn to the influence of the biological environment in which nanoparticles are applied on their interactions at the organismal and cellular levels. Due to their high surface-free energy, nanoparticles absorb biomolecules when they contact a biological or abiotic environment, and the interaction environment changes the bio-identity of intact nanoparticles. This coating layer formed by biomolecules in biological fluids was named protein corona by KA Dawson in 2007. In recent years, the term biomolecular corona has also been used due to the complexity of biological fluids [71, 72].

Formation of protein corona: in addition to the physicochemical properties of nanoparticles such as shape, size, and surface charge, it also depends on the ratio of nanoparticles and proteins, the type of medium, and the different molecules that mix between nanoparticles and proteins [73].

This layer, called the protein corona, which is formed by the incorporation of nanoparticles into biological fluids, consists of many biomolecules such as albumin, complementary proteins, and apolipoproteins, which will recreate the biological identity of nanoparticles. Protein corona formation occurs rapidly, randomly, and dynamically when it interacts with the biological fluid, as determined by the physicochemical properties of the nanoparticles, and is irreversible. Due to the change in bio-identity of the nanoparticles, their biological fate, and therapeutic effects such as circulation time, biodistribution, stability, immune system activation, cellular uptake, and targeting effect will also change [74, 75].

The main purpose of nanoparticles developed as drug delivery systems is to show the maximum effect by using the minimum dose in the targeted tissue and to reduce the drug distribution in normal tissues. Two basic methods, active and passive targeting approaches, are used for targeting nanoparticles. To briefly mention, in passive targeting, nanoparticles are transported to the target site through natural physiological processes and passive factors. The physicochemical properties of nanoparticles, such as size, charge, and shape, are crucial in passive targeting. As a result of various modifications made on the structure of the drug carrier nanoparticle, its delivery to specific cells, tissues, and organs is defined as active targeting. In active targeting nanoparticles, strategies such as chemical (pH, reactive oxygen species, proteases) and physical factors (heat, magnetic field, ultrasound) or targeting with cell-specific binding are utilized [76 - 78].

The protein corona on the nanoparticle surface, the first part encountered by the biological system, can directly affect the targeting abilities of nanoparticles. Nanoparticles coated with proteins called opsonins can be easily recognized by the mononuclear phagocyte system and easily removed from the blood by activating the immune system. Another negative effect of protein corona is the inhibition of nanoparticle targeting. The surface of nanoparticles is modified for targeting using various ligands, but the formation of protein corona may lose the targeting ability with this ligand. Protein corona may appear as a disease mechanism as well as reducing its therapeutic effects by affecting the targeting methods of nanoparticles. Studies show that the specific composition of the protein corona may mediate a pathogenic process in a clinically relevant disease [79].

Besides the existing and proven adverse effects of protein corona on the therapeutic effects of nanoparticles, there are ways that can be used to enhance the therapeutic effects of nanoparticles. Recent studies have shown that the mechanism underlying the success of selective organ targeting is the specific protein corona around the nanoparticles. Controlling protein corona formation and composition increases the circulation time of nanoparticles, further demonstrating the potential to avoid non-specific cellular uptake. When proteins with affinity for specific receptors can be incorporated into the protein corona, nanoparticles can overcome the weaknesses of current targeting approaches, giving nanoparticles tremendous targeting capability [77, 80].

Considering the creation of a protein corona resulting from nanoparticles' engagement with biological fluids, it is viable to create nanoparticles that possess biocompatibility, biosafety, suitable biodistribution and increased therapeutic effectiveness. Currently, the use of protein corona for specific therapeutic purposes is a promising strategy. Studies involving the biomimetic approach in nanomedicine by utilizing the protein corona have increased recently [81].

The most widely used biomimetic approach using the protein corona is cell membrane decoration. Red blood cells (RBC), white blood cells (WBC), platelets, and exosomes are the main cell types used for cell membrane decoration. With this method, nanoparticles imitate the functionality of various cell types, allowing them to stay in circulation for a longer time and reduce unwanted immune responses [82].

Endogenous protein coating is another approach performed using the protein corona in the biomimetic approach. The most abundant component of the protein corona is protein. The endogenous protein coating allows for great control biological the formation of protein corona to achieve several specific therapeutic effects. The endogenous coating increases the targeting ability of the nanoparticles as well as allows them to stay in the circulation longer and show lower immune activation [82].

Modification of protein corona components with biomolecules without changing their functions is another widely used method. Many biomolecules are used in this method. Biomimetic peptic modification is one of the most widely used methods. The objective of biomolecule modification is to confer selectivity and targeting capabilities on nanoparticles in biological fluids. [82].

CONCLUDING REMARKS

This section is a review of nano-drug delivery systems inspired by biology and defined as biomimetics. With the introduction of nanotechnology into our lives and its application in the field of medicine, many nanomedicines have been developed. While nanomedicine was mostly in the form of static structures in its early period, more dynamic and organism-sensitive systems are being developed today. As our knowledge of biological systems increases, more rational designs can be made that are stimulus-sensitive, targeted, and able to evade the immune system. Although polymeric nanoparticles are also early-time bio-inspired structures, information and examples of emerging technologies are presented due to the development of more recent technologies. We hope readers in this field will benefit from this chapter.

REFERENCES

[1]   ISO/DIS 80004-1(en), "Nanotechnologies – Vocabulary — Part 1: Core terms and definitions", Available From: https://www.iso.org/obp/ui/#iso:std:iso:80004:-1:dis:ed-1:v1:en/ (Accessed 08.12.2022).

[2]   ISO/TS 80004-5:2011(en), "Nanotechnologies — Vocabulary — Part 5: Nano/bio interface",

Available From: https://www.iso.org/obp/ui/#iso:std:iso:80004:-1:dis:ed-1:v1:en/ (Accessed 08.12.2022).

[3] K.G. Gareev, D.S. Grouzdev, V.V. Koziaeva, N.O. Sitkov, H. Gao, T.M. Zimina, and M. Shevtsov, "Biomimetic nanomaterials: Diversity, technology, and biomedical applications", *Nanomaterials,* vol. 12, no. 14, pp. 2485-2515, 2022.
[http://dx.doi.org/10.3390/nano12142485] [PMID: 35889709]

[4] C. Vauthier, and D. Labarre, "Modular biomimetic drug delivery systems", *J. Drug Deliv. Sci. Technol.,* vol. 18, no. 1, pp. 59-68, 2008.
[http://dx.doi.org/10.1016/S1773-2247(08)50008-6]

[5] D. Guimarães, A. Cavaco-Paulo, and E. Nogueira, "Design of liposomes as drug delivery system for therapeutic applications", *Int. J. Pharm.,* vol. 601, p. 120571, 2021.
[http://dx.doi.org/10.1016/j.ijpharm.2021.120571] [PMID: 33812967]

[6] L. Sercombe, T. Veerati, F. Moheimani, S.Y. Wu, A.K. Sood, and S. Hua, "Advances and challenges of liposome assisted drug delivery", *Front. Pharmacol.,* vol. 6, p. 286, 2015.
[http://dx.doi.org/10.3389/fphar.2015.00286] [PMID: 26648870]

[7] A. Botet-Carreras, M.T. Montero, J. Sot, Ò. Domènech, and J.H. Borrell, "Engineering and development of model lipid membranes mimicking the HeLa cell membrane", *Colloids Surf. A Physicochem. Eng. Asp.,* vol. 630, p. 127663, 2021.
[http://dx.doi.org/10.1016/j.colsurfa.2021.127663]

[8] B.D. Henry, D.R. Neill, K.A. Becker, S. Gore, L. Bricio-Moreno, R. Ziobro, M.J. Edwards, K. Mühlemann, J. Steinmann, B. Kleuser, L. Japtok, M. Luginbühl, H. Wolfmeier, A. Scherag, E. Gulbins, A. Kadioglu, A. Draeger, and E.B. Babiychuk, "Engineered liposomes sequester bacterial exotoxins and protect from severe invasive infections in mice", *Nat. Biotechnol.,* vol. 33, no. 1, pp. 81-88, 2015.
[http://dx.doi.org/10.1038/nbt.3037] [PMID: 25362245]

[9] D.A. Bricarello, E.J. Mills, J. Petrlova, J.C. Voss, and A.N. Parikh, "Ganglioside embedded in reconstituted lipoprotein binds cholera toxin with elevated affinity", *J. Lipid Res.,* vol. 51, no. 9, pp. 2731-2738, 2010.
[http://dx.doi.org/10.1194/jlr.M007401] [PMID: 20472870]

[10] J. Yang, A. Bahreman, G. Daudey, J. Bussmann, R.C.L. Olsthoorn, and A. Kros, "Drug delivery *via* cell membrane fusion using lipopeptide modified liposomes", *ACS Cent. Sci.,* vol. 2, no. 9, pp. 621-630, 2016.
[http://dx.doi.org/10.1021/acscentsci.6b00172] [PMID: 27725960]

[11] M. Farid, T. Faber, D. Dietrich, and A. Lamprecht, "Cell membrane fusing liposomes for cytoplasmic delivery in brain endothelial cells", *Colloids Surf. B Biointerfaces,* vol. 194, p. 111193, 2020.
[http://dx.doi.org/10.1016/j.colsurfb.2020.111193] [PMID: 32592944]

[12] T. Zheng, Y. Chen, Y. Shi, and H. Feng, "High efficiency liposome fusion induced by reducing undesired membrane peptides interaction", *Open Chem.,* vol. 17, no. 1, pp. 31-42, 2019.
[http://dx.doi.org/10.1515/chem-2019-0004]

[13] F. Xu, Q. Xia, and P. Wang, "Rationally designed DNA nanostructures for drug delivery", *Front Chem.,* vol. 8, pp. 751-764, 2020.
[http://dx.doi.org/10.3389/fchem.2020.00751] [PMID: 33195016]

[14] S. Raniolo, F. Iacovelli, V. Unida, A. Desideri, and S. Biocca, "*In Silico* and in cell analysis of openable DNA nanocages for miRNA silencing", *Int. J. Mol. Sci.,* vol. 21, no. 1, pp. 61-72, 2019.
[http://dx.doi.org/10.3390/ijms21010061] [PMID: 31861821]

[15] L. Yang, Y. Zhao, X. Xu, K. Xu, M. Zhang, K. Huang, H. Kang, H. Lin, Y. Yang, and D. Han, "An Intelligent DNA Nanorobot for Autonomous Anticoagulation", *Angew. Chem. Int. Ed.,* vol. 59, no. 40, pp. 17697-17704, 2020.
[http://dx.doi.org/10.1002/anie.202007962] [PMID: 32573062]

[16]   S. Li, Q. Jiang, S. Liu, Y. Zhang, Y. Tian, C. Song, J. Wang, Y. Zou, G.J. Anderson, J.Y. Han, Y. Chang, Y. Liu, C. Zhang, L. Chen, G. Zhou, G. Nie, H. Yan, B. Ding, and Y. Zhao, "A DNA nanorobot functions as a cancer therapeutic in response to a molecular trigger *in vivo*", *Nat. Biotechnol.,* vol. 36, no. 3, pp. 258-264, 2018.
[http://dx.doi.org/10.1038/nbt.4071] [PMID: 29431737]

[17]   W. Ma, Y. Zhan, Y. Zhang, X. Shao, X. Xie, C. Mao, W. Cui, Q. Li, J. Shi, J. Li, C. Fan, and Y. Lin, "An Intelligent DNA Nanorobot with *in vitro* Enhanced Protein Lysosomal Degradation of HER2", *Nano Lett.,* vol. 19, no. 7, pp. 4505-4517, 2019.
[http://dx.doi.org/10.1021/acs.nanolett.9b01320] [PMID: 31185573]

[18]   S. Liu, Q. Jiang, X. Zhao, R. Zhao, Y. Wang, Y. Wang, J. Liu, Y. Shang, S. Zhao, T. Wu, Y. Zhang, G. Nie, and B. Ding, "A DNA nanodevice-based vaccine for cancer immunotherapy", *Nat. Mater.,* vol. 20, no. 3, pp. 421-430, 2021.
[http://dx.doi.org/10.1038/s41563-020-0793-6] [PMID: 32895504]

[19]   G. Zhu, L. Mei, H.D. Vishwasrao, O. Jacobson, Z. Wang, Y. Liu, B.C. Yung, X. Fu, A. Jin, G. Niu, Q. Wang, F. Zhang, H. Shroff, and X. Chen, "Intertwining DNA-RNA nanocapsules loaded with tumor neoantigens as synergistic nanovaccines for cancer immunotherapy", *Nat. Commun.,* vol. 8, no. 1, pp. 1482-1495, 2017.
[http://dx.doi.org/10.1038/s41467-017-01386-7] [PMID: 29133898]

[20]   European Medicines Agency (EU), *Guideline on development, production, characterization and specification for monoclonal antibodies and related products.,* 2016.

[21]   M. Liu, L. Li, D. Jin, and Y. Liu, "Nanobody—A versatile tool for cancer diagnosis and therapeutics", *Wiley Interdiscip. Rev. Nanomed. Nanobiotechnol.,* vol. 13, no. 4, p. e1697, 2021.
[http://dx.doi.org/10.1002/wnan.1697] [PMID: 33470555]

[22]   World Health Organization (WHO), "Guidelines for the production and quality control of monoclonal antibodies and related products intended for medicinal use", *Replacement of Annex 3 of WHO Technical Report Series,* vol. 822, 2022.

[23]   R Luo, H Liu, and Z. Cheng, "Protein scaffolds: Antibody alternatives for cancer diagnosis and therapy", *RSC chemical biology ,* vol. 3, no. 7, pp. 830-847, 2022.
[http://dx.doi.org/10.1039/D2CB00094F]

[24]   S. Ståhl, T. Gräslund, A. Eriksson Karlström, F.Y. Frejd, P.Å. Nygren, and J. Löfblom, "Affibody Molecules in Biotechnological and Medical Applications", *Trends Biotechnol.,* vol. 35, no. 8, pp. 691-712, 2017.
[http://dx.doi.org/10.1016/j.tibtech.2017.04.007] [PMID: 28514998]

[25]   F.Y. Frejd, and K.T. Kim, "Affibody molecules as engineered protein drugs", *Exp. Mol. Med.,* vol. 49, no. 3, p. e306, 2017.
[http://dx.doi.org/10.1038/emm.2017.35] [PMID: 28336959]

[26]   J.P. Salvador, L. Vilaplana, and M.P. Marco, "Nanobody: outstanding features for diagnostic and therapeutic applications", *Anal. Bioanal. Chem.,* vol. 411, no. 9, pp. 1703-1713, 2019.
[http://dx.doi.org/10.1007/s00216-019-01633-4] [PMID: 30734854]

[27]   S.T. Yurkin, and Z. Wang, "Cell membrane-derived nanoparticles: emerging clinical opportunities for targeted drug delivery", *Nanomedicine (Lond.),* vol. 12, no. 16, pp. 2007-2019, 2017.
[http://dx.doi.org/10.2217/nnm-2017-0100] [PMID: 28745122]

[28]   M. Zhang, Y. Du, S. Wang, and B. Chen, "A Review of Biomimetic Nanoparticle Drug Delivery Systems Based on Cell Membranes", *Drug Des. Devel. Ther.,* vol. 14, pp. 5495-5503, 2020.
[http://dx.doi.org/10.2147/DDDT.S282368] [PMID: 33363358]

[29]   Ş. Hamarat Baysal, and A.H. Uslan, "ENCAPSULATION OF PEG-UREASE/PEG-AlaDH ENZYME SYSTEM IN ERYTHROCYTE", *Artif. Cells Blood Substit. Immobil. Biotechnol.,* vol. 29, no. 5, pp. 405-412, 2001.

[http://dx.doi.org/10.1081/BIO-100106924] [PMID: 11708664]

[30]   G Ak, H Yilmaz, A Guneş, and S. Hamarat Sanlier, *In vitro and in vivo evaluation of folate receptor-targeted a novel magnetic drug delivery system for ovarian cancer therapy.*, vol. 46, no. sup 1, pp. 926-937, 2018.
[http://dx.doi.org/10.1080/21691401.2018.1439838]

[31]   J.S. Brenner, D.C. Pan, J.W. Myerson, O.A. Marcos-Contreras, C.H. Villa, P. Patel, H. Hekierski, S. Chatterjee, J.Q. Tao, H. Parhiz, K. Bhamidipati, T.G. Uhler, E.D. Hood, R.Y. Kiseleva, V.S. Shuvaev, T. Shuvaeva, M. Khoshnejad, I. Johnston, J.V. Gregory, J. Lahann, T. Wang, E. Cantu, W.M. Armstead, S. Mitragotri, and V. Muzykantov, "Red blood cell-hitchhiking boosts delivery of nanocarriers to chosen organs by orders of magnitude", *Nat. Commun.,* vol. 9, no. 1, pp. 2684-2698, 2018.
[http://dx.doi.org/10.1038/s41467-018-05079-7] [PMID: 29992966]

[32]   V. Lenders, R. Escudero, X. Koutsoumpou, L. Armengol Álvarez, J. Rozenski, S.J. Soenen, Z. Zhao, S. Mitragotri, P. Baatsen, K. Allegaert, J. Toelen, and B.B. Manshian, "Modularity of RBC hitchhiking with polymeric nanoparticles: testing the limits of non-covalent adsorption", *J. Nanobiotechnology,* vol. 20, no. 1, pp. 333-345, 2022.
[http://dx.doi.org/10.1186/s12951-022-01544-0] [PMID: 35842697]

[33]   G. Lu, X. Wang, F. Li, S. Wang, J. Zhao, J. Wang, J. Liu, C. Lyu, P. Ye, H. Tan, W. Li, G. Ma, and W. Wei, "Engineered biomimetic nanoparticles achieve targeted delivery and efficient metabolism-based synergistic therapy against glioblastoma", *Nat. Commun.,* vol. 13, no. 1, pp. 4214-4231, 2022.
[http://dx.doi.org/10.1038/s41467-022-31799-y] [PMID: 35864093]

[34]   A.S. Gdowski, J.B. Lampe, V.J.T. Lin, R. Joshi, Y.C. Wang, A. Mukerjee, J.K. Vishwanatha, and A.P. Ranjan, "Bioinspired Nanoparticles Engineered for Enhanced Delivery to the Bone", *ACS Appl. Nano Mater.,* vol. 2, no. 10, pp. 6249-6257, 2019.
[http://dx.doi.org/10.1021/acsanm.9b01226] [PMID: 33585803]

[35]   H. Lin, C. Yang, Y. Luo, M. Ge, H. Shen, X. Zhang, and J. Shi, "Biomimetic nanomedicine-triggered in situ vaccination for innate and adaptive immunity activations for bacterial osteomyelitis treatment", *ACS Nano,* vol. 16, no. 4, pp. 5943-5960, 2022.
[http://dx.doi.org/10.1021/acsnano.1c11132] [PMID: 35316599]

[36]   Y. Guo, S.Z. Wang, X. Zhang, H.R. Jia, Y.X. Zhu, X. Zhang, G. Gao, Y.W. Jiang, C. Li, X. Chen, S.Y. Wu, Y. Liu, and F.G. Wu, "In situ generation of micrometer-sized tumor cell-derived vesicles as autologous cancer vaccines for boosting systemic immune responses", *Nat. Commun.,* vol. 13, no. 1, pp. 6534-6554, 2022.
[http://dx.doi.org/10.1038/s41467-022-33831-7] [PMID: 36319625]

[37]   M. Holay, Z. Guo, J. Pihl, J. Heo, J.H. Park, R.H. Fang, and L. Zhang, "Bacteria-Inspired Nanomedicine", *ACS Appl. Bio Mater.,* vol. 4, no. 5, pp. 3830-3848, 2021.
[http://dx.doi.org/10.1021/acsabm.0c01072] [PMID: 34368643]

[38]   T. Harimoto, J. Hahn, Y.Y. Chen, J. Im, J. Zhang, N. Hou, F. Li, C. Coker, K. Gray, N. Harr, S. Chowdhury, K. Pu, C. Nimura, N. Arpaia, K.W. Leong, and T. Danino, "A programmable encapsulation system improves delivery of therapeutic bacteria in mice", *Nat. Biotechnol.,* vol. 40, no. 8, pp. 1259-1269, 2022.
[http://dx.doi.org/10.1038/s41587-022-01244-y] [PMID: 35301496]

[39]   F.P. Canale, C. Basso, G. Antonini, M. Perotti, N. Li, A. Sokolovska, J. Neumann, M.J. James, S. Geiger, W. Jin, J.P. Theurillat, K.A. West, D.S. Leventhal, J.M. Lora, F. Sallusto, and R. Geiger, "Metabolic modulation of tumours with engineered bacteria for immunotherapy", *Nature,* vol. 598, no. 7882, pp. 662-666, 2021.
[http://dx.doi.org/10.1038/s41586-021-04003-2] [PMID: 34616044]

[40]   J.D. Cecil, N. Sirisaengtaksin, N.M. O'Brien-Simpson, and A.M. Krachler, "Outer Membrane Vesicle-Host Cell Interactions", *Microbiol. Spectr.,* vol. 7, no. 1, p. 7.1.06, 2019.
[http://dx.doi.org/10.1128/microbiolspec.PSIB-0001-2018] [PMID: 30681067]

[41]   A.T. Jan, "Outer Membrane Vesicles (OMVs) of Gram-negative Bacteria: A Perspective Update", *Front. Microbiol.,* vol. 8, pp. 1053-1064, 2017.
[http://dx.doi.org/10.3389/fmicb.2017.01053] [PMID: 28649237]

[42]   Y. Huang, M.P. Nieh, W. Chen, and Y. Lei, "Outer membrane vesicles (OMVs) enabled bio-applications: A critical review", *Biotechnol. Bioeng.,* vol. 119, no. 1, pp. 34-47, 2022.
[http://dx.doi.org/10.1002/bit.27965] [PMID: 34698385]

[43]   K. Kuerban, X. Gao, H. Zhang, J. Liu, M. Dong, L. Wu, R. Ye, M. Feng, and L. Ye, "Doxorubicin-loaded bacterial outer-membrane vesicles exert enhanced anti-tumor efficacy in non-small-cell lung cancer", *Acta Pharm. Sin. B,* vol. 10, no. 8, pp. 1534-1548, 2020.
[http://dx.doi.org/10.1016/j.apsb.2020.02.002] [PMID: 32963948]

[44]   W. Huang, Q. Zhang, W. Li, M. Yuan, J. Zhou, L. Hua, Y. Chen, C. Ye, and Y. Ma, "Development of novel nanoantibiotics using an outer membrane vesicle-based drug efflux mechanism", *J. Control. Release,* vol. 317, pp. 1-22, 2020.
[http://dx.doi.org/10.1016/j.jconrel.2019.11.017] [PMID: 31738965]

[45]   R. Li, and Q. Liu, "Engineered Bacterial Outer Membrane Vesicles as Multifunctional Delivery Platforms", *Front. Mater.,* vol. 7, pp. 202-220, 2020.
[http://dx.doi.org/10.3389/fmats.2020.00202]

[46]   S. Wang, J. Gao, and Z. Wang, "Outer membrane vesicles for vaccination and targeted drug delivery", *Wiley Interdiscip. Rev. Nanomed. Nanobiotechnol.,* vol. 11, no. 2, p. e1523, 2019.
[http://dx.doi.org/10.1002/wnan.1523] [PMID: 29701017]

[47]   Y. Wang, Y. Wang, N. Kang, Y. Liu, W. Shan, S. Bi, L. Ren, and G. Zhuang, "Construction and Immunological Evaluation of CpG-Au@HBc Virus-Like Nanoparticles as a Potential Vaccine", *Nanoscale Res. Lett.,* vol. 11, no. 1, pp. 338-347, 2016.
[http://dx.doi.org/10.1186/s11671-016-1554-y] [PMID: 27435343]

[48]   K.D. Brune, D.B. Leneghan, I.J. Brian, A.S. Ishizuka, M.F. Bachmann, S.J. Draper, S. Biswas, and M. Howarth, "Plug-and-Display: decoration of Virus-Like Particles via isopeptide bonds for modular immunization", *Sci. Rep.,* vol. 6, no. 1, p. 19234, 2016.
[http://dx.doi.org/10.1038/srep19234] [PMID: 26781591]

[49]   M. Fusciello, F. Fontana, S. Tähtinen, C. Capasso, S. Feola, B. Martins, J. Chiaro, K. Peltonen, L. Ylösmäki, E. Ylösmäki, F. Hamdan, O.K. Kari, J. Ndika, H. Alenius, A. Urtti, J.T. Hirvonen, H.A. Santos, and V. Cerullo, "Artificially cloaked viral nanovaccine for cancer immunotherapy", *Nat. Commun.,* vol. 10, no. 1, pp. 5747-5760, 2019.
[http://dx.doi.org/10.1038/s41467-019-13744-8] [PMID: 31848338]

[50]   E. Shoeb, and K. Hefferon, "Future of cancer immunotherapy using plant virus-based nanoparticles", *Future Sci. OA,* vol. 5, no. 7, p. FSO401, 2019.
[http://dx.doi.org/10.2144/fsoa-2019-0001] [PMID: 31428448]

[51]   A. Nuñez-Rivera, P.G.J. Fournier, D.L. Arellano, A.G. Rodriguez-Hernandez, R. Vazquez-Duhalt, and R.D. Cadena-Nava, "Brome mosaic virus-like particles as siRNA nanocarriers for biomedical purposes", *Beilstein J. Nanotechnol.,* vol. 11, pp. 372-382, 2020.
[http://dx.doi.org/10.3762/bjnano.11.28] [PMID: 32175217]

[52]   W. Shan, D. Zhang, Y. Wu, X. Lv, B. Hu, X. Zhou, S. Ye, S. Bi, L. Ren, and X. Zhang, "Modularized peptides modified HBc virus-like particles for encapsulation and tumor-targeted delivery of doxorubicin", *Nanomedicine,* vol. 14, no. 3, pp. 725-734, 2018.
[http://dx.doi.org/10.1016/j.nano.2017.12.002] [PMID: 29275067]

[53]   S. Shapira, I. Boustanai, D. Kazanov, M. Ben Shimon, A. Fokra, and N. Arber, "Innovative dual system approach for selective eradication of cancer cells using viral-based delivery of natural bacterial toxin–antitoxin system", *Oncogene,* vol. 40, no. 31, pp. 4967-4979, 2021.
[http://dx.doi.org/10.1038/s41388-021-01792-8] [PMID: 34172933]

[54]   G. Battogtokh, Y.Y. Cho, J.Y. Lee, H.S. Lee, and H.C. Kang, "Mitochondrial-Targeting Anticancer Agent Conjugates and Nanocarrier Systems for Cancer Treatment", *Front. Pharmacol.,* vol. 9, pp. 922-942, 2018.
[http://dx.doi.org/10.3389/fphar.2018.00922] [PMID: 30174604]

[55]   S.R. Bonam, F. Wang, and S. Muller, "Lysosomes as a therapeutic target", *Nat. Rev. Drug Discov.,* vol. 18, no. 12, pp. 923-948, 2019.
[http://dx.doi.org/10.1038/s41573-019-0036-1] [PMID: 31477883]

[56]   F. Geisslinger, M. Müller, A.M. Vollmar, and K. Bartel, "Targeting Lysosomes in Cancer as Promising Strategy to Overcome Chemoresistance—A Mini Review", *Front. Oncol.,* vol. 10, pp. 1156-1163, 2020.
[http://dx.doi.org/10.3389/fonc.2020.01156] [PMID: 32733810]

[57]   X. Guo, X. Wei, Z. Chen, X. Zhang, G. Yang, and S. Zhou, "Multifunctional nanoplatforms for subcellular delivery of drugs in cancer therapy", *Prog. Mater. Sci.,* vol. 107, p. 100599, 2020.
[http://dx.doi.org/10.1016/j.pmatsci.2019.100599]

[58]   Y. Liu, H.R. Jia, X. Han, and F.G. Wu, "Endoplasmic reticulum-targeting nanomedicines for cancer therapy", *Smart Materials in Medicine,* vol. 2, pp. 334-349, 2021.
[http://dx.doi.org/10.1016/j.smaim.2021.09.001]

[59]   P. Gao, W. Pan, N. Li, and B. Tang, "Boosting Cancer Therapy with Organelle-Targeted Nanomaterials", *ACS Appl. Mater. Interfaces,* vol. 11, no. 30, pp. 26529-26558, 2019.
[http://dx.doi.org/10.1021/acsami.9b01370] [PMID: 31136142]

[60]   R.A.J.F. Oerlemans, S.B.P.E. Timmermans, and J.C.M. van Hest, "Artificial Organelles: Towards Adding or Restoring Intracellular Activity", *ChemBioChem,* vol. 22, no. 12, pp. 2051-2078, 2021.
[http://dx.doi.org/10.1002/cbic.202000850] [PMID: 33450141]

[61]   T. Einfalt, D. Witzigmann, C. Edlinger, S. Sieber, R. Goers, A. Najer, M. Spulber, O. Onaca-Fischer, J. Huwyler, and C.G. Palivan, "Biomimetic artificial organelles with *in vitro* and *in vivo* activity triggered by reduction in microenvironment", *Nat. Commun.,* vol. 9, no. 1, pp. 1127-1139, 2018.
[http://dx.doi.org/10.1038/s41467-018-03560-x] [PMID: 29555899]

[62]   L.M.P.E. van Oppen, L.K.E.A. Abdelmohsen, S.E. van Emst-de Vries, P.L.W. Welzen, D.A. Wilson, J.A.M. Smeitink, W.J.H. Koopman, R. Brock, P.H.G.M. Willems, D.S. Williams, and J.C.M. van Hest, "Biodegradable Synthetic Organelles Demonstrate ROS Shielding in Human-Complex-I-Deficient Fibroblasts", *ACS Cent. Sci.,* vol. 4, no. 7, pp. 917-928, 2018.
[http://dx.doi.org/10.1021/acscentsci.8b00336]

[63]   X. Ren, D. Chen, Y. Wang, H. Li, Y. Zhang, H. Chen, X. Li, and M. Huo, "Nanozymes-recent development and biomedical applications", *J. Nanobiotechnology,* vol. 20, no. 1, pp. 92-110, 2022.
[http://dx.doi.org/10.1186/s12951-022-01295-y] [PMID: 35193573]

[64]   M. Liang, and X. Yan, "Nanozymes: From New Concepts, Mechanisms, and Standards to Applications", *Acc. Chem. Res.,* vol. 52, no. 8, pp. 2190-2200, 2019.
[http://dx.doi.org/10.1021/acs.accounts.9b00140] [PMID: 31276379]

[65]   C. Cao, T. Zhang, N. Yang, X. Niu, Z. Zhou, J. Wang, D. Yang, P. Chen, L. Zhong, X. Dong, and Y. Zhao, "POD Nanozyme optimized by charge separation engineering for light/pH activated bacteria catalytic/photodynamic therapy", *Signal Transduct. Target. Ther.,* vol. 7, no. 1, pp. 86-97, 2022.
[http://dx.doi.org/10.1038/s41392-022-00900-8] [PMID: 35342192]

[66]   F. Gao, T. Shao, Y. Yu, Y. Xiong, and L. Yang, "Surface-bound reactive oxygen species generating nanozymes for selective antibacterial action", *Nat. Commun.,* vol. 12, no. 1, pp. 745-763, 2021.
[http://dx.doi.org/10.1038/s41467-021-20965-3] [PMID: 33531505]

[67]   G. Ak, Ü.F. Bozkaya, H. Yılmaz, Ö. Sarı Turgut, İ. Bilgin, C. Tomruk, Y. Uyanıkgil, and Ş. Hamarat Şanlıer, "An intravenous application of magnetic nanoparticles for osteomyelitis treatment: An efficient alternative", *Int. J. Pharm.,* vol. 592, p. 119999, 2021.

[http://dx.doi.org/10.1016/j.ijpharm.2020.119999] [PMID: 33190790]

[68]   A. Zeybek, G. Şanlı-Mohamed, G. Ak, H. Yılmaz, and Ş.H. Şanlier, "*In vitro* evaluation of doxorubicin-incorporated magnetic albumin nanospheres", *Chem. Biol. Drug Des.,* vol. 84, no. 1, pp. 108-115, 2014.
[http://dx.doi.org/10.1111/cbdd.12300] [PMID: 24524300]

[69]   Ş. Hamarat Şanlier, G. Ak, H. Yılmaz, A. Ünal, Ü.F. Bozkaya, G. Tanıyan, Y. Yıldırım, and G. Yıldız Türkyılmaz, "Development of ultrasoundtriggered and magnetic-targeted nanobubble system for dual-drug delivery", *J. Pharm. Sci.,* vol. 108, no. 3, pp. 1272-1283, 2019.
[http://dx.doi.org/10.1016/j.xphs.2018.10.030] [PMID: 30773203]

[70]   H. Liu, J. Wang, M. Wang, Y. Wang, S. Shi, X. Hu, Q. Zhang, D. Fan, and P. Xu, "Biomimetic nanomedicine coupled with neoadjuvant chemotherapy to suppress breast cancer metastasis via tumor microenvironment remodeling", *Adv. Funct. Mater.,* vol. 31, no. 25, p. 2100262, 2021.
[http://dx.doi.org/10.1002/adfm.202100262]

[71]   M.P. Monopoli, C. Åberg, A. Salvati, and K.A. Dawson, "Biomolecular coronas provide the biological identity of nanosized materials", *Nat. Nanotechnol.,* vol. 7, no. 12, pp. 779-786, 2012.
[http://dx.doi.org/10.1038/nnano.2012.207] [PMID: 23212421]

[72]   Y. Wang, R. Cai, and C. Chen, "The nano–bio interactions of nanomedicines: Understanding the biochemical driving forces and redox reactions", *Acc. Chem. Res.,* vol. 52, no. 6, pp. 1507-1518, 2019.
[http://dx.doi.org/10.1021/acs.accounts.9b00126] [PMID: 31149804]

[73]   S. Tenzer, D. Docter, J. Kuharev, A. Musyanovych, V. Fetz, R. Hecht, F. Schlenk, D. Fischer, K. Kiouptsi, C. Reinhardt, K. Landfester, H. Schild, M. Maskos, S.K. Knauer, and R.H. Stauber, "Rapid formation of plasma protein corona critically affects nanoparticle pathophysiology", *Nat. Nanotechnol.,* vol. 8, no. 10, pp. 772-781, 2013.
[http://dx.doi.org/10.1038/nnano.2013.181] [PMID: 24056901]

[74]   M. Lundqvist, J. Stigler, G. Elia, I. Lynch, T. Cedervall, and K.A. Dawson, "Nanoparticle size and surface properties determine the protein corona with possible implications for biological impacts", *Proc. Natl. Acad. Sci.,* vol. 105, no. 38, pp. 14265-14270, 2008.
[http://dx.doi.org/10.1073/pnas.0805135105] [PMID: 18809927]

[75]   V. Francia, K. Yang, S. Deville, C. Reker-Smit, I. Nelissen, and A. Salvati, "Corona composition can affect the mechanisms cells use to internalize nanoparticles", *ACS. Nano.,* vol. 13, no. 10, pp. 11107-11121, 2019.
[http://dx.doi.org/10.1021/acsnano.9b03824] [PMID: 31525954]

[76]   Q. Xiao, M. Zoulikha, M. Qiu, C. Teng, C. Lin, X. Li, M.A. Sallam, Q. Xu, and W. He, "The effects of protein corona on *in vivo* fate of nanocarriers", *Adv. Drug Deliv. Rev.,* vol. 186, p. 114356, 2022.
[http://dx.doi.org/10.1016/j.addr.2022.114356] [PMID: 35595022]

[77]   Z. Li, Y. Wang, J. Zhu, Y. Zhang, W. Zhang, M. Zhou, C. Luo, Z. Li, B. Cai, S. Gui, Z. He, and J. Sun, "Emerging well-tailored nanoparticulate delivery system based on in situ regulation of the protein corona", *J. Control. Rel.,* vol. 320, pp. 1-18, 2020.
[http://dx.doi.org/10.1016/j.jconrel.2020.01.007] [PMID: 31931050]

[78]   R.J.C. Bose, K. Ha, and J.R. McCarthy, "Bio-inspired nanomaterials as novel options for the treatment of cardiovascular disease", *Drug Discov. Today,* vol. 26, no. 5, pp. 1200-1211, 2021.
[http://dx.doi.org/10.1016/j.drudis.2021.01.035] [PMID: 33561512]

[79]   Z. Wang, C. Wang, S. Liu, W. He, L. Wang, J. Gan, Z. Huang, Z. Wang, H. Wei, J. Zhang, and L. Dong, "Specifically formed corona on silica nanoparticles enhances transforming growth factor $\beta 1$ activity in triggering lung fibrosis", *ACS Nano,* vol. 11, no. 2, pp. 1659-1672, 2017.
[http://dx.doi.org/10.1021/acsnano.6b07461] [PMID: 28085241]

[80]   S. Schöttler, G. Becker, S. Winzen, T. Steinbach, K. Mohr, K. Landfester, V. Mailänder, and F.R. Wurm, "Protein adsorption is required for stealth effect of poly(ethylene glycol)- and poly(phosphoester)-coated nanocarriers", *Nat. Nanotechnol.,* vol. 11, no. 4, pp. 372-377, 2016.

[http://dx.doi.org/10.1038/nnano.2015.330] [PMID: 26878141]

[81]   X. Wang, and W. Zhang, "The janus of protein corona on nanoparticles for tumor targeting, immunotherapy and diagnosis", *J. Control. Release,* vol. 345, pp. 832-850, 2022.
[http://dx.doi.org/10.1016/j.jconrel.2022.03.056] [PMID: 35367478]

[82]   Z. Chen, X. Chen, J. Huang, J. Wang, and Z. Wang, "Harnessing protein corona for biomimetic nanomedicine design", *Biomimetics,* vol. 7, no. 3, p. 126, 2022.
[http://dx.doi.org/10.3390/biomimetics7030126] [PMID: 36134930]

CHAPTER 2

Lipid-Based Nanocarriers and Applications in Medicine

Ümmühan Fulden Aydın[1,*] and **Abdullah Tuli**[1]

[1] *Çukurova University, Faculty of Medicine, Medical Biochemistry Department, Adana, Turkey*

Abstract: Lipid nanocarriers have recently arisen with a wide range of uses and research areas, with the advantages they offer in virtue of their unique properties. They are easily synthesized, scaled up, biodegradable, proper to transport many bioactive components, have a high loading capacity, and are convenient for various routes of administration (parenteral, oral, dermal, ocular, *etc.*). These carriers overcome the problems of bioactive substances such as low solubility, plasma half-life and bioavailability, and side effects, as well as providing controlled release, local delivery, and targeting. Lipid-based nanoparticular systems can be categorized into two basic classes, vesicular and non-vesicular. While liposomes are the most widely used vesicular structures, solid lipid nanoparticles and nano-structured lipid carriers are non-vesicular nanocarriers. These nanocarriers have many medical uses, such as cancer therapy, gene therapy, photodynamic therapy, treatment of infectious diseases and neurodegenerative diseases, vaccines, imaging, *etc*. It is essential that the synthesis method of lipid-based nanocarriers and the components from which they are composed are selected in accordance with the medical application area and characterization studies are carried out. In this article, liposomes, solid lipid nanoparticles and nano-structured lipid carriers will be discussed as lipid-based nanocarriers, synthesis and characterization methods will be emphasized and examples from medical applications will be given.

Keywords: Cancer Therapy, Characterization, Controlled Release, Drug Delivery, Emulsions, Entrapment Efficiency, High-Pressure Homogenization, Liposomes, Loading Capacity, Medical Applications, Nano-Structured Lipid Carriers, Particle Size, Polydispersity Index, Solid Lipid Nanoparticles, Solvent Evaporation, Solvent Injection, Targeted Therapy, Thin Film Hydration, Vaccines, Zeta Potential.

---
* **Corresponding author Ümmühan Fulden Aydın:** Çukurova University, Faculty of Medicine, Medical Biochemistry Department, Adana, Turkey; Tel: +90 322 338 60 60/ 3466 ; E-mail:bozkayafulden@gmail.com

Habibe Yılmaz (Ed.)

INTRODUCTION

Lipid-based nanocarriers have been in the limelight in recent years. These nanocarriers are preferred due to their dedicated physical properties as well as their unique size and shape properties. They are easily synthesized, scaled up, biodegradable, proper to transport many bioactive components, have a high loading capacity, and are convenient for various routes of administration (parenteral, oral, dermal, ocular, *etc.*). There is a need for lipid-based nano-drug carriers to increase the low biological efficiency of drugs with low plasma solubility and bioavailability due to their lipophilicity and reduce the side effects they cause by providing controlled drug release and local drug delivery. Like other nano drug carriers, lipid-based nanocarriers are designed to solve the problems of conventional therapies, such as low bioavailability, high toxicity, low plasma stability, and thus reduced half-life [1 - 3].

Lipid-based nanoparticular systems can be categorized into two basic classes, vesicular and non-vesicular. The discovery of vesicular liposomes, one of the most extensively used nanoparticles in the pharmaceutical field, was initiated in 1961 by hematologist Dr. Alec D. Bangham. Solid lipid nanoparticles (SLNs), discovered in 1991 as a choice to liposomes, and nano-structured lipid carriers (NLCs), which are superior to them, are non-vesicular lipid-based carriers [4].

In this section, the structural properties, synthesis, and characterization methods of liposomes, SLNs, and NLCs will be discussed and their medical applications will be mentioned.

LIPOSOMES

Liposomes are bilayer vesicles ranging in size from 20 nm to 2.5 μm, consisting of cholesterol and amphiphilic natural or synthetic phospholipids, allowing loading of both lipophilic and hydrophilic compounds into them. The properties of liposomes also differ according to the synthesis method, the type, composition, and ratio of lipids, and, therefore the size and surface charge that they affect [4 - 6].

Liposomes are generally classified according to their lamellarity and size. Liposomes consisting of a single bilayer are called unilamellar vesicles and are also subclassed according to their size as giant unilamellar vesicles (GUV, >1μm), large unilamellar vesicles (LUV, >100 nm) and small unilamellar vesicles (SUV, <100 nm). However, the structure in which 2-5 and >5 bilayers are sequentially intertwined is called oligolamellar vesicles (OLV, 100-1000 nm) and multilamellar vesicles (MLV, >0.5 μm) respectively, and the structure formed by the presence of bilayer vesicles in a large vesicle is called multivesicular vesicles

(MVV, >1 μm) as shown in Fig. (**1**) [7]. The sizes of liposomes can vary depending on the synthesis method, and at the same time, the area of application is an important factor in deciding the type of liposome. SUV is mostly preferred for its long circulation time, passive targeting to the desired area, and cell uptake. For example, Doxil®, one of the FDA (Food and Drug Administration)-approved liposomal drugs, is SUV, an IV-administered liposomal formulation loaded with doxorubicin used for ovarian cancer, Kaposi's sarcoma, and myeloid melanoma. However, Arikayce® is loaded with amikacin, which is used for lung diseases; its size is approximately 300 nm, and is a liposomal formulation applied with a nebulizer. Whether the liposome is unilamellar or multilamellar also affects the choice of liposome type since it is an important parameter in the release of the loaded substance. Unilamellar ones release their content more easily than multilamellar ones but have a higher loading capacity for hydrophilic agents [8 - 10].

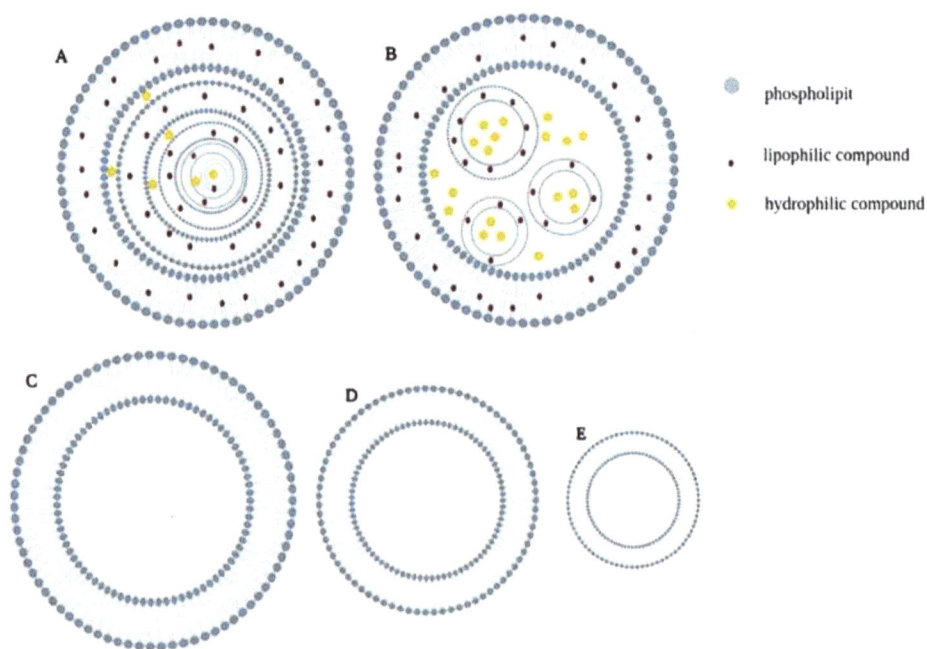

Fig. (1). Classification of liposomes according to lamelarity and size. **A**) Multilamellar liposome (MLV) **B**) Multivesicular liposome (MVV) **C**) Giant unilamellar liposome (GUV) **D**) Large unilamellar liposome (LUV) **E**) Small unilamellar liposome (SUV).

Formation and Composition of Liposomes

As it is known, phospholipids, the basic building blocks of liposomes, are amphiphilic, causing them to self-assemble in the aqueous environment due to the polar head groups and tails formed by the apolar fatty acid chains. As the polar phosphate groups interact with the water molecules in the aqueous medium, the hydrophobic tails begin to orient toward each other. This orientation causes them to form bilayer spheres called liposomes. Lipids, which take part in the formation of liposomes that need energy input thermodynamically by methods such as sonication, homogenization, *etc.*, tend to form these vesicles with stable and minimum surface tension in aqueous media [11].

The lipid composition affects the liposome's charge, stability, flexibility, and size. Phospholipids used in liposome production can be glycerophospholipids or sphingolipids. The most commonly used phospholipids are phosphatidylcholine (PC), phosphatidylethanolamine (PE), and phosphatidylserine (PS). Since these phospholipids are a component of the membrane structure of normal cells, they can be obtained primarily from egg yolk and soybean. The fatty acid component in natural phospholipids, whether PC, PE, or PS, can be palmitic acid, stearic acid, oleic acid, linoleic acid, and arachidonic acid. Egg yolk phospholipids contain more arachidonic and docosahexaenoic acids than soybean phospholipids and are more resistant to oxidative stress because they have more saturated fatty acids. While the saturated fatty acids in the egg yolk phospholipids are in the sn-1 position and the unsaturated ones are in the sn-2 position (*e.g.*, egg PC comprises about 40% 1-palmitoyl-2-oleoyl-phosphatidylcholine), the saturated and unsaturated fatty acids in soybean phospholipids can be in both positions [12 - 15]. In addition to natural phospholipids, synthetic phospholipids are often used in liposome production. For example; 1,2-distearoyl-sn-glycero-3-phospholetha-nolamine (DSPE) and 1,2-distearoyl-sn-glycero-3-phosphocholine (DSPC) are stearic acid-based synthetic phospholipids, 1,2-dipalmitoil-sn-glycero-3-phos-phocholine (DPPC) and 1,2-dipalmitoil-sn-glycero-3-phosphoetanolamine (DPPE) are palmitic acid-based synthetic phospholipids, and 1,2-dioleoyl-sn-glycero-3-phosphoethanolamine (DOPE), 1,2-dioleoyl-3-trimethylammonium-propane (DOTAP), 1,2-dioleoyl-sn-glycero-3-phosphocholine (DOPC), 1-palmitoyl-2-oleoyl-sn-glycero-3-phosphocholine (POPC), and 1,2-dimyristoyl-sn-glycero-3-phosphocholine (DMPC) are different types of synthetic phos-pholipids [13]. González-Cela-Casamayor MA *et al.* [16] developed liposomes produced with DOPC and DMPC for the treatment of glaucoma and simultaneously ocular surface conservation. On the other hand, Leone M *et al.* [17] loaded lovastatin into liposomes generated with DOPC and DOPE to inhibit 3-hydroxy-3-methyl-glutaryl coenzyme A reductase. In addition, polyethylene glycol (PEG) conjugated forms of synthetic phospholipids are also often used in the formation of liposomes. In another study, Yang T *et al.* [18, 19] performed liposome synthesis using PEGylated DSPE and soybean phosphatidylcholine to improve the solubility and stability of Taxol® and accomplished *in vivo* and *in vitro* studies of this nanocarrier system.

Sphingomyelin, one of the sphingolipid structures used in forming liposomes, has a sphingosine skeleton instead glycerol and is naturally found in the membrane structure of animal cells. They contain fatty acid chains with longer chain lengths than their phosphatidylcholine structures and have the capability of making intramolecular and intermolecular hydrogen bonds, and for this reason, the characteristic features of the liposomes constituted differentiate from those formed by glycerophospholipids. In addition, sphingomyelin has more saturation than phosphatidylcholine and interacts more tightly with cholesterol [12]. Lim EU *et al.* [20] investigated the release kinetics of ibuprofen and acetaminophen loaded by forming sphingomyelin-cholesterol liposomes in their study. Besides, Carter K *et al.* [21] put emphasis on that using sphingomyelin instead of DSPE in liposomes created as irinotecan carriers, increasing cholesterol, and removing PEG enhance the serum stability of liposomes and that these liposomes they obtained are promising for chemophototherapy.

Most of the liposome structures contain cholesterol, as well as phospholipids. Although cholesterol itself cannot form a bilayer, it is easily incorporated into the bilayer with phospholipids by the orientation of the polar hydroxyl group and the apolar part [14]. Cholesterol plays a substantial role in the stability and permeability of the lipid bilayer. In addition, it is notified that it increases phospholipid packaging and therefore has a critical role in the retention of content and, in particular, inhibits phospholipid exchange with lipoproteins. Although many researchers have stated that the ratio of cholesterol in liposomes should be 2:1 or 1:1 (lipid:cholesterol), the reason has not yet been explained [22]. Briuglia *et al.* [23], in their study, performed liposome synthesis by combining DMPC, DPPC, and DSPC with cholesterol at different percentages and accomplished various characterization studies on the drug loading efficiency and stability of these liposomes. According to the results obtained, they mentioned that the ratio of 70:30% lipid:cholesterol is optimal, correlation is observed for various lipids, and the ratio of 2:1 in the literature is supported.

Preparation Methods of Liposomes

Different synthesis protocols, described below in detail, are as follows: thin-film hydration, injection, reverse-phase evaporation, detergent removal, freeze-thaw, a French pressure cell, and microfluidics.

Thin-Film Hydration

It is the most frequently utilized technique for liposome synthesis. In this procedure, the organic solvent is removed in a round bottomed environment after the lipids are dissolved. If a lipophilic agent is desired to be loaded into liposomes, it must be added at this stage. The resulting lipid film layer is hydrated with the aqueous medium and the energy input mentioned above is provided, usually with the help of a sonic bath. Loading of hydrophilic agents takes place during the hydration phase and the length of the hydration time affects the loading capacity [24]. The thin film method is schematized in Fig. (**2**).

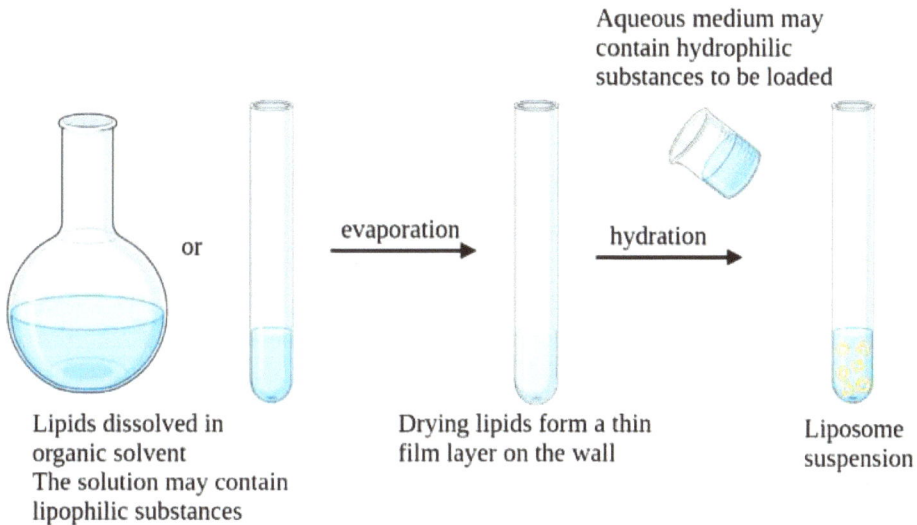

Fig. (2). Liposome synthesis by thin film hydration method.

While the most common organic solvents in this method are chloroform and ethanol, the solutions used for hydration are usually distilled water, buffer solutions, 0.9% NaCl, PBS (phosphate buffered saline), 10% sucrose, and 5% dextrose. The volume used for hydration is also a parameter that should be taken into account in liposome preparation [24, 25]. In this method, MLVs can be obtained using a large-volume hydration solution, and MLVs can then be converted into SUVs by lamellarity and size regulation by methods such as extrusion or probe-type sonication. Extrusion is often used to reduce liposome sizes with different pore diameters and make them more uniform using polycarbonate membranes [6, 26].

Injection

The injection method is based on injecting lipids dissolved in ether or ethanol into the aqueous medium and afterward removing the organic solvent. Lipids are dissolved in heated diethyl ether or diethyl ether-methanol mixture in the ether injection method. It is subsequently injected into the aqueous medium at a certain pressure with the help of a fine needle. Similarly, lipids dissolved in ethanol are injected into a very large volume of the aqueous medium in ethanol injection. In both methods, while lipophilic substances are added into the organic phase to be injected, hydrophilic substances are inserted into the aqueous medium. After injection, organic solvents are removed by a technique such as a dialysis and filtration, and SUV is obtained.

The most important advantage of the injection method is that it is easy to scale up. The liposomes are diluted and have a high polydispersity index (PDI) [4, 6, 13, 27].

Reverse-Phase Evaporation

Lipids dissolved in an organic solvent are mixed directly with the aqueous medium. First, inverted micelles are formed in the emulsion after sonication. Then, the inverse micelles in the gel structure, which are formed by the evaporation of the organic solvent, begin to evolve into liposomes. The resulting liposome sizes can be decreased by extrusion. This method is not recommended for the synthesis of liposomes in which structures such as therapeutic peptides and DNA will be encapsulated, which can be degraded by the use of sonication due to the organic phase consisting of various mixtures of diethyl ether, isopropyl ether, and chloroform [6, 13].

Detergent Removal

In this procedure, lipids are dissolved in a detergent solution and mixed, and therefore mixed micelles (detergent-lipid micelles) are formed. Then, by removing the detergent, the lipids form a bilayer and turn into LUVs. The most common detergents are Triton X-100, sodium cholate, sodium deoxycholate, and alkyl glycoside as surfactants with a high critical micelle concentration. The removal of detergent is accomplished by dialysis, chromatography, impregnation with hydrophobic beads, or dilution [9, 13, 27].

Freeze-Thaw

The freeze-thaw method is the process of making liposomes more unilamellar and uniform and improving encapsulation efficiency by performing fast freezing and slow thawing cycles at $-196\,°C$ in liquid nitrogen [6, 13, 28].

French Pressure Cell

MLVs are reshaped by passing them through a small high-pressure orifice, thus creating polydisperse SUVs. However, in the process of french pressure, degradation of phospholipids and the component to be encapsulated may occur [27].

Microfluidics

Microfluidics has recently become one of the most preferred liposome synthesis methods. Due to the advantages such as large-scale production, high uniformity, and adjustable size (from nano to micro), many researchers opt for the microfluidic method. The organic phase in which the lipids dissolve and the aqueous medium are passed through different grooves and mixed with high pressure on a chip, therefore, the stable liposomes are formed when the mixture reaches equilibrium. The fact that conventional liposome formation methods have multiple stages is one of the reasons why the microfluidics method is preferred [29 - 31].

Characterization of Liposomes

Characterization studies should be performed to determine the properties of the synthesized liposomes. These studies include measurement of size, determination of morphology (shape, lamellae) and loading efficiency, and measurement of zeta potential.

Dynamic light scattering (DLS), size exclusion chromatography (SEC), and microscopy techniques are used for size measurement. Negative staining transmission electron microscopy (TEM), the most used cryogenic-TEM (cryo-TEM), freeze-fracture TEM, and scanning electron microscopy (SEM) are types that are useful for providing information on the size and morphology of liposomes. Atomic force microscopy (AFM), another type of microscopy, is suitable for stability studies as well as for providing information on the morphology, size, and stiffness of liposomes [32, 33]. HPLC (high-performance liquid chromatography)-SEC enables the purification of liposomes by time and hydrodynamic size [34]. The DLS method is a method frequently used in hydrodynamic size measurement as well as providing PDI with the same measurement. In the use of lipid-based nanocarrier systems, including liposomes, as drug carriers, a PDI value of ≤ 0.3 is considered an indicator of the homogeneous nanoparticle population [35].

Zeta potential defines the surface charge of particles in colloidal dispersions. In the measurement performed in devices based on light scattering, the liposome suspension is loaded into a special cuvette containing a pair of electrodes, and voltage is implemented by placing the cuvette into the device. Zeta potentials are determined by measuring the light scattering created by the migration of particles to the oppositely charged electrode at a certain speed. Liposomes can be negatively charged, positively charged, or neutral. If they are less than -30 mV or higher than +30 mV, it means that they are extremely anionic and extremely cationic, respectively [36, 37].

The loading/encapsulation/entrapment efficiency is determined by calculating the percentage ratio of the determined amount of the compound planned to be loaded in the liposome to the initial amount, as shown in Equation 1. For the calculation, firstly the amount of loaded and non-entrapped components must be determined. For this purpose, the unloaded components are removed from the liposomes by dialysis, exclusion chromatography, or centrifugation. The quantification of the compound can be carried out by several methods such as HPLC, GC-MS (gas chromatography-mass spectroscopy), LC-MS (liquid chromatography-mass spectroscopy), and visible/ultraviolet/fluorometric spectroscopy, *etc.*, which are suitable for its analysis [36, 38].

$$Entrapment\ Efficiency\ \% = \frac{Loaded\ amount\ of\ bioactive\ compound}{Total\ amount\ of\ active\ compound} x100 \quad (1)$$

Medical Applications of Liposomes

Liposomes have a wide range of medical applications, including cancer treatment, vaccines, infectious diseases, diagnostics, and photodynamic therapy (PDT) [22]. Nowadays, it is possible to find many publications and studies related to the applications of liposomes. Although some of them have not yet been produced and used on an industrial scale, some of them have received FDA and EMA (European Medicines Agency) approval, and some are still in the research phase.

Nanoparticle drug delivery systems can be targeted by passive targeting for cancer therapy. In passive targeting, there is increased permeability, mainly due to the larger pore sizes among the endothelial cells in the tumor region compared to normal tissue. Nanoparticles with a size of <400 nm are allowed to pass into tumor tissue by using this principle, called the enhanced permeability and retention (EPR) effect, whereas not being allowed to pass into normal tissues [39]. For example, the previously mentioned Doxil® is a PEGylated formulation containing doxorubicin for anticancer treatment. LipoDox also has the same formulation as Doxil® and is manufactured to address the lack of supply. Myocet® is an EMA-approved non-PEGylated formulation of doxorubicin used for metastatic breast cancer [40]. Another example of a liposomal drug, Marqibo®, is FDA-approved sphingomyelin-cholesterol-based, contains vincristine, and is used in the therapy of acute lymphoblastic leukemia [41]. In another targeting strategy, which is active targeting, the surfaces of the nanocarriers are functionalized with various agents for tissue-specific or overexpressed receptors, thereby increasing the accumulation of the nanocarrier system in the targeted tissue compared to passive targeting [42].

Aptamers, peptides, proteins, or antibodies, as well as ligands such as folic acid, can be used for active targeting [43]. Chaini *et al.* [44] loaded bleomycin into folic acid-bound and PEGylated liposomes to target cervical and breast cancer and examined its anticancer activity. Depending on their results, folate-bound liposomal bleomycin shows much more cytotoxicity in cervical cancer cells with folate receptors than folate-free liposomal bleomycin, while it did not show in breast cancer cells without folate receptors.

Liposomal vaccine formulations have the ability to trigger both humoral and cell-mediated immunity. Antigens are loaded into liposomes and captured by antigen-presenting cells. T cells are also stimulated because antigen-presenting cells produce cytokines and chemokines. Additionally, a controlled, slow release of antigen from liposomes also provides a long-lasting dependent immune response [45]. An important benefit of liposomes as a vaccine system is that they are versatile. Liposomes can be loaded with DNA, RNA, or peptide/protein antigens, as well as a wide variety of adjuvants [46]. Cruz *et al.* [47] developed a liposomal vaccine to enhance the immunological response against tumor-associated antigens, prepared by co-loading the ESO-1 peptide and immunostimulatory adjuvants targeted with PEG-conjugated Fcγ. With this targeted liposomal vaccine, antigen delivery to dendritic cells would be accomplished and immunotherapeutic efficacy would be possible. Epaxal® is an example of an FDA-approved Hepatitis A vaccine. Hemagglutinin and neuraminidase were anchored to liposomes derived from phosphatidylcholine and phosphati-dylethanolamine, and inactivated hepatitis A virus was adsorbed onto the liposome structure to obtain the final vaccine formulation [48]. Liposomal formulations can be used for the treatment of infections as well as for prevention with vaccines. Ambisome®, Abelcet®, and Amphotec® are liposomal antifungals containing amphotericin B used to treat infections such as leishmaniasis, aspergillosis, *etc* [49, 50].

Some limitations of imaging methods such as magnetic resonance imaging (MRI), traditional positron emission tomography (PET), computed tomography (CT), or single-photon emission computed tomography (SPECT) have been tried to become through nanoparticular systems. The ability of nanoparticles to be targeted and accumulated in the desired region brings innovation. The detection of this accumulation and uptake into cells makes it possible to diagnose and make dose adjustments before treatment. Mahakian and colleagues [51] compared PET imaging of ^{64}Cu liposomes and 2-deoxy-2-[^{18}F]fluoro-D-glucose (^{18}F-FDG) in head and neck cancers. In these cancers, especially lesions called mild dysplasia can be visualized with ^{64}Cu liposomes, but not with ^{18}F-FDG. Further, they determined that the accumulation of ^{64}Cu liposome is higher than ^{18}F-FDG with a value of 2,12-4,4%ID/cc (percentage of injected dose per cubic centimeter). In another study, the researchers created a new formulation to be used for both PET and MRI imaging in solid tumors such as glioblastoma by loading magnetic nanoparticles into liposomes that they had created from glucose-modified and ^{68}Ga-chelating phospholipids [52].

PDT is a type of treatment that aims to kill cancer cells without damaging healthy cells in superficial cancers. However, due to the non-tumor selectivity of many photosensitizers, liposomal systems have been considered because of their ability to increase their high loading capacity and flexibility to enhance the safety and efficacy of PDT [53]. Visudyne® is an approved liposomal photosensitizing formulation used in the treatment of subfoveal choroidal neovascularization, pathological myopia, and ocular histoplasmosis syndrome. The solution came from loading into the liposomal system because the photosensitizer in its content is hydrophobic and spontaneously forms aggregates in the aqueous medium [54].

SOLID LIPID NANOPARTICLES (SLNS)

SLNs are colloidal lipid nanoparticles with a size of 40-1000 nm, obtained by dispersing solid lipids in water (0.1-30% (w/w)) or in an aqueous solution of surfactant (0.5-5% (w/w)) [4, 5, 55]. The lipids used to form SLNs are non-polar and solid at body and room temperature. Solid lipids allow sustainable drug release because of the fact that they allow the decreased mobility of loaded drugs compared to liquid lipids. SLNs are suitable for loading hydrophobic materials as well as hydrophilic ones [56].

SLNs can be classified into three types according to morphological patterns (Fig. **3**): A) homogeneous matrix model, B) drug-enriched shell model C) drug-enriched core model.

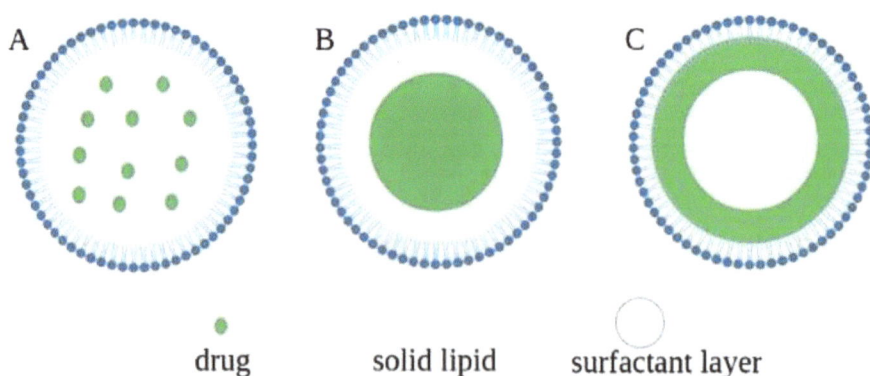

drug   solid lipid   surfactant layer

Fig. (3). Types of SLNs **A)** Homogeneous matrix model **B)** Drug-enriched core model **C)** Drug-enriched shell model.

In the homogeneous matrix model, drugs are dispersed as amorphous crystals in the lipid matrix. These SLNs are achieved by the cold homogenization method. If a lipophilic drug is to be loaded, it is added before the surfactants. The drugs crystallize evenly during cooling, without forming phases with lipid clusters. In the enriched shell model, the drug is located inside the shell, but there is little or no drug in the center. Such SLNs are suitable for burst-style release. After hot homogenization, lipids precipitate at crystallization temperatures in the core when cooled, while drugs move to the outer surface of the lipids due to the reduced solubility in the surfactant solution. In the enriched core model, on the other hand, drugs are aggregated in the center with the opposite mechanism. The pre-emulsion, which is in the form of a supersaturated drug in the lipid, is put through rapid cooling. Throughout the time of this rapid cooling, the drug begins to crystallize in the center. After further cooling, the lipids aggregate in the shell [3, 57].

Formation and Composition of SLNs

Among the structures that make up SLNs, fatty acids, triglycerides, partial glycerides, waxes, and surfactants can be counted [57]. These lipids are generally recognized as safe (GRAS, a definition of the FDA), biocompatible, biodegradable, and possess melting points above 40 °C and are selected based on loading efficiency, characteristics of the material to be loaded, particle size, and storage stability [58].

Stearic acid, palmitic acid, decanoic acid, behenic acid, and myristic acid, which are saturated fatty acids and are in solid form at room temperature, are fatty acids that are often used in the preparation of SLNs. Whilst stearic acid provides the highest entrapment efficiency, SLNs including stearic acid and palmitic acid has a higher degradation rate and particle size than others [57]. Fatty acid alcohols such as stearyl alcohol and cetyl alcohol, which are further metabolized by the fatty alcohol dehydrogenase [4, 57], and the saturated fatty acid esters glyceryl tristearate (Dynasan® 118), glycerol tripalmitate (Dynasan® 116) and glycerol trimyristate (Dynasan® 114), are also used in the production of SLNs [57, 59]. Partial glycerides, on the other hand, are used for the stability of SLNs by acting as a surfactant, and also affect the entrapment efficiency. While glyceryl monostearate (GMS) and glyceryl monooleate (GMO) [4] are generally preferred in topical application, glyceryl palmitostearate (Precirol® ATO 5) and glyceryl behenate (Compritol® ATO 888) in oral pharmaceuticals [57, 60]. However, Hippalgaonkar *et al.* [61] tested different lipids such as Compritol 888 ATO, Precirol® ATO 5, Dynasan® 116, Dynasan® 118, or Softisan® 154 for SLN synthesis in order to transport indomethacin ocularly and chose Compritol® 888 ATO due to its high solubility and high loading efficiency.

Cetyl palmitate, a hydrophobic wax with long fatty acid chains, and beeswax and carnauba wax are often used in manufacturing SLNs [57]. In a study comparing wax and glyceride solid lipid nanoparticles, wax SLNs were noted to be more stable and had excellent particle size distributions, whereas SLNs comprising glycerides had a higher loading capacity [62]. In another study, the tacrolimus loading capacity of beeswax-SLNs compared to stearic acid-SLNs increased from 2.3% to 2.9%, and the particle size from 167.3 nm to 274.9 nm [63].

Surfactants play a role in the formation of SLN, during the dispersion of lipid melts in aqueous media and in ensuring their steric stability by surrounding them after cooling [58]. Surfactants could be phospholipids, steroids, the mentioned above partial glycerides, non-ionic polyoxyethylene sorbitan copolymers (*e.g.*, polysorbate 80, also known as Tween® 80), polyethylene glycol/polyoxypropylene copolymers, and polyoxyethylene alkyl/aryl ethers, cationic dimethyldioc-tadecylammonium bromide, and DOTAP, or anionic sodium lauryl sulfate, and bile salts. Ionic surfactants provide steric repulsion stability, meanwhile, non-ionic surfactants provide electrostatic stability. Non-ionic surfactants are recommended for parenteral or oral administration and are more suitable for loading hydrophobic drugs. Toxicity should be taken into account in the selection of surfactants. The descending order of toxicity is as follows: cationic, anionic, non-ionic, and amphoteric [57, 58]. Pizzol CD *et al.* [64] underline that the selection of lipid and surfactant combinations may influence the toxicity of SLNs.

The formation of a pre-emulsion derived from lipids and surfactants, followed by the reduction of the size by methods such as homogenization and ultrasonication, are common steps in the preparation of SLNs. The choice of synthesis method varies according to the size and the nature of the material to be loaded [56].

Preparation Methods of SLNs

The preparation methods of SLNs are as follows: hot homogenization, cold homogenization, ultrasonication/high-speed homogenization, solvent emulsi-fication-evaporation, solvent emulsification-diffusion, microemulsion, double emulsion, supercritical fluid, and spray drying.

Hot Homogenization

The hot homogenization method is one of the high-pressure homogenization methods. The lipids and pharmaceutical compounds are heated 5-10 °C above the melting temperature. The pre-emulsion formed by mixing lipid melt and aqueous surfactant solution is then passed through a high-pressure homogenizer (*e.g.*, ultra-turrax). When the emulsion is cooled to room temperature, crystallization occurs. Particles are usually small in size and have a narrow particle distribution.

Supercooling may be required, as crystallization may be delayed by virtue of the small size and the presence of surfactant. At homogenization, 3-5 cycles at 500-1500 bar are sufficient. Enhancing the homogenization cycle can lead to an increase in particle size. One of the biggest disadvantages of this method is the ability to break down heat-sensitive compounds [65 - 69].

Cold Homogenization

The cold homogenization method is also the high-pressure homogenization technique. The initial step is the same as the previous method and it is useful in eliminating the disadvantages of this method, such as drug deterioration and supercooling. The drug-containing lipids are first melted and then quickly cooled with liquid nitrogen or dry ice. 50-100 μm sized particles are obtained by milling the cooled emulsion. It is then dispersed at room temperature by high-pressure homogenization in the cold aqueous surfactant solution, resulting in larger-sized SLNs than hot homogenization [66 - 68].

Ultrasonication/High-Speed Homogenization

In this method, lipids and the substance to be loaded are melted and mixed with a hot surfactant solution at high speed. The resulting pre-emulsion is subjected to ultrasound until the desired particle size is achieved. Both bath and probe-type sonication can be used, but probe-type is often preferred for ultrasonication. Nevertheless, this can lead to metal contamination [67, 70].

Solvent Emulsification-Evaporation

Lipids and the thermolabile, lipophilic drug are completely dissolved in an organic solvent such as cyclohexane, chloroform, and toluene, which is immiscible with water. Nanodispersions are then formed by mixing with a high-pressure homogenizer in an aqueous solution of surfactant. SLNs are formed by the evaporation of the solvent and the complete separation of lipid contents from the aqueous media through sintered disc filter funnels. The size of the SLNs obtained by this method is around 40-100 nm, with a narrow distribution width. Increasing lipid content can increase lipid size [71, 72].

Solvent Emulsification-Diffusion

In this method, the drug does not need to be dissolved directly in the lipid. Organic solvents such as methyl-, ethyl-, isopropyl-acetate, and benzyl alcohol, which are partially miscible with water, are saturated with water. Lipids and drugs are then added to this saturated system and mixed with an aqueous surfactant solution to form an oil/water (o/w) emulsion. Water is added to this emulsion in a ratio of 1:5 or 1:10, the organic solvent begins to diffuse into the aqueous medium, and SLNs are formed spontaneously. The disadvantage of both solvent emulsification-evaporation and solvent emulsification-diffusion methods is that they require the use of toxic organic solvents and advanced solvent removal such as ultrafiltration [71, 72]. SLNs' sizes below 100 nm are produced by this method [4].

Microemulsion

Lipids, surfactants, co-surfactants, and water are mixed at 65-70 °C to form a transparent solution. Lipid crystallization then occurs when this hot microemulsion is diluted 1:25 or 1:50 with cold water at 2-3 °C to provide a high-temperature difference [4, 66, 67]. The microemulsion method may be preferred to load mechanical stress-sensitive compounds such as phenolics [73].

Double Emulsion

With this method, hydrophilic drugs, proteins, and peptides can be encapsulated into SLNs. An o/w emulsion is originated by mixing melted lipids with an aqueous surfactant solution containing the active compound to be entrapped. The o/w emulsion is added to the aqueous surfactant and the co-surfactant medium to form a clear w/o/w emulsion. SLNs are obtained by ultrafiltration or centrifugation at 4 °C for 30 minutes at 12000 g after mixing this warm medium with a cold dispersion medium [67, 71].

Supercritical Fluid

A supercritical fluid is a liquid at a temperature and pressure above the supercritical point, where the liquid and gas phases cannot be distinguished, to be used instead of organic solvents [74]. Carbon dioxide (CO_2) is often used in the production of SLNs. Lipids and the drug are dissolved with the help of a water-immiscible, and in the presence of a surfactant, an emulsion is formed with the help of a homogenizer. After adding w/o emulsion to the extraction column from above, supercritical CO_2 is supplied to the extraction column by a countercurrent. SLNs are formed after the solvent has been removed quickly and completely [71, 72].

Spray-Drying

It is used to obtain dry and more stable SLNs from aqueous dispersion as a cheaper alternative to lyophilization and is recommended for lipids with melting temperatures above 70 °C [67, 71].

Characterization of SLNs

Characterization studies of SLNs include determining size, PDI, morphology, and crystallinity, measuring zeta potential, and calculating entrapment efficiency.

DLS is used to measure the size of SLNs. However, laser diffraction (LD), also known as static light scattering, is also preferred. Although SLNs can be in the 40-1000 nm range in size, SLNs with a size of 50-300 nm are mainly used to deliver chemotherapeutic agents and target the central nervous system (CNS). The size of SLNs varies according to the synthesis method, the properties of the lipids, and the temperature. Low surfactant concentrations can lead to an increase in size. Nanoparticles with a PDI value less than 0.5 are considered to show homogeneous size distribution, while those greater than 0.5 show heterogeneous distribution. PDI value less than 0.3 is suitable for SLNs [70]. The high-pressure homogenization method and the solvent evaporation method generally enable the production of SLN less than 100 nm and PDI less than 0.2 [75].

SEM and TEM analyses are also used for size measurement and reveal morphology. SLN is not always perfectly spherical. They can be stick-form, pellet-form, or anisotropic. After electron bombardment is applied while taking images in electron microscopes, lipids may melt and therefore SLNs may deteriorate. This problem is overcome with cryogenic electron microscopes. AFM provides better resolution 3D topophragic images, whereas SEM provides 3D and TEM 2D images.

The zeta potential of SLNs determines their fate *in vivo*. Generally, positively charged SLNs have a longer circulation time and are stable at 20-40 mV. Positively charged SLNs interact with serum components, thus they are protected from phagocytosis but are indicated to destabilize cells [75, 76]. On the other hand, it is also stated that it is appropriate for SLNs to have a zeta potential higher than -60 mV to have good stability [2]. However, near-neutral particles are preferred, not strong charges, for systemic drug delivery. SLNs with low zeta potential are obtained with the use of a nonionic surfactant. Wherefore, it is possible to increase the zeta potential of SLNs by coating with materials such as PEG or by surrounding with nonionic surfactants [77].

The crystallinity of SLNs affects the loading efficiency of the molecules [78]. Differential scanning calorimetry (DSC), X-ray scattering, X-ray diffraction (XRD), FTIR, and Raman spectroscopy are methods used to determine lipid crystallinity [4, 76]. DSC is a technique used to analyze polymorphic transitions and changes in lipid physical properties. It also provides information about melting points and melting enthalpies of lipids [75, 76]. XRD is frequently utilized to investigate the polymorphic behavior and crystallinity of the constituents of SLNs [75]. Raman spectroscopy and FT-IR are also often used to detect polymorphic transitions of lipids [76].

For SLNs, the encapsulation efficiency is found by the calculation made using equation 1 following the determination of the loaded amount after the removal of the unloaded active compound. The free active compound is collected by extraction with an organic solvent, microfiltration, ultrafiltration, (ultra-) centrifugation, and size exclusion chromatography and then determined by HPLC, MS, and UV/visible spectroscopy [70].

Medical Applications of SLNs

SLNs are also widely used in targeted cancer therapy, gene therapy, infections, and vaccines. In addition, SLNs can be used to cross biological barriers such as the blood-brain barrier (BBB) as well as for local drug delivery by pulmonary, oral, topical, rectal, ocular, *etc.* administration [79, 80].

Ak G *et al.* [81] synthesized dual drug-loaded functionalized SLNs to target glioblastoma multiforme. They loaded temozolomide and carmustine into SLNs functionalized with beta-hydroxybutyrate and stearyl amine to cross the BBB and be targeted to the MCT-1 receptor. Accordingly, they found SLNs targeted to MCT-1, capable of controlled release, 220.9 ± 35.73 nm in size. Bhatt R *et al.* [82] prepared SLNs loaded with rosmarinic acid for intranasal administration in order to reduce oxidative stress in the CNS in Huntington's disease. They found that these SLNs with a size of about 150 nm regressed hunting-like behaviors in rats, but also significantly reduced oxidative stress in the striatum. They also stated that SLNs given intranasally are more effective than those given intravenously and their accumulation in the brain is higher. In another study, Morsi NM *et al.* [83] succeeded in loading the low water-soluble lipophilic vinpocetine alkaloid, which is used in brain ischemia but has a very low oral bioavailability of 6.7% and is eliminated from the body in approximately 8 hours, into SLNs for the purpose of passing through the BBB.

Khallaf RA *et al.* [84] performed the topical application of SLNs. For this purpose, they loaded 5-fluorouracil into shell-enriched SLNs and made it into a gel formulation, thus the effectiveness of the drug for skin carcinoma in mice is increased and they have also obtained a potential topical delivery system formulation for other hydrophilic drugs. In another study, SLNs which were measured 223.7 ± 4.6 nm in size, were loaded with epirubicin for pulmonary application. While empty SLNs did not cause any toxic effect in the A549 cell line, epirubicin-loaded SLNs were found to be statistically significantly more cytotoxic in A549 cells than free epirubicin. In addition, according to pharmacokinetic studies, after inhalation of epirubicin-loaded SLNs to rats, the concentration of the drug was found to be higher in the lungs than in the plasma [85].

In addition to cancer treatment, SLNs are also used in the treatment of various infections. Desai J. and Thakkar H. [86] targeted SLNs loaded with darunavir against the HIV reservoir. They found that SLNs grafted with a peptide designed to target the CD4 receptor possessed by HIV host cells were more abundant than those without the CD4 receptor in HIV host cells. As a result, they produced actively targeted SLNs against HIV with higher oral bioavailability than the free drug. Another study includes loading miconazole to SLNs to increase its oral bioavailability and investigating its antifungal activity and determining its pharmacokinetics in rabbits. Accordingly, 23 nm sized SLNs increased the bioavailability of the antifungal agent 2.5 times, as well as improved its activity compared to the free-form drug [87]. Pandey R and Ashler GK [88] produced SLNs co-loaded with rifampicin, isoniazid, and pyrazinamide. They administered these SLNs to guinea pigs with tuberculosis model by inhaler. According to this, drug concentrations in the delivery system remained in the plasma for 5 days after a single dose inhalation application. They determined those tubercle bacilli were completely cleared after 7 doses of administration. In the oral application of conventional drugs, they had to apply 46 doses to achieve this result.

Vicente-Pascual M *et al.* [89] administered SLNs loaded with a non-viral vector inducing IL-10 production to IL-10 knockout mice as eye drops for gene therapy of corneal inflammation. After 3 doses of topical inoculation, vectors became transfected into corneal epithelial cells. Han Y *et al.* [90] worked with SLNs, which they actively targeted with transferrin and covered with PEG, in order to create both gene therapy and chemotherapeutic effects for the treatment of lung cancer. The transfection efficiency of the plasmid in the SLN was significantly greater than the transfection efficiency of the naked plasmid, and the tumor volume of mice treated with doxorubicin (the chemotherapeutic agent) and plasmid-loaded SLNs shrank much more than in other groups (such as free doxorubicin).

SLNs have been studied for many years to transport peptide/protein compounds. Ansari MJ *et al.* [91], in their study to solve the problems of insulin-dependent diabetic patients due to insulin injections many times a day, found that when they loaded insulin into SLNs and administered to diabetic rats, its oral bioavailability was 5 times higher than that of free insulin solution. Another study on the production of protein-containing SLN was conducted for hepatitis B vaccination. Saraf S *et al.* [92] used the SLNs containing the hepatitis B antigen for a vaccination with intranasal and intramuscular administration by taking advantage of the adjuvant effect of SLNs themselves. Although there are many studies with lipid nanocarriers on vaccine production, the importance of the use of them in the field of vaccination has been demonstrated once again with the COVID-19 pandemic. Many scientists have benefited from lipid nanoparticles in the production of the SARS-CoV-2 vaccine [93 - 95].

NANO-STRUCTURED LIPID CARRIERS (NLCS)

Nanostructured lipid carriers were developed by Müller and Lucks [96] in the late 1990s to solve the problems of SLNs, such as lower loading capacity and high drug expulsion, which are 10-1000 nm in size and defined as the second generation of solid lipid nanoparticles [2, 97 - 99]. Although NLCs are also in solid form at room and body temperature like SLNs, they consist not only of solid lipids but also of oils, which will make the matrix of the nanoparticle more imperfect. This imperfect crystal matrix structure provides NLCs with enhanced properties than SLNs such as enhanced drug loading capacity and improved stability [99, 100].

NLCs are classified into 3 types: 1) the imperfect type 2) the amorphous type 3) the multiple oil-in-solid fat-in-water (O/F/W) type.

The imperfect type is obtained by mixing lipids such as glycerides with different saturation and chain-length fatty acids, and this imperfection provides more room for the drug. In the amorphous type, a structureless amorphous matrix is formed by mixing special lipids such as hydroxyoctacosanyl hydroxystearate or isopropyl myristate with the solid lipid, and drug expulsion is preserved from the more ordered crystalline structure. Multiple O/F/W type is a solid matrix with oil particles. The emulsion consisting of these oil compartments in a solid matrix allows more drugs to be loaded and prolonged drug release [4, 99].

Formation and Composition of NLCs

NLCs consist of solid lipids, liquid lipids, surfactants, and water. All lipids in the formation of NLCs are also defined as GRAS. Solid lipids are mixed with liquid lipids in a ratio of 70:30 to 99.9:0.1, while the surfactant ratio can vary in the range of 0.5-5% [99, 101].

The solid lipids and surfactants used in the manufacture of NLCs may be the same as those used for SLNs (*e.g.*, stearic acid, GMS, carnauba wax, cetyl palmitate, Precirol® ATO 5, Compritol® 888 ATO, *etc.* for solid lipids; Tween® 80, Tween® 20, lecithin, poloxamer 188, poloxamer 407, Cremophor® RH40 (PEG-40 hydrogenated castor oil), Kolliphor® EL (Polyoxyl castor oil), *etc.* for surfactants) [4, 99 - 103]. Glyceryl behenate can provide high loading capacity due to its imperfect crystal lattice. In addition, it should be noted that GMS works as a stabilizer, nonionic emulsifier, and plasticizer for pharmaceuticals [4].

Miglyol® 812 and 810, Softison 378 (Medium chain triglycerides (MCT), oleic acid, alpha-tocopheryl acetate, squalene, Labrasol® (PEG-8 caprylic/capric glycerides), LabrafacTM PG (propylene glycol dicaprylocaprate), soy lecithin, paraffin oil, 2-octyl dodecanol, isopropyl myristate, and Gelucire®44/14 (lauroyl polyoxylglycerides), some botanic oils such as limonene, corn oil, soybean oil, and sunflower oil, Mediterranean essential oils, Siberian pure seed oil, *etc.* are frequently used in the formation of NLC as liquid lipids [4, 101 - 104]. NLCs are mainly derived from oleic acid as an emulsifying agent and penetration enhancer and MCTs, which have high stability against oxidation and provide drugs with high solubility. The fatty acids of MCT can be composed of capric acid, caproic acid, caprylic acid, lauric acid, and myristic acid. Botanical oils are important in protecting from oxidative damage by means of their antioxidant properties [4, 104].

Liquid and solid lipids should be selected according to the properties of the active substance to be transported, the targeted organism, tissue, or organ. At the same time, the ratio of liquid lipid to solid lipid should be determined well. It should be noted that this ratio may affect the size, shape, drug loading capacity, storage stability, or drug release of NLCs [3, 105, 106].

Preparation Methods of NLCs

The techniques used in NLC synthesis are the same as for SLNs. While many researchers give their synthesis methods, instead of giving SLN and NLC in separate categories, they gather both under a single heading and explain them in that way. As a result, it can be stated that the synthesis methods of NLCs are also hot or cold homogenization, ultrasonication/high-speed homogenization, solvent emulsification-evaporation, solvent emulsification-diffusion, microemulsion, double emulsion, supercritical fluid, and spray drying which are given under the preparation methods of SLNs [2 - 4, 55, 100, 102 - 105, 107].

In general, NLC production is obtained by melting solid and liquid lipids and emulsifying them in aqueous surfactant solution. Among the mentioned methods, the high-pressure homogenization methods are often preferred due to its ease of application and providing large-scale production. In addition, probe sonication is often preferred for size reduction [2, 3, 100 - 103, 107].

Characterization of NLCs

Characterization studies are undoubtedly a very critical step for the medical application of NLCs. NLCs, just as SLNs, are mainly characterized by particle size, PDI, morphology, zeta potential, crystallinity, and entrapment efficiency [2].

Both DLS and LD methods can be used for particle size and PDI measurement for NLCs. PDI values in the range of 0.1-0.25 indicate a very narrow size distribution. A PDI value less than 0.3 is generally admissible, also, values less than 0.5 are likely to be considered homogeneous [2, 3, 99]. The presence of liquid lipids also reduces the PDI value of the particles. Furthermore, compared to SLNs, the liquid lipids allow NLCs to form smooth and smaller-sized particles with lower surface tension as well as reduce the viscosity of the particles [108]. NLCs loaded with chemotherapeutic agents should be 50-300 nm in size. It is also stated that NLCs with a size of <100 nm reach tumor sites effectively due to the EPR effect and are eliminated more slowly by reticuloendothelial system organs than larger carriers [103]. DLS and LD reveal the size of particles assuming they are spherical. However, NLCs do not always show spherical morphology and may be anisometric. Anisometric NLCs have more surface area than spherical NLCs and need more surfactant for stabilization [99, 103]. Techniques that can determine the particle size, as well as the morphology of NLCs, are again SEM, TEM, and AFM. In addition, field emission SEM (FESEM) is a technique that can be more effective than SEM in determining the nanoscale. Cryo-TEM, -SEM, or even cryo-FESEM can also be preferred for characterization [4, 103, 109].

In the case of NLCs, a zeta potential higher than -30mV is required for good physical stabilization. It is stated that the use of non-ionic surfactant decreases the zeta potential, while the increase in the oil content rises it [2].

The crystallinity measurement of NLCs is also performed by DSC, XRD, FTIR, and Raman spectroscopy methods. The crystallinity of NLCs is reduced by adding oils to solid lipids [2, 4].

Medical Applications of NLCs

NLCs, as the other lipid carriers mentioned, have a great variety of medical applications in virtue of their even more advanced properties.

Liu D *et al.* [110] synthesized NLCs coated with chitosan copolymer func-tionalized with N-acetyl-L-cysteine (CS-NAC) to enhance the low intraocular penetration of anti-inflammatory, anti-microbial, anti-oxidant, but low-soluble curcumin. The NLCs obtained had a size of 88.64±1.25nm, a zeta potential of 22.51±0.34 mV, and an entrapment efficiency of > 90%. In this study, curcumin, whose solubility was increased, was made suitable for topical application with an NLC formulation whose clearance was decreased by CS-NAC. In another study on the topical application of NLCs, Rosita N *et al.* [111] compared the skin penetration of para-methoxycinnamic acid (PMCA) loaded NLCs with SLNs and simple PMCA cream. They found that after 4.5 hours of application into the at skin, NLC PMCA, which has a smaller size than the others and a higher loading efficiency than the SLN formulation, penetrates deeper compared to PMCA SLN and PMCA Cream. Although SLNs are also known to be quite powerful topical carriers, the improved properties of NLCs lead to their more effective results. Taking a different approach to diabetic ulcer healing, Sun D *et al.* [112] have impregnated 20(S)-protopanaxadiol (PPD) loaded NLCs into silicone elastomer networks. Compared with the PPD solution, PPD-loaded NLCs ($p < 0.01$) and PPD-loaded NLC-impregnated silicone elastomers (PPD-NS, $p < 0.05$) showed good anti-inflammatory potential with significant NO inhibition. At the same time, PPD-NS led to a pro-angiogenesis activity, and these two significant effects were suggested as important indicators of *in vivo* recovery.

NLCs are very popular in research to cross the BBB. In a study aimed at the treatment of insomnia, temazepam-loaded NLCs produced with oleic acid, Compritol® 888 ATO, and Poloxamer® 407 were found to have $75.2 \pm 0.1\%$ loading efficiency, 306.6 ± 49.6 nm average size, -10.2 ± 0.3 mV zeta potential. According to the results of this study, which was carried out to ensure that orally taken temazepam reaches the desired effective concentration in the brain, the relative bioavailability of temazepam transported with NLCs was found to be 292.7% compared to the temazepam suspension, and the concentration of the temazepam loaded-NLC formulation in the brain was shown to be much higher according to the scintigraphy images [113]. In another interesting study, Battaglini M and colleagues [114] examined NLCs loaded with CeO_2 nanoparticles (nanoceria, NC), which draw attention to the treatment of various neurological diseases such as Alzheimer's, Parkinson's, and stroke, on the BBB that they constructed with neurons, astrocytes and endothelial cells in 3-dimensions (3-D). They showed that spherical NLCs with a diameter of 100–200 nm, together with smaller structures around 20 nm, have antioxidant properties, can pass the 3-D BBB model, and have neuroprotective and pro-neurogenic effects. Sharma P *et al.* [115] loaded cytarabine into NLCs for the treatment of meningeal leukemia. These spherical and smooth NLCs were stable at 2-8 °C for up to 3 months, showed sustained release for up to 72 hours ($89.9 \pm 1.11\%$), and induced greater cytotoxicity than free form cytarabine in cell culture studies at 48 and 72 h.

As mentioned before, the use of lipid nanoparticles in vaccines is numerous. Gerhardt A *et al.* [116] produced an RNA-containing NLC vaccine to improve the storage conditions of RNA vaccines for SARS-CoV-2. It is known that current SARS-CoV-2 vaccines need to be stored at -20 °C or -70 °C and transported by the cold chain. Whereas, Gerhardt A *et al.* stated that these thermostable NLC-RNA complexes they produced maintain their *in vivo* activity when stored at room temperature for 8 months and in the refrigerator for 21 months. Moa M *et al.* [117], with their study, aimed to perform both dendritic cell (DC) vaccination for gastric cancer immunotherapy and photodynamic therapy with chlorin e6 (Ce6). For this purpose, they loaded Ce6 into actively targeted NLCs by functionalization with folic acid. At the same time, it is thought that reactive oxygen species to be produced by the action of Ce6 in NLCs that are about 100 nm in size and thus benefit from the EPR effect, will cause the release of tumor-associated antigens that will lead to DC vaccination.

Another study on the applications of NLCs was designed to allow simultaneous diagnosis and treatment. These NLCs produced for theranostic purposes were loaded with doxorubicin and docosahexaenoic acid, which is stated to increase the anticancer activity of doxorubicin, and labeled with Tc-99m for scintigraphy imaging [118]. These theranostic NLCs with a size of about 70 nm were found to have a longer circulatory life and to be accumulated more in the tumor area in virtue of the EPR effect than the free drug, and they also showed increased anticancer activity against breast cancer. Zhu X *et al.* [119] loaded superparamagnetic iron oxide into NLCs for MR imaging of the liver and conjugated them with galactose to target the overexpressed asialoglycoprotein receptors in hepatocytes. Based on their results, they concluded that the NLCs are good targeted imaging agents for the diagnosis of liver diseases.

CONCLUDING REMARKS

Lipid-based nanocarriers are very important nanotechnological materials that have been studied for many years, are used with approval today, and whose clinical research and development studies continue. It is also obvious that they can be used in a wide variety of areas due to their capacity to carry bioactive materials in many different structures. At the same time, they continue to be preferred because they significantly increase the bioavailability of these materials, as they are both biocompatible and non-toxic, and reduce the side effects of the drugs. Synthesis of these carriers is also not very difficult and can be easily scaled up. The fact that they serve more than one purpose, such as the transport of more than one bioactive product at the same time, or the simultaneous treatment and diagnosis, makes them attractive. As a result, it is clear that the utilization and research of these lipid-based nanocarriers, which have already been studied and used for many years, will continue.

REFERENCES

[1]   A. Patidar, D.S. Thakur, P. Kumar, and J. Verma, "A review on novel lipid based nanocarriers", *Int. J. Pharm. Pharm. Sci.,* vol. 2, no. 4, pp. 30-35. 2010.

[2]   N. Dhiman, R. Awasthi, B. Sharma, H. Kharkwal, and G.T. Kulkarni, "Lipid nanoparticles as carriers for bioactive delivery", *Front Chem.,* vol. 9, p. 580118. 2021.
[http://dx.doi.org/10.3389/fchem.2021.580118] [PMID: 33981670]

[3]   S.U. Rawal, and M.M. Patel, Lipid nanoparticulate systems: Modern versatile drug carriers. *Lipid Nanocarriers for Drug Targeting.,* William Andrew, pp. 49-138. 2018.
[http://dx.doi.org/10.1016/B978-0-12-813687-4.00002-5]

[4]   T. Shukla, N. Upmanyu, P.S. Pandey, and D. Gosh, Lipid nanocarriers.*Lipid Carriers for Drug Targeting.,* William Andrew, pp. 1-47. 2018.
[http://dx.doi.org/10.1016/B978-0-12-813687-4.00001-3]

[5]   M. Mehanna, A. Motawaa, and M. Samaha, "Pharmaceutical particulate carriers: Lipid: Based carriers", *Natl. J. Physiol. Pharm. Pharmacol.,* vol. 2, no. 1, pp. 10-22. 2012.

[6]   A. Akbarzadeh, R. Rezaei-Sadabady, S. Davaran, S.W. Joo, N. Zarghami, Y. Hanifehpour, M. Samiei, M. Kouhi, and K. Nejati-Koshki, "Liposome: Classification, preparation, and applications", *Nanoscale Res. Lett.,* vol. 8, no. 1, p. 102. 2013.
[http://dx.doi.org/10.1186/1556-276X-8-102] [PMID: 23432972]

[7]   C. Has, and P. Sunthar, "A comprehensive review on recent preparation techniques of liposomes", *J. Liposome Res.,* vol. 30, no. 4, pp. 336-365. 2020.
[http://dx.doi.org/10.1080/08982104.2019.1668010] [PMID: 31558079]

[8]   P. Liu, G. Chen, and J. Zhang, "A review of liposomes as a drug delivery system: current status of approved products, regulatory environments, and future perspectives", *Molecules,* vol. 27, no. 4, p. 1372. 2022.
[http://dx.doi.org/10.3390/molecules27041372] [PMID: 35209162]

[9]   D. Lombardo, and M.A. Kiselev, "Methods of liposome preparation: Formation and control factors of versatile nanocarriers for biomedical and nanomedicine application", *Pharmaceutics,* vol. 14, no. 3, p. 543. 2022.
[http://dx.doi.org/10.3390/pharmaceutics14030543] [PMID: 35335920]

[10] M. Shirley, "Amikacin liposome inhalation suspension: A review in mycobacterium avium complex lung disease", *Drugs,* vol. 79, no. 5, pp. 555-562. 2019.
[http://dx.doi.org/10.1007/s40265-019-01095-z] [PMID: 30877642]

[11] D. Yadav, K. Sandeep, D. Pandey, and R.K. Dutta, "Liposomes for drug delivery", *J. Biotechnol. Biomater.,* vol. 7, no. 4, p. 4. 2017.
[http://dx.doi.org/10.4172/2155-952X.1000276]

[12] J. Li, X. Wang, T. Zhang, C. Wang, Z. Huang, X. Luo, and Y. Deng, "A review on phospholipids and their main applications in drug delivery systems", *Asian J. Pharm.,* vol. 10, no. 2, pp. 81-98. 2015.
[http://dx.doi.org/10.1016/j.ajps.2014.09.004]

[13] H. Nsairat, D. Khater, U. Sayed, F. Odeh, A. Al Bawab, and W. Alshaer, "Liposomes: Structure, composition, types, and clinical applications", *Heliyon,* vol. 8, no. 5, p. e09394. 2022.
[http://dx.doi.org/10.1016/j.heliyon.2022.e09394] [PMID: 35600452]

[14] H. Daraee, A. Etemadi, M. Kouhi, S. Alimirzalu, and A. Akbarzadeh, "Application of liposomes in medicine and drug delivery", *Artif. Cells Nanomed. Biotechnol.,* vol. 44, no. 1, pp. 381-391. 2016.
[http://dx.doi.org/10.3109/21691401.2014.953633] [PMID: 25222036]

[15] P. van Hoogevest, and A. Wendel, "The use of natural and synthetic phospholipids as pharmaceutical excipients", *Eur. J. Lipid Sci. Technol.,* vol. 116, no. 9, pp. 1088-1107. 2014.
[http://dx.doi.org/10.1002/ejlt.201400219] [PMID: 25400504]

[16] M.A. González-Cela-Casamayor, J.J. López-Cano, I. Bravo-Osuna, V. Andrés-Guerrero, M. Vicario-de-la-Torre, M. Guzmán-Navarro, J.M. Benítez-del-Castillo, R. Herrero-Vanrell, and I.T. Molina-Martínez, "Novel osmoprotective DOPC-DMPC liposomes loaded with antihypertensive drugs as potential strategy for glaucoma treatment", *Pharmaceutics,* vol. 14, no. 7, p. 1405. 2022.
[http://dx.doi.org/10.3390/pharmaceutics14071405] [PMID: 35890300]

[17] G. Leone, M. Consumi, C. Franzi, G. Tamasi, S. Lamponi, A. Donati, A. Magnani, C. Rossi, and C. Bonechi, "Development of liposomal formulations to potentiate natural lovastatin inhibitory activity towards 3-hydroxy-3-methyl-glutaryl coenzyme A (HMG-CoA) reductase", *J. Drug Deliv. Sci. Technol.,* vol. 43, pp. 107-112. 2018.
[http://dx.doi.org/10.1016/j.jddst.2017.09.019]

[18] O. Ishida, K. Maruyama, H. Tanahashi, M. Iwatsuru, K. Sasaki, M. Eriguchi, and H. Yanagie, "Liposomes bearing polyethyleneglycol-coupled transferrin with intracellular targeting property to the solid tumors *in vivo*", *Pharm. Res.,* vol. 18, no. 7, pp. 1042-1048. 2001.
[http://dx.doi.org/10.1023/A:1010960900254] [PMID: 11496943]

[19] T. Yang, F.D. Cui, M.K. Choi, J.W. Cho, S.J. Chung, C.K. Shim, and D.D. Kim, "Enhanced solubility and stability of PEGylated liposomal paclitaxel: *In vitro* and *in vivo* evaluation", *Int. J. Pharm.,* vol. 338, no. 1-2, pp. 317-326. 2007.
[http://dx.doi.org/10.1016/j.ijpharm.2007.02.011] [PMID: 17368984]

[20] E.B. Lim, S. Haam, and S.W. Lee, "Sphingomyelin-based liposomes with different cholesterol contents and polydopamine coating as a controlled delivery system", *Colloids Surf. A Physicochem. Eng. Asp.,* vol. 618, p. 126447. 2021.
[http://dx.doi.org/10.1016/j.colsurfa.2021.126447]

[21] K.A. Carter, D. Luo, A. Razi, J. Geng, S. Shao, J. Ortega, and J.F. Lovell, "Sphingomyelin liposomes containing porphyrin– phospholipid for irinotecan chemophototherapy", *Theranostics,* vol. 6, no. 13, pp. 2329-2336. 2016.
[http://dx.doi.org/10.7150/thno.15701] [PMID: 27877238]

[22] P. Nakhaei, R. Margiana, D.O. Bokov, W.K. Abdelbasset, M.A. Jadidi Kouhbanani, R.S. Varma, F. Marofi, M. Jarahian, and N. Beheshtkhoo, "Liposomes: Structure, biomedical applications, and stability parameters with emphasis on cholesterol", *Front. Bioeng. Biotechnol.,* vol. 9, p. 705886. 2021.
[http://dx.doi.org/10.3389/fbioe.2021.705886] [PMID: 34568298]

[23] M.L. Briuglia, C. Rotella, A. McFarlane, and D.A. Lamprou, "Influence of cholesterol on liposome stability and on *in vitro* drug release", *Drug Deliv. Transl. Res.,* vol. 5, no. 3, pp. 231-242. 2015.
[http://dx.doi.org/10.1007/s13346-015-0220-8] [PMID: 25787731]

[24] D.E. Large, R.G. Abdelmessih, E.A. Fink, and D.T. Auguste, "Liposome composition in drug delivery design, synthesis, characterization, and clinical application", *Adv. Drug Deliv. Rev.,* vol. 176, p. 113851. 2021.
[http://dx.doi.org/10.1016/j.addr.2021.113851] [PMID: 34224787]

[25] JS Dua, AC Rana, and AK Bhandari, "Liposome: Methods of preparation and applications", *Int. J. Pharm. Sci. Res.,* vol. 3, pp. 14-20. 2012.

[26] M.R. Mozafari, "Liposomes: An overview of manufacturing techniques", *Cell. Mol. Biol. Lett.,* vol. 10, no. 4, pp. 711-719. 2005.
[PMID: 16341279]

[27] Y.P. Patil, and S. Jadhav, "Novel methods for liposome preparation", *Chem. Phys. Lipids,* vol. 177, pp. 8-18. 2014.
[http://dx.doi.org/10.1016/j.chemphyslip.2013.10.011] [PMID: 24220497]

[28] A. Gonzalez Gomez, C. Xu, and Z. Hosseinidoust, "Preserving the efficacy of glycopeptide antibiotics during nanoencapsulation in liposomes", *ACS Infect. Dis.,* vol. 5, no. 10, pp. 1794-1801. 2019.
[http://dx.doi.org/10.1021/acsinfecdis.9b00232] [PMID: 31397146]

[29] V.M. Shah, D.X. Nguyen, P. Patel, B. Cote, A. Al-Fatease, Y. Pham, M.G. Huynh, Y. Woo, and A.W.G. Alani, "Liposomes produced by microfluidics and extrusion: A comparison for scale-up purposes", *Nanomedicine,* vol. 18, pp. 146-156. 2019.
[http://dx.doi.org/10.1016/j.nano.2019.02.019] [PMID: 30876818]

[30] D. van Swaay, and A. deMello, "Microfluidic methods for forming liposomes", *Lab Chip,* vol. 13, no. 5, pp. 752-767. 2013.
[http://dx.doi.org/10.1039/c2lc41121k] [PMID: 23291662]

[31] K.K. Ajeeshkumar, P.A. Aneesh, N. Raju, M. Suseela, C.N. Ravishankar, and S. Benjakul, "Advancements in liposome technology: Preparation techniques and applications in food, functional foods, and bioactive delivery: A review", *Compr. Rev. Food Sci. Food Saf.,* vol. 20, no. 2, pp. 1280-1306. 2021.
[http://dx.doi.org/10.1111/1541-4337.12725] [PMID: 33665991]

[32] K. Edwards, and A. Baeumner, "Analysis of liposomes", *Talanta,* vol. 68, no. 5, pp. 1432-1441. 2006.
[http://dx.doi.org/10.1016/j.talanta.2005.08.031] [PMID: 18970482]

[33] C. Chen, S. Zhu, T. Huang, S. Wang, and X. Yan, "Analytical techniques for single-liposome characterization", *Anal. Methods,* vol. 5, no. 9, p. 2150. 2013.
[http://dx.doi.org/10.1039/c3ay40219c]

[34] J. Cauzzo, N. Jayakumar, B.S. Ahluwalia, A. Ahmad, and N. Škalko-Basnet, "Characterization of liposomes using quantitative phase microscopy (QPM)", *Pharmaceutics,* vol. 13, no. 5, p. 590. 2021.
[http://dx.doi.org/10.3390/pharmaceutics13050590] [PMID: 33919040]

[35] M. Danaei, M. Dehghankhold, S. Ataei, F. Hasanzadeh Davarani, R. Javanmard, A. Dokhani, S. Khorasani, and M. Mozafari, "Impact of particle size and polydispersity index on the clinical applications of lipidic nanocarrier systems", *Pharmaceutics,* vol. 10, no. 2, p. 57. 2018.
[http://dx.doi.org/10.3390/pharmaceutics10020057] [PMID: 29783687]

[36] O. Popovska, J. Simonovska, Z. Kavrakovski, and V. Rafajlovska, "An Overview: Methods for preparation and characterization of liposomes as drug delivery systems", *Int. J. Pharm. Phytopharmacol. Res.,* vol. 3, no. 3, pp. 182-189. 2013.

[37] J.D. Clogston, and A.K. Patri, "Zeta potential measurement", In: McNeil S. Eds. Characterization of nanoparticles intended for drug delivery, Methods in Molecular Biology 697 2011.
[http://dx.doi.org/10.1007/978-1-60327-198-1_6]

[38] Creative-biostructure.com, "Liposome encapsulation efficiency determination", Available From: https://www.creative-biostructure.com/ (Accessed 30th November 2022).

[39] M. Alavi, and M. Hamidi, "Passive and active targeting in cancer therapy by liposomes and lipid nanoparticles", *Drug Metab. Pers. Ther.,* vol. 34, no. 1, p. 20180032. 2019.
[http://dx.doi.org/10.1515/dmpt-2018-0032] [PMID: 30707682]

[40] H. Pandey, R. Rani, and V. Agarwal, "Liposome and their applications in cancer therapy", *Braz. Arch. Biol. Technol.,* vol. 59, no. 0, p. 59. 2016.
[http://dx.doi.org/10.1590/1678-4324-2016150477]

[41] J.A. Silverman, and S.R. Deitcher, "Marqibo® (vincristine sulfate liposome injection) improves the pharmacokinetics and pharmacodynamics of vincristine", *Cancer Chemother. Pharmacol.,* vol. 71, no. 3, pp. 555-564. 2013.
[http://dx.doi.org/10.1007/s00280-012-2042-4] [PMID: 23212117]

[42] M.F. Attiaa, N. Antona, J. Wallyna, Z. Ziad Omrand, and T.F. Vandammea, "An overview of active and passive targeting strategies toimprove the nanocarriers efficiency to tumour sites, royal pharmaceutical society", *J. Pharm. Pharmacol.,* vol. 71, pp. 1185-1198. 2019.
[http://dx.doi.org/10.1111/jphp.13098] [PMID: 31049986]

[43] W. Yan, S.S.Y. Leung, and K.K.W. To, "Updates on the use of liposomes for active tumor targeting in cancer therapy", *Nanomedicine,* vol. 15, no. 3, pp. 303-318. 2020.

[http://dx.doi.org/10.2217/nnm-2019-0308] [PMID: 31802702]

[44]　M. Chiani, D. Norouzian, M.A. Shokrgozar, K. Azadmanesh, A. Najmafshar, M.R. Mehrabi, and A. Akbarzadeh, "Folic acid conjugated nanoliposomes as promising carriers for targeted delivery of bleomycin", *Artif. Cells Nanomed. Biotechnol.,* vol. 46, no. 4, pp. 757-763. 2018.
[http://dx.doi.org/10.1080/21691401.2017.1337029] [PMID: 28643525]

[45]　N. Marasini, K.A. Ghaffar, M. Skwarczynski, and I. Toth, Liposomes as a vaccine delivery system.*Micro and Nano Technologies.,* Elsevier, pp. 221-239. 2017.

[46]　R.A. Schwendener, "Liposomes as vaccine delivery systems: A review of the recent advances", *Ther. Adv. Vaccines,* vol. 2, no. 6, pp. 159-182. 2014.
[http://dx.doi.org/10.1177/2051013614541440] [PMID: 25364509]

[47]　N. Lamichhane, T. Udayakumar, W. D'Souza, C. Simone II, S. Raghavan, J. Polf, and J. Mahmood, "Liposomes: Clinical applications and potential for image-guided drug delivery", *Molecules,* vol. 23, no. 2, p. 288. 2018.
[http://dx.doi.org/10.3390/molecules23020288] [PMID: 29385755]

[48]　P.A. Bovier, "Epaxal® is an example of an FDA-approved hepatitis A vaccine", *Expert Rev. Vaccines,* vol. 7, no. 8, pp. 1141-1150. 2008.
[http://dx.doi.org/10.1586/14760584.7.8.1141] [PMID: 18844588]

[49]　R. Nisini, N. Poerio, S. Mariotti, F. De Santis, and M. Fraziano, "The multirole of liposomes in therapy and prevention of infectious diseases", *Front. Immunol.,* vol. 9, p. 155. 2018.
[http://dx.doi.org/10.3389/fimmu.2018.00155] [PMID: 29459867]

[50]　Ambisome, "Mechanism of action of AmBisome® (amphotericin B) liposome for injection", Available From: https://www.ambisome.com/ (Accessed on cited:30th November 2022).

[51]　L.M. Mahakian, D.G. Farwell, H. Zhang, J.W. Seo, B. Poirier, S.P. Tinling, A.M. Afify., E.M. Haynam, D. Shaye, and K.W. Ferrara, "Comparison of PET imaging with 64Cu-liposomes and 18F-FDG in the 7,12-dimethylbenz[a]anthracene (DMBA)-induced hamster buccal pouch model of oral dysplasia and squamous cell carcinoma", *Mol. Imaging Biol.,* vol. 16, no. 2, pp. 284-292. 2014.
[http://dx.doi.org/10.1007/s11307-013-0676-1] [PMID: 24019092]

[52]　J. Malinge, B. Géraudie, P. Savel, V. Nataf, A. Prignon, C. Provost, Y. Zhang, P. Ou, K. Kerrou, J.N. Talbot, J.M. Siaugue, M. Sollogoub, and C. Ménager, "Liposomes for PET and MR imaging and for dual targeting (magnetic field/glucose moiety): Synthesis, properties and *in vivo* studies", *Mol. Pharm.,* vol. 14, no. 2, pp. 406-414. 2017.
[http://dx.doi.org/10.1021/acs.molpharmaceut.6b00794] [PMID: 28029258]

[53]　A. Derycke, and P.A. de Witte, "Liposomes for photodynamic therapy", *Adv. Drug Deliv. Rev.,* vol. 56, no. 1, pp. 17-30. 2004.
[http://dx.doi.org/10.1016/j.addr.2003.07.014] [PMID: 14706443]

[54]　S. Ghosh, K.A. Carter, and J.F. Lovell, "Liposomal formulations of photosensitizers", *Biomaterials,* vol. 218, p. 119341. 2019.
[http://dx.doi.org/10.1016/j.biomaterials.2019.119341] [PMID: 31336279]

[55]　S. Mohammadi-Samani, and P. Ghasemiyeh, "Solid lipid nanoparticles and nanostructured lipid carriers as novel drug delivery systems: Applications, advantages and disadvantages", *Res. Pharm. Sci.,* vol. 13, no. 4, pp. 288-303. 2018.
[http://dx.doi.org/10.4103/1735-5362.235156] [PMID: 30065762]

[56]　Y. Mirchandani, V.B. Patravale, and B. S, "Solid lipid nanoparticles for hydrophilic drugs", *J. Control. Release,* vol. 335, pp. 457-464. 2021.
[http://dx.doi.org/10.1016/j.jconrel.2021.05.032] [PMID: 34048841]

[57]　S. Khatak, and H. Dureja, "Structural composition of solid lipid nanoparticles for invasive and noninvasive drug delivery", *Curr. Nanomater.,* no. 2, pp. 129-153. 2017.

[58]   R.A. Hernández-Esquivel, G. Navarro-Tovar, E. Zárate-Hernández, and P. Aguirre-Bañuelos, Solid lipid nanoparticles.*Nanocomposite Materials for Biomedical and Energy Storage Applications.,* intechopen, pp. 1-340. 2022.
[http://dx.doi.org/10.5772/intechopen.102536]

[59]   A.E.B. Yassin, M.K. Anwer, H.A. Mowafy, I.M. El-Bagory, M.A. Bayomi, and I.A. Alsarra, "Optimization of 5-fluorouracil solid-lipid nanoparticles: A preliminary study to treat colon cancer", *Int. J. Med. Sci.,* vol. 7, no. 6, pp. 398-408. 2010.
[http://dx.doi.org/10.7150/ijms.7.398] [PMID: 21103076]

[60]   M.H. Aburahma, and S.M. Badr-Eldin, "Compritol 888 ATO: A multifunctional lipid excipient in drug delivery systems and nanopharmaceuticals", *Expert Opin. Drug Deliv.,* vol. 11, no. 12, pp. 1865-1883. 2014.
[http://dx.doi.org/10.1517/17425247.2014.935335] [PMID: 25152197]

[61]   K. Hippalgaonkar, G.R. Adelli, K. Hippalgaonkar, M.A. Repka, and S. Majumdar, "Indomethacin-loaded solid lipid nanoparticles for ocular delivery: Development, characterization, and *in vitro* evaluation", *J. Ocul. Pharmacol. Ther.,* vol. 29, no. 2, pp. 216-228. 2013.
[http://dx.doi.org/10.1089/jop.2012.0069] [PMID: 23421502]

[62]   V. Jenning, and S. Gohla, "Comparison of wax and glyceride solid lipid nanoparticles (SLN®)", *Int. J. Pharm.,* vol. 196, no. 2, pp. 219-222. 2000.
[http://dx.doi.org/10.1016/S0378-5173(99)00426-3] [PMID: 10699722]

[63]   I.L. Dantas, K.T.S. Bastos, M. Machado, J.G. Galvão, A.D. Lima, J.K.M.C. Gonsalves, E.D.P. Almeida, A.A.S. Araújo, C.T. de Meneses, V.H.V. Sarmento, R.S. Nunes, and A.A.M. Lira, "Influence of stearic acid and beeswax as solid lipid matrix of lipid nanoparticles containing tacrolimus", *J. Therm. Anal. Calorim.,* vol. 132, no. 3, pp. 1557-1566. 2018.
[http://dx.doi.org/10.1007/s10973-018-7072-7]

[64]   C. Pizzol, F. Filippin-Monteiro, J. Restrepo, F. Pittella, A. Silva, P. Alves de Souza, A. Machado de Campos, and T. Creczynski-Pasa, "Influence of surfactant and lipid type on the physicochemical properties and biocompatibility of solid lipid nanoparticles", *Int. J. Environ. Res. Public. Health.,* vol. 11, no. 8, pp. 8581-8596. 2014.
[http://dx.doi.org/10.3390/ijerph110808581] [PMID: 25141003]

[65]   N. Naseri, H. Valizadeh, and P. Zakeri-Milani, "Solid lipid nanoparticles and nanostructured lipid carriers: Structure, preparation and application", *Adv. Pharm. Bull.,* vol. 5, no. 3, pp. 305-313. 2015.
[http://dx.doi.org/10.15171/apb.2015.043] [PMID: 26504751]

[66]   N. Yadav, S. Khatak, and U.V.S. Sara, "Solid lipid nanoparticles: A review", *Int J App Pharm,* vol. 5, no. 2, pp. 8-18. 2013.

[67]   J. Ezzati Nazhad Dolatabadi, H. Valizadeh, and H. Hamishehkar, "Solid lipid nanoparticles as efficient drug and gene delivery systems: Recent breakthroughs", *Adv. Pharm. Bull.,* vol. 5, no. 2, pp. 151-159. 2015.
[http://dx.doi.org/10.15171/apb.2015.022] [PMID: 26236652]

[68]   S. Mukherjee, S. Ray, and R.S. Thakur, "Solid lipid nanoparticles: A modern formulation approach in drug delivery system", *Indian J. Pharm. Sci.,* vol. 71, no. 4, pp. 349-358. 2009.
[http://dx.doi.org/10.4103/0250-474X.57282] [PMID: 20502539]

[69]   U. Numanoğlu, and N. Tarımcı, "Characterization of solid lipid nanoparticles (SLNTM) and their pharmaceutical and cosmetic applications", *J.Fac. Pharm. Ankara.,* vol. 35, no. 3, pp. 211-235. 2006.

[70]   H. Nsairat, D. Khater, F. Odeh, F. Al-Adaileh, S. Al-Taher, A.M. Jaber, W. Alshaer, A. Al Bawab, and M.S. Mubarak, "Lipid nanostructures for targeting brain cancer", *Heliyon,* vol. 7, no. 9, p. e07994. 2021.
[http://dx.doi.org/10.1016/j.heliyon.2021.e07994] [PMID: 34632135]

[71] S.V. Khairnar, P. Pagare, A. Thakre, A.R. Nambiar, V. Junnuthula, M.C. Abraham, P. Kolimi, D. Nyavanandi, and S. Dyawanapelly, "Review on the scale-up methods for the preparation of solid lipid nanoparticles", *Pharmaceutics,* vol. 14, no. 9, p. 1886. 2022.
[http://dx.doi.org/10.3390/pharmaceutics14091886] [PMID: 36145632]

[72] V.A. Duong, T.T.L. Nguyen, and H.J. Maeng, "Preparation of solid lipid nanoparticles and nanostructured lipid carriers for drug delivery and the effects of preparation parameters of solvent injection method", *Molecules,* vol. 25, no. 20, p. 4781. 2020.
[http://dx.doi.org/10.3390/molecules25204781] [PMID: 33081021]

[73] A. Borges, V. de Freitas, N. Mateus, I. Fernandes, and J. Oliveira, "Solid lipid nanoparticles as carriers of natural phenolic compounds", *Antioxidants,* vol. 9, no. 10, p. 998. 2020.
[http://dx.doi.org/10.3390/antiox9100998] [PMID: 33076501]

[74] J.S. Al, S.A. Alkhoori, and L.F. Yousef, "Phenolic acids from plants: Extraction and application to human health, In: Atta-ur-Rahman Eds", *Stud. Nat. Prod. Chem.,* vol. 48, pp. 389-417. 2018.

[75] V. Andonova, and P. Peneva, "Characterization methods for solid lipid nanoparticles (SLN) and nanostructured lipid carriers (NLC)", *Curr. Pharm. Des.,* vol. 23, no. 43, pp. 6630-6642. 2018.
[http://dx.doi.org/10.2174/1381612823666171115105721] [PMID: 29141534]

[76] N. Kathe, B. Henriksen, and H. Chauhan, "Physicochemical characterization techniques for solid lipid nanoparticles: Principles and limitations", *Drug Dev. Ind. Pharm.,* vol. 40, no. 12, pp. 1565-1575. 2014.
[http://dx.doi.org/10.3109/03639045.2014.909840] [PMID: 24766553]

[77] L. Xu, X. Wang, Y. Liu, G. Yang, R.J. Falconer, and C.X. Zhao, "Lipid nanoparticles for drug delivery", *Adv. NanoBiomed Res.,* vol. 2, no. 2, p. 2100109. 2022.
[http://dx.doi.org/10.1002/anbr.202100109] [PMID: 35179344]

[78] A.A. Attama, and C.C. Müller-Goymann, "Effect of beeswax modification on the lipid matrix and solid lipid nanoparticle crystallinity", *Colloids Surf. A Physicochem. Eng. Asp.,* vol. 315, no. 1-3, pp. 189-195. 2008.
[http://dx.doi.org/10.1016/j.colsurfa.2007.07.035]

[79] A. Deshpande, M. Mohamed, S.B. Daftardar, M. Patel, S.H.S. Boddu, and J. Nesamony, Solid lipid nanoparticles in drug delivery: Opportunities and challenges.*Emerging Nanotechnologies for Diagnostics, Drug Delivery and Medical Devices.,* Elsevier, pp. 291-330. 2017.

[80] R.K. Tekade, R. Maheshwari, M. Tekade, and M.B. Chougule, Solid lipid nanoparticles for targeting and delivery of drugs and genes. *Nanotechnology-Based Approaches for Targeting and Delivery of Drugs and Genes.,* Academic Press, pp. 256-286. 2017.
[http://dx.doi.org/10.1016/B978-0-12-809717-5.00010-5]

[81] G. Ak, A. Ünal, T. Karakayalı, B. Özel, N. Selvi Günel, and Ş. Hamarat Şanlıer, "Brain-targeted, drug-loaded solid lipid nanoparticles against glioblastoma cells in culture", *Colloids Surf. B Biointerfaces,* vol. 206, p. 111946. 2021.
[http://dx.doi.org/10.1016/j.colsurfb.2021.111946] [PMID: 34216850]

[82] R. Bhatt, D. Singh, A. Prakash, and N. Mishra, "Development, characterization and nasal delivery of rosmarinic acid-loaded solid lipid nanoparticles for the effective management of Huntington's disease", *Drug Deliv.,* vol. 22, no. 7, pp. 931-939. 2015.
[http://dx.doi.org/10.3109/10717544.2014.880860] [PMID: 24512295]

[83] N.M. Morsi, D.M. Ghorab, and H.A. Badie, "Brain targeted solid lipid nanoparticles for brain ischemia: Preparation and *in vitro* characterization", *Pharm. Dev. Technol.,* vol. 18, no. 3, pp. 736-744. 2013.
[http://dx.doi.org/10.3109/10837450.2012.734513] [PMID: 23477526]

[84] R.A. Khallaf, H.F. Salem, and A. Abdelbary, "5-Fluorouracil shell-enriched solid lipid nanoparticles (SLN) for effective skin carcinoma treatment", *Drug Deliv.,* vol. 23, no. 9, pp. 3452-3460. 2016.

[http://dx.doi.org/10.1080/10717544.2016.1194498] [PMID: 27240935]

[85]   L. Hu, Y. Jia, and WenDing, "Preparation and characterization of solid lipid nanoparticles loaded with epirubicin for pulmonary delivery", *Pharmazie,* vol. 65, no. 8, pp. 585-587. 2010.
[PMID: 20824958]

[86]   J. Desai, and H. Thakkar, "Darunavir-loaded lipid nanoparticles for targeting to HIV reservoir", *AAPS. Pharm.Sci.Tech.,* vol. 19, no. 2, pp. 648-660. 2018.
[http://dx.doi.org/10.1208/s12249-017-0876-0] [PMID: 28948564]

[87]   B. Aljaeid, and K.M. Hosny, "Miconazole-loaded solid lipid nanoparticles: Formulation and evaluation of a novel formula with high bioavailability and antifungal activity", *Int. J. Nanomed.,* vol. 11, pp. 441-447. 2016.
[http://dx.doi.org/10.2147/IJN.S100625] [PMID: 26869787]

[88]   R. Pandey, and G.K. Khuller, "Solid lipid particle-based inhalable sustained drug delivery system against experimental tuberculosis", *Tuberculosis,* vol. 85, no. 4, pp. 227-234. 2005.
[http://dx.doi.org/10.1016/j.tube.2004.11.003] [PMID: 15922668]

[89]   M. Vicente-Pascual, I. Gómez-Aguado, J. Rodríguez-Castejón, A. Rodríguez-Gascón, E. Muntoni, L. Battaglia, A. del Pozo-Rodríguez, and M.Á. Solinís Aspiazu, "Topical administration of SLN-based gene therapy for the treatment of corneal inflammation by de novo IL-10 production", *Pharmaceutics,* vol. 12, no. 6, p. 584. 2020.
[http://dx.doi.org/10.3390/pharmaceutics12060584] [PMID: 32586018]

[90]   Y Han, P Zhang, Y Chen, J Sun, and F Kong, "Co-delivery of plasmid DNA and doxorubicin by solid lipid nanoparticles for lung cancer therapy", *Int. J. Mol. Med.,* vol. 34, no. 1, pp. 191-196. 2014.

[91]   M.J. Ansari, M.K. Anwer, S. Jamil, R. Al-Shdefat, B.E. Ali, M.M. Ahmad, and M.N. Ansari, "Enhanced oral bioavailability of insulin-loaded solid lipid nanoparticles: Pharmacokinetic bioavailability of insulin-loaded solid lipid nanoparticles in diabetic rats", *Drug Deliv.,* vol. 23, no. 6, pp. 1972-1979. 2016.
[http://dx.doi.org/10.3109/10717544.2015.1039666] [PMID: 26017100]

[92]   S. Saraf, D. Mishra, A. Asthana, R. Jain, S. Singh, and N.K. Jain, "Lipid microparticles for mucosal immunization against hepatitis B", *Vaccine,* vol. 24, no. 1, pp. 45-56. 2006.
[http://dx.doi.org/10.1016/j.vaccine.2005.07.053] [PMID: 16122855]

[93]   X. Hou, T. Zaks, R. Langer, and Y. Dong, "Lipid nanoparticles for mRNA delivery", *Nat. Rev. Mater.,* vol. 6, no. 12, pp. 1078-1094. 2021.
[http://dx.doi.org/10.1038/s41578-021-00358-0] [PMID: 34394960]

[94]   Science.org, "Better fat bubbles could power a new generation of mRNA vaccines", Available From: https://www.science.org/ (Accessed on cited:10th December 2022),

[95]   B. Wilson, and K.M. Geetha, "Lipid nanoparticles in the development of mRNA vaccines for COVID-19", *J. Drug Deliv. Sci. Technol.,* vol. 74, p. 103553. 2022.
[http://dx.doi.org/10.1016/j.jddst.2022.103553] [PMID: 35783677]

[96]   R.H. Müller, and J.S. Lucks, "Arzneistofftrager aus festen Lipidteilchen, Feste Lipidnanospha¨ren (SLN)", patent EP0605497. 1996.

[97]   D.K. Mishra, R. Shandilya, and P.K. Mishra, "Lipid based nanocarriers: A translational perspective", *Nanomedicine,* vol. 14, no. 7, pp. 2023-2050. 2018.
[http://dx.doi.org/10.1016/j.nano.2018.05.021] [PMID: 29944981]

[98]   R.H. Müller, M. Radtke, and S.A. Wissing, "Nanostructured lipid matrices for improved microencapsulation of drugs", *Int. J. Pharm.,* vol. 242, no. 1-2, pp. 121-128. 2002.
[http://dx.doi.org/10.1016/S0378-5173(02)00180-1] [PMID: 12176234]

[99]   A. Khosa, S. Reddi, and R.N. Saha, "Nanostructured lipid carriers for site-specific drug delivery", *Biomed. Pharmacother.,* vol. 103, pp. 598-613. 2018.
[http://dx.doi.org/10.1016/j.biopha.2018.04.055] [PMID: 29677547]

[100] C. Tapeinos, M. Battaglini, and G. Ciofani, "Advances in the design of solid lipid nanoparticles and nanostructured lipid carriers for targeting brain diseases", *J. Control. Release,* vol. 264, pp. 306-332. 2017.
[http://dx.doi.org/10.1016/j.jconrel.2017.08.033] [PMID: 28844756]

[101] A. Beloqui, M.Á. Solinís, A. Rodríguez-Gascón, A.J. Almeida, and V. Préat, "Nanostructured lipid carriers: Promising drug delivery systems for future clinics", *Nanomedicine,* vol. 12, no. 1, pp. 143-161. 2016.
[http://dx.doi.org/10.1016/j.nano.2015.09.004] [PMID: 26410277]

[102] T. Waghule, V.K. Rapalli, S. Gorantla, R.N. Saha, S.K. Dubey, A. Puri, and G. Singhvi, "Nanostructured lipid carriers as potential drug delivery systems for skin disorders", *Curr. Pharm. Des.,* vol. 26, no. 36, pp. 4569-4579. 2020.
[http://dx.doi.org/10.2174/1381612826666200614175236] [PMID: 32534562]

[103] M. Haider, S.M. Abdin, L. Kamal, and G. Orive, "Nanostructured lipid carriers for delivery of chemotherapeutics: A review", *Pharmaceutics,* vol. 12, no. 3, p. 288. 2020.
[http://dx.doi.org/10.3390/pharmaceutics12030288] [PMID: 32210127]

[104] E.B. Souto, I. Baldim, W.P. Oliveira, R. Rao, N. Yadav, F.M. Gama, and S. Mahant, "SLN and NLC for topical, dermal, and transdermal drug delivery", *Expert Opin. Drug Deliv.,* vol. 17, no. 3, pp. 357-377. 2020.
[http://dx.doi.org/10.1080/17425247.2020.1727883] [PMID: 32064958]

[105] M. Elmowafy, and M.M. Al-Sanea, "Nanostructured lipid carriers (NLCs) as drug delivery platform: Advances in formulation and delivery strategies", *Saudi Pharm. J.,* vol. 29, no. 9, pp. 999-1012. 2021.
[http://dx.doi.org/10.1016/j.jsps.2021.07.015] [PMID: 34588846]

[106] A. Puri, K. Loomis, B. Smith, J.H. Lee, A. Yavlovich, E. Heldman, and R. Blumenthal, "Lipid-based nanoparticles as pharmaceutical drug carriers: From concepts to clinic", *Crit. Rev. Ther. Drug Carrier Syst.,* vol. 26, no. 6, pp. 523-580. 2009.
[http://dx.doi.org/10.1615/CritRevTherDrugCarrierSyst.v26.i6.10] [PMID: 20402623]

[107] C. Tapeinos, M. Battaglini, and G. Ciofani, "Advances in the design of solid lipid nanoparticles and nanostructured lipid carriers for targeting brain diseases", *J. Control. Release,* vol. 264, no. 264, pp. 306-332. 2017.
[http://dx.doi.org/10.1016/j.jconrel.2017.08.033] [PMID: 28844756]

[108] K Pathak, L Keshri, and M Shah, "lipid nanocarriers: Influence of lipids on product development and pharmacokinetics, critical reviews™ in therapeutic drug carrier systems", *Crit. Rev. Ther. Drug. Carrier. Syst.,* vol. 28, no. 4, pp. 357-393. 2011.

[109] M.A. Iqbal, S. Md, J.K. Sahni, S. Baboota, S. Dang, and J. Ali, "Nanostructured lipid carriers system: Recent advances in drug delivery", *J. Drug Target.,* vol. 20, no. 10, pp. 813-830. 2012.
[http://dx.doi.org/10.3109/1061186X.2012.716845] [PMID: 22931500]

[110] D. Liu, J. Li, H. Pan, F. He, Z. Liu, Q. Wu, C. Bai, S. Yu, and X. Yang, "Potential advantages of a novel chitosan-N-acetylcysteine surface modified nanostructured lipid carrier on the performance of ophthalmic delivery of curcumin", *Sci. Rep.,* vol. 6, no. 1, p. 28796. 2016.
[http://dx.doi.org/10.1038/srep28796] [PMID: 27350323]

[111] N. Rosita, A.A. Sultani, and D.M. Hariyadi, "Penetration study of p-methoxycinnamic acid (PMCA) in nanostructured lipid carrier, solid lipid nanoparticles, and simple cream into the rat skin", *Sci. Rep.,* vol. 12, no. 1, p. 19365. 2022.
[http://dx.doi.org/10.1038/s41598-022-23514-0] [PMID: 36371457]

[112] D. Sun, S. Guo, L. Yang, Y. Wang, X. Wei, S. Song, Y. Yang, Y. Gan, and Z. Wang, "Silicone elastomer gel impregnated with 20(S)-protopanaxadiol-loaded nanostructured lipid carriers for ordered diabetic ulcer recovery", *Acta Pharmacol. Sin.,* vol. 41, no. 1, pp. 119-128. 2020.
[http://dx.doi.org/10.1038/s41401-019-0288-7] [PMID: 31534201]

[113]   N. E Eleraky, M. M Omar, H. A Mahmoud, and H. A Abou-Taleb, "Nanostructured lipid carriers to mediate brain delivery of temazepam: Design and *in vivo* study", *Pharmaceutics,* vol. 12, no. 5, p. 451. 2020.
[http://dx.doi.org/10.3390/pharmaceutics12050451] [PMID: 32422903]

[114]   M. Battaglini, C. Tapeinos, I. Cavaliere, A. Marino, A. Ancona, N. Garino, V. Cauda, F. Palazon, D. Debellis, and G. Ciofani, "Design, fabrication, and *in vitro* evaluation of nanoceria-loaded nanostructured lipid carriers for the treatment of neurological diseases", *ACS Biomater. Sci. Eng.,* vol. 5, no. 2, pp. 670-682. 2019.
[http://dx.doi.org/10.1021/acsbiomaterials.8b01033] [PMID: 33405830]

[115]   P. Sharma, B. Dube, and K. Sawant, "Development and evaluation of nanostructured lipid carriers of cytarabine for treatment of meningeal leukemia", *J. Nanosci. Nanotechnol.,* vol. 11, no. 8, pp. 6676-6682. 2011.
[http://dx.doi.org/10.1166/jnn.2011.4235] [PMID: 22103067]

[116]   A. Gerhardt, E. Voigt, M. Archer, S. Reed, E. Larson, N. Van Hoeven, R. Kramer, C. Fox, and C. Casper, "A flexible, thermostable nanostructured lipid carrier platform for RNA vaccine delivery", *Mol. Ther. Methods Clin. Dev.,* vol. 25, pp. 205-214. 2022.
[http://dx.doi.org/10.1016/j.omtm.2022.03.009] [PMID: 35308783]

[117]   M. Mao, S. Liu, Y. Zhou, G. Wang, J. Deng, and L. Tian, "Nanostructured lipid carrier delivering chlorins e6 as in situ dendritic cell vaccine for immunotherapy of gastric cancer", *J. Mater. Res.,* vol. 35, no. 23-24, pp. 3257-3264. 2020.
[http://dx.doi.org/10.1557/jmr.2020.227] [PMID: 33424109]

[118]   RS Fernandes, JO Silva, and SV Mussi, "Nanostructured lipid carrier co-loaded with doxorubicin and docosahexaenoic acid as a theranostic agent: Evaluation of biodistribution and antitumor activity in experimental model", *Mol. Imaging. Biol.,* vol. 20, p. 437Y447. 2018.

[119]   X. Zhu, X. Deng, C. Lu, Y. Chen, L. Jie, Q. Zhang, W. Li, Z. Wang, Y. Du, and R. Yu, "SPIO-loaded nanostructured lipid carriers as liver-targeted molecular T2-weighted MRI contrast agent", *Quant. Imaging Med. Surg.,* vol. 8, no. 8, pp. 770-780. 2018.
[http://dx.doi.org/10.21037/qims.2018.09.03] [PMID: 30306057]

CHAPTER 3

Metallic Nanoparticles: Synthesis and Applications in Medicine

Şeref Akay[1,*] and **Sultan Eda Kuş**[1]

¹ Faculty of Engineering, Department of Genetics and Bioengineering, Alanya Alaaddin Keykubat University, Antalya, Turkey

Abstract: The progress in nanoscience and advances in the fabrication, characterization, and modification of materials at the nanoscale have paved the way for the production and use of nanoparticles with different properties. Today, the chemical agents used in many therapies cannot achieve the desired effectiveness due to dose-dependent toxicity, low solubility and bioavailability, damage to non-target organs and tissues due to non-specificity, and side effects. Nanoparticle systems produced in different forms and compositions are one of the main approaches used to eliminate the negative aspects of conventional chemical agents. Among these nanoparticle systems, metallic nanoparticles represent a promising approach. During the last two decades, metallic nanoparticles (MNPs) have drawn great attention due to their optical, electrical, and physicochemical properties as well as their size-dependent properties. The large surface to volume ratio and surface reactivity of metallic nanoparticles provide great potential for combining them with different biological/chemical agents, as well as they can also be formulated as a bioactive nanoplatform alone. In this regard, the present chapter summarizes the general aspects of metallic nanoparticles, common methods for synthesis, and various applications in the biomedical field.

Keywords: Antibacterial, Antitumor, Biogenic, Bottom-up, Cancer, Chemical reduction, Diagnostic, Drug delivery, Gold nanoparticles, Green synthesis, Chemotherapy, Iron oxide, Laser ablation, Magnetic nanoparticles metallic nanoparticles, Nanomaterial, Nanotechnology, Silver nanoparticles, Targeting, Top-down.

INTRODUCTION

The ability to detect the Earth's magnetic field and navigate through it is widespread in nature. The most well-researched organisms are *magnetotactic bacteria*, abundant aquatic microorganisms, which produce distinctive organelles called magnetosomes to accomplish *magnetotaxis* and to seek for zones that

---

* **Corresponding author Şeref Akay:** Faculty of Engineering, Department of Genetics and Bioengineering, Alanya Alaaddin Keykubat University, Antalya, Turkey; Tel: 90 242 510 61 20; E-mail: seref.akay@alanya.edu.tr

Habibe Yılmaz (Ed.)

promote growth in chemically diverse water layers and sediments [1]. *Magnetospirillum gryphiswaldense* is one of the renowned microbes representing magnetotactic bacteria.

These strains use the Earth's magnetic field for navigating owing to magnetosomes, which consist of magnetic iron nanocrystals enclosed in a membrane [2]. The nanocrystals in the magnetosomes must be aligned in a chain to create effective magnetoreception, as shown in Fig. (**1**), instead of agglomerated forms. Thus, the magnetic moments of each nanocrystal in the chain produce a dipole potent enough to line up the entire bacterium within the low geomagnetic field [1, 3]. Empty membrane capsules are formed first in the biosynthesis of magnetosomes, and large amounts of iron are transferred and magnetites (Fe_3O_4) are produced inside these vesicles [2]. These nanosized particles can work like a navigation system because they show single-field magnetic properties.

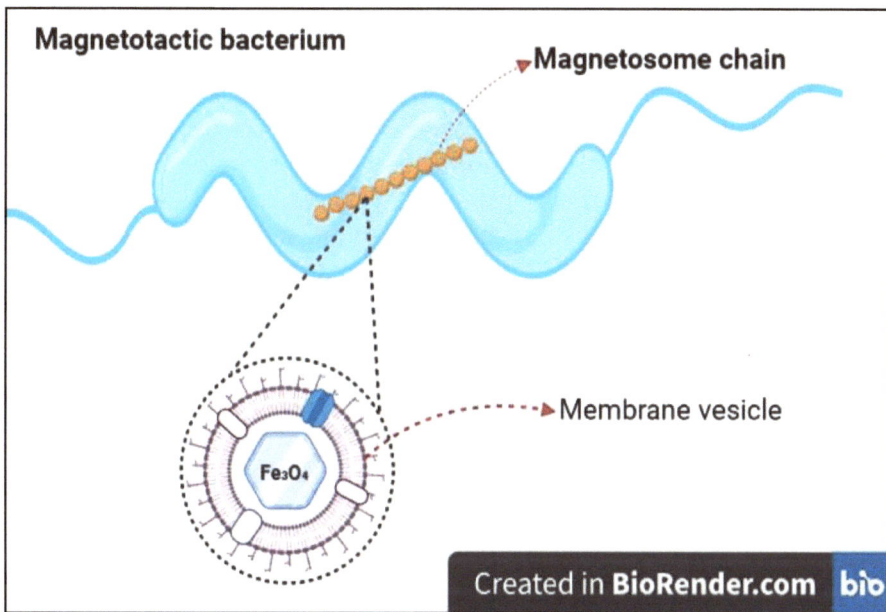

Fig. (1). Magnetosomes alignment to help bacteria floating in the water to navigate.

Instead, if the bacteria produced a single bulk material with the size of the individual particle chain, the magnetic field would become ineffective, and the bacteria would not be able to align according to the magnetic field inclination of the region [3]. This example is a very good example of the changing properties of metallic materials at the nanoscale and the benefit of nature from these changing properties.

Configurations of nanometric materials cause changes in their physical, chemical, and biological behaviors. This effect, which is defined as the quantum confinement effect, states that the properties of materials depend on their size at the nanoscale. The size-dependent properties include chemical reactivity, melting point, electrical conductivity, magnetic penetration, and alter as a function of particle size. The most substantial outcome of quantum effect is providing the ability to researchers for adjusting the properties of the material for required purposes [4]. Unpredictable alterations in surface properties due to the size at the nanoscale have drawn the attention of different fields of nanomaterials. Because the surface area and reactivity establish great potential for many applications [5].

Metallic or metal nanoparticles (MNPs) usually have a metal core consisting of an inorganic metal or metal oxide coated with a shell of organic or inorganic material. Mostly copper, gold, silver, iron, aluminum, lead, and zinc metals are used.

MNPs have emerged as a new type of nanomaterial in the field of nanotechnology in the last few years. Noble metals with useful effects on health, such as gold, silver, and platinum, are used for the production of nanoparticles. Significant advances have appeared in colloid science over the past century, including the introduction of pioneering methodologies for the production of metals, nano forms of metal oxides, and organic products. Especially with the use of MNPs as drug release and carrier systems in cancer treatment, substantial progress has been made in this area. Today, MNPs and related nanostructures are investigated more closely because of their outstanding properties which are beneficial for catalysis, material science, detection and treatment of disease, and sensor technology [5, 6].

MNPs are considered flexible and interesting nanostructures due to both the simple synthesis and the wide range of properties they present. This flexible platform provides high applicability in many fields, such as biomolecular ultrasensitive sensing, protein and cell labeling, targeted delivery of intracellular therapeutic agents, and hyperthermal therapy for cancer. In terms of comprehending their optical and electronic features, MNPs also offer an advantage over other nanoplatforms [6 - 8].

Many MNPs in commercial goods such as personal care products (cosmetics, toothpaste, *etc.*), detergents, drugs, and pharmaceuticals are in direct contact with the human body. Even with the concerns about the toxicity of some MNPs, they have been mostly proven advantageous with appropriate size and dosage. Besides the most desirable properties, such as high surface area and antimicrobial effect, it is thought that the simplicity of functionalization of MNPs can improve their biomedical capacity and enhance their therapeutic performance [9, 10].

The interaction ability of MNPs with most components of the immune system provides them with advanced bioactivity and the morphology of the nanoparticles is the key for the desired activity. This interaction ability and related bioactivity have resulted in increasing interest and potential applications of MNPs. MNPs have also been applied in many fields due to their impressive physicochemical properties and high surface-to-volume ratios compared to their bulk forms and their resistance to harsh environments. Another useful property of MNPs is their ultrafine size, which makes them penetrate and be absorbed more easily through the biological membrane and selectively delivered to certain affected cells [11 - 13]. The key attributions that make MNPs beneficial are highlighted in Fig. (2).

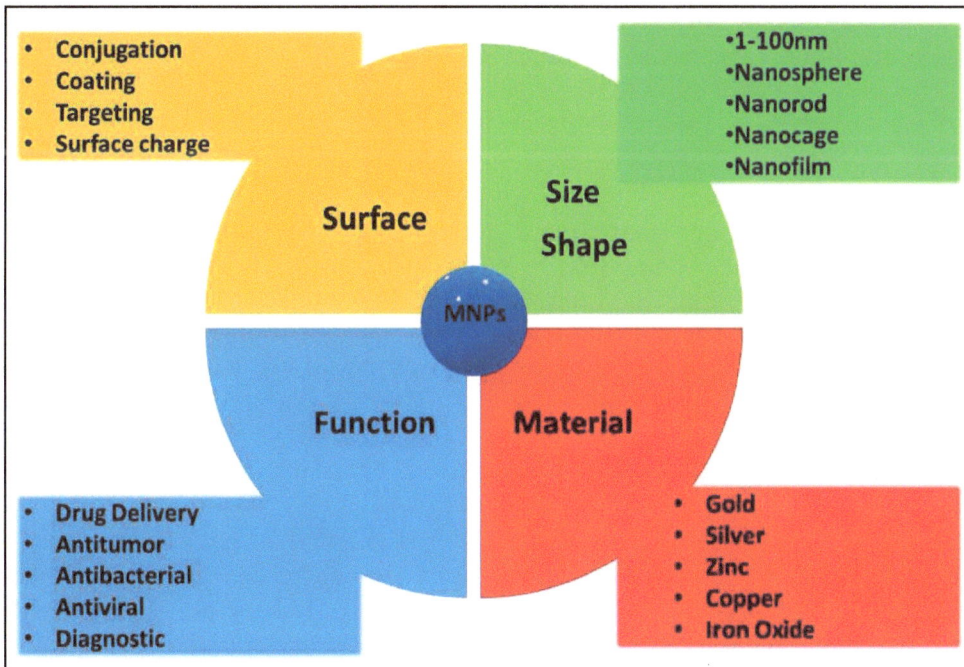

- Conjugation
- Coating
- Targeting
- Surface charge

•1-100nm
•Nanosphere
•Nanorod
•Nanocage
•Nanofilm

Surface

Size

Shape

MNPs

Function

Material

- Drug Delivery
- Antitumor
- Antibacterial
- Antiviral
- Diagnostic

- Gold
- Silver
- Zinc
- Copper
- Iron Oxide

Fig. (2). General highlights of MNPs.

Gold nanoparticles (AuNPs) and other MNPs, such as silver (AgNPs) and iron oxide nanoparticles (IONPs), are widely used for numerous biomedical applications including diagnostic, treatment, and drug delivery purposes. AgNPs are renowned for their antimicrobial and inflammatory capabilities. This property is utilized selectively to promote healing, and many commercial wound dressings, implant coatings, and pharmaceutical dosages are formulated with a combination of AgNPs. The use of MNPs is persistently rising globally in biomedicine and related fields [5].

Colloidal gold is attracting increasing attention in nanotechnological progress and development due to its nanosized dimensions. Optical properties and color of AuNPs change depending on the size and the morphology of particles. For example, spherical AuNPs smaller than the size of 30 nm are red, while the particles up to 100 nm color change to pink and larger sizes appear darker [14]. Major properties of AuNPs that make them very popular are their easy synthesis, low toxicity, and interaction ability with targeted biomolecules [15].

The shape and size of AuNPs can be readily controlled in the synthesis process. The usual size is in the range of 1 to 150 nm and with variations in size and shape, unique physiochemical and optical/electrical properties can be achieved. AuNPs have the highest stability, and their characteristics are almost the same even though their concentrations are changed. Their biocompatibility is also an important reason for being widely used in biomedical applications such as tumor detection and therapy [16]. AuNPs are excellent candidates for electron-dense labeling agents, thus, they have applications in the areas of histochemistry and cytochemistry. Moreover, AuNPs provide a functional and large surface for conjugating and immobilizing biomolecules such as proteins, enzymes, and antibodies for biocatalysis and biosensing applications [14]. The functionalization of AuNPs through ligands and alterations in their features as a result of analyte ligand binding, provide a versatile platform for various applications, such as self-assembly, transformation of particles into DNA templates, bioelectronics-based detection techniques [17].

AgNPs which were synthesized in 1880, are potent antiseptic and antibacterial agents used in the environmental and biomedical field. Nowadays, AgNPs are usually used in antibacterial applications due to the lethal effects of high silver ion release [14]. On the other hand, AgNPs also have large surface area, which can be functionalized through interaction with different polymers and biomolecules. AgNPs are used in a variety of applications, such as wound dressings, drug administration, and surface coating for surgical instruments and implants. Because of their antibacterial and antiviral properties, they have been employed in the development of tissue scaffolds [9].

SYNTHESIS OF METALLIC NANOPARTICLES

Nanoparticles are commonly synthesized using a variety of procedures, including physical, chemical, and biological methods. The choice of metal nanoparticle preparation is important, since the mechanism of interaction of metal ions with the reducing agent significantly affects the structure, size, and stability of the resulting nanoparticles [5].

Production methods of MNPs are categorized into two main types as *Bottom-Up* and *Top-Down Methods* as represented in Fig. (**3**). The main difference between both methods is the form of starting material; in Top-Down techniques, the synthesis starts with bulk materials, whereas in Bottom-Up, the synthesis is initiated at atomic levels.

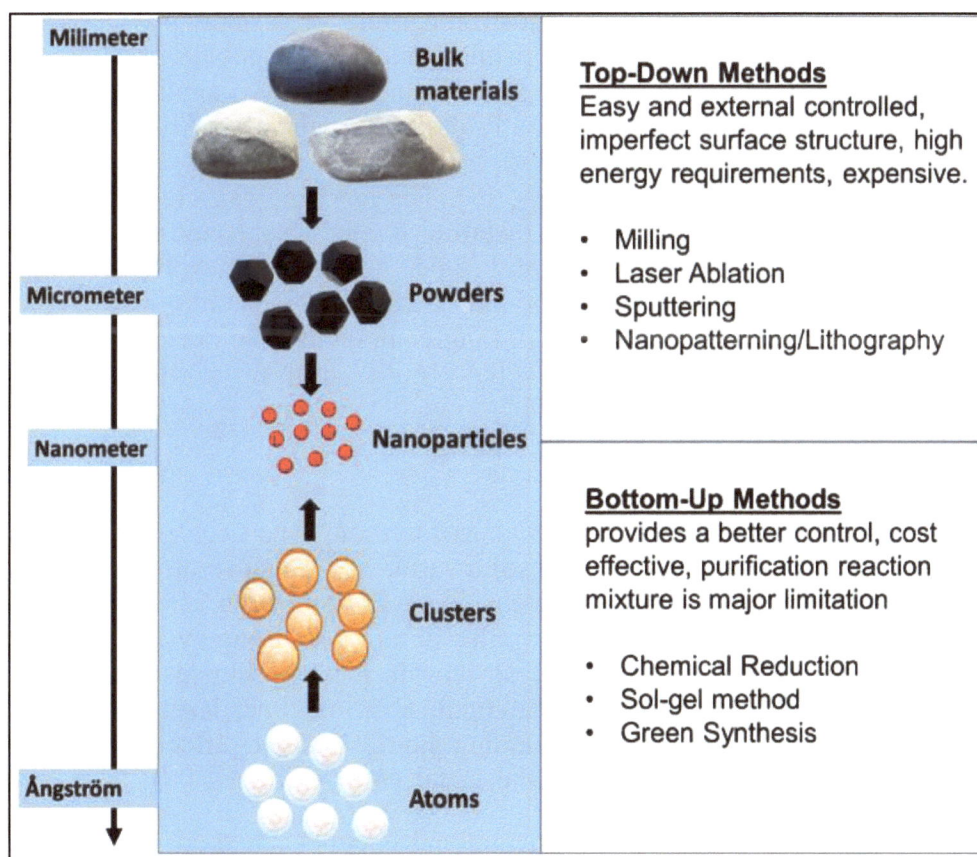

Fig. (3). Schematic representation of Top-Down and Bottom-Up methods for nanoparticle synthesis.

Top-Down

In this method, the bulk material is converted into nanosized particles by applying different physical and chemical processes [18]. Although Top-Down methods are easy to apply, they are not suitable for preparing particles of informal shape and very small size. The main constraint with Top-Down approach is generating nanoparticles with imperfect surfaces and structures which significantly alter their physicochemical properties. These methods, moreover, involve high-pressure and high-temperature processes which increase the cost of manufacturing due to the large energy requirement [12].

Milling

It is the most concrete Top-Down example as it involves the direct breakdown of bulk materials into micro/nano structures. The working principle of mechanical grinding is to reduce the grain size by high energy ball grinding. Depending on the mechanical energy induced in the powder mixture, it is classified as low-energy and high-energy grinding. Nanosized particles are usually produced using a high-energy ball milling process. In this method, bulk powder is added to a container with several heavy balls. High mechanical energy is applied to the bulk powder material with the help of the ball rotating at high speed. Since the kinetic energy generated by the balls depends on mass and velocity, heavy materials such as ceramics are favored for milling [6, 19]. The resulting particle size is in the range of 3-25 nm if a single phase of metallic powder is subjected to mechanical milling [19]. Parameters such as mill type, grinding atmosphere, grinding medium, density, time, and temperature play a crucial role in controlling the shape and size of MNPs. Various types of milling equipment have been developed to overcome the limitations, including agitator mills, drum mills, vibratory mills, abrasive mills, and planetary mills [12].

Laser Ablation

In this method, strong laser irradiation is used to remove particles from the target materials. As the laser beams hit the solid surface of the substrate, the particles evaporate from the surface due to the charge energy of the laser beam and result in nanoparticle formation. The laser pulse duration and energy determine the relative amount of ablated atoms and particles formed. Various parameters such as duration of the laser, pulse, wavelength, ablation time, laser fluence, and presence of surfactant in the surrounding liquid medium affect the ablation efficiency and the characteristics of the metal particle formed [19].

Laser ablation is preferred over conventional chemical methods due to time-saving processing, improving control of the size and shape of particles with higher yields, and providing long-term stability [12].

During the processing, no solvent is needed, thus, the absence of chemical contamination is another advantage of this method [13]. However, the high energy requirements for generating a laser beam and, increasing and controlling the temperature of both the target and surrounding media limits the application of laser ablation [20].

Sputtering

Sputtering is a process that uses high-energy gas or plasma particles to erode target materials and deposit them on a surface to create thin films of nanomaterials. The process involves highly energetic bombarding that causes the physical subtraction of atoms or clusters due to the momentum transfer between the surface and particles [19]. The whole process is carried out in an evacuated chamber with low pressure, where the sputtering gas is manipulated by applying a high voltage to generate gas ions.

Different sources can be used for sputtering, such as radio frequency, magnetron, and DC diodes. The sputtering is attractive because the resulting nanoparticles have the same composition as the substrate material [18]. The success of sputtering; depends on parameters including layer thickness, type of substrate and gas, temperature, and size and shape of nanoparticles [12].

Nanopatterning/Lithography

Lithography is a practical technique that uses a light beam or electrons to selectively remove nanostructures from the substrate material. The method relies on the patterning of a material surface through exposure to light or electrons, and an etching and/or decomposition stage that selectively removes the undesired material to fabricate the actual pattern [18, 21]. This approach is regarded as an imprinting process in which a substrate is cut or formed into the desired pattern structure by an electron or light beams. Lithography can be applied in two ways according to the exposure method. If the process requires a mask for transferring nanopatterns, the technique is called mask nanolithography. On the contrary, it is maskless lithography if the process involves direct printing or writing of the desired nanostructures on the surface by probe lithography, focused ion beam lithography, or electron beam lithography. Lithography is conducted at mild temperature and pressure and uses metal precursor inks for the synthesis of MNPs [12, 18].

Bottom-Up Methods

Nanoparticle synthesis methods using the Bottom-Up approach rely on the generation of nanoparticles from smaller molecules, such as the assembly of atoms, molecules, or small particles. In this method, nanostructured building blocks of nanoparticles are first generated and then combined to produce the ultimate nanostructure using chemical and biological procedures. Bottom-Up approaches provide more precise control for particle formation of uniform size, shape, and composition. These methods are advantageous also due to the reduced cost of production procedures [5, 22].

Chemical Reduction

One of the most widely applied Bottom-Up synthesis methods is chemical reduction. Due to low-cost processing and simplicity, it can be employed for the synthesis of different MNPs [6]. This approach involves a redox reaction in which first, ionic salts of the metal precursors have formed, and the metal ions in solution are reduced by diverse chemical agents under controlled conditions [5, 6, 12]. Various types of reducing agents such as sodium borohydrate, hydrazine, lithium aluminum hydride, sodium citrate, tannic acid, alcohols, *etc.* can be utilized in the chemical reduction of metal precursors to form MNPs [12]. It is also important to prevent agglomeration during nanoparticle synthesis by chemical reduction. Therefore, protective stabilizing agents are used for avoiding particles from agglomeration or sedimentation. Stabilizers with different functionalities interact with the particle or are adsorbed on the surface of the particles and stabilize the nanoparticle growth. In general, low molecular weight compounds such as citrate, SDS, or chitosan and high-molecular-weight polymers such as starch, Polyvinylpyrrolidone (PVP), and Polyethylene glycol (PEG), are employed as stabilizing agents [6]. Despite the advantages and simplicity of the chemical reduction approach, there are some factors that limit the application of chemically produced MNPs in biomedical fields. As the synthesis involves toxic chemicals and solvents, and due to the toxicity of stabilizing agents in the latter stage of synthesis, concerns have arisen about the biocompatibility of the as-produced nanomaterials. The toxic side products of chemical reduction methods lead to environmental issues as well [6]. To achieve uniform particle size and desired shape, the chemical reduction can be regulated by adjusting the reaction parameters such as temperature, pressure, duration, mixing, and the ratio of reducing and stabilizing agents [12].

Sol-Gel Method

A wet chemical approach called the sol-gel method is widely employed for the formation of nanomaterials. This technique is utilized to create a variety of metal-oxide-based nanomaterials. This process is called as a sol-gel method due to the formation of two phases: a solution and a gel phase. The metal precursors are first obtained in solution form, and then the sol form is converted to a gel which consists of a network structure [5]. The sol-gel procedure involves several steps in the formation of nanostructures. Metal oxides are hydrolyzed in the first stage to form a solution. Next, the solution is allowed to evaporate, and the removal of solvents leads to the formation of a viscous solution. At the same time, the polycondensation reactions occur and result in the formation of hydroxo-(M–OH–M) or oxo-(M–O–M) bridges. In the subsequent steps, particle growth and agglomeration of the particles occur. Then the particles are left for aging and drying to remove water and solvent. Lastly, the particles are calcinated to obtain nanoparticles. Particle size and morphology can be controlled by systematic monitoring of reaction parameters such as hydrolysis and aging duration, pH, and concentration of precursor. The sol-gel approach provides several advantages including homogenous particle formation, applicability in low temperatures, and being a versatile method for complex nanostructures [18].

Biogenic Method

Although physical and chemical approaches are more commonly used in nanoparticle manufacturing, the usage of harmful substances severely limits their biological applications. Therefore, to expand its biomedical applications; biocompatible, non-toxic, environmentally friendly methods had to be developed. Besides the eco-friendly nature, biological approaches provide many advantages, including sustainable resources, low-cost strategies, and equipment. Furthermore, green synthesis does not require the use of high pressure, temperature, and toxic chemicals [23]. In this perspective, green synthesis utilizes nontoxic and benign materials including bacteria, enzymes, fungi, yeast, and plants, and extracts derived from these biological sources. The reduction mechanism is similar to the chemical methods, however, instead of using synthetic toxic chemicals in the green synthesis of MNPs, metal ions are reduced by biological molecules such as proteins, phenolic compounds, alkaloids, and flavonoids [13]. The various contents of biomolecules in biological environments play a key role in not only reducing metal ions to nanoparticles but are also important for subsequent capping and stabilizing steps [24].

The green production of MNPs by plants is the most preferred route for synthesis due to low risks and mild conditions. The ability of herbal extracts to reduce metallic precursors is due to their rich phytochemical composition that consists of various compounds such as alkaloids, phenolic acids, carboxylic acids, polyphenols, flavonoids, proteins, sugars, terpenoids, ketones, aldehydes, and amides. The procedure for the green route is a facile method, accomplished by contacting the metal solution with a plant extract at room temperature. The color change of the mixture solution after a short while indicates the formation of MNPs [6, 25]. Various parts of plants including leaves, plants, roots, barks, and fruits can be utilized as a source of reducing media. There are number of research that report the formation of MNPs using extracts derived from different plants such as Ginkgo [26], Nimtree [27], mallow [28], pumpkin [29], and Tulsi [30].

Microbial cells are able to reduce the metal ions that they accumulate from the surrounding environment and convert them into nanoparticle form. The conversion is generally performed by the enzymatic action of microorganisms and the responsible enzyme for this reaction is mostly reductase. The microbial synthesis of MNPs can be extracellular in which the nanoparticle formation occurs at the cell membrane surface, or intracellular in which nanoparticles are generated inside the cell [12]. *Pseudomonas stutzeri* AG259 isolated from a silver mine, which is the first bacteria demonstrated to accumulate silver and form AgNPs with the size 35 to 46 nm [31]. Similarly, bacteria-mediated synthesis of MNPs was also reported in *Bacillus thuringiensis, Corynebacterium, Bacillus cereus, Ureibacillus thermosphaericus, Escherichia coli* and *Klebsiella pneumonia* [6, 12, 24, 25].

Fungi have recently emerged as better alternatives for the biogenic creation of MNPs because of their high intracellular metal absorption capability, ability to create nanoparticles with various chemical compositions, and ability to produce a vast quantity of proteins [12]. Mukherjee *et al.*, (2002) demonstrated the production of AgNP in the fungus *Verticillium* by contacting the fungal biomass with silver solution [32]. In another study, the synthesis of iron oxide nanoparticles has been reported in the endophyte *Penicillium oxalicum* [33].

Algae are considered another important biological source for the biological synthesis of MNPs due to their capability of accumulating heavy metals. They also comprise fats, carbohydrates, proteins, minerals, and carotenoids which are significant biomolecules in reducing, stabilizing, or capping of MNPs. *Sargassum wightii, Chlorella vulgaris, Spirogyra insignis,* and *Fucus vesiculosus* are some algae that are reported to involve in the biological synthesis of MNPs [6, 34].

BIOMEDICAL APPLICATIONS OF MNPS

Drug Delivery

Drug delivery systems (DDS) are the most promising developments for health care applications and despite the enormous advancement in the field, it is still continuously progressing. This is because of the challenge and the difficulties in conventional drug administration routes to achieve the desired target and dose [35, 36]. Drug delivery systems have been progressing in multidisciplinary research field for more than two decades and made an effective impact on the treatment of different medical issues [35, 37]. Nanoscale materials offer almost infinite possibility to tailor the fundamental properties of the drug molecule including release kinetics, solubility, half-life, and bioavailability. Regarding the diversity of physiological processes occurring at nanoscales, the relatively smaller size compared to cells and organelles, makes these materials a promising nominee for DDS [35].

Metallic Nanoparticles (MNPs) have been selected and handled as smart DDS because of their direct targeting ability and thereby reducing the side effects. In recent years, many types of MNPs have been developed for cancer treatment as well as many others have been utilized as drug delivery carriers for various agents such as antibodies, peptides, nucleic acids, and chemotherapy agents. Functionalization of MNPs also enhances the capabilities of carrying more and various drugs and effective inside delivery with minimum adverse effects. MNPs are unsurpassed compared to other nanoparticles due to a higher surface, high pore size and volume, and adjustable optical properties. Moreover, their surface can be easily modified or coated to harbor targeting agents and active biomolecules, other than to improve biocompatibility and reduce toxicity. In addition, MNPs and their conjugated forms allow multiple drug loading, enhance the solubility of hydrophobic agents, and improve circulation time, thereby preventing rapid renal excretion [35].

The design of MNPs to provide slow and sustained release of drugs and the ability to deliver the drug payload to the target site without affecting the other cells are two important parameters for the efficiency of MNPs-based DDS. These two factors can be satisfied by the size and functionalization possibility of MNPs. MNPs have gained great focus because of the distinctive material and size-dependent physicochemical properties which organic NPs do not pose [35].

Miller-Kleinhenz *et al.* (2018) designed ultra-small magnetic iron oxide nanoparticle (IONP) drug carriers functionalized with peptides to target and inhibit cancerous cells. It has been shown that the developed targeted FeO-Doxorubicin nanoparticles inhibited breast cancer cell invasion compared to conventional treatment [38].

Various uses of metallic nanoparticles as active biomolecule carriers, pave the way for efficient therapy for many diseases [36]. Gold nanoparticles (AuNPs) have been utilized widely as potential drug carriers among all types of noble metallic nanoparticles. Producing AuNPs with various size ranges (1-100nm) and shapes (nanospheres, nanorods, *etc.*) is rather simple. AuNPs can be biofunctionalized to modify the surface properties in order to achieve desired characteristics such as biocompatibility, stability, and biodistribution [36]. In a recent study, the design of AS-1411 and chitosan functionalized AuNPs to deliver methotrexate (MTX) into cancer cells has been reported. The developed drug delivery system with an average size of 62 ± 2.4 nm showed that they are able to accumulate through AS-1411 targeting mechanism, penetrate breast cancer cells, and release the therapeutic drug [39].

Valdivia *et al.* (2022) reported the production of gold NPs and iron oxide NPs with micellar functionalization for the delivery of dexamethasone, a corticosteroid frequently used to treat a variety of diseases such as arthritis, allergic reactions, and inflammation. The micellar form was developed through the self-assembly of surfactant molecules, and MNPs and drugs were encapsulated in the micellar form. The systems have been shown to carry up to 74% of drugs with reduced toxicity, high stability, and homogeneity in aqueous solution [40].

MNPs have been suggested as efficient vehicles for the delivery of antibodies, peptides, and RNA/DNA to target different types of cells. The latest studies revealed that different types of AuNPs prevent the degradation of nucleic acids by nuclease. Oligonucleotides conjugated AuNPs have been shown promising agents for the delivery of gene regulatory compounds [36, 41].

DNA-conjugated AuNPs developed by Yokomori *et al.* (2022) have demonstrated the successful retaining of encapsulated CRISPR/Cas9 ribonucleoproteins for genome editing purposes. DNA–AuNPs functionalization through biotin moieties has been shown effective in retaining the target molecule. The study revealed the potential of MNP-DNA conjugation as carriers for direct protein/nucleic acid delivery [42]. Some specific examples MNPs developed for drug delivery and treatment purpose are presented in Table **1**.

Table 1. Some MNPs, related synthesis, and applications.

Type of MNPs	Production Method	Particle Size nm	Drug and Functionalization	Application and Outcomes	Refs.
AuNPs	chemical reduction with oleylamine	240.7 ± 7.4	oleic acid based surface encapsulation /dexamethasone	hybrid drug delivery nanomaterials with high stability and monodispersity	[12]
ZnO	chemical reduction with oleylamine	21–39	ZnO-quercetin	inhibition of the growth of breast cancer cells (MCF-7) with high biocompatibility	[43]
AuNPs	chemical reduction with trisodium citrate	62 ± 2.4	AS-1411 aptamer and Chitosan conjugated/methotrexate	successfully target breast cancer cells and release the therapeutic drug	[39]
AuNPs	chemical reduction with sodium citrate	50 -100	gelatin coated methotrexate	cytotoxic effect on the MCF-7 breast cancer cell line	[44]

(Table 1) cont.....

Type of MNPs	Production Method	Particle Size nm	Drug and Functionalization	Application and Outcomes	Refs.
AuNPs	reduction by sodium citrate	72.56	PVP-coated Folic acid functionalized curcumin	folate-based tumor-specific targeting and delivery to the breast without harming normal cells	[45]
AuNPs	reduction by sodium black tea extracts	19	theaflavin conjugated	internalization and apoptotic ability of theaflavin enhanced against ovarian teratocarcinoma cell line PA-1	[46]
AgNPs	reducing with NaBH4	21.14 ± 9.48	encapsulated in dipalmitoyl-phosphatidyl choline (DPPC)	stable nanocapsule with enhanced cytotoxicity at a lower dose of AgNPs	[47]
AgNPs	reducing with β-cyclodextrin	9	–	toxicity through membrane depolarization and ROS generation	[48]
AgNPs	reduction by extract of Ginkgo biloba leaves	40.2 ± 1.2	–	inhibit tumor proliferation, ROS generation and mitochondrial caspase apoptotic pathway	[26]
IONP (Fe_3O_4)	crystallization of the Fe3O4 shells with oleic acid	4-12	dihydroxyphenyl)propionic acid (DHCA)- and F56 peptide conjugated	exhibited T1–T2 dual-mode MRI imaging and tumor-targeting performance	[49]

Anticancer Therapy

Cancer is the leading threat with increasing frequency worldwide for healthcare due to high mortality and reducing life quality as a consequence of conventional chemotherapy. Tumors are composed of cancer cells surrounded by the extracellular matrix, which contains various types of normal cells, including immune and adipocyte cells, and a unique and complex tumor microenvironment (TME) [50, 51]. Despite the great development in cancer therapy that has been recorded so far, classical approaches are still challenging because of side effects and low success of recovery. For many antitumor treatments, the success of the therapy depends on several parameters such as tumor type, size, aggression, and localization. The foremost challenges usually observed in antitumor therapy are the lack of precise therapy and low drug accumulation inside the tumor microenvironment. Consequently, untargeted healthy tissues are affected through adverse effects of non-specific therapy. Besides, the complex TME itself imposes a lot of challenges for treatment procedures [51 - 53]. Therefore, antitumor therapies are now interested in engineered approaches that aim at patients with particular disease profiles and characteristics. Nanomedicine, which emerged as a combination of nanotechnology, biomaterials, and medicine has arisen as a promising method for combating these obstacles. MNPs have been offered as an alternative attempt for the more effective cancer therapy. MNPs are favorable in cancer treatment due to narrow size distribution, having various shapes, and the possibility to modify surfaces. Furthermore, MNPs can readily penetrate tumor cells or TME thanks to their small size and high density. Besides, they can be used in combination with external stimuli such as light and ultrasound for more effective and localized therapies [51, 54].

Despite the frequently applied cancer therapies such as chemo/radiotherapy, and surgery, nanomedicine-based strategies are providing targeting ability and non-invasive approaches for the treatment of certain cancer types [51].

MNPs can make the tumor microenvironment (TME) reachable by converting inconvenient conditions to favorable ones. Particularly, the capability of targeting and sensitizing the target tissues to external stimuli, such as light, sound, and magnetic fields, can be improved by using MNPs [23]. Some types of MNPs do not require external stimulation, yet they are naturally responsive to internal tumor conditions such as pH, hypoxia, and redox potential. Furthermore, their properties can be adjusted to respond to certain stimulus mechanisms by modifying the surface with different biomolecules and coatings. Considering their beneficial properties, MNPs and metal-derived nanostructures have been increasingly investigated in a number of studies as novel techniques to fight cancer. The main mechanisms of MNPs for inhibiting the proliferative effects of tumor cells include the generation of reactive oxygen species (ROS), changing mitochondrial membrane activation, caspase expression, and blocking anti-apoptotic protein expression [54].

The fundamental approaches utilizing MNPs currently used in cancer therapy are summarized in Fig. (4).

The attractive advantages of AuNPs such as reduced toxicity, biocompatibility, superior stability, and improved retention capacity have made them the most investigated MNPs amongst others. Moreover, surface modification and engineering through numerous biological and chemical manufacturing methods allow the design and synthesis of AuNPs with desired size and structures. For example, nanospheres with tiny surface areas are expedient for the fabrication of cytotoxic agents, whereas nanoshells with large surface areas and internal pores are favorable for drug delivery applications. Hence, the flexible and adjustable nature of AuNPs puts them forward as adequate vehicles that are capable of targeting and inhibiting cancer cells. The other adjustable optical properties, photothermal properties, and surface plasmon resonance (SPR) are beneficial for different cancer therapy modalities such as tools in phototherapy and photoimaging including magnetic resonance imaging (MRI), photoacoustic imaging (PAI), and X-ray scatter imaging. Hyperthermal treatment is a good example of using AuNPs in cancer, where they produce local heat as a response to specific frequencies [54].

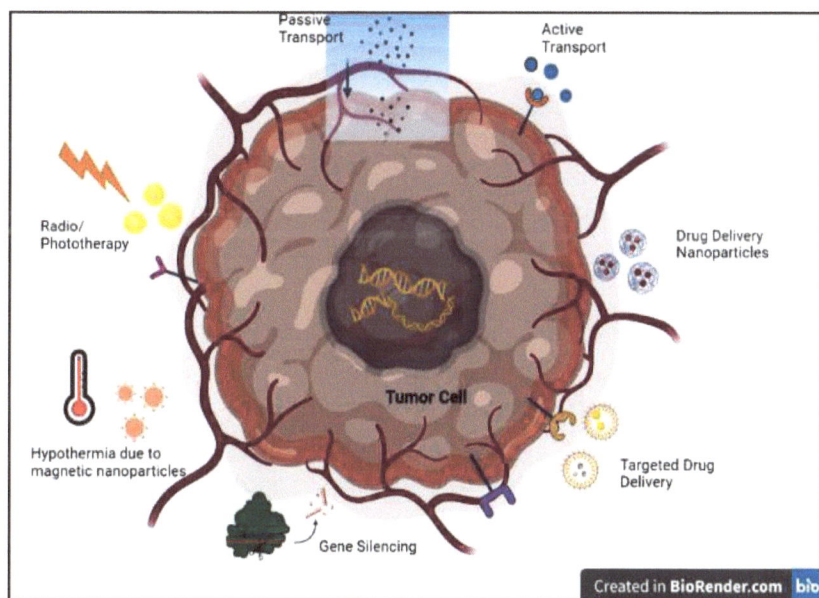

Fig. (4). MNPs based approaches for cancer therapy.

There are many applications that include AuNPs in cancer therapy such as gene silencing and radiotherapy. For instance, Majumdar *et al.*, (2019) have reported the antitumor potential of green synthesized AuNPs against liver cancer and breast cancer [55]. Strong anticarcinogenic activity of AuNPs has also been demonstrated against colon cancer [56], pancreatic cancer [46, 57], ovarian cancer [58], and cervix carcinoma [59, 60].

Silver nanoparticles (AgNPs) are another extremely considered nanomaterial in medicine. AgNPs have been widely investigated due to their incomparable properties such as surface-to-volume ratio, biocompatibility, superb SPR, and smooth functionalization. Dominant antibacterial and antitumor activities of AgNPs have also made them attractive biomedical vehicles [54]. Oncological researches have also demonstrated that AgNPs are able to regulate the autophagy of tumor cells, either by showing cytotoxic activity or in combined with carrying agents in other medications. Pertaining to the anticancer mechanisms, AgNPs affect the cell membrane which leads to access and accumulation of nanoparticles in tumor cells, eventually resulting in cell death or hampering their proliferation. AgNPs can also play a role in modulating signaling metabolism, generating oxidative stress through Ag^+ cations, decreasing ATP in cancer cells, and ROS production. Interaction of Ag^+ ions with nucleic acids causes DNA fragmentation in mitochondria and nuclei and results in cell death. Similar pathways have been attributed to other metal-based nanoparticles such as Platinum, Palladium, iron oxide, and manganese [54, 61]. Significant antitumor activity of AgNPs have been reported by Yusuf and Casey (2020). The synthesis of nanoparticles has been achieved by reducing $NaBH_4$ and desired toxicities have been observed against human leukemia monocytic cell lines (THP-1) [62]. Another study has demonstrated that AgNPs taken up by triple–negative breast cancer (TNBC) cells cause extensive DNA damage without similar damage in normal cells [63]. AgNPs are also shown to be cytotoxic to various tumors, including ovary [28, 64], glioblastoma [65], cervical [26], and pancreas [66].

Iron Oxide based magnetic nanoparticles exhibit functional properties which offer number of advantages for medical applications [67]. In addition to their biocompatibility and chemical inertness, the magnetic properties of iron oxide nanoparticles have nominated them as unique agents for combined cancer therapy such as photothermal and photodynamic therapy. Iron oxide nanoparticles are one of the metallic nanoformulations approved by FDA for cancer diagnostics and imaging [54]. For example, NanoTherm® (MagForce AG, Germany) which includes 15 nm nanoparticles of iron oxide distributed in water, licensed by the European Union for the treatment of brain cancer.

Antibacterial and Antiviral Applications

MNPs, such as silver (Ag), gold (Au), and zinc (Zn), have proposed as effective tools to combat bacterial resistance. Metal nanoparticles can inhibit bacterial growth *via* multiple pathways, broadening the range of antimicrobial activity. For example, silver nanoparticles (AgNPs) have been shown to kill bacteria by neutralizing the membrane surface and altering penetrability. The other major pathways underlying nanomaterial antibacterial actions are reactive oxygen species (ROS) generation, which leads to cell membrane deformation, and intracellular effects, such as interactions with proteins and nucleic acids [68 - 70].

Particularly considering the rising burden of infections due to biofilm formation and increased resistance, the current antibacterial strategies are inadequate to eradicate infections. Increasing conventional antibiotic concentration or dose frequency contributes to the development of multidrug-resistant strains. Moreover, most of these antibiotic molecules are failed to achieve the source of infection which is surrounded by a complex biofilm matrix [71, 72]. Therefore, MNPs have been found promising in controlling infections.

Silver is a well-known antimicrobial agent that is effective against both Gram-positive and Gram-negative bacteria. As a result of their antimicrobial and catalytic properties, AgNPs are increasingly being used in biomedical applications. Because of their small size and high surface-to-volume ratio, AgNPs have exciting catalytic and antibacterial effects on bacterial cell membranes. Nanoparticles, due to their small size, can pass through the cell membrane and act as a catalyst to deactivate enzymes required for bacterial metabolism [29]. The antimicrobial mechanism of silver nanoparticles is attributed to several mechanisms, the majority of which are based on Ag^+ ions. In aqueous solution, Ag^+ ions can form bonds with negatively charged fragments of the cell membrane, resulting in the formation of holes in the membrane. The most advantageous property of AgNPs is that, due to different mechanisms of antimicrobial action, resistance to them is rare and slow in comparison to antibiotic resistance [69, 73]. Silver nanoparticles have not only been used as antimicrobial agents, but they have also been shown to be an excellent vehicle for preventing viruses from binding to host cells [69, 74]. Recent studies have investigated and reported the antiviral potency of AgNPs against SARS-CoV-2 [75, 76].

CONCLUDING REMARKS

The rise in the number of diseases in the biomedical field and, the inadequacy and ineffectiveness of the developed therapies against them increase the need for new tools and methods. On the other hand, instead of developing new treatment methods, studies to improve existing treatments have been accelerated. With the advances in nanotechnology, nanoparticle systems have become a promising path for many drug-involved applications. Numerous methods have been developed for both the delivery of existing active drug components and, the controlled and effective use of nanoparticles with similar activity in these treatments. Besides the advantages of small size, MNPs exhibit different optical, electrical, and magnetic properties compared to other forms of nanoparticle systems. In addition, their surface properties allow them to be conjugated with different components, enabling the development of multifunctional tools. Thus, MNPs have been used in many biomedical fields, including drug delivery platforms, different cancer treatments, antibacterial resistance applications, and imaging. The beneficial contributions of MNPs, especially in cancer therapy, have shown that these tools will be an effective method for many diseases in the future.

REFERENCES

[1] M. Toro-Nahuelpan, G. Giacomelli, O. Raschdorf, S. Borg, J.M. Plitzko, M. Bramkamp, D. Schüler, and F.D. Müller, "MamY is a membrane-bound protein that aligns magnetosomes and the motility axis of helical magnetotactic bacteria", *Nat. Microbiol.,* vol. 4, no. 11, pp. 1978-1989, 2019.
[http://dx.doi.org/10.1038/s41564-019-0512-8] [PMID: 31358981]

[2] D. Schüler, C.L. Monteil, and C.T. Lefevre, "Magnetospirillum gryphiswaldense", *Trends Microbiol.,* vol. 28, no. 11, pp. 947-948, 2020.
[http://dx.doi.org/10.1016/j.tim.2020.06.001] [PMID: 32674989]

[3] C. Binns, Size Matters.*Introduction to Nanoscience and Nanotechnology.* John Wiley & Sons, Inc.: Hoboken, NJ, USA, 2010, pp. 11-32.
[http://dx.doi.org/10.1002/9780470618837.ch1]

[4] C. Daruich De Souza, B. Ribeiro Nogueira, and M.E.C.M. Rostelato, "Review of the methodologies used in the synthesis gold nanoparticles by chemical reduction", *J. Alloys Compd.,* vol. 798, pp. 714-740, 2019.
[http://dx.doi.org/10.1016/j.jallcom.2019.05.153]

[5] P.G. Jamkhande, N.W. Ghule, A.H. Bamer, and M.G. Kalaskar, "Metal nanoparticles synthesis: An overview on methods of preparation, advantages and disadvantages, and applications", *J. Drug Deliv. Sci. Technol.,* vol. 53, p. 101174, 2019.
[http://dx.doi.org/10.1016/j.jddst.2019.101174]

[6] A.J. Shnoudeh, I. Hamad, R.W. Abdo, L. Qadumii, A.Y. Jaber, H.S. Surchi, and S.Z. Alkelany, Synthesis, characterization, and applications of metal nanoparticles.*Biomat.Bionanotech.* Elsevier Inc., 2019, pp. 527-612.
[http://dx.doi.org/10.1016/B978-0-12-814427-5.00015-9]

[7] D. Feldheim, and C. Foss, *Metal nanoparticles. synthesis, characterization and applications.* Marcel Dekker: New York, 2002.

[8] N. Joudeh, and D. Linke, "Nanoparticle classification, physicochemical properties, characterization, and applications: A comprehensive review for biologists", *J. Nanobiotechnology,* vol. 20, no. 1, p. 262, 2022.
[http://dx.doi.org/10.1186/s12951-022-01477-8] [PMID: 35672712]

[9] R. Eivazzadeh-Keihan, E. Bahojb Noruzi, K. Khanmohammadi Chenab, A. Jafari, F. Radinekiyan, S.M. Hashemi, F. Ahmadpour, A. Behboudi, J. Mosafer, A. Mokhtarzadeh, A. Maleki, and M.R. Hamblin, "Metal–based nanoparticles for bone tissue engineering", *J. Tissue Eng. Regen. Med.,* vol. 14, no. 12, pp. 1687-1714, 2020.
[http://dx.doi.org/10.1002/term.3131] [PMID: 32914573]

[10] Y.H. Luo, L.W. Chang, and P. Lin, "Metal-based nanoparticles and the immune system: Activation, inflammation, and potential applications", *BioMed Res. Int.,* vol. 2015, pp. 1-12, 2015.
[http://dx.doi.org/10.1155/2015/143720] [PMID: 26125021]

[11] N. Kumar, and S. Kumbhat, Introduction. In; Essentials in Nanoscience and Nanotechnology, Wiley 2016.
[http://dx.doi.org/10.1002/9781119096122]

[12] G. Habibullah, J. Viktorova, and T. Ruml, "Current strategies for noble metal nanoparticle synthesis", *Nanoscale Res. Lett.,* vol. 16, no. 1, p. 47, 2021.
[http://dx.doi.org/10.1186/s11671-021-03480-8] [PMID: 33721118]

[13] L. Marinescu, D. Ficai, O. Oprea, A. Marin, A. Ficai, E. Andronescu, and A.M. Holban, "Optimized synthesis approaches of metal nanoparticles with antimicrobial applications", *J. Nanomater.,* vol. 2020, pp. 1-14, 2020.
[http://dx.doi.org/10.1155/2020/6651207]

[14] Kumar N., and Kumbhat S., *Nanomaterials. in: Essentials in nanoscience and nanotechnology.* Wiley,

2016, pp. 149-188.
[http://dx.doi.org/10.1002/9781119096122.ch2]

[15]   S. Thota, and D.C. Crans, *Introduction. In: Metal nanoparticles.* Wiley-VCH Verlag GmbH & Co. KGaA: Weinheim, Germany, 2017, pp. 1-14.

[16]   G.R. Rudramurthy, and M.K. Swamy, "Potential applications of engineered nanoparticles in medicine and biology: An update", *J. Biol. Inorg. Chem.,* vol. 23, no. 8, pp. 1185-1204, 2018.
[http://dx.doi.org/10.1007/s00775-018-1600-6] [PMID: 30097748]

[17]   R. Sakthi Devi, A. Girigoswami, M. Siddharth, and K. Girigoswami, "Applications of gold and silver nanoparticles in theranostics", *Appl. Biochem. Biotechnol.,* vol. 194, no. 9, pp. 4187-4219, 2022.
[http://dx.doi.org/10.1007/s12010-022-03963-z] [PMID: 35551613]

[18]   N. Baig, I. Kammakakam, and W. Falath, "Nanomaterials: A review of synthesis methods, properties, recent progress, and challenges", *Mat. Adv.,* vol. 2, no. 6, pp. 1821-1871, 2021.
[http://dx.doi.org/10.1039/D0MA00807A]

[19]   N Kumar, and S Kumbhat, "Nanomaterials: General Synthetic Approaches", In: *Essentials in nanoscience and nanotechnology.* Wiley Online Library, 2016, pp. 29-76.
[http://dx.doi.org/10.1002/9781119096122]

[20]   C.K. Ghosh, Synthesis of noble metal nanoparticles: Chemical and physical routes.*Nanotechnology.* CRC Press, 2017, pp. 107-129.
[http://dx.doi.org/10.1201/9781315116730-6]

[21]   M. Tulinski, and M. Jurczyk, "Nanomaterials synthesis methods", In: *In: Metrology and Standardization of Nanotechnology* Wiley-VCH Verlag GmbH & Co. KGaA: Weinheim, Germany, 2017, pp. 75-98.
[http://dx.doi.org/10.1002/9783527800308.ch4]

[22]   S. Thota, and D.C. Crans, *Metal Nanoparticles.* Wiley-VCH Verlag GmbH & Co. KGaA: Weinheim, Germany, 2018.

[23]   E. Tinajero-Díaz, D. Salado-Leza, C. Gonzalez, M. Martínez Velázquez, Z. López, J. Bravo-Madrigal, P. Knauth, F.Y. Flores-Hernández, S.E. Herrera-Rodríguez, R.E. Navarro, A. Cabrera-Wrooman, E. Krötzsch, Z.Y.G. Carvajal, and R. Hernández-Gutiérrez, "Green metallic nanoparticles for cancer therapy: Evaluation models and cancer applications", *Pharmaceutics,* vol. 13, no. 10, p. 1719, 2021.
[http://dx.doi.org/10.3390/pharmaceutics13101719] [PMID: 34684012]

[24]   K.R.B. Singh, V. Nayak, J. Singh, A.K. Singh, and R.P. Singh, "Potentialities of bioinspired metal and metal oxide nanoparticles in biomedical sciences", *RSC Advances,* vol. 11, no. 40, pp. 24722-24746, 2021.
[http://dx.doi.org/10.1039/D1RA04273D] [PMID: 35481029]

[25]   H. Chopra, S. Bibi, I. Singh, M.M. Hasan, M.S. Khan, Q. Yousafi, A.A. Baig, M.M. Rahman, F. Islam, T.B. Emran, and S. Cavalu, "Green metallic nanoparticles: Biosynthesis to applications", *Front. Bioeng. Biotechnol.,* vol. 10, p. 874742, 2022.
[http://dx.doi.org/10.3389/fbioe.2022.874742] [PMID: 35464722]

[26]   Z. Xu, Q. Feng, M. Wang, H. Zhao, Y. Lin, and S. Zhou, "Green biosynthesized silver nanoparticles with aqueous extracts of ginkgo biloba induce apoptosis via mitochondrial pathway in cervical cancer cells", *Front. Oncol.,* vol. 10, p. 575415, 2020.
[http://dx.doi.org/10.3389/fonc.2020.575415] [PMID: 33194686]

[27]   N.S. Alharbi, and N.S. Alsubhi, "Green synthesis and anticancer activity of silver nanoparticles prepared using fruit extract of Azadirachta indica", *J. Radiat. Res. Appl. Sci.,* vol. 15, no. 3, pp. 335-345, 2022.
[http://dx.doi.org/10.1016/j.jrras.2022.08.009]

[28]   S. Abbasi, A. İlhan, H. Jabbari, P. Javidzade, M. Safari, and F.A. Zadeh, "Cytotoxicity evaluation of synthesized silver nanoparticles by a green method against ovarian cancer cell lines", *Nanomedicine.*

Res. J., vol. 7, pp. 156-164, 2022.
[http://dx.doi.org/10.22034/NMRJ.2022.02.005]

[29]   C. Krishnaraj, B.J. Ji, S.L. Harper, and S.I. Yun, "Plant extract-mediated biogenic synthesis of silver, manganese dioxide, silver-doped manganese dioxide nanoparticles and their antibacterial activity against food- and water-borne pathogens", *Bioprocess. Biosyst. Eng.,* vol. 39, no. 5, pp. 759-772, 2016.
[http://dx.doi.org/10.1007/s00449-016-1556-2] [PMID: 26857369]

[30]   S. Jain, and M.S. Mehata, "Medicinal plant leaf extract and pure flavonoid mediated green synthesis of silver nanoparticles and their enhanced antibacterial property", *Sci. Rep.,* vol. 7, no. 1, p. 15867, 2017.
[http://dx.doi.org/10.1038/s41598-017-15724-8] [PMID: 29158537]

[31]   R.M. Slawson, M.I. Van Dyke, H. Lee, and J.T. Trevors, "Germanium and silver resistance, accumulation, and toxicity in microorganisms", *Plasmid,* vol. 27, no. 1, pp. 72-79, 1992.
[http://dx.doi.org/10.1016/0147-619X(92)90008-X] [PMID: 1741462]

[32]   P. Mukherjee, A. Ahmad, D. Mandal, S. Senapati, S.R. Sainkar, M.I. Khan, R. Parishcha, P.V. Ajaykumar, M. Alam, R. Kumar, and M. Sastry, "Fungus-mediated synthesis of silver nanoparticles and their immobilization in the mycelial matrix: A novel biological approach to nanoparticle synthesis", *Nano Lett.,* vol. 1, no. 10, pp. 515-519, 2001.
[http://dx.doi.org/10.1021/nl0155274]

[33]   P. Mathur, S. Saini, E. Paul, C. Sharma, and P. Mehtani, "Endophytic fungi mediated synthesis of iron nanoparticles: Characterization and application in methylene blue decolorization", *Curr. Opin. Green Sustain. Chem.,* vol. 4, p. 100053, 2021.
[http://dx.doi.org/10.1016/j.crgsc.2020.100053]

[34]   E.A. Shalaby, Algae-mediated silver nanoparticles: Synthesis, properties, and biological activities.*Green Synthesis of Silver Nanomaterials.* Elsevier, 2022, pp. 525-545.
[http://dx.doi.org/10.1016/B978-0-12-824508-8.00009-5]

[35]   V. Chandrakala, V. Aruna, and G. Angajala, "Review on metal nanoparticles as nanocarriers: Current challenges and perspectives in drug delivery systems", *Emer. Mat.,* vol. 5, no. 6, pp. 1593-1615, 2022.
[http://dx.doi.org/10.1007/s42247-021-00335-x] [PMID: 35005431]

[36]   B. Klębowski, J. Depciuch, M. Parlińska-Wojtan, and J. Baran, "Applications of noble metal-based nanoparticles in medicine", *Int. J. Mol. Sci.,* vol. 19, no. 12, p. 4031, 2018.
[http://dx.doi.org/10.3390/ijms19124031] [PMID: 30551592]

[37]   R. Tietze, J. Zaloga, H. Unterweger, S. Lyer, R.P. Friedrich, C. Janko, M. Pöttler, S. Dürr, and C. Alexiou, "Magnetic nanoparticle-based drug delivery for cancer therapy", *Biochem. Biophys. Res. Commun.,* vol. 468, no. 3, pp. 463-470, 2015.
[http://dx.doi.org/10.1016/j.bbrc.2015.08.022] [PMID: 26271592]

[38]   J. Miller-Kleinhenz, X. Guo, W. Qian, H. Zhou, E.N. Bozeman, L. Zhu, X. Ji, Y.A. Wang, T. Styblo, R. O'Regan, H. Mao, and L. Yang, "Dual-targeting Wnt and uPA receptors using peptide conjugated ultra-small nanoparticle drug carriers inhibited cancer stem-cell phenotype in chemo-resistant breast cancer", *Biomaterials,* vol. 152, pp. 47-62, 2018.
[http://dx.doi.org/10.1016/j.biomaterials.2017.10.035] [PMID: 29107218]

[39]   M. Shahidi, O. Abazari, P. Dayati, A. Bakhshi, A. Rasti, F. Haghiralsadat, S.M. Naghib, and D. Tofighi, "Aptamer-functionalized chitosan-coated gold nanoparticle complex as a suitable targeted drug carrier for improved breast cancer treatment", *Nanotechnol. Rev.,* vol. 11, no. 1, pp. 2875-2890, 2022.
[http://dx.doi.org/10.1515/ntrev-2022-0479]

[40]   V. Valdivia, R. Gimeno-Ferrero, M. Pernia Leal, C. Paggiaro, A.M. Fernández-Romero, M.L. González-Rodríguez, and I. Fernández, "Biologically relevant micellar nanocarrier systems for drug encapsulation and functionalization of metallic nanoparticles", *Nanomaterials.,* vol. 12, no. 10, p. 1753, 2022.

[http://dx.doi.org/10.3390/nano12101753] [PMID: 35630975]

[41]   A.A. Yaqoob, H. Ahmad, T. Parveen, A. Ahmad, M. Oves, I.M.I. Ismail, H.A. Qari, K. Umar, and M.N. Mohamad Ibrahim, "Recent advances in metal decorated nanomaterials and their various biological applications: A review", *Front. Chem.,* vol. 8, p. 341, 2020.
[http://dx.doi.org/10.3389/fchem.2020.00341] [PMID: 32509720]

[42]   M. Yokomori, H. Suzuki, A. Nakamura, S.S. Sugano, and M. Tagawa, "DNA-functionalized colloidal crystals for macromolecular encapsulation", *Soft. Matter.,* vol. 18, no. 36, pp. 6954-6964, 2022.
[http://dx.doi.org/10.1039/D2SM00949H] [PMID: 36063070]

[43]   P. Sathishkumar, Z. Li, R. Govindan, R. Jayakumar, C. Wang, and F. Long Gu, "Zinc oxide-quercetin nanocomposite as a smart nano-drug delivery system: Molecular-level interaction studies", *Appl. Surf. Sci.,* vol. 536, p. 147741, 2021.
[http://dx.doi.org/10.1016/j.apsusc.2020.147741]

[44]   B. Khodashenas, M. Ardjmand, A.S. Rad, and M.R. Esfahani, "Gelatin-coated gold nanoparticles as an effective pH-sensitive methotrexate drug delivery system for breast cancer treatment", *Mater. Today Chem.,* vol. 20, p. 100474, 2021.
[http://dx.doi.org/10.1016/j.mtchem.2021.100474]

[45]   S. Mahalunkar, A.S. Yadav, M. Gorain, V. Pawar, R. Braathen, S. Weiss, B. Bogen, S.W. Gosavi, and G.C. Kundu, "Functional design of pH-responsive folate-targeted polymer-coated gold nanoparticles for drug delivery and *in vivo* therapy in breast cancer", *Int. J. Nanomedicine,* vol. 14, pp. 8285-8302, 2019.
[http://dx.doi.org/10.2147/IJN.S215142] [PMID: 31802866]

[46]   S. Saha, X. Xiong, P.K. Chakraborty, K. Shameer, R.R. Arvizo, R.A. Kudgus, S.K.D. Dwivedi, M.N. Hossen, E.M. Gillies, J.D. Robertson, J.T. Dudley, R.A. Urrutia, R.G. Postier, R. Bhattacharya, and P. Mukherjee, "Gold nanoparticle reprograms pancreatic tumor microenvironment and inhibits tumor growth", *ACS Nano,* vol. 10, no. 12, pp. 10636-10651, 2016.
[http://dx.doi.org/10.1021/acsnano.6b02231] [PMID: 27758098]

[47]   A. Yusuf, A. Brophy, B. Gorey, and A. Casey, "Liposomal encapsulation of silver nanoparticles enhances cytotoxicity and causes induction of reactive oxygen species-independent apoptosis", *J. Appl. Toxicol.,* vol. 38, no. 5, pp. 616-627, 2018.
[http://dx.doi.org/10.1002/jat.3566] [PMID: 29181855]

[48]   S. Dey, L. Fageria, A. Sharma, S. Mukherjee, S. Pande, R. Chowdhury, and S. Chowdhury, "Silver nanoparticle-induced alteration of mitochondrial and ER homeostasis affects human breast cancer cell fate", *Toxicol. Rep.,* vol. 9, pp. 1977-1984, 2022.
[http://dx.doi.org/10.1016/j.toxrep.2022.10.017] [PMID: 36518460]

[49]   D. Liu, J. Li, C. Wang, L. An, J. Lin, Q. Tian, and S. Yang, "Ultrasmall Fe@Fe3O4 nanoparticles as T1–T2 dual-mode MRI contrast agents for targeted tumor imaging", *Nanomedicine.,* vol. 32, p. 102335, 2021.
[http://dx.doi.org/10.1016/j.nano.2020.102335] [PMID: 33220508]

[50]   C Soica, I Pinzaru, C Trandafirescu, F Andrica, C Danciu, M Mioc, D Coricovac, C Sitaru, and C Dehelean, "Silver-, gold-, and iron-based metallic nanoparticles: Biomedical applications as theranostic agents for cancer", *Des. Nanostructures. Theranostics. Appl.,* pp. 161-242, 2018.
[http://dx.doi.org/10.1016/B978-0-12-813669-0.00005-1]

[51]   R. Khursheed, K. Dua, S. Vishwas, M. Gulati, N.K. Jha, G.M. Aldhafeeri, F.G. Alanazi, B.H. Goh, G. Gupta, K.R. Paudel, P.M. Hansbro, D.K. Chellappan, and S.K. Singh, "Biomedical applications of metallic nanoparticles in cancer: Current status and future perspectives", *Biomed. Pharmacother.,* vol. 150, p. 112951, 2022.
[http://dx.doi.org/10.1016/j.biopha.2022.112951] [PMID: 35447546]

[52]   X. He, S. Chen, and X. Mao, "Utilization of metal or non-metal-based functional materials as efficient composites in cancer therapies", *RSC. Adv.,* vol. 12, no. 11, pp. 6540-6551, 2022.

[http://dx.doi.org/10.1039/D1RA08335J] [PMID: 35424648]

[53]   M. Vinardell, and M. Mitjans, "Antitumor activities of metal oxide nanoparticles", *Nanomaterials.,* vol. 5, no. 2, pp. 1004-1021, 2015.
[http://dx.doi.org/10.3390/nano5021004] [PMID: 28347048]

[54]   D.N. Păduraru, D. Ion, A.G. Niculescu, F. Muşat, O. Andronic, A.M. Grumezescu, and A. Bolocan, "Recent developments in metallic nanomaterials for cancer therapy, diagnosing and imaging applications", *Pharma.,* vol. 14, no. 2, p. 435, 2022.
[http://dx.doi.org/10.3390/pharmaceutics14020435] [PMID: 35214167]

[55]   M. Majumdar, S.C. Biswas, R. Choudhury, P. Upadhyay, A. Adhikary, D.N. Roy, and T.K. Misra, "Synthesis of gold nanoparticles using citrus macroptera fruit extract: Anti–biofilm and anticancer activity", *ChemistrySelect.,* vol. 4, no. 19, pp. 5714-5723, 2019.
[http://dx.doi.org/10.1002/slct.201804021]

[56]   P.E. Costantini, M. Di Giosia, L. Ulfo, A. Petrosino, R. Saporetti, C. Fimognari, P.P. Pompa, A. Danielli, E. Turrini, L. Boselli, and M. Calvaresi, "Spiky gold nanoparticles for the photothermal eradication of colon cancer cells", *Nanomaterials.,* vol. 11, no. 6, p. 1608, 2021.
[http://dx.doi.org/10.3390/nano11061608] [PMID: 34207455]

[57]   Y. Huai, Y. Zhang, X. Xiong, S. Das, R. Bhattacharya, and P. Mukherjee, "Gold nanoparticles sensitize pancreatic cancer cells to gemcitabine", *Cell. Stress.,* vol. 3, no. 8, pp. 267-279, 2019.
[http://dx.doi.org/10.15698/cst2019.08.195] [PMID: 31440741]

[58]   R. Maity, M. Chatterjee, A. Banerjee, A. Das, R. Mishra, S. Mazumder, and N. Chanda, "Gold nanoparticle-assisted enhancement in the anti-cancer properties of theaflavin against human ovarian cancer cells", *Mater. Sci. Eng. C,* vol. 104, p. 109909, 2019.
[http://dx.doi.org/10.1016/j.msec.2019.109909] [PMID: 31499983]

[59]   L. Qian, W. Su, Y. Wang, M. Dang, W. Zhang, and C. Wang, "Synthesis and characterization of gold nanoparticles from aqueous leaf extract of Alternanthera sessilis and its anticancer activity on cervical cancer cells", *Artif. Cells. Nanomed. Biotech.,* vol. 47, no. 1, pp. 1173-1180, 2019.
[http://dx.doi.org/10.1080/21691401.2018.1549064]

[60]   J Lopes-Nunes, AS Agonia, T Rosado, E Gallardo, R Palmeira-De-oliveira, A Palmeira-De-oliveira, J Martinez-De-oliveira, J Fonseca-Moutinho, MPC Campello, A Paiva, A Paulo, A Vulgamott, AD Ellignton, PA Oliveira, and C Cruz, "Aptamer-functionalized gold nanoparticles for drug delivery to gynecological carcinoma cells cancers", *Cancers.,* vol. 13, pp. 4038-4038, 2021.
[http://dx.doi.org/10.3390/cancers13164038]

[61]   J. Xu, W. Han, P. Yang, T. Jia, S. Dong, H. Bi, A. Gulzar, D. Yang, S. Gai, F. He, J. Lin, and C. Li, "Tumor microenvironment-responsive mesoporous MnO 2 -coated upconversion nanoplatform for self-enhanced tumor theranostics", *Adv. Funct. Mater.,* vol. 28, no. 36, p. 1803804, 2018.
[http://dx.doi.org/10.1002/adfm.201803804]

[62]   A. Yusuf, and A. Casey, "Evaluation of silver nanoparticle encapsulation in DPPC-based liposome by different methods for enhanced cytotoxicity", *Int. J. Polym. Mater.,* vol. 69, no. 13, pp. 860-871, 2020.
[http://dx.doi.org/10.1080/00914037.2019.1626390]

[63]   J. Swanner, C.D. Fahrenholtz, I. Tenvooren, B.W. Bernish, J.J. Sears, A. Hooker, C.M. Furdui, E. Alli, W. Li, G.L. Donati, K.L. Cook, P.A. Vidi, and R. Singh, "Silver nanoparticles selectively treat triple–negative breast cancer cells without affecting non–malignant breast epithelial cells *in vitro* and *in vivo*", *FASEB Bioadv.,* vol. 1, no. 10, pp. 639-660, 2019.
[http://dx.doi.org/10.1096/fba.2019-00021] [PMID: 32123812]

[64]   K. Lavudi, V.S. Harika, R.R. Kokkanti, S. Patchigolla, A. Sinha, S. Patnaik, and J. Penchalaneni, "2-Dimensional *in vitro* culture assessment of ovarian cancer cell line using cost effective silver nanoparticles from *Macrotyloma uniflorum* seed extracts", *Front. Bioeng. Biotechnol.,* vol. 10, p. 978846, 2022.
[http://dx.doi.org/10.3389/fbioe.2022.978846] [PMID: 36051584]

[65]   K. Urbańska, B. Pająk, A. Orzechowski, J. Sokołowska, M. Grodzik, E. Sawosz, M. Szmidt, and P. Sysa, "The effect of silver nanoparticles (AgNPs) on proliferation and apoptosis of *in ovo* cultured glioblastoma multiforme (GBM) cells", *Nanoscale Res. Lett.,* vol. 10, no. 1, p. 98, 2015.
[http://dx.doi.org/10.1186/s11671-015-0823-5] [PMID: 25852394]

[66]   E. Zielinska, A. Zauszkiewicz-Pawlak, M. Wojcik, and I. Inkielewicz-Stepniak, "Silver nanoparticles of different sizes induce a mixed type of programmed cell death in human pancreatic ductal adenocarcinoma", *Oncotarget,* vol. 9, no. 4, pp. 4675-4697, 2018.
[http://dx.doi.org/10.18632/oncotarget.22563] [PMID: 29435134]

[67]   J.J. Xu, W.C. Zhang, Y.W. Guo, X.Y. Chen, and Y.N. Zhang, "Metal nanoparticles as a promising technology in targeted cancer treatment", *Drug Deliv.,* vol. 29, no. 1, pp. 664-678, 2022.
[http://dx.doi.org/10.1080/10717544.2022.2039804] [PMID: 35209786]

[68]   H. Kotrange, A. Najda, A. Bains, R. Gruszecki, P. Chawla, and M.M. Tosif, "Metal and metal oxide nanoparticle as a novel antibiotic carrier for the direct delivery of antibiotics", *Int. J. Mol. Sci.,* vol. 22, no. 17, p. 9596, 2021.
[http://dx.doi.org/10.3390/ijms22179596] [PMID: 34502504]

[69]   R. Thomas, P. Jishma, S. Snigdha, K.R. Soumya, J. Mathew, and E.K. Radhakrishnan, "Enhanced antimicrobial efficacy of biosynthesized silver nanoparticle based antibiotic conjugates", *Inorg. Chem. Commun.,* vol. 117, p. 107978, 2020.
[http://dx.doi.org/10.1016/j.inoche.2020.107978]

[70]   A. Ahsan, M.A. Farooq, A. Ahsan Bajwa, and A. Parveen, "Green synthesis of silver nanoparticles using parthenium hysterophorus: Optimization, characterization and *in vitro* therapeutic evaluation", *Molecules,* vol. 25, no. 15, p. 3324, 2020.
[http://dx.doi.org/10.3390/molecules25153324] [PMID: 32707950]

[71]   S. Fulaz, S. Vitale, L. Quinn, and E. Casey, "Nanoparticle–biofilm interactions: The role of the EPS matrix", *Trends. Microbiol.,* vol. 27, no. 11, pp. 915-926, 2019.
[http://dx.doi.org/10.1016/j.tim.2019.07.004] [PMID: 31420126]

[72]   K.M. Rubey, and J.S. Brenner, "Nanomedicine to fight infectious disease", *Adv. Drug Deliv. Rev.,* vol. 179, p. 113996, 2021.
[http://dx.doi.org/10.1016/j.addr.2021.113996] [PMID: 34634395]

[73]   R.Y. Pelgrift, and A.J. Friedman, "Nanotechnology as a therapeutic tool to combat microbial resistance", *Adv. Drug Deliv. Rev.,* vol. 65, no. 13-14, pp. 1803-1815, 2013.
[http://dx.doi.org/10.1016/j.addr.2013.07.011] [PMID: 23892192]

[74]   G. Thirumurugan, J.V.L.N. Seshagiri Rao, and M.D. Dhanaraju, "Elucidating pharmacodynamic interaction of silver nanoparticle: Topical deliverable antibiotics", *Sci. Rep.,* vol. 6, no. 1, p. 29982, 2016.
[http://dx.doi.org/10.1038/srep29982] [PMID: 27427207]

[75]   S.S. Jeremiah, K. Miyakawa, T. Morita, Y. Yamaoka, and A. Ryo, "Potent antiviral effect of silver nanoparticles on SARS-CoV-2", *Biochem. Biophys. Res. Commun.,* vol. 533, no. 1, pp. 195-200, 2020.
[http://dx.doi.org/10.1016/j.bbrc.2020.09.018] [PMID: 32958250]

[76]   Q. He, J. Lu, N. Liu, W. Lu, Y. Li, C. Shang, X. Li, L. Hu, and G. Jiang, "Antiviral properties of silver nanoparticles against SARS-CoV-2: Effects of surface coating and particle size", *Nanomaterials.,* vol. 12, no. 6, p. 990, 2022.
[http://dx.doi.org/10.3390/nano12060990] [PMID: 35335803]

Recent Advances in Biotechnology, 2023, Vol. 8, 81-116

CHAPTER 4

Photodynamic Therapy and Applications in Cancer

Ceren Sarı[1] and **Figen Celep Eyüpoğlu**[1,*]

[1] *Department of Medical Biology, Faculty of Medicine, Karadeniz Technical University, Trabzon, Turkey*

Abstract: The idea of using light as a therapeutic tool has been popular for thousands of years. Scientific discoveries in line with technological innovations have contributed to the advancement of photodynamic therapy as a therapeutic modality. Photodynamic therapy is based on the generation of highly reactive species that alter the molecular systematics of cells through interactions between light, photosensitizer, and molecular oxygen. It has a minimally invasive protocol that can be combined with other clinical methods or can be stand-alone. The development of photosensitizers with the integration of nanotechnological approaches has provided favorable results over the years in malignant and non-malignant diseases by facilitating target-site action, selectivity, and controllable drug release. This chapter presents a review of photodynamic therapy with its important aspects; history, mechanism of action, cellular effects, integration into nanoscale drug delivery systems, and combinational therapeutic approaches in cancer.

Keywords: Anticancer therapy, Apoptosis, Autophagy, Cancer, Tumor, Cell death, Clinical application, Combination therapy, Drug delivery systems, Immunogenic cell death, Light therapy, Nanotechnology, Necroptosis, Necrosis, Neoplasms, Photodynamic therapy, Photosensitizer, Reactive oxygen species, Singlet oxygen, Vascular endothelium.

INTRODUCTION

Photodynamic therapy (PDT) is a minimally invasive and highly selective therapeutic modality in which photosensitizer molecules, light, and molecular oxygen are combined to destroy unwanted cells and tissues. Photosensitizer molecules, which do not exert toxic effects under normal conditions, are activated by light. Following activation, they transfer energy to molecular oxygen through photochemical reactions, producing highly toxic reactive oxygen species (ROS) that lead to cell demise [1, 2].

---

* **Corresponding author Figen Celep Eyüpoğlu**: Department of Medical Biology, Faculty of Medicine, Karadeniz Technical University, Trabzon, Turkey; Tel: +90 462 377 7941; E-mail: fcelep@ktu.edu.tr

Habibe Yılmaz (Ed.)

PDT has yielded effective results in various clinical fields, such as dermatology, immunology, cardiology, ophthalmology, urology, and dentistry [2 - 11]. The most comprehensive field of study of PDT is cancer therapy. Antitumoral PDT studies have documented promising results in different types of cancer (breast, skin, brain, lung, head and neck, gastrointestinal, gynecological, prostate, *etc.*) [12 - 19].

PDT has various advantages compared to conventional therapeutic approaches such as surgery, chemotherapy, and radiotherapy. The only route of activation of photosensitizer is co-administration with light, resulting in minimal systemic toxicity because exposure time and site can be managed. Numerous potential applications of PDT are conceivable, as ROS can damage cells from different origins. Besides, PDT may be recommended as adjuvant therapy to overcome some of the difficulties of common therapeutic models or to support their outcomes [20 - 23].

BRIEF HISTORY OF LIGHT THERAPY

Light has a special meaning in ancient civilizations [24]. Sun worship represents many ancient religions in Asia and Europe. Ancient cultures believed the sun had a healing potential for various ailments. In the times of ancient Egypt, extracts from different plants were applied to skin lesions and exposed to sunlight as a treatment protocol. In other ancient civilizations, such as China, Greece, and India, phototherapy applications were used to treat different diseases such as vitiligo, psoriasis, skin malignancies, and even psychosis [24 - 26].

The development of modern phototherapy dates back to the early 1900s. Arnold Rikli, considered to be one of the pioneers of modern phototherapy, mentioned sunbathing as a treatment method and helped to re-emerge the therapeutic effects of sunlight. Phototherapy was popularized by Niels Finsen, who won the Nobel Prize in 1903 for his study applying carbon arc phototherapy in the treatment of lupus vulgaris. In the early 20th century, Oscar Raab, a student of Professor Herman von Tappeiner, conducted experiments to analyze the toxic effects of acridine on paramecia. He noticed that the toxic effects of acridine varied with daylight and were minimal on stormy days. Further experiments have confirmed that the toxicity of acridine is dose- and light-dependent, thus, the combination of the two has shown highly toxic effects. Continuing Raab's research, Von Tappeiner, together with dermatologist Albert Jesionek, focused on the implementation of eosin as a photosensitizer in skin cancer, skin lupus, and condylomas of the female genitals. In 1904, von Tappeiner and Albert Jodlbauer stated that oxygen is an exigency for the process of photosensitization. These studies were collected in a book in 1907, in which von Tappeiner used the term "photodynamic action" to delineate the phenomenon of oxygen-induced photosensitization [25 - 27].

EXCITING JOURNEY OF PHOTODYNAMIC THERAPY

Von Tappeiner's clear prediction of the phototherapeutic application of photosensitizers has accelerated scientific studies in this field. The use of hematoporphyrin, a derivative of porphyrin, has greatly contributed to the development of PDT [26]. Hausmann reported that the combination of hematoporphyrin and light, which he applied to paramecium and red blood cells, significantly killed the cells. He also observed skin reactions in mice exposed to light after being treated with hematoporphyrin [25, 28]. In 1913, Friedrich Meyer-Betz self-administered hematoporphyrin to observe how it would work in humans. He was the first scientist to use porphyrins as photosensitizers in humans, observing edema and prolonged pain in light-exposed areas [28, 29].

The concept of modern PDT began in the 1960s with the studies of Samuel Schwartz and Richard Lipson. Schwartz successfully developed a hematoporphyrin derivative (HpD), which has higher phototoxicity than hematoporphyrin, and Lipson demonstrated tumoral accumulation and therapeutic effects of HpD in patients with different lesions [28 - 30]. In 1975, an important wall of PDT studies was built by Thomas Dougherty's study, which combined HpD and red light to destruct mammary tumors in mice [31]. That same year, J. F. Kelly demonstrated that HpD-PDT eradicates bladder carcinoma in mice [32]. In 1976, another breakthrough in the development of PDT occurred with the first human bladder cancer study to demonstrate the therapeutic effects of HpD-PDT by M. E. Snell and J. F. Kelly [33]. Over the years, several studies have been conducted and different types of tumors (breast, colorectal, pancreas, head and neck, brain, cholangiocarcinoma, mesothelioma, *etc.*) were treated with PDT [34 - 47].

Photofrin, one of the most common HpDs, received its first healthcare approval for use in the prophylactic treatment of bladder cancer in 1993. The Food and Drug Administration (FDA) approved Photofrin in 1995 for the treatment of esophageal cancer. Later, Photofrin was approved for usage in early-stage lung cancer in 1998 [48]. Thus, PDT has strengthened its spot in the clinical literature as a novel therapeutic modality.

DEVELOPMENT OF PHOTOSENSITIZERS

Photosensitizers are an essential part of PDT procedures. Various photosensitizers have been studied over the years, and some of them have been approved for clinical use [49]. As a result of intensive studies on PDT, it has been understood that ideal photosensitizers should have important properties such as high levels of chemical purity, room-temperature stability, easy synthetic routes, efficient photochemical reactivity, phototoxicity in the presence of a specific wavelength, minimum cytotoxicity in the dark, high solubility in tissues and selectivity to neoplastic tissues [50 - 53]. Porphyrins and their derivatives (such as hemato-porphyrin and Photofrin) constitute first-generation photosensitizers. Although these photosensitizers are believed to exhibit favorable photodynamic activity, they have non-negligible disadvantages such as low chemical purity, poor penetration into tissues, and prolonged skin sensitivity to light [30, 53]. Second-generation photosensitizers such as 5-aminolevulinic acid, texaphyrin, benzoporphyrin derivatives, thiopurine derivatives, chlorine, and phthalocyanines have been designed to overcome the disadvantages of first-generation photosensitizers [30, 53 - 57]. They are characterized by higher chemical purity, higher singlet oxygen yield, and deep-seated tissue penetration due to strong absorption in the wavelength range of 650-800 nm. Besides, they show fewer side effects due to their high selectivity to tumoral tissues and quick removal from the body [30, 49, 53, 58]. Despite the improved photochemical properties, second-generation photosensitizers have disadvantages such as low stability and poor water solubility [49, 53]. Different approaches have been considered to maximize the qualified features of the second-generation photosensitizers and to minimize their disadvantages, resulting in third-generation photosensitizers. They mostly contain second-generation photosensitizers that are conjugated to a photo-activated drug or encapsulated in nanoparticles, which cause minimal damage in the surrounding tissue and exhibit a higher affinity to targeted lesions. This minimizes off-target effects and reduces dark toxicity [59 - 61]. Although different properties of third-generation photosensitizers have been improved, difficulties with the procedure of parenteral administration restrict the widespread clinical application of PDT. Therefore, it is necessary to design drug-delivery tools that will facilitate the bioavailability of photodynamic applications.

LIGHT SOURCES

Light is essential in PDT, as the photodynamic reaction relies on the activation of the photosensitizer by light. The light sources used in the early stages of PDT were non-coherent light sources (*e.g.*, conventional arc lamps) that were safe, easy to use, and inexpensive. However, it is difficult to control the light dose in conventional lamps. Also, they may cause thermal effects that should be avoided in PDT [28, 62]. Alternatives to non-coherent light sources are lasers and light-emitting diodes (LED). Lasers are commonly used in superficial and interstitial PDT. They produce monochromatic coherent light with a narrow bandwidth and provide a specific wavelength for photosensitizers. Since they generate a very narrow beam of light, optical fibers can be coupled to lasers and routed to the target site for endoscopic or interstitial application [62, 63]. On the other hand,

LEDs are cheaper and less hazardous light sources that can produce high-energy light of specific wavelengths. These light sources do not cause thermal destruction and require remarkably less energy than lasers. Their ease of use due to being assembled in various sizes and shapes is important for the effective application of PDT [62, 64 - 67].

Since light is an important part of PDT, it should be noted that tissue-specific clinical efficacy is highly correlated with light fluence rate, exposure time, and mode of light delivery. Low light fluence rates are better for PDT because high rates can deplete tissue oxygen and lead to inefficient degradation of photosensitizers [28, 68]. The depth of light penetration into tissues varies at different wavelengths (Fig. **1**). In the spectral range of 600-1300 nm, called the therapeutic window or optical window, light has the capacity to penetrate tissues at maximum depth. Shorter wavelengths have less tissue penetration and may cause skin photosensitivity. In contrast, longer wavelengths are ineffective at generating ROS due to the lack of sufficient free energy to excite oxygen [1, 28, 63].

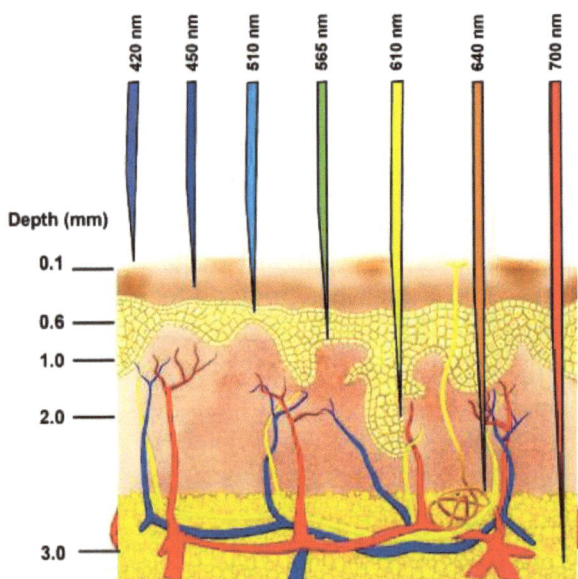

Fig. (1). Tissue penetration depths of light wavelengths.

ADMINISTRATION OF PHOTOSENSITIZERS

Administration of photosensitizers can be completed topically in cream form or locally/systemically by subcutaneous or intravenous injection. The uptake of photosensitizers takes approximately 6-48 hours following the administration. At the end of this period, light is applied to the relevant site of the body. The light dose or quantity of energy depends on the light fluence rate, the molar extinction coefficient of the photosensitizer, and the concentration of the photosensitizer [69].

PRINCIPLES OF PHOTODYNAMIC THERAPY

PDT requires the combined action of three non-toxic components; a photosensitizer, light of a specific wavelength, and molecular oxygen. Photosensitizers are in an inactive state until exposure to light and mostly accumulate in tumor cells. When activated by light, they transfer energy to oxygen molecules (O_2) to generate ROS, such as singlet oxygen ($^1O_2$), hydroxyl radical ($^.OH$), superoxide radical ($O_2^{.-}$), and hydrogen peroxide (H_2O_2). These cytotoxic products initiate a series of biochemical events, causing inevitable cell damage and, eventually, cell death [61, 67, 70].

The photodynamic reaction can occur by two mechanisms (type I and type II) whose initial stages are similar. Following the irradiation of the photosensitizer, it is converted from the ground (singlet) state (S_0) to the excited singlet state (S_1) due to photon absorption. S_1 is a short-lived (nanosecond) and unstable state where the photosensitizer can return to the S_0 state by losing its energy through light emission (fluorescence) or heat production. Alternatively, the S_1 state can undergo intersystem crossing and drive to a more stable, long-lived (microseconds) excited triplet state (T_1). Photosensitizers can decay to the S_0 state from the T_1 state by emitting light (phosphorescence) or by energy transferring to another molecule [53, 71]. On the other hand, excited molecules (T_1) can undergo two kinds of reactions that culminate in cellular death. In type I reaction, the energy of T_1-state photosensitizer is transferred to biomolecules *via* electron or hydrogen transfer between photosensitizer and surrounding tumor tissue, thereby producing ROS and causing damage. In type II reactions, excited photosensitizer molecules transfer energy directly to molecular oxygen. Thus, highly active singlet oxygen molecules are produced and interact with cellular components such as proteins, lipids, and nucleic acids, ultimately resulting in cell death (Fig. **2**) [53, 61, 72]. Type II reaction is usually predominant in PDT. However, the contribution of both reactions to PDT efficiency and extent of damage depends on several factors, such as oxygen concentration, light fluence rate, chemical structure of the photosensitizer, its total concentration, and intracellular localization. It is also noteworthy that as oxygen is depleted, type I reaction begins to prevail [28, 29, 73, 74].

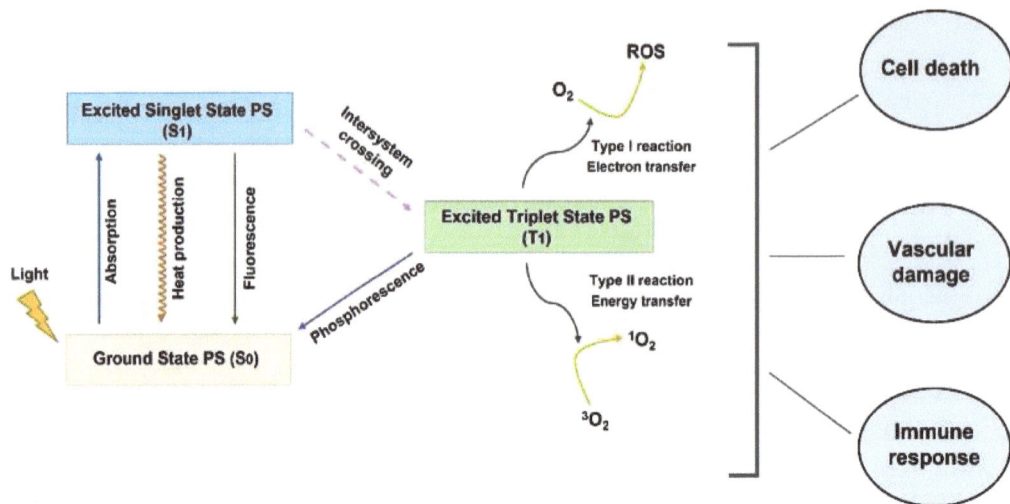

Fig. (2). Mechanism of action and anti-tumor responses of PDT. PS; photosensitizer.

Cellular Uptake

Due to the limited lifespan of reactive species produced by excited photosensitizers, cellular damage is highly correlated with the uptake and localization of photosensitizers. The overall charge and lipophilicity of photosensitizers are important parameters for cellular uptake, transport, and intracellular localization.

Studies have revealed that negative charges are unfavorable for the passive transport of photosensitizers across cell membranes, especially if photosensitizers do not have sufficient lipophilicity. Photosensitizers with more than two negative charges tend to be too polar to diffuse. Therefore, they are taken into cells by endocytosis and mostly accumulate in lysosomes [71, 75, 76]. On the other hand, positively charged photosensitizers are easily transported into cells and electrostatically attracted by negatively charged components of mitochondrial membranes. The affinity of cationic photosensitizers to anionic regions of proteins is thought to play an important role in their selective accumulation in tumors [71, 77].

The uptake and localization of photosensitizers also depend on the structural balance between hydrophobicity/hydrophilicity. Most hydrophobic photosensitizers have sufficient affinity for membranes, however, hydrophilic photosensitizers enter cells by pinocytosis and/or endocytosis. The fact that hydrophilic photosensitizers do not aggregate in aqueous media increases the effects of PDT because aggregation causes internal conversion to the ground state (S_0), thereby reducing the efficiency of photosensitizer [78]. In addition, hydrophilic photosensitizers are quickly eliminated from the body, resulting in fewer side effects [79]. Although hydrophobic photosensitizers are easier to pass through membranes than hydrophilic ones, they may need to be used with designated delivery systems due to their tendency to aggregate in aqueous media [80]. These observations have lightened amphiphilic photosensitizers in PDT studies. Amphiphilic photosensitizers are advantageous because the hydrophilic part makes them water-soluble and facilitates drug administration, while the hydrophobic part enables entry and accumulation into the cell [75, 81, 82].

Subcellular Distribution and Intracellular Targets

Subcellular distribution of photosensitizers following cellular uptake is another important factor influencing PDT activity. Intracellular targets of PDT-induced photodamage alter the type of cellular response by activating multiple signaling pathways. This ultimately affects PDT-mediated cell death.

Lysosomes

Preliminary studies have suggested that lysosomes are critical intracellular targets for PDT. Photosensitizers localized in lysosomes generally predispose cells to necrosis by delaying or blocking apoptosis [71]. The cell is thought to be digested by hydrolytic enzymes released from impaired lysosomes caused by photodamage. However, further studies indicated that lysosome-localized photosensitizers exhibited significantly lower efficacy compared to those localized in mitochondria and other organelles. The underlying reason may be the inactivation of lysosomal enzymes due to PDT or cytosolic inhibitors [83].

Mitochondria

Mitochondria are critical subcellular targets for PDT, as mitochondrial dysfunction due to photodamage can rapidly induce an apoptotic response. Since mitochondria are one of the apoptotic regulators and produce most of the energy, photosensitizers localized in mitochondria are reported to be effective in killing cells [71, 84]. These photosensitizers especially induce apoptotic cell death accompanied by oxidative stress. This is due to oxidative damage caused by ROS, as well as superoxide anions produced as the secondary product by photodamage of the electron transport chain [71, 85].

Endoplasmic Reticulum

Endoplasmic reticulum (ER) has a critical role in signal transduction, calcium homeostasis, and protein processing in cells [86]. ROS production induced by ER stress is known to trigger immunogenic cell death. Therefore, photosensitizers localized in ER are considered to be potent eradicators of tumors [87 - 89]. Several studies have shown that photosensitizers directly localized in ER modulate high levels of ROS production and robust immune responses [90, 91]. However, most photosensitizers do not tend to accumulate in ER. To achieve this, they should possess hydrophobic or amphiphilic properties [89].

Biomembranes

Membrane binding of photosensitizer is directly proportional to PDT efficiency [92]. Singlet oxygen is more likely to react with membrane components if a photosensitizer is anchored in the plasma membrane. However, singlet oxygen mostly fails to oxidize cellular components if the photosensitizer remains at the membrane boundary without contacting the membrane [93]. Apoptotic cell death is usually induced when membrane-localized photosensitizers cause mild oxidative damage. On the other hand, in case of excessive damage, the integrity of the membrane is completely destroyed, thus, necrosis occurs [71, 94]. The vital role of biomembranes in cell organization, transport mechanisms and survival may explain the high PDT efficacy of membrane-localized photosensitizers. Even low levels of membrane oxidation caused by PDT can lead to membrane barrier disruption, functional alteration of signaling cascades, and cell death [71, 95, 96].

PDT-Induced Cell Death Pathways

A thorough understanding of the cellular mechanisms of PDT is important in terms of therapeutic outcomes. PDT destroys cancer cells directly by different cell death pathways or indirectly by impairing vascularization that provides nutrients and oxygen to tumor cells. Different forms of PDT-induced cell death occasionally overlap or contribute to each other's activation. Therefore, detailed studies of light dosimetry, intracellular localization of photosensitizer, and genotype of target tissue are needed in many cases [1, 28].

Apoptosis

Apoptosis is the main cell death pathway induced by PDT. It is a programmed cell death classified into intrinsic (mitochondrial) and extrinsic (death receptor) pathways. Intrinsic apoptosis can be caused by various factors such as intracellular ROS production, ER stress, DNA damage *etc* [97]. Mitochondria play a critical role in the intrinsic pathway. The mitochondrial membrane becomes permeable due to alterations in membrane potential and facilitates the cytosolic release of proapoptotic molecules such as cytochrome c, thus leading to oligomerization of apoptotic peptide activating factor 1 (Apaf-1). This results in the recruitment of procaspases and activation of effector caspases that can cleave cellular substrates. On the other hand, the extrinsic pathway is triggered when death ligands bind to plasma membrane receptors. These receptors combine to form a death-inducing signaling complex (DISC) that activates procaspases. This leads to apoptosis by activation of effector caspases [98].

Intrinsic apoptosis is the most common PDT-mediated apoptotic pathway. It is a tightly regulated pathway dominated by anti-apoptotic and pro-apoptotic proteins of the B-cell lymphoma 2 (Bcl-2) family. Several studies have shown that mitochondria-localized photosensitizers impair the function of anti-apoptotic proteins without damaging proapoptotic proteins [1, 99, 100]. It has also been shown that some of the photosensitizers localized in lysosomes induce apoptosis through lysosomal photodamage. The release of lysosomal proteases facilitates proteasomal cleavage of Bid, one of the proapoptotic members of the Bcl-2 family. Translocation of truncated Bid (tBid) to mitochondria causes cytochrome c release, leading to apoptosis [101 - 103]. Moreover, caspase-independent death effectors such as apoptosis inducing factor (AIF) can be released from mitochondria along with a second mitochondria-derived activator of caspase/direct inhibitor of apoptosis-binding protein with low pI (Smac/DIABLO). This indicates that caspase-dependent and independent pathways can be activated simultaneously in PDT-mediated cell death [104].

Necrosis

Necrosis is a form of cell death in which energy is not required, leading to a pyknotic nucleus, cellular swelling, and cytoplasmic membrane ruptures [1]. In necrosis, cells are rapidly damaged due to different factors such as radiation, exposure to chemicals, and hypoxia. A typical feature of necrotic cell death is the inflammatory response triggered by the release of cellular contents into the extracellular environment. For decades, necrosis was thought to be the most common type of PDT-induced cell death. The main reason for this was the assumption that cellular fragments formed in late apoptotic cells are associated with necrotic cell death. In recent years, a comprehensive understanding of cell death mechanisms has revealed that PDT can trigger different cell death pathways other than necrosis [97]. Necrosis is not the primarily targeted death pathway in most PDT protocols, as it may exhibit adverse effects on normal cells and tissues [103].

In the context of PDT, a combination of photosensitizer and light fluence at high doses can result in necrotic cell death. In a study investigating the effects of cytotoxic doses, it was shown that the PDT protocol which causes less than 70% cytotoxicity, mainly induces apoptosis, while higher doses that cause 99% cytotoxicity trigger necrotic cell death [105]. Necrosis is mostly induced by photosensitizers that lead to loss of membrane integrity. Another important factor leading to necrosis is the incubation period of photosensitizers. In a study using zinc (II) phthalocyanine (ZnPc), the short incubation time (2 hours) of ZnPc triggered the early loss of membrane integrity, causing necrosis. When the incubation period was increased to 24 hours, partial inhibition of the membrane was observed, and cells mostly underwent apoptosis [106].

Autophagy

Autophagy is a process that facilitates the recycling of cellular components under starvation. However, in some conditions, it helps to ensure cell death in apoptosis-lacking cells [107 - 109]. While we still need extensive studies to understand whether autophagy causes survival or death following PDT, up-to-date studies have provided bright insight so far. Accordingly, PDT-mediated autophagy may cause cell survival or death, depending on the type of photosensitizer, ROS production, the presence of apoptotic mechanism, and the extent of photodamage during treatment [1, 97].

Several studies have reported increased phototoxicity when autophagy is inhibited. Therefore, autophagy has been proposed as a survival mechanism to deal with photo-induced stress rather than a cell death pathway [110, 111]. However, the pro-survival effect of autophagy is thought to occur up to a certain threshold of phototoxicity [98, 109, 112]. PDT-induced autophagy-associated cell death occurs mainly in apoptosis-deficient or apoptosis-resistant cells [1, 108]. It has been shown that photosensitizers localized in lysosomes and mitochondria are capable of triggering autophagic cell death. Autophagic cell death can be observed in the presence of intense photodamage in which the function of lysosomes is impaired. A pro-survival response may occur in cases of mild PDT, even if the lysosomes are partially affected. However, parallel damage in mitochondria and lysosomes enables this response to be switched to autophagic cell death [113, 114].

Necroptosis

Necroptosis is a regulated type of necrosis in which death receptors are involved [115]. It is triggered by the detection of stress signals *via* death receptors (*e.g.*, TNFR1, FAS, TRAIL), especially in the case of caspase-8 inhibition. Recognition of death signals by receptors normally leads to cell death *via* extrinsic apoptosis. However, in cells where caspase-8 is inhibited, the mechanism of cell death is switched to necroptosis due to the activation of receptor-interacting protein kinases RIPK1, RIPK3, and mixed lineage kinase domain-like protein (MKLK) [116 - 121]. One of the first studies on PDT-induced necroptosis showed that RIPK3-dependent necroptosis can be activated in human glioblastoma cells with the help of the necrosome complex composed of RIPK1 and RIPK3 [122].

Subsequent studies have revealed that tumor type and photosensitizer concentration are critical factors in PDT-induced necroptosis. For instance, in a study examining the PDT-mediated effects of hiporfin on various osteosarcoma cell lines (DLM-8, 143B, and HOS), low hiporfin concentrations have activated autophagy as a protective mechanism in 143B and HOS cell lines. Moreover, the expression of RIP1, a necroptosis modulator, has increased in DLM-8 cells. More interestingly, pre-incubation of these cells with a necroptosis inhibitor has increased cell viability, whereas no increase has been observed when an apoptosis inhibitor was used. This indicated that necroptosis was the main cell death mechanism in DLM-8 cells [123]. In another study, low concentrations of talaporfin sodium were shown to activate necroptosis in glioblastoma cells following PDT, while high concentrations induced non-necroptotic necrosis [124]. All of these studies demonstrate that PDT-mediated necroptosis, like in other types of cell death, is dependent on multiple factors [125].

Paraptosis

Paraptosis is a type of programmed cell death characterized by cytoplasmic vacuolation that involves the swelling of ER or mitochondria [126]. Paraptosis is usually related to the presence of misfolded proteins in ER. Therefore, targeting photosensitizers in ER can cause paraptosis [127]. Different studies have shown that photosensitizers that mostly accumulate in ER may trigger paraptosis [128 - 132]. However, extensive studies are needed to analyze the mechanism and factors of PDT-induced paraptosis.

Immunogenic Cell Death

For many years, it was believed that apoptosis was immunologically silent, while necrosis was considered a highly inflammatory cell death. The main reason is the engulfing of apoptotic cells by phagocytic cells without releasing their intracellular contents. In contrast, intracellular content containing inflammatory molecules is rapidly released during necrosis. These molecules are known as damage-associated molecular patterns (DAMPs) and are recognized by pattern recognition receptors (PRRs) in immune cells [98]. It is now known that, as a type of apoptosis, an immune response against pathogens or tumor antigens can be activated in dying cells. This type of cell death is immunogenic cell death, a distinct form of apoptosis that can be induced by a variety of therapeutic modalities including photodynamic therapy [133].

Immunogenic cell death involves the stimulation of immune system components by DAMPs. The molecular patterns of DAMPs may differ depending on the treatment protocol and the genotypic origin of cancer cells. In PDT-induced immunogenic cell death, calreticulin (CRT), a calcium-binding protein, is usually observed to be exposed on the outer surface of the plasma membrane as a DAMP. The translocation of CRT from ER to the outer surface of the cell membrane acts as the "eat me" signal for antigen-presenting cells, thereby promoting immune response [89].

PDT-MEDIATED IMMUNE RESPONSE

Studies have revealed that PDT can activate both innate and adaptive immunity, thereby helping to establish long-term immunological memory. Activation of the immune system as a result of the photodynamic effect was first described with macrophage activation due to lipid peroxidation of lymphocyte membranes as a result of ROS formation [134]. Subsequently, PDT-treated cells were shown to secrete different cytokines, including tumor necrosis factor, interleukin (IL)-1β, and IL-6, which are involved in the recruitment of various myeloid cells [135]. Further studies have shown that PDT-treated tumor cell lysates can stimulate dendritic cells and modulate an antitumor immune response [136]. The first clinical case of immune-responsive PDT was studied in the treatment of multifocal angiosarcoma of the head and neck. According to this study, PDT succeeded in local control of the targeted tumor, meanwhile, untreated distant tumors were observed to undergo spontaneous remission accompanied by increased immune cell infiltration [137].

The current knowledge of the immunology of tumor cells provides a solid background for future research on PDT-activated immune response [138]. Although PDT-induced immune responses still need to be investigated, it is now clear that PDT-mediated oxidative stress can activate an acute inflammatory response to maintain homeostasis and elicit both innate and adaptive immunity [89].

ANTIVASCULAR EFFECTS OF PDT

The vascular system is the essential biological tool for the delivery of nutrients and oxygen to physiological tissues, including tumors. Tumor angiogenesis is the creation of a vascular network that will provide oxygen and nutrients to sustain tumor growth [139]. Most photosensitizers exhibit affinity to the endothelium. Therefore, one of the expected outcomes of PDT is vascular destruction [25]. PDT-induced vascular damage is thought to be crucial, as it prevents cancer cells from getting the necessary nutrients through the vascular network [140]. The antivascular response of PDT is highly influenced by the type and dose of photosensitizer, drug-light interval, and light influence. For example, a study using verteporfin has demonstrated that high doses induced vascular occlusion, while lower doses promoted the hyperpermeable response [141]. A different study has shown that the intensity of vascular damage caused by PDT correlated with the protoporphyrin IX levels in the vessel walls [142]. In addition, a study on the PDT sensitivity of endothelial cells has shown that these cells are more sensitive to PDT than smooth muscle cells or fibroblasts [143]. In another study investigating PDT-mediated effects in bovine aortic endothelial cells and human colon adenocarcinoma cells, it was documented that photosensitizers mostly accumulate in endothelial cells, and these cells are more sensitive to PDT than cancer cells [144]. The common interpretation of these studies indicates that PDT is an efficient therapeutic modality for tumor vascularity.

PDT IN CANCER TREATMENT

PDT arouses interest with its successful results reflected in oncological and non-oncological studies. Anticancer effects of PDT may occur through direct photodamage of tumor cells, destruction of tumor vasculature, and/or activation of the immune system [145]. The characteristics of tumor cells, such as high metabolic activity, hyperpermeable tumor neovasculature, poor lymphatic drainage, and low pH are thought to allow photosensitizers to accumulate in tumor tissues selectively. Besides, limiting the light application to the treatment site increases the anticancer efficacy of PDT [146]. In addition to tumor cells, PDT can affect cellular and non-cellular components of the tumor microenvironment (fibroblasts, immune cells, extracellular matrix, *etc.*) [147].

PDT has proven to be largely successful in some precancerous skin lesions and non-melanoma skin cancers [148, 149]. Besides, many tumor types, such as skin, head and neck, breast, lung, bladder, prostate, pancreas, biliary tract, and brain, have been shown to provide favorable outcomes in experimental and clinical settings [49, 148]. While PDT does not require an invasive protocol in superficial oncological lesions, deeply located tumors can be reached using light sources integrated into fiber optic systems [150].

PDT can be combined with other therapeutic modalities to increase treatment efficacy. Considering the fact that molecular targets of conventional therapies and PDT would differ, combination strategies may provide better outcomes [151]. PDT may overcome the resistance of cancer cells to standard treatment methods. In addition, administration of lower doses in combination therapy models may cause fewer side effects [30, 152].

Combination with Chemotherapy

Chemotherapy is a widely used method in clinical cancer therapy. Chemotherapeutic drugs do not exert selective toxicity to cancer cells, they can be severely damaging to other cells in the body. Moreover, possible resistance to these drugs limits the efficacy of treatment. It is suggested that a combination with PDT can overcome such undesirable effects. In the case of combined application, synergistic effects will provide better therapeutic outcomes. Chemotherapeutic drugs may increase the susceptibility of cancer cells to ROS, while PDT may overcome multidrug resistance [30, 153].

Numerous studies have revealed the high efficacy of combinational therapy. For instance, the combination of doxorubicin with nickel oxide nanoparticles has been shown to result in higher cell death [154]. In a study using a $^1O_2$-sensitive nanocarrier delivery system, it was shown that doxorubicin can easily reach the tumor site, destroy cancer cells, and minimize side effects [155]. Another study noted that cisplatin-induced genotoxicity could be blunted or reduced when riboflavin was used as a photosensitizer [156]. In a study, doxorubicin-loaded nanoparticles-Chlorine e6-encapsulating microbubbles complex (DOX-NPs/Ce--MBs) was designed to overcome anticancer multidrug resistance. It has been shown that ROS, produced as a result of laser activation of Ce6, inhibits DOX effusion by damaging the efflux pumps, which is the main characteristic of multidrug resistant cells [157]. Multidrug resistance is usually related to overexpression of P-glycoprotein on the cell surface [153]. Different studies have shown that PDT-induced ROS reduces P-glycoprotein levels and drug efflux in cancer cells, thereby reversing drug resistance [158, 159]. In conclusion, promising results of the chemo-PDT combination provide theoretical background for further studies and thus raise the expectancy for high-level therapeutic efficacy.

Combination with Radiotherapy

Radiotherapy is a treatment method based on the application of ionizing radiation in some diseases, especially cancer. It causes cell death by increasing genotoxicity and DNA damage [160]. Above 50% of patients with malignant tumors are recommended to receive radiation therapy. In some cancer cases (*e.g.*, skin cancer, cervical cancer and lymphoma), the therapeutic efficiency of radiotherapy alone can be greater than 90%. However, the non-specificity of radiotherapy causes damage to healthy tissues. Besides, hypoxic cells in tumors may resist radiotherapy [153, 161, 162].

PDT combined radiotherapy can significantly reduce symptoms and prolong survival in patients with advanced malignancies. It has been stated that various radiation-resistant tumor cells are sensitive to PDT [163]. Moreover, photocytotoxic damage produces singlet oxygen that primarily acts on cell membranes and cellular substrates, while nuclear DNA is the main target of ionizing radiation [164]. Therefore, the combination of PDT and radiotherapy can achieve tumor control by causing damage to different targets [160].

Several studies have shown favorable results of the PDT-radiotherapy combination. For instance, the combination of PDT and gamma-irradiation using 5-aminolevulinic acid has been reported to cause cytotoxicity by creating an additive effect [165]. Another study reported that Bowen's disease can be treated using combination therapy with 5-aminolevulinic acid-mediated PDT [166]. Inhibition of tumor growth was reported in a study examining the effect of hematoporphyrin dimethyl ether-mediated PDT and radiotherapy [167]. In a comprehensive study investigating the effect of PDT in combination with radiotherapy, a total of 90 cases with mid-to-late malignant tumors (rectal cancer, gastric cancer, bladder cancer, esophageal cancer, cervical cancer, and superficial tumors) were examined. According to the results, a combination of PDT with intensity modulated radiotherapy (IMRT) significantly reduced symptoms and improved the quality of life of patients [153, 168].

Combination with Immunotherapy

The PDT-induced immune response plays an important role in controlling tumor growth and preventing metastasis. These processes involve complex mechanisms that influence each other, and both innate and adaptive immunity are involved in the PDT-induced response. Therefore, almost every point and player of the immune system can be targeted for treatment.

PDT triggers immunotherapeutic response through different mechanisms: (1) PDT-induced immunogenic cell death enhances tumor destruction with contributions of local immune cells; (2) tumor-specific antigens released during immunogenic cell death act as an *In situ* vaccine; (3) DAMPs boost the immunogenicity of tumor antigens; and (4) pro-inflammatory cytokines provide the immune system activation [169, 170]. These effects can be potentiated by different immunotherapy agents that increase tumor immunogenicity (such as immunoadjuvants) or reduce immunosuppression (such as immune checkpoint inhibitors) [148].

Several studies have demonstrated that PDT-generated tumor cell lysates may have vaccine potential. A study analyzing the combination of dendritic cell immunotherapy and hypericin-mediated PDT noted that immunogenic cell death-based vaccines may be useful in clinical settings [171]. In another study, it has been reported that PDT-generated murine tumor cell lysates may be effective vaccines [136]. In addition, a different study showed that 5-aminolevulinic acid-mediated PDT tumor lysates can increase the antigen-presenting capacity of *ex vivo*-generated dendritic cells [172]. The use of surgically removed tumor tissues to obtain PDT-based vaccines is also among the approaches recommended in the literature [173]. Another approach is to combine PDT with monoclonal antibodies. In a study, the tumor was destroyed by providing deep tissue penetration with photosensitizers conjugated to monoclonal antibodies, which target epidermal growth factor receptors (EGFRs) [174]. Furthermore, co-administration of PDT with immune checkpoint inhibitors has been shown to induce a potent tumor-specific immune response, mediating regression of both irradiated primary tumors and non-irradiated distant tumors [175].

NANOSCALE DRUG DELIVERY SYSTEMS

Drug delivery strategies play a major role in the therapeutic success of any treatment, including PDT. Since most photosensitizers are hydrophobic, their tendency to aggregate in aqueous media negatively affects PDT. Various drug delivery strategies have been designed to increase the specificity of photosensitizers and reduce side effects [1].

Nanoparticles

Nanoparticles are one of the most prominent drug delivery systems in PDT with their advantages such as easy penetration into cells due to submicron size, allowing the release of photosensitizer at constant concentration, biocompatibility, and photostability [176, 177]. Polymeric nanoparticles (PNPs), solid lipid nanoparticles (SLNs), nanostructured lipid carriers (NLCs), and metallic nanoparticles (MNs) are mostly preferred nanoparticle types in PDT studies.

PNPs are nanostructures that physically trap hydrophobic photosensitizers through hydrophobic or electrostatic interactions between the photosensitizer and the polymer. Their surface properties, sizes, and shapes can be optimized to enable polymer degradation and control drug release [178]. They are generally prepared from natural or synthetic polymers such as polycaprolactone (PCL), polylactic acid (PLA), poly-D, L-lactide-co-glycolide (PLGA), gelatin, chitosan, *etc* [177, 179, 180]. SLNs are colloidal particles prepared from lipids and surfactants. They have a wide range of applications due to their biocompatibility. However, several disadvantages, such as insufficient encapsulation capacity, tendency to crystallize, and uncontrollable drug leakage during storage, limit their efficiency [181 - 183]. NLCs have been developed to cope with these problems. They have an irregular matrix of solid and oil lipid phases, which prevents solid lipid crystallization and promotes increased drug encapsulation [183, 184]. Another type of nanoparticle used in PDT is metallic nanoparticles. Their long-term activity and stability in different biological conditions, such as pH, ionic strength, and pressure, provide advantages for their application in medical research. Gold nanoparticles are one of the most popular metallic nanoparticles used in biology and medicine [1, 185, 186]. They can be conjugated with different biological ligands to target overexpressed cell surface receptors in cancer cells.

They also increase the production of singlet oxygen and provide controlled drug release [187]. Different studies using all the specified nanoparticles as a drug delivery system have demonstrated the effective results of combining photosensitizers with nanoparticles, revealing the great potential of nanoparticles in future studies of PDT [188 - 193].

Liposomes

Liposomes are nanometric-sized unilamellar or multilamellar phospholipid vesicles that can be formulated to encapsulate hydrophilic or hydrophobic drugs. They are one of the most studied drug delivery systems due to their simple structure and controllable size, as well as being biocompatible and biodegradable. Hydrophobic photosensitizers dissolve in the phospholipid bilayer, while hydrophilic photosensitizers are encapsulated in the inner core. Therefore, liposomal delivery increases phototoxicity by reducing photosensitizer aggregation [1, 178, 194 - 197]. To increase specificity, liposomes can also be functionalized with monoclonal antibodies or antibody fragments (immuno-liposome). Such liposomes increase the efficacy of treatment by targeting cells that express cell surface antigens specific to these antibodies [198].

Hydrogels

Hydrogels are three-dimensional polymeric systems that can absorb and retain water. These systems are generally used for hydrophilic drugs because they can be dispersed in the matrix [177]. Hydrogels easily penetrate tissues and their nanoscale formulation helps them to diffuse and retain photosensitizers into tumors [199]. Their formulations can be developed by adding functional moieties, this improves their cellular uptake and targeting capabilities [200]. Hydrogel scaffolds can prevent premature photosensitizer release and accumulation in non-specific tissues, thereby increasing PDT efficacy [201]. Besides being encapsulated in the hydrogel matrix, photosensitizer may also be used as a component (*e.g.*, cross-linker) in the construction of hydrogel [202]. The photosensitizer, which is the covalent component of hydrogel, transfers its energy to molecular oxygen when exposed to light, resulting in the formation of singlet oxygen and ROS [203, 204].

LIMITATIONS AND FUTURE DIRECTIONS

PDT is a multicomponent therapeutic approach. In this light-based method, the photosensitizer does not produce singlet oxygen or ROS in the absence of light. The procedure has a variety of applications, as the timing and irradiation area can be optimized. However, light does not penetrate deep tissues, thus limiting the treatment efficacy of deep-seated tumors [205, 206]. Light penetration mostly depends on the depth of the target tissue and the light wavelength. Inhomogeneous regions in or between tissues can cause light to be scattered, transmitted, reflected, or absorbed [207, 208]. This indicates that the therapeutic window (600-1300 nm) should be considered to ensure maximum penetration [206].

The period of time between drug administration and light exposure is known as a drug-light interval. An ideal drug-light interval may differ from lesion to lesion due to variables such as targeted tissue, location, tumor size and depth [30, 209]. This complicates the implementation of a standard treatment protocol. Other important limitations of PDT are the difficulties of utilization in treatments of larger tumor masses and its inability to be implemented in the whole body in advanced cancers [210, 211]. Furthermore, the half-life of singlet oxygen, which is the main cause of destructive effects in PDT, is very short. Therefore, the expected therapeutic efficacy of PDT may be limited to the photosensitization area in some cases [212]. Since the efficacy of PDT is also dependent on molecular oxygen, tumor hypoxia may negatively affect the therapeutic outcomes of PDT [30, 213].

Future strategies for perfecting PDT mainly involve the improvement of photosensitizers with an extended period of action and efficiency. This can be achieved by the development of novel photosensitizers with strong absorption in the near-infrared (NIR) region, which can accomplish sufficient depth of treatment [71, 214]. These developments can be modulated by nanotechnological approaches. The specificity of photosensitizers is provided by different modalities such as antibody conjugation, synthesis of special photosensitizers containing different molecules (*e.g.*, DNA or peptide-based binders), or designing photosensitizers that attach to magnetic nanoparticles and directing them to target tissues with external magnetic field [50, 215]. In addition to advances in functional photosensitizers, the development of endoscopic techniques and devices will contribute to the clinical use of PDT in the near future [216].

CONCLUDING REMARKS

PDT promises undeniably influential results in experimental studies and clinical cases, but it has not been considered a standard treatment yet [71]. In light of advances in laser technology, nanotechnology, photochemistry, and photobiology, PDT can be effectively integrated into clinical studies, and the aforementioned limitations can be overcome. These advances will lead to the re-establishment of existing protocols with better equipment and improved technology. Therefore, PDT is likely to become more centralized as part of a multimodal approach or as a stand-alone therapy, which can be a powerful alternative to conventional therapeutic modalities [207].

REFERENCES

[1]   U. Chilakamarthi, and L. Giribabu, "Photodynamic therapy: Past, present and future", *Chem. Rec.,* vol. 17, no. 8, pp. 775-802, 2017.
[http://dx.doi.org/10.1002/tcr.201600121] [PMID: 28042681]

[2]   M. Tampa, M.I. Sarbu, C. Matei, C.I. Mitran, M.I. Mitran, C. Caruntu, C. Constantin, M. Neagu, and S.R. Georgescu, "Photodynamic therapy: A hot topic in dermato-oncology (Review)", *Oncol. Lett.,* vol. 17, no. 5, pp. 4085-4093, 2019.
[http://dx.doi.org/10.3892/ol.2019.9939] [PMID: 30944601]

[3]   T. Kimura, S. Takatsuki, S. Miyoshi, K. Fukumoto, M. Takahashi, E. Ogawa, A. Ito, T. Arai, S. Ogawa, and K. Fukuda, "Nonthermal cardiac catheter ablation using photodynamic therapy", *Circ. Arrhythm. Electrophysiol.,* vol. 6, no. 5, pp. 1025-1031, 2013.
[http://dx.doi.org/10.1161/CIRCEP.113.000810] [PMID: 23995252]

[4]   K. Fujita, Y. Imamura, K. Shinoda, C.S. Matsumoto, Y. Mizutani, K. Hashizume, A. Mizota, and M. Yuzawa, "One-year outcomes with half-dose verteporfin photodynamic therapy for chronic central serous chorioretinopathy", *Ophthalmology,* vol. 122, no. 3, pp. 555-561, 2015.
[http://dx.doi.org/10.1016/j.ophtha.2014.09.034] [PMID: 25444637]

[5]   S.G. Rockson, P. Kramer, M. Razavi, A. Szuba, S. Filardo, P. Fitzgerald, J.P. Cooke, S. Yousuf, A.R. DeVault, M.F. Renschler, and D.C. Adelman, "Photoangioplasty for human peripheral atherosclerosis: Results of a phase I trial of photodynamic therapy with motexafin lutetium (Antrin)", *Circulation,* vol. 102, no. 19, pp. 2322-2324, 2000.
[http://dx.doi.org/10.1161/01.CIR.102.19.2322] [PMID: 11067782]

[6]   R. Falk-Mahapatra, and S.O. Gollnick, "Photodynamic therapy and immunity: An update", *Photochem. Photobiol.,* vol. 96, no. 3, pp. 550-559, 2020.
[http://dx.doi.org/10.1111/php.13253] [PMID: 32128821]

[7]   H. Gursoy, C. Ozcakir-Tomruk, J. Tanalp, and S. Yilmaz, "Photodynamic therapy in dentistry: A literature review", *Clin. Oral Investig.,* vol. 17, no. 4, pp. 1113-1125, 2013.
[http://dx.doi.org/10.1007/s00784-012-0845-7] [PMID: 23015026]

[8]   G. Bozzini, P. Colin, N. Betrouni, P. Nevoux, A. Ouzzane, P. Puech, A. Villers, and S. Mordon, "Photodynamic therapy in urology: What can we do now and where are we heading?", *Photodiagn. Photodyn. Ther.,* vol. 9, no. 3, pp. 261-273, 2012.
[http://dx.doi.org/10.1016/j.pdpdt.2012.01.005] [PMID: 22959806]

[9]   A.D. Singh, P.K. Kaiser, J.E. Sears, M. Gupta, P.A. Rundle, and I.G. Rennie, "Photodynamic therapy of circumscribed choroidal haemangioma", *Br. J. Ophthalmol.,* vol. 88, no. 11, pp. 1414-1418, 2004.
[http://dx.doi.org/10.1136/bjo.2004.044396] [PMID: 15489484]

[10]   M.A. Blasi, A.C. Tiberti, A. Scupola, A. Balestrazzi, E. Colangelo, P. Valente, and E. Balestrazzi, "Photodynamic therapy with verteporfin for symptomatic circumscribed choroidal hemangioma: Five-year outcomes", *Ophthalmology,* vol. 117, no. 8, pp. 1630-1637, 2010.
[http://dx.doi.org/10.1016/j.ophtha.2009.12.033] [PMID: 20417564]

[11]   D.K. Newman, "Photodynamic therapy: Current role in the treatment of chorioretinal conditions", *Eye,* vol. 30, no. 2, pp. 202-210, 2016.
[http://dx.doi.org/10.1038/eye.2015.251] [PMID: 26742867]

[12]   E. Ostańska, D. Aebisher, and D. Bartusik-Aebisher, "The potential of photodynamic therapy in current breast cancer treatment methodologies", *Biomed. Pharmacother.,* vol. 137, p. 111302, 2021.
[http://dx.doi.org/10.1016/j.biopha.2021.111302] [PMID: 33517188]

[13]   S.M. Fien, and A.R. Oseroff, "Photodynamic therapy for non-melanoma skin cancer", *J. Natl. Compr. Canc. Netw.,* vol. 5, no. 5, pp. 531-540, 2007.
[http://dx.doi.org/10.6004/jnccn.2007.0046] [PMID: 17509255]

[14]   K. Moghissi, K. Dixon, J.A.C. Thorpe, M. Stringer, and C. Oxtoby, "Photodynamic therapy (PDT) in early central lung cancer: A treatment option for patients ineligible for surgical resection", *Thorax,* vol. 62, no. 5, pp. 391-395, 2007.
[http://dx.doi.org/10.1136/thx.2006.061143] [PMID: 17090572]

[15]   S.W. Cramer, and C.C. Chen, "Photodynamic therapy for the treatment of glioblastoma", *Front. Surg.,* vol. 6, p. 81, 2020.
[http://dx.doi.org/10.3389/fsurg.2019.00081] [PMID: 32039232]

[16]   K.H. Nelke, W. Pawlak, J. Leszczyszyn, and H. Gerber, "Photodynamic therapy in head and neck cancer", *Postepy Hig. Med. Dosw.,* vol. 68, pp. 119-128, 2014.
[http://dx.doi.org/10.5604/17322693.1088044] [PMID: 24491903]

[17]   A.K. Kubba, "Role of photodynamic therapy in the management of gastrointestinal cancer", *Digestion,* vol. 60, no. 1, pp. 1-10, 1999.
[http://dx.doi.org/10.1159/000007582] [PMID: 9892792]

[18]   Y. Matoba, K. Banno, I. Kisu, and D. Aoki, "Clinical application of photodynamic diagnosis and photodynamic therapy for gynecologic malignant diseases: A review", *Photodiagn. Photodyn. Ther.,* vol. 24, pp. 52-57, 2018.
[http://dx.doi.org/10.1016/j.pdpdt.2018.08.014] [PMID: 30172075]

[19]   M. Osuchowski, D. Bartusik-Aebisher, F. Osuchowski, and D. Aebisher, "Photodynamic therapy for prostate cancer: A narrative review", *Photodiagn. Photodyn. Ther.,* vol. 33, p. 102158, 2021.
[http://dx.doi.org/10.1016/j.pdpdt.2020.102158] [PMID: 33352313]

[20]   X. Li, S. Lee, and J. Yoon, "Supramolecular photosensitizers rejuvenate photodynamic therapy", *Chem. Soc. Rev.,* vol. 47, no. 4, pp. 1174-1188, 2018.
[http://dx.doi.org/10.1039/C7CS00594F] [PMID: 29334090]

[21]   A. Nanashima, H. Yamaguchi, S. Shibasaki, N. Ide, T. Sawai, T. Tsuji, S. Hidaka, Y. Sumida, T. Nakagoe, and T. Nagayasu, "Adjuvant photodynamic therapy for bile duct carcinoma after surgery: A preliminary study", *J. Gastroenterol.,* vol. 39, no. 11, pp. 1095-1101, 2004.
[http://dx.doi.org/10.1007/s00535-004-1449-z] [PMID: 15580404]

[22]   X. Wang, G. Ramamurthy, A.A. Shirke, E. Walker, J. Mangadlao, Z. Wang, Y. Wang, L. Shan, M.D. Schluchter, Z. Dong, S.M. Brady-Kalnay, N.K. Walker, M. Gargesha, G. MacLennan, D. Luo, R. Sun, B. Scott, D. Roy, J. Li, and J.P. Basilion, "Photodynamic therapy is an effective adjuvant therapy for image-guided surgery in prostate cancer", *Cancer Res.,* vol. 80, no. 2, pp. 156-162, 2020.
[http://dx.doi.org/10.1158/0008-5472.CAN-19-0201] [PMID: 31719100]

[23]   N.R. Rigual, G. Shafirstein, J. Frustino, M. Seshadri, M. Cooper, G. Wilding, M.A. Sullivan, and B. Henderson, "Adjuvant intraoperative photodynamic therapy in head and neck cancer", *JAMA Otolaryngol. Head Neck Surg.,* vol. 139, no. 7, pp. 706-711, 2013.
[http://dx.doi.org/10.1001/jamaoto.2013.3387] [PMID: 23868427]

[24]   R. Ackroyd, C. Kelty, N. Brown, and M. Reed, "The history of photodetection and photodynamic therapy", *Photochem. Photobiol.,* vol. 74, no. 5, pp. 656-669, 2001.
[http://dx.doi.org/10.1562/0031-8655(2001)074<0656:THOPAP>2.0.CO;2] [PMID: 11723793]

[25]   M.D. Daniell, and J.S. Hill, "A history of photodynamic therapy", *ANZ J. Surg.,* vol. 61, no. 5, pp. 340-348, 1991.
[http://dx.doi.org/10.1111/j.1445-2197.1991.tb00230.x] [PMID: 2025186]

[26]   M.H. Abdel-kader, *CHAPTER 1 The Journey of PDT Throughout History: PDT from Pharos to Present. Photodynamic Medicine: From Bench to Clinic.* The Royal Society of Chemistry, 2016, pp. 1-21.

[27]   R-M. Szeimies, J. Dräger, C. Abels, and M. Landthaler, "History of photodynamic therapy in dermatology", In: *Comprehensive Series in Photosciences. 2.*, P. Calzavara-Pinton, R-M. Szeimies, B. Ortel, Eds., Elsevier, 2001, pp. 3-15.

[28]   J.H. Correia, J.A. Rodrigues, S. Pimenta, T. Dong, and Z. Yang, "Photodynamic therapy review: Principles, photosensitizers, applications, and future directions", *Pharmaceutics*, vol. 13, no. 9, p. 1332, 2021.
[http://dx.doi.org/10.3390/pharmaceutics13091332] [PMID: 34575408]

[29]   D.E.J.G.J. Dolmans, D. Fukumura, and R.K. Jain, "Photodynamic therapy for cancer", *Nat. Rev. Cancer*, vol. 3, no. 5, pp. 380-387, 2003.
[http://dx.doi.org/10.1038/nrc1071] [PMID: 12724736]

[30]   G. Gunaydin, M.E. Gedik, and S. Ayan, "Photodynamic therapy for the treatment and diagnosis of Cancer–A review of the current clinical status", *Front Chem.*, vol. 9, p. 686303, 2021.
[http://dx.doi.org/10.3389/fchem.2021.686303] [PMID: 34409014]

[31]   T.J. Dougherty, G.B. Grindey, R. Fiel, K.R. Weishaupt, and D.G. Boyle, "Photoradiation therapy. II. cure of animal tumors with hematoporphyrin and light", *J. Natl. Cancer Inst.*, vol. 55, no. 1, pp. 115-121, 1975.
[http://dx.doi.org/10.1093/jnci/55.1.115] [PMID: 1159805]

[32]   J.F. Kelly, M.E. Snell, and M.C. Berenbaum, "Photodynamic destruction of human bladder carcinoma", *Br. J. Cancer*, vol. 31, no. 2, pp. 237-244, 1975.
[http://dx.doi.org/10.1038/bjc.1975.30] [PMID: 1164470]

[33]   J.F. Kelly, and M.E. Snell, "Hematoporphyrin derivative: A possible aid in the diagnosis and therapy of carcinoma of the bladder", *J. Urol.*, vol. 115, no. 2, pp. 150-151, 1976.
[http://dx.doi.org/10.1016/S0022-5347(17)59108-9] [PMID: 1249866]

[34]   T.J. Dougherty, G. Lawrence, J.H. Kaufman, D. Boyle, K.R. Weishaupt, and A. Goldfarb, "Photoradiation in the treatment of recurrent breast carcinoma", *J. Natl. Cancer Inst.*, vol. 62, no. 2, pp. 231-237, 1979.
[PMID: 283259]

[35]   T.S. Mang, R. Allison, G. Hewson, W. Snider, and R. Moskowitz, "A phase II/III clinical study of tin ethyl etiopurpurin (Purlytin)-induced photodynamic therapy for the treatment of recurrent cutaneous metastatic breast cancer", *Cancer J. Sci. Am.*, vol. 4, no. 6, pp. 378-384, 1998.
[PMID: 9853137]

[36]   A. Dimofte, T.C. Zhu, S.M. Hahn, and R.A. Lustig, "*In vivo* light dosimetry for motexafin lutetium-mediated PDT of recurrent breast cancer", *Lasers Surg. Med.*, vol. 31, no. 5, pp. 305-312, 2002.
[http://dx.doi.org/10.1002/lsm.10115] [PMID: 12430147]

[37]   H. Barr, N. Krasner, P.B. Boulos, P. Chatlani, and S.G. Bown, "Photodynamic therapy for colorectal cancer: A quantitative pilot study", *Br. J. Surg.*, vol. 77, no. 1, pp. 93-96, 2005.
[http://dx.doi.org/10.1002/bjs.1800770132] [PMID: 2302524]

[38]   P. Mĺkvy, H. Messmann, J. Regula, M. Conio, M. Pauer, C.E. Millson, A.J. MacRobert, and S.G. Bown, "Photodynamic therapy for gastrointestinal tumors using three photosensitizers--ALA induced PPIX, Photofrin and MTHPC. A pilot study", *Neoplasma*, vol. 45, no. 3, pp. 157-161, 1998.
[PMID: 9717528]

[39]   S.G. Bown, A.Z. Rogowska, D.E. Whitelaw, W.R. Lees, L.B. Lovat, P. Ripley, L. Jones, P. Wyld, A. Gillams, and A.W. Hatfield, "Photodynamic therapy for cancer of the pancreas", *Gut*, vol. 50, no. 4, pp. 549-557, 2002.
[http://dx.doi.org/10.1136/gut.50.4.549] [PMID: 11889078]

[40]   V.G. Schweitzer, "Photodynamic therapy for treatment of head and neck cancer", *Otolaryngol. Head Neck Surg.*, vol. 102, no. 3, pp. 225-232, 1990.
[http://dx.doi.org/10.1177/019459989010200304] [PMID: 2108409]

[41]   M.A. Biel, "Photodynamic therapy and the treatment of head and neck neoplasia", *Laryngoscope,* vol. 108, no. 9, pp. 1259-1268, 1998.
[http://dx.doi.org/10.1097/00005537-199809000-00001] [PMID: 9738739]

[42]   D.R. Sandeman, "Photodynamic therapy in the management of malignant gliomas: A review", *Lasers Med. Sci.,* vol. 1, no. 3, pp. 163-174, 1986.
[http://dx.doi.org/10.1007/BF02040233]

[43]   J.S. Hill, A.H. Kaye, W.H. Sawyer, G. Morstyn, P.D. Megison, and S.S. Stylli, "Selective uptake of hematoporphyrin derivative into human cerebral glioma", *Neurosurgery,* vol. 26, no. 2, pp. 248-254, 1990.
[http://dx.doi.org/10.1227/00006123-199002000-00011] [PMID: 2137904]

[44]   E.A. Popovic, A.H. Kaye, and J.S. Hill, "Photodynamic therapy of brain tumors", *J. Clin. Laser Med. Surg.,* vol. 14, no. 5, pp. 251-261, 1996.
[http://dx.doi.org/10.1089/clm.1996.14.251] [PMID: 9612191]

[45]   M.A. Rosenthal, B. Kavar, J.S. Hill, D.J. Morgan, R.L. Nation, S.S. Stylli, R.L. Basser, S. Uren, H. Geldard, M.D. Green, S.B. Kahl, and A.H. Kaye, "Phase I and pharmacokinetic study of photodynamic therapy for high-grade gliomas using a novel boronated porphyrin", *J. Clin. Oncol.,* vol. 19, no. 2, pp. 519-524, 2001.
[http://dx.doi.org/10.1200/JCO.2001.19.2.519] [PMID: 11208846]

[46]   M.A.E.J. Ortner, J. Liebetruth, S. Schreiber, M. Hanft, U. Wruck, V. Fusco, J.M. Müller, H. Hörtnagl, and H. Lochs, "Photodynamic therapy of nonresectable cholangiocarcinoma", *Gastroenterology,* vol. 114, no. 3, pp. 536-542, 1998.
[http://dx.doi.org/10.1016/S0016-5085(98)70537-2] [PMID: 9496944]

[47]   H.I. Pass, T.F. DeLaney, Z. Tochner, P.E. Smith, B.K. Temeck, H.W. Pogrebniak, K.C. Kranda, A. Russo, W.S. Friauf, J.W. Cole, J.B. Mitchell, and G. Thomas, "Intrapleural photodynamic therapy: Results of a phase I trial", *Ann. Surg. Oncol.,* vol. 1, no. 1, pp. 28-37, 1994.
[http://dx.doi.org/10.1007/BF02303538] [PMID: 7834425]

[48]   T.J. Dougherty, C.J. Gomer, B.W. Henderson, G. Jori, D. Kessel, M. Korbelik, J. Moan, and Q. Peng, "Photodynamic therapy", *J. Natl. Cancer Inst.,* vol. 90, no. 12, pp. 889-905, 1998.
[http://dx.doi.org/10.1093/jnci/90.12.889] [PMID: 9637138]

[49]   R. Baskaran, J. Lee, and S.G. Yang, "Clinical development of photodynamic agents and therapeutic applications", *Biomater. Res.,* vol. 22, no. 1, p. 25, 2018.
[http://dx.doi.org/10.1186/s40824-018-0140-z] [PMID: 30275968]

[50]   P. Agostinis, K. Berg, K.A. Cengel, T.H. Foster, A.W. Girotti, S.O. Gollnick, S.M. Hahn, M.R. Hamblin, A. Juzeniene, D. Kessel, M. Korbelik, J. Moan, P. Mroz, D. Nowis, J. Piette, B.C. Wilson, and J. Golab, "Photodynamic therapy of cancer: An update", *CA Cancer J. Clin.,* vol. 61, no. 4, pp. 250-281, 2011.
[http://dx.doi.org/10.3322/caac.20114] [PMID: 21617154]

[51]   J. Kou, D. Dou, and L. Yang, "Porphyrin photosensitizers in photodynamic therapy and its applications", *Oncotarget,* vol. 8, no. 46, pp. 81591-81603, 2017.
[http://dx.doi.org/10.18632/oncotarget.20189] [PMID: 29113417]

[52]   E.S. Nyman, and P.H. Hynninen, "Research advances in the use of tetrapyrrolic photosensitizers for photodynamic therapy", *J. Photochem. Photobiol. B,* vol. 73, no. 1-2, pp. 1-28, 2004.
[http://dx.doi.org/10.1016/j.jphotobiol.2003.10.002] [PMID: 14732247]

[53]   S. Kwiatkowski, B. Knap, D. Przystupski, J. Saczko, E. Kędzierska, K. Knap-Czop, J. Kotlińska, O. Michel, K. Kotowski, and J. Kulbacka, "Photodynamic therapy: Mechanisms, photosensitizers and combinations", *Biomed. Pharmacother.,* vol. 106, pp. 1098-1107, 2018.
[http://dx.doi.org/10.1016/j.biopha.2018.07.049] [PMID: 30119176]

[54] I. Yoon, J.Z. Li, and Y.K. Shim, "Advance in photosensitizers and light delivery for photodynamic therapy", *Clin. Endosc.,* vol. 46, no. 1, pp. 7-23, 2013.
[http://dx.doi.org/10.5946/ce.2013.46.1.7] [PMID: 23423543]

[55] R.R. Allison, G.H. Downie, R. Cuenca, X.H. Hu, C.J.H. Childs, and C.H. Sibata, "Photosensitizers in clinical PDT", *Photodiagn. Photodyn. Ther.,* vol. 1, no. 1, pp. 27-42, 2004.
[http://dx.doi.org/10.1016/S1572-1000(04)00007-9] [PMID: 25048062]

[56] U. Bazylińska, J. Pietkiewicz, J. Saczko, M. Nattich-Rak, J. Rossowska, A. Garbiec, and K.A. Wilk, "Nanoemulsion-templated multilayer nanocapsules for cyanine-type photosensitizer delivery to human breast carcinoma cells", *Eur. J. Pharm. Sci.,* vol. 47, no. 2, pp. 406-420, 2012.
[http://dx.doi.org/10.1016/j.ejps.2012.06.019] [PMID: 22796218]

[57] S. Duchi, G. Sotgiu, E. Lucarelli, M. Ballestri, B. Dozza, S. Santi, A. Guerrini, P. Dambruoso, S. Giannini, D. Donati, C. Ferroni, and G. Varchi, "Mesenchymal stem cells as delivery vehicle of porphyrin loaded nanoparticles: Effective photoinduced *in vitro* killing of osteosarcoma", *J. Control. Release,* vol. 168, no. 2, pp. 225-237, 2013.
[http://dx.doi.org/10.1016/j.jconrel.2013.03.012] [PMID: 23524189]

[58] Y. Sun, X. Geng, Y. Wang, X. Su, R. Han, J. Wang, X. Li, P. Wang, K. Zhang, and X. Wang, "Highly efficient water-soluble photosensitizer based on chlorin: Synthesis, characterization, and evaluation for photodynamic therapy", *ACS Pharmacol. Transl. Sci.,* vol. 4, no. 2, pp. 802-812, 2021.
[http://dx.doi.org/10.1021/acsptsci.1c00004] [PMID: 33860203]

[59] F. Setaro, J.W.H. Wennink, P.I. Mäkinen, L. Holappa, P.N. Trohopoulos, S. Ylä-Herttuala, C.F. van Nostrum, A. de la Escosura, and T. Torres, "Amphiphilic phthalocyanines in polymeric micelles: A supramolecular approach toward efficient third-generation photosensitizers", *J. Mater. Chem. B Mater. Biol. Med.,* vol. 8, no. 2, pp. 282-289, 2020.
[http://dx.doi.org/10.1039/C9TB02014D] [PMID: 31803886]

[60] M.S. Gualdesi, J. Vara, V. Aiassa, C.I. Alvarez Igarzabal, and C.S. Ortiz, "New poly(acrylamide) nanoparticles in the development of third generation photosensitizers", *Dyes Pigments,* vol. 184, p. 108856, 2021.
[http://dx.doi.org/10.1016/j.dyepig.2020.108856]

[61] A.G. Niculescu, and A.M. Grumezescu, "Photodynamic therapy—an up-to-date review", *Appl. Sci.,* vol. 11, no. 8, p. 3626, 2021.
[http://dx.doi.org/10.3390/app11083626]

[62] Z. Huang, "A review of progress in clinical photodynamic therapy", *Technol. Cancer Res. Treat.,* vol. 4, no. 3, pp. 283-293, 2005.
[http://dx.doi.org/10.1177/153303460500400308] [PMID: 15896084]

[63] M.M. Kim, and A. Darafsheh, "Light sources and dosimetry techniques for photodynamic therapy", *Photochem. Photobiol.,* vol. 96, no. 2, pp. 280-294, 2020.
[http://dx.doi.org/10.1111/php.13219] [PMID: 32003006]

[64] M.H. Schmidt, D.M. Bajic, K.W. Reichert II, T.S. Martin, G.A. Meyer, and H.T. Whelan, "Light-emitting diodes as a light source for intraoperative photodynamic therapy", *Neurosurgery,* vol. 38, no. 3, pp. 552-556, 1996.
[PMID: 8837808]

[65] R.A. Lustig, T.J. Vogl, D. Fromm, R. Cuenca, R. Alex Hsi, A.K. D'Cruz, Z. Krajina, M. Turić, A. Singhal, and J.C. Chen, "A multicenter Phase I safety study of intratumoral photoactivation of talaporfin sodium in patients with refractory solid tumors", *Cancer,* vol. 98, no. 8, pp. 1767-1771, 2003.
[http://dx.doi.org/10.1002/cncr.11708] [PMID: 14534895]

[66] J. Chen, L. Keltner, J. Christophersen, F. Zheng, M. Krouse, A. Singhal, and S. Wang, "New technology for deep light distribution in tissue for phototherapy", *Cancer J.,* vol. 8, no. 2, pp. 154-163, 2002.

[http://dx.doi.org/10.1097/00130404-200203000-00009] [PMID: 11999949]

[67]   J. Dobson, G.F. de Queiroz, and J.P. Golding, "Photodynamic therapy and diagnosis: Principles and comparative aspects", *Vet. J.,* vol. 233, pp. 8-18, 2018.
[http://dx.doi.org/10.1016/j.tvjl.2017.11.012] [PMID: 29486883]

[68]   B.A. Hartl, H. Hirschberg, L. Marcu, and S.R. Cherry, "Characterizing low fluence thresholds for *in vitro* photodynamic therapy", *Biomed. Opt. Express,* vol. 6, no. 3, pp. 770-779, 2015.
[http://dx.doi.org/10.1364/BOE.6.000770] [PMID: 25798302]

[69]   J.D. Breskey, S.E. Lacey, B.J. Vesper, W.A. Paradise, J.A. Radosevich, and M.D. Colvard, "Photodynamic therapy: Occupational hazards and preventative recommendations for clinical administration by healthcare providers", *Photomed. Laser Surg.,* vol. 31, no. 8, pp. 398-407, 2013.
[http://dx.doi.org/10.1089/pho.2013.3496] [PMID: 23859750]

[70]   J.F. Algorri, M. Ochoa, P. Roldán-Varona, L. Rodríguez-Cobo, and J.M. López-Higuera, "Photodynamic therapy: A compendium of latest reviews", *Cancers,* vol. 13, no. 17, p. 4447, 2021.
[http://dx.doi.org/10.3390/cancers13174447] [PMID: 34503255]

[71]   L. Benov, "Photodynamic therapy: Current status and future directions", *Med. Princ. Pract.,* vol. 24, no. Suppl 1, suppl. Suppl. 1, pp. 14-28, 2015.
[http://dx.doi.org/10.1159/000362416] [PMID: 24820409]

[72]   A. Nowak-Stepniowska, P. Pergoł, and A. Padzik-Graczyk, "[Photodynamic method of cancer diagnosis and therapy--mechanisms and applications]", *Postepy Biochem.,* vol. 59, no. 1, pp. 53-63, 2013.
[PMID: 23821943]

[73]   A.P. Castano, T.N. Demidova, and M.R. Hamblin, "Mechanisms in photodynamic therapy: Part two—cellular signaling, cell metabolism and modes of cell death", *Photodiagn. Photodyn. Ther.,* vol. 2, no. 1, pp. 1-23, 2005.
[http://dx.doi.org/10.1016/S1572-1000(05)00030-X] [PMID: 25048553]

[74]   W.P. Li, C.J. Yen, B.S. Wu, and T.W. Wong, "Recent advances in photodynamic therapy for deep-seated tumors with the aid of nanomedicine", *Biomedicines,* vol. 9, no. 1, p. 69, 2021.
[http://dx.doi.org/10.3390/biomedicines9010069] [PMID: 33445690]

[75]   A.P. Castano, T.N. Demidova, and M.R. Hamblin, "Mechanisms in photodynamic therapy: Part one—photosensitizers, photochemistry and cellular localization", *Photodiagn. Photodyn. Ther.,* vol. 1, no. 4, pp. 279-293, 2004.
[http://dx.doi.org/10.1016/S1572-1000(05)00007-4] [PMID: 25048432]

[76]   R.W. Boyle, and D. Dolphin, "Structure and biodistribution relationships of photodynamic sensitizers", *Photochem. Photobiol.,* vol. 64, no. 3, pp. 469-485, 1996.
[http://dx.doi.org/10.1111/j.1751-1097.1996.tb03093.x] [PMID: 8806226]

[77]   T.J. Jensen, M.G.H. Vicente, R. Luguya, J. Norton, F.R. Fronczek, and K.M. Smith, "Effect of overall charge and charge distribution on cellular uptake, distribution and phototoxicity of cationic porphyrins in HEp2 cells", *J. Photochem. Photobiol. B,* vol. 100, no. 2, pp. 100-111, 2010.
[http://dx.doi.org/10.1016/j.jphotobiol.2010.05.007] [PMID: 20558079]

[78]   X.F. Zhang, and H.J. Xu, "Influence of halogenation and aggregation on photosensitizing properties of zinc phthalocyanine (ZnPC)", *J. Chem. Soc., Faraday Trans.,* vol. 89, no. 18, pp. 3347-3351, 1993.
[http://dx.doi.org/10.1039/ft9938903347]

[79]   K. Sakamoto, T. Kato, T. Kawaguchi, E. Ohno-Okumura, T. Urano, T. Yamaoka, S. Suzuki, and M.J. Cook, "Photosensitizer efficacy of non-peripheral substituted alkylbenzopyridoporphyrazines for photodynamic therapy of cancer", *J. Photochem. Photobiol. Chem.,* vol. 153, no. 1-3, pp. 245-253, 2002.
[http://dx.doi.org/10.1016/S1010-6030(02)00292-7]

[80]   N. Mehraban, and H. Freeman, "Developments in PDT sensitizers for increased selectivity and singlet oxygen production", *Materials,* vol. 8, no. 7, pp. 4421-4456, 2015.
[http://dx.doi.org/10.3390/ma8074421] [PMID: 28793448]

[81]   M. Ethirajan, Y. Chen, P. Joshi, and R.K. Pandey, "The role of porphyrin chemistry in tumor imaging and photodynamic therapy", *Chem. Soc. Rev.,* vol. 40, no. 1, pp. 340-362, 2011.
[http://dx.doi.org/10.1039/B915149B] [PMID: 20694259]

[82]   N. Malatesti, I. Munitic, and I. Jurak, "Porphyrin-based cationic amphiphilic photosensitisers as potential anticancer, antimicrobial and immunosuppressive agents", *Biophys. Rev.,* vol. 9, no. 2, pp. 149-168, 2017.
[http://dx.doi.org/10.1007/s12551-017-0257-7] [PMID: 28510089]

[83]   N.L. Oleinick, and H.H. Evans, "The photobiology of photodynamic therapy: Cellular targets and mechanisms", *Radiat. Res.,* vol. 150, no. 5, pp. S146-S156, 1998.
[http://dx.doi.org/10.2307/3579816] [PMID: 9806617]

[84]   S.M. Mahalingam, J.D. Ordaz, and P.S. Low, "Targeting of a photosensitizer to the mitochondrion enhances the potency of photodynamic therapy", *ACS. Omega.,* vol. 3, no. 6, pp. 6066-6074, 2018.
[http://dx.doi.org/10.1021/acsomega.8b00692] [PMID: 30023938]

[85]   S. Wu, F. Zhou, Y. Wei, W.R. Chen, Q. Chen, and D. Xing, "Cancer phototherapy via selective photoinactivation of respiratory chain oxidase to trigger a fatal superoxide anion burst", *Antioxid. Redox Signal.,* vol. 20, no. 5, pp. 733-746, 2014.
[http://dx.doi.org/10.1089/ars.2013.5229] [PMID: 23992126]

[86]   A. Spang, "The endoplasmic reticulum-the caring mother of the cell", *Curr. Opin. Cell Biol.,* vol. 53, pp. 92-96, 2018.
[http://dx.doi.org/10.1016/j.ceb.2018.06.004] [PMID: 30006039]

[87]   A.D. Garg, and P. Agostinis, "ER stress, autophagy and immunogenic cell death in photodynamic therapy-induced anti-cancer immune responses", *Photochem. Photobiol. Sci.,* vol. 13, no. 3, pp. 474-487, 2014.
[http://dx.doi.org/10.1039/c3pp50333j] [PMID: 24493131]

[88]   W. Li, J. Yang, L. Luo, M. Jiang, B. Qin, H. Yin, C. Zhu, X. Yuan, J. Zhang, Z. Luo, Y. Du, Q. Li, Y. Lou, Y. Qiu, and J. You, "Targeting photodynamic and photothermal therapy to the endoplasmic reticulum enhances immunogenic cancer cell death", *Nat. Commun.,* vol. 10, no. 1, p. 3349, 2019.
[http://dx.doi.org/10.1038/s41467-019-11269-8] [PMID: 31350406]

[89]   R. Alzeibak, T.A. Mishchenko, N.Y. Shilyagina, I.V. Balalaeva, M.V. Vedunova, and D.V. Krysko, "Targeting immunogenic cancer cell death by photodynamic therapy: Past, present and future", *J. Immunother. Cancer,* vol. 9, no. 1, p. e001926, 2021.
[http://dx.doi.org/10.1136/jitc-2020-001926] [PMID: 33431631]

[90]   A.D. Garg, D.V. Krysko, T. Verfaillie, A. Kaczmarek, G.B. Ferreira, T. Marysael, N. Rubio, M. Firczuk, C. Mathieu, A.J.M. Roebroek, W. Annaert, J. Golab, P. de Witte, P. Vandenabeele, and P. Agostinis, "A novel pathway combining calreticulin exposure and ATP secretion in immunogenic cancer cell death", *EMBO J.,* vol. 31, no. 5, pp. 1062-1079, 2012.
[http://dx.doi.org/10.1038/emboj.2011.497] [PMID: 22252128]

[91]   I. Adkins, J. Fucikova, A.D. Garg, P. Agostinis, and R. Špíšek, "Physical modalities inducing immunogenic tumor cell death for cancer immunotherapy", *OncoImmunology,* vol. 3, no. 12, p. e968434, 2014.
[http://dx.doi.org/10.4161/21624011.2014.968434] [PMID: 25964865]

[92]   C. Pavani, A.F. Uchoa, C.S. Oliveira, Y. Iamamoto, and M.S. Baptista, "Effect of zinc insertion and hydrophobicity on the membrane interactions and PDT activity of porphyrin photosensitizers", *Photochem. Photobiol. Sci.,* vol. 8, no. 2, pp. 233-240, 2009.
[http://dx.doi.org/10.1039/b810313e] [PMID: 19247516]

[93]   F. Engelmann, S. Rocha, H. Toma, K. Araki, and M. Baptista, "Determination of n-octanol/water partition and membrane binding of cationic porphyrins", *Int. J. Pharm.,* vol. 329, no. 1-2, pp. 12-18, 2007.
[http://dx.doi.org/10.1016/j.ijpharm.2006.08.008] [PMID: 16979860]

[94]   P. Agostinis, H. Breyssens, E. Buytaert, and N. Hendrickx, "Regulatory pathways in photodynamic therapy induced apoptosis", *Photochem. Photobiol. Sci.,* vol. 3, no. 8, pp. 721-729, 2004.
[http://dx.doi.org/10.1039/b315237e] [PMID: 15295626]

[95]   D.A. Al-Mutairi, J.D. Craik, I. Batinic-Haberle, and L.T. Benov, "Photosensitizing action of isomeric zinc N -methylpyridylporphyrins in human carcinoma cells", *Free Radic. Res.,* vol. 40, no. 5, pp. 477-483, 2006.
[http://dx.doi.org/10.1080/10715760600577849] [PMID: 16551574]

[96]   A. Uzdensky, "Signal transduction and photodynamic therapy", *Curr. Signal Transduct. Ther.,* vol. 3, no. 1, pp. 55-74, 2008.
[http://dx.doi.org/10.2174/157436208783334277]

[97]   D.R. Mokoena, B.P. George, and H. Abrahamse, "Photodynamic therapy induced cell death mechanisms in breast cancer", *Int. J. Mol. Sci.,* vol. 22, no. 19, p. 10506, 2021.
[http://dx.doi.org/10.3390/ijms221910506] [PMID: 34638847]

[98]   C. Donohoe, M.O. Senge, L.G. Arnaut, and L.C. Gomes-da-Silva, "Cell death in photodynamic therapy: From oxidative stress to anti-tumor immunity", *Biochim. Biophys. Acta Rev. Cancer,* vol. 1872, no. 2, p. 188308, 2019.
[http://dx.doi.org/10.1016/j.bbcan.2019.07.003] [PMID: 31401103]

[99]   D. Kessel, and M. Castelli, "Evidence that bcl-2 is the target of three photosensitizers that induce a rapid apoptotic response", *Photochem. Photobiol.,* vol. 74, no. 2, pp. 318-322, 2001.
[http://dx.doi.org/10.1562/0031-8655(2001)074<0318:ETBITT>2.0.CO;2] [PMID: 11547571]

[100] L. Xue, S. Chiu, and N.L. Oleinick, "Photochemical destruction of the Bcl-2 oncoprotein during photodynamic therapy with the phthalocyanine photosensitizer Pc 4", *Oncogene,* vol. 20, no. 26, pp. 3420-3427, 2001.
[http://dx.doi.org/10.1038/sj.onc.1204441] [PMID: 11423992]

[101] J.J. Reiners Jr, J.A. Caruso, P. Mathieu, B. Chelladurai, X-M. Yin, and D. Kessel, "Release of cytochrome c and activation of pro-caspase-9 following lysosomal photodamage involves bid cleavage", *Cell Death Differ.,* vol. 9, no. 9, pp. 934-944, 2002.
[http://dx.doi.org/10.1038/sj.cdd.4401048] [PMID: 12181744]

[102] S. Ichinose, J. Usuda, T. Hirata, T. Inoue, K. Ohtani, S. Maehara, M. Kubota, K. Imai, Y. Tsunoda, Y. Kuroiwa, K. Yamada, H. Tsutsui, K. Furukawa, T. Okunaka, N. Oleinick, and H. Kato, "Lysosomal cathepsin initiates apoptosis, which is regulated by photodamage to Bcl-2 at mitochondria in photodynamic therapy using a novel photosensitizer, ATX-s10 (Na)", *Int. J. Oncol.,* vol. 29, no. 2, pp. 349-355, 2006.
[http://dx.doi.org/10.3892/ijo.29.2.349] [PMID: 16820876]

[103] D. Kessel, and N.L. Oleinick, "Cell death pathways associated with photodynamic therapy: An update", *Photochem. Photobiol.,* vol. 94, no. 2, pp. 213-218, 2018.
[http://dx.doi.org/10.1111/php.12857] [PMID: 29143339]

[104] D.J. Granville, B.A. Cassidy, D.O. Ruehlmann, J.C. Choy, C. Brenner, G. Kroemer, C. van Breemen, P. Margaron, D.W. Hunt, and B.M. McManus, "Mitochondrial release of apoptosis-inducing factor and cytochrome c during smooth muscle cell apoptosis", *Am. J. Pathol.,* vol. 159, no. 1, pp. 305-311, 2001.
[http://dx.doi.org/10.1016/S0002-9440(10)61696-3] [PMID: 11438477]

[105] S. Nagata, A. Obana, Y. Gohto, and S. Nakajima, "Necrotic and apoptotic cell death of human malignant melanoma cells following photodynamic therapy using an amphiphilic photosensitizer, ATX-S10(Na)", *Lasers Surg. Med.,* vol. 33, no. 1, pp. 64-70, 2003.

[http://dx.doi.org/10.1002/lsm.10190] [PMID: 12866123]

[106] C. Fabris, G. Valduga, G. Miotto, L. Borsetto, G. Jori, S. Garbisa, and E. Reddi, "Photosensitization with zinc (II) phthalocyanine as a switch in the decision between apoptosis and necrosis", *Cancer Res.,* vol. 61, no. 20, pp. 7495-7500, 2001.
[PMID: 11606385]

[107] Z. Yin, C. Pascual, and D. Klionsky, "Autophagy: Machinery and regulation", *Microb. Cell,* vol. 3, no. 12, pp. 588-596, 2016.
[http://dx.doi.org/10.15698/mic2016.12.546] [PMID: 28357331]

[108] J.J. Reiners Jr, P. Agostinis, K. Berg, N.L. Oleinick, and D.H. Kessel, "Assessing autophagy in the context of photodynamic therapy", *Autophagy,* vol. 6, no. 1, pp. 7-18, 2010.
[http://dx.doi.org/10.4161/auto.6.1.10220] [PMID: 19855190]

[109] D. Kessel, and J.J. Reiners Jr, "Apoptosis and autophagy after mitochondrial or endoplasmic reticulum photodamage", *Photochem. Photobiol.,* vol. 83, no. 5, pp. 1024-1028, 2007.
[http://dx.doi.org/10.1111/j.1751-1097.2007.00088.x] [PMID: 17880495]

[110] D. Kessel, M.G.H. Vicente, and J.J. Reiners Jr, "Initiation of apoptosis and autophagy by photodynamic therapy", *Autophagy,* vol. 2, no. 4, pp. 289-290, 2006.
[http://dx.doi.org/10.4161/auto.2792] [PMID: 16921269]

[111] M. Dewaele, W. Martinet, N. Rubio, T. Verfaillie, P.A. de Witte, J. Piette, and P. Agostinis, "Autophagy pathways activated in response to PDT contribute to cell resistance against ROS damage", *J. Cell. Mol. Med.,* vol. 15, no. 6, pp. 1402-1414, 2011.
[http://dx.doi.org/10.1111/j.1582-4934.2010.01118.x] [PMID: 20626525]

[112] M. Andrzejak, M. Price, and D.H. Kessel, "Apoptotic and autophagic responses to photodynamic therapy in 1c1c7 murine hepatoma cells", *Autophagy,* vol. 7, no. 9, pp. 979-984, 2011.
[http://dx.doi.org/10.4161/auto.7.9.15865] [PMID: 21555918]

[113] W.K. Martins, N.F. Santos, C.S. Rocha, I.O.L. Bacellar, T.M. Tsubone, A.C. Viotto, A.Y. Matsukuma, A.B.P. Abrantes, P. Siani, L.G. Dias, and M.S. Baptista, "Parallel damage in mitochondria and lysosomes is an efficient way to photoinduce cell death", *Autophagy,* vol. 15, no. 2, pp. 259-279, 2019.
[http://dx.doi.org/10.1080/15548627.2018.1515609] [PMID: 30176156]

[114] W.K. Martins, R. Belotto, M.N. Silva, D. Grasso, M.D. Suriani, T.S. Lavor, R. Itri, M.S. Baptista, and T.M. Tsubone, "Autophagy regulation and photodynamic therapy: Insights to improve outcomes of cancer treatment", *Front. Oncol.,* vol. 10, p. 610472, 2021.
[http://dx.doi.org/10.3389/fonc.2020.610472] [PMID: 33552982]

[115] T.V. Berghe, A. Linkermann, S. Jouan-Lanhouet, H. Walczak, and P. Vandenabeele, "Regulated necrosis: The expanding network of non-apoptotic cell death pathways", *Nat. Rev. Mol. Cell Biol.,* vol. 15, no. 2, pp. 135-147, 2014.
[http://dx.doi.org/10.1038/nrm3737] [PMID: 24452471]

[116] D. Vercammen, R. Beyaert, G. Denecker, V. Goossens, G. Van Loo, W. Declercq, J. Grooten, W. Fiers, and P. Vandenabeele, "Inhibition of caspases increases the sensitivity of L929 cells to necrosis mediated by tumor necrosis factor", *J. Exp. Med.,* vol. 187, no. 9, pp. 1477-1485, 1998.
[http://dx.doi.org/10.1084/jem.187.9.1477] [PMID: 9565639]

[117] N. Holler, R. Zaru, O. Micheau, M. Thome, A. Attinger, S. Valitutti, J.L. Bodmer, P. Schneider, B. Seed, and J. Tschopp, "Fas triggers an alternative, caspase-8–independent cell death pathway using the kinase RIP as effector molecule", *Nat. Immunol.,* vol. 1, no. 6, pp. 489-495, 2000.
[http://dx.doi.org/10.1038/82732] [PMID: 11101870]

[118] Y. Cho, S. Challa, D. Moquin, R. Genga, T.D. Ray, M. Guildford, and F.K.M. Chan, "Phosphorylation-driven assembly of the RIP1-RIP3 complex regulates programmed necrosis and virus-induced inflammation", *Cell,* vol. 137, no. 6, pp. 1112-1123, 2009.
[http://dx.doi.org/10.1016/j.cell.2009.05.037] [PMID: 19524513]

[119] D.W. Zhang, J. Shao, J. Lin, N. Zhang, B.J. Lu, S.C. Lin, M.Q. Dong, and J. Han, "RIP3, an energy metabolism regulator that switches TNF-induced cell death from apoptosis to necrosis", *Science,* vol. 325, no. 5938, pp. 332-336, 2009.
[http://dx.doi.org/10.1126/science.1172308] [PMID: 19498109]

[120] X. Sun, J. Lee, T. Navas, D.T. Baldwin, T.A. Stewart, and V.M. Dixit, "RIP3, a novel apoptosis-inducing kinase", *J. Biol. Chem.,* vol. 274, no. 24, pp. 16871-16875, 1999.
[http://dx.doi.org/10.1074/jbc.274.24.16871] [PMID: 10358032]

[121] Y.K. Dhuriya, and D. Sharma, "Necroptosis: A regulated inflammatory mode of cell death", *J. Neuroinflammation,* vol. 15, no. 1, p. 199, 2018.
[http://dx.doi.org/10.1186/s12974-018-1235-0] [PMID: 29980212]

[122] I. Coupienne, G. Fettweis, N. Rubio, P. Agostinis, and J. Piette, "5-ALA-PDT induces RIP3-dependent necrosis in glioblastoma", *Photochem. Photobiol. Sci.,* vol. 10, no. 12, pp. 1868-1878, 2011.
[http://dx.doi.org/10.1039/c1pp05213f] [PMID: 22033613]

[123] M. Sun, C. Zhou, H. Zeng, N. Puebla-Osorio, E. Damiani, J. Chen, H. Wang, G. Li, F. Yin, L. Shan, D. Zuo, Y. Liao, Z. Wang, L. Zheng, Y. Hua, and Z. Cai, "Hiporfin-mediated photodynamic therapy in preclinical treatment of osteosarcoma", *Photochem. Photobiol.,* vol. 91, no. 3, pp. 533-544, 2015.
[http://dx.doi.org/10.1111/php.12424] [PMID: 25619546]

[124] Y. Miki, J. Akimoto, K. Moritake, C. Hironaka, and Y. Fujiwara, "Photodynamic therapy using talaporfin sodium induces concentration-dependent programmed necroptosis in human glioblastoma T98G cells", *Lasers Med. Sci.,* vol. 30, no. 6, pp. 1739-1745, 2015.
[http://dx.doi.org/10.1007/s10103-015-1783-9] [PMID: 26109138]

[125] T. Mishchenko, I. Balalaeva, A. Gorokhova, M. Vedunova, and D.V. Krysko, "Which cell death modality wins the contest for photodynamic therapy of cancer?", *Cell Death Dis.,* vol. 13, no. 5, p. 455, 2022.
[http://dx.doi.org/10.1038/s41419-022-04851-4] [PMID: 35562364]

[126] F. Fontana, M. Raimondi, M. Marzagalli, A. Di Domizio, and P. Limonta, "The emerging role of paraptosis in tumor cell biology: Perspectives for cancer prevention and therapy with natural compounds", *Biochim. Biophys. Acta Rev. Cancer,* vol. 1873, no. 2, p. 188338, 2020.
[http://dx.doi.org/10.1016/j.bbcan.2020.188338] [PMID: 31904399]

[127] D. Kessel, "Apoptosis, paraptosis and autophagy: Death and survival pathways associated with photodynamic therapy", *Photochem. Photobiol.,* vol. 95, no. 1, pp. 119-125, 2019.
[http://dx.doi.org/10.1111/php.12952] [PMID: 29882356]

[128] D. Kessel, "Exploring modes of photokilling by hypericin", *Photochem. Photobiol.,* vol. 96, no. 5, pp. 1101-1104, 2020.
[http://dx.doi.org/10.1111/php.13275] [PMID: 32343412]

[129] D. Kessel, "Hypericin accumulation as a determinant of PDT efficacy", *Photochem. Photobiol.,* vol. 96, no. 5, pp. 1144-1147, 2020.
[http://dx.doi.org/10.1111/php.13302] [PMID: 32599667]

[130] I. Rizvi, G. Obaid, S. Bano, T. Hasan, and D. Kessel, "Photodynamic therapy: Promoting *in vitro* efficacy of photodynamic therapy by liposomal formulations of a photosensitizing agent", *Lasers Surg. Med.,* vol. 50, no. 5, pp. 499-505, 2018.
[http://dx.doi.org/10.1002/lsm.22813] [PMID: 29527710]

[131] D. Kessel, and J.J. Reiners Jr, "Effects of combined lysosomal and mitochondrial photodamage in a non-small-cell lung cancer cell line: The role of paraptosis", *Photochem. Photobiol.,* vol. 93, no. 6, pp. 1502-1508, 2017.
[http://dx.doi.org/10.1111/php.12805] [PMID: 28696570]

[132] W.J. Cho, D. Kessel, J. Rakowski, B. Loughery, A.J. Najy, T. Pham, S. Kim, Y.T. Kwon, I. Kato, H.E. Kim, and H.R.C. Kim, "Photodynamic therapy as a potent radiosensitizer in head and neck squamous

cell carcinoma", *Cancers,* vol. 13, no. 6, p. 1193, 2021.
[http://dx.doi.org/10.3390/cancers13061193] [PMID: 33801879]

[133] B. Montico, A. Nigro, V. Casolaro, and J. Dal Col, "Immunogenic apoptosis as a novel tool for anticancer vaccine development", *Int. J. Mol. Sci.,* vol. 19, no. 2, p. 594, 2018.
[http://dx.doi.org/10.3390/ijms19020594] [PMID: 29462947]

[134] N. Yamamoto, S. Homma, T.W. Sery, L.A. Donoso, and J. Kenneth Hoober, "Photodynamic immunopotentiation: *in vitro* activation of macrophages by treatment of mouse peritoneal cells with haematoporphyrin derivative and light", *Eur. J. Cancer Clin. Oncol.,* vol. 27, no. 4, pp. 467-471, 1991.
[http://dx.doi.org/10.1016/0277-5379(91)90388-T] [PMID: 1827722]

[135] S.O. Gollnick, X. Liu, B. Owczarczak, D.A. Musser, and B.W. Henderson, "Altered expression of interleukin 6 and interleukin 10 as a result of photodynamic therapy *in vivo*", *Cancer Res.,* vol. 57, no. 18, pp. 3904-3909, 1997.
[PMID: 9307269]

[136] S.O. Gollnick, L. Vaughan, and B.W. Henderson, "Generation of effective antitumor vaccines using photodynamic therapy", *Cancer Res.,* vol. 62, no. 6, pp. 1604-1608, 2002.
[PMID: 11912128]

[137] P.S.P. Thong, K.W. Ong, N.S.G. Goh, K.W. Kho, V. Manivasager, R. Bhuvaneswari, M. Olivo, and K.C. Soo, "Photodynamic-therapy-activated immune response against distant untreated tumours in recurrent angiosarcoma", *Lancet Oncol.,* vol. 8, no. 10, pp. 950-952, 2007.
[http://dx.doi.org/10.1016/S1470-2045(07)70318-2] [PMID: 17913664]

[138] V. Vithanage, C. Jayasinghe, P. Costa, and S. Rajendram, "Photodynamic therapy: An overview and insights into a prospective mainstream anticancer therapy", *J. Turk. Chem. Soc. Sect A: Chem.,* vol. 9, pp. 821-848, 2022.
[http://dx.doi.org/10.18596/jotcsa.1000980]

[139] S.M. Weis, and D.A. Cheresh, "Tumor angiogenesis: Molecular pathways and therapeutic targets", *Nat. Med.,* vol. 17, no. 11, pp. 1359-1370, 2011.
[http://dx.doi.org/10.1038/nm.2537] [PMID: 22064426]

[140] D.E. Dolmans, A. Kadambi, J.S. Hill, C.A. Waters, B.C. Robinson, J.P. Walker, D. Fukumura, and R.K. Jain, "Vascular accumulation of a novel photosensitizer, MV6401, causes selective thrombosis in tumor vessels after photodynamic therapy", *Cancer Res.,* vol. 62, no. 7, pp. 2151-2156, 2002.
[PMID: 11929837]

[141] C. He, P. Agharkar, and B. Chen, "Intravital microscopic analysis of vascular perfusion and macromolecule extravasation after photodynamic vascular targeting therapy", *Pharm. Res.,* vol. 25, no. 8, pp. 1873-1880, 2008.
[http://dx.doi.org/10.1007/s11095-008-9604-5] [PMID: 18446275]

[142] T.A. Middelburg, H.S. de Bruijn, L. Tettero, A. van der Ploeg van den Heuvel, H.A.M. Neumann, E.R.M. de Haas, and D.J. Robinson, "Topical hexylaminolevulinate and aminolevulinic acid photodynamic therapy: Complete arteriole vasoconstriction occurs frequently and depends on protoporphyrin IX concentration in vessel wall", *J. Photochem. Photobiol. B,* vol. 126, pp. 26-32, 2013.
[http://dx.doi.org/10.1016/j.jphotobiol.2013.06.014] [PMID: 23892187]

[143] C.J. Gomer, N. Rucker, and A.L. Murphree, "Differential cell photosensitivity following porphyrin photodynamic therapy", *Cancer Res.,* vol. 48, no. 16, pp. 4539-4542, 1988.
[PMID: 2969280]

[144] C.M.L. West, D.C. West, S. Kumar, and J.V. Moore, "A comparison of the sensitivity to photodynamic treatment of endothelial and tumour cells in different proliferative states", *Int. J. Radiat. Biol.,* vol. 58, no. 1, pp. 145-156, 1990.
[http://dx.doi.org/10.1080/09553009014551501] [PMID: 1973432]

[145] A.F. Dos Santos, D.R.Q. De Almeida, L.F. Terra, M.S. Baptista, and L. Labriola, "Photodynamic therapy in cancer treatment: An update review", *J. Cancer Metastasis Treat.,* vol. 2019, pp. 5-25, 2019.
[http://dx.doi.org/10.20517/2394-4722.2018.83]

[146] A.P. Castano, T.N. Demidova, and M.R. Hamblin, "Mechanisms in photodynamic therapy: Part three—Photosensitizer pharmacokinetics, biodistribution, tumor localization and modes of tumor destruction", *Photodiagn. Photodyn. Ther.,* vol. 2, no. 2, pp. 91-106, 2005.
[http://dx.doi.org/10.1016/S1572-1000(05)00060-8] [PMID: 25048669]

[147] A.J. Sorrin, M. Kemal Ruhi, N.A. Ferlic, V. Karimnia, W.J. Polacheck, J.P. Celli, H.C. Huang, and I. Rizvi, "Photodynamic therapy and the biophysics of the tumor microenvironment", *Photochem. Photobiol.,* vol. 96, no. 2, pp. 232-259, 2020.
[http://dx.doi.org/10.1111/php.13209] [PMID: 31895481]

[148] X. Li, J.F. Lovell, J. Yoon, and X. Chen, "Clinical development and potential of photothermal and photodynamic therapies for cancer", *Nat. Rev. Clin. Oncol.,* vol. 17, no. 11, pp. 657-674, 2020.
[http://dx.doi.org/10.1038/s41571-020-0410-2] [PMID: 32699309]

[149] M.T. Wan, and J.Y. Lin, "Current evidence and applications of photodynamic therapy in dermatology", *Clin. Cosmet. Investig. Dermatol.,* vol. 7, pp. 145-163, 2014.
[PMID: 24899818]

[150] S. Mallidi, S. Anbil, A.L. Bulin, G. Obaid, M. Ichikawa, and T. Hasan, "Beyond the barriers of light penetration: Strategies, perspectives and possibilities for photodynamic therapy", *Theranostics,* vol. 6, no. 13, pp. 2458-2487, 2016.
[http://dx.doi.org/10.7150/thno.16183] [PMID: 27877247]

[151] A. Juzeniene, Q. Peng, and J. Moan, "Milestones in the development of photodynamic therapy and fluorescence diagnosis", *Photochem. Photobiol. Sci.,* vol. 6, no. 12, pp. 1234-1245, 2007.
[http://dx.doi.org/10.1039/b705461k] [PMID: 18046478]

[152] J.O. Yoo, and K.S. Ha, "New insights into the mechanisms for photodynamic therapy-induced cancer cell death", *Int. Rev. Cell Mol. Biol.,* vol. 295, pp. 139-174, 2012.
[http://dx.doi.org/10.1016/B978-0-12-394306-4.00010-1] [PMID: 22449489]

[153] Q. Zhang, and L. Li, "Photodynamic combinational therapy in cancer treatment", *J. BUON,* vol. 23, no. 3, pp. 561-567, 2018.
[PMID: 30003719]

[154] S. Nazir, S. Bano, S. Munir, M. Fahad Al-Ajmi, M. Afzal, and K. Mazhar, "Smart nickel oxide based core–shell nanoparticles for combined chemo and photodynamic cancer therapy", *Int. J. Nanomedicine,* vol. 11, pp. 3159-3166, 2016.
[http://dx.doi.org/10.2147/IJN.S106533] [PMID: 27471383]

[155] X. Wang, G. Meng, S. Zhang, and X. Liu, "A reactive 1o2 - responsive combined treatment system of photodynamic and chemotherapy for cancer", *Sci. Rep.,* vol. 6, no. 1, p. 29911, 2016.
[http://dx.doi.org/10.1038/srep29911] [PMID: 27443831]

[156] E. Husain, and I. Naseem, "Riboflavin-mediated cellular photoinhibition of cisplatin-induced oxidative DNA breakage in mice epidermal keratinocytes", *Photodermatol. Photoimmunol. Photomed.,* vol. 24, no. 6, pp. 301-307, 2008.
[http://dx.doi.org/10.1111/j.1600-0781.2008.00380.x] [PMID: 19000187]

[157] D. Kim, S. Park, H. Yoo, S. Park, J. Kim, K. Yum, K. Kim, and H. Kim, "Overcoming anticancer resistance by photodynamic therapy-related efflux pump deactivation and ultrasound-mediated improved drug delivery efficiency", *Nano Converg.,* vol. 7, no. 1, p. 30, 2020.
[http://dx.doi.org/10.1186/s40580-020-00241-8] [PMID: 32897469]

[158] J. Shi, Y. Su, W. Liu, J. Chang, and Z. Zhang, "A nanoliposome-based photoactivable drug delivery system for enhanced cancer therapy and overcoming treatment resistance", *Int. J. Nanomedicine,* vol.

12, pp. 8257-8275, 2017.
[http://dx.doi.org/10.2147/IJN.S143776] [PMID: 29180864]

[159] C.Y. Zhao, R. Cheng, Z. Yang, and Z.M. Tian, "Nanotechnology for cancer therapy based on chemotherapy", *Molecules,* vol. 23, no. 4, p. 826, 2018.
[http://dx.doi.org/10.3390/molecules23040826] [PMID: 29617302]

[160] J. Xu, J. Gao, and Q. Wei, "Combination of photodynamic therapy with radiotherapy for cancer treatment", *J. Nanomater.,* vol. 2016, pp. 1-7, 2016.
[http://dx.doi.org/10.1155/2016/8507924]

[161] K. Graham, and E. Unger, "Overcoming tumor hypoxia as a barrier to radiotherapy, chemotherapy and immunotherapy in cancer treatment", *Int. J. Nanomedicine,* vol. 13, pp. 6049-6058, 2018.
[http://dx.doi.org/10.2147/IJN.S140462] [PMID: 30323592]

[162] A. Menegakis, R. Klompmaker, C. Vennin, A. Arbusà, M. Damen, B. van den Broek, D. Zips, J. van Rheenen, L. Krenning, and R.H. Medema, "Resistance of hypoxic cells to ionizing radiation is mediated in part via hypoxia-induced quiescence", *Cells,* vol. 10, no. 3, p. 610, 2021.
[http://dx.doi.org/10.3390/cells10030610] [PMID: 33801903]

[163] A.R. Montazerabadi, A. Sazgarnia, M.H. Bahreyni-Toosi, A. Ahmadi, and A. Aledavood, "The effects of combined treatment with ionizing radiation and indocyanine green-mediated photodynamic therapy on breast cancer cells", *J. Photochem. Photobiol. B,* vol. 109, pp. 42-49, 2012.
[http://dx.doi.org/10.1016/j.jphotobiol.2012.01.004] [PMID: 22325306]

[164] A. Colasanti, A. Kisslinger, M. Quarto, and P. Riccio, "Combined effects of radiotherapy and photodynamic therapy on an *in vitro* human prostate model", *Acta Biochim. Pol.,* vol. 51, no. 4, pp. 1039-1046, 2004.
[PMID: 15625575]

[165] R. Allman, P. Cowburn, and M. Mason, "Effect of photodynamic therapy in combination with ionizing radiation on human squamous cell carcinoma cell lines of the head and neck", *Br. J. Cancer,* vol. 83, no. 5, pp. 655-661, 2000.
[http://dx.doi.org/10.1054/bjoc.2000.1328] [PMID: 10944608]

[166] A. Nakano, D. Watanabe, Y. Akita, T. Kawamura, Y. Tamada, and Y. Matsumoto, "Treatment efficiency of combining photodynamic therapy and ionizing radiation for Bowen's disease", *J. Eur. Acad. Dermatol. Venereol.,* vol. 25, no. 4, pp. 475-478, 2011.
[http://dx.doi.org/10.1111/j.1468-3083.2010.03757.x] [PMID: 20569287]

[167] Z. Luksiene, A. Kalvelyte, and R. Supino, "On the combination of photodynamic therapy with ionizing radiation", *J. Photochem. Photobiol. B,* vol. 52, no. 1-3, pp. 35-42, 1999.
[http://dx.doi.org/10.1016/S1011-1344(99)00098-6] [PMID: 10643073]

[168] S. Na, Ed., *Photodynamic therapy combined with imcrt for cancer :A clinical study.*, 2009.

[169] M.F. Naylor, W.R. Chen, T.K. Teague, L.A. Perry, and R.E. Nordquist, "In situ photoimmunotherapy: A tumour-directed treatment for melanoma", *Br. J. Dermatol.,* vol. 155, no. 6, pp. 1287-1292, 2006.
[http://dx.doi.org/10.1111/j.1365-2133.2006.07514.x] [PMID: 17107404]

[170] P. Mroz, J.T. Hashmi, Y.Y. Huang, N. Lange, and M.R. Hamblin, "Stimulation of anti-tumor immunity by photodynamic therapy", *Expert Rev. Clin. Immunol.,* vol. 7, no. 1, pp. 75-91, 2011.
[http://dx.doi.org/10.1586/eci.10.81] [PMID: 21162652]

[171] A.D. Garg, L. Vandenberk, C. Koks, T. Verschuere, L. Boon, S.W. Van Gool, and P. Agostinis, "Dendritic cell vaccines based on immunogenic cell death elicit danger signals and T cell–driven rejection of high-grade glioma", *Sci. Transl. Med.,* vol. 8, no. 328, p. 328ra27, 2016.
[http://dx.doi.org/10.1126/scitranslmed.aae0105] [PMID: 26936504]

[172] J. Ji, Z. Fan, F. Zhou, X. Wang, L. Shi, H. Zhang, P. Wang, D. Yang, L. Zhang, W.R. Chen, and X. Wang, "Improvement of DC vaccine with ALA-PDT induced immunogenic apoptotic cells for skin squamous cell carcinoma", *Oncotarget,* vol. 6, no. 19, pp. 17135-17146, 2015.

[http://dx.doi.org/10.18632/oncotarget.3529] [PMID: 25915530]

[173] M. Korbelik, B. Stott, and J. Sun, "Photodynamic therapy-generated vaccines: Relevance of tumour cell death expression", *Br. J. Cancer,* vol. 97, no. 10, pp. 1381-1387, 2007.
[http://dx.doi.org/10.1038/sj.bjc.6604059] [PMID: 17971767]

[174] K. Ito, M. Mitsunaga, T. Nishimura, H. Kobayashi, and H. Tajiri, "Combination photoimmunotherapy with monoclonal antibodies recognizing different epitopes of human epidermal growth factor receptor 2: An assessment of phototherapeutic effect based on fluorescence molecular imaging", *Oncotarget,* vol. 7, no. 12, pp. 14143-14152, 2016.
[http://dx.doi.org/10.18632/oncotarget.7490] [PMID: 26909859]

[175] C. He, X. Duan, N. Guo, C. Chan, C. Poon, R.R. Weichselbaum, and W. Lin, "Core-shell nanoscale coordination polymers combine chemotherapy and photodynamic therapy to potentiate checkpoint blockade cancer immunotherapy", *Nat. Commun.,* vol. 7, no. 1, p. 12499, 2016.
[http://dx.doi.org/10.1038/ncomms12499] [PMID: 27530650]

[176] G. Calixto, B. Fonseca-Santos, M. Chorilli, and J. Bernegossi, "Nanotechnology-based drug delivery systems for treatment of oral cancer: A review", *Int. J. Nanomedicine,* vol. 9, pp. 3719-3735, 2014.
[http://dx.doi.org/10.2147/IJN.S61670] [PMID: 25143724]

[177] G. Calixto, J. Bernegossi, L. de Freitas, C. Fontana, and M. Chorilli, "Nanotechnology-based drug delivery systems for photodynamic therapy of cancer: A review", *Molecules,* vol. 21, no. 3, p. 342, 2016.
[http://dx.doi.org/10.3390/molecules21030342] [PMID: 26978341]

[178] L. Li, and K.M. Huh, "Polymeric nanocarrier systems for photodynamic therapy", *Biomater. Res.,* vol. 18, no. 1, p. 19, 2014.
[http://dx.doi.org/10.1186/2055-7124-18-19] [PMID: 26331070]

[179] A. Kumari, S.K. Yadav, and S.C. Yadav, "Biodegradable polymeric nanoparticles based drug delivery systems", *Colloids Surf. B Biointerfaces,* vol. 75, no. 1, pp. 1-18, 2010.
[http://dx.doi.org/10.1016/j.colsurfb.2009.09.001] [PMID: 19782542]

[180] A. Manmode, D. Sakarkar, and N. Mahajan, "Nanoparticles-tremendous therapeutic potential: A review", *Int. J. Pharm. Tech. Res.,* p. 1, 2009.

[181] S. Mukherjee, S. Ray, and R.S. Thakur, "Solid lipid nanoparticles: A modern formulation approach in drug delivery system", *Indian J. Pharm. Sci.,* vol. 71, no. 4, pp. 349-358, 2009.
[http://dx.doi.org/10.4103/0250-474X.57282] [PMID: 20502539]

[182] R. López-García, and A. Ganem-Rondero, "Solid lipid nanoparticles (SLN) and nanostructured lipid carriers (NLC): Occlusive effect and penetration enhancement ability", *J. Cos. Derma. Sci. Appl.,* vol. 5, no. 2, pp. 62-72, 2015.
[http://dx.doi.org/10.4236/jcdsa.2015.52008]

[183] I. Chauhan, M. Yasir, M. Verma, and A.P. Singh, "Nanostructured lipid carriers: A groundbreaking approach for transdermal drug delivery", *Adv. Pharm. Bull.,* vol. 10, no. 2, pp. 150-165, 2020.
[http://dx.doi.org/10.34172/apb.2020.021] [PMID: 32373485]

[184] P. Jain, P. Rahi, V. Pandey, S. Asati, and V. Soni, "Nanostructure lipid carriers: A modish contrivance to overcome the ultraviolet effects", *Egypt. j. basic appl. sci.,* vol. 4, no. 2, pp. 89-100, 2017.
[http://dx.doi.org/10.1016/j.ejbas.2017.02.001]

[185] P. Tharkar, A.U. Madani, A. Lasham, A.N. Shelling, and R. Al-Kassas, "Nanoparticulate carriers: An emerging tool for breast cancer therapy", *J. Drug Target.,* vol. 23, no. 2, pp. 97-108, 2015.
[http://dx.doi.org/10.3109/1061186X.2014.958844] [PMID: 25230853]

[186] L. Shang, X. Zhou, J. Zhang, Y. Shi, and L. Zhong, "Metal nanoparticles for photodynamic therapy: A potential treatment for breast cancer", *Molecules,* vol. 26, no. 21, p. 6532, 2021.
[http://dx.doi.org/10.3390/molecules26216532] [PMID: 34770941]

[187]   P. García Calavia, G. Bruce, L. Pérez-García, and D.A. Russell, "Photosensitiser-gold nanoparticle conjugates for photodynamic therapy of cancer", *Photochem. Photobiol. Sci.,* vol. 17, no. 11, pp. 1534-1552, 2018.
[http://dx.doi.org/10.1039/c8pp00271a] [PMID: 30118115]

[188]   T.Y. Ohulchanskyy, I. Roy, L.N. Goswami, Y. Chen, E.J. Bergey, R.K. Pandey, A.R. Oseroff, and P.N. Prasad, "Organically modified silica nanoparticles with covalently incorporated photosensitizer for photodynamic therapy of cancer", *Nano Lett.,* vol. 7, no. 9, pp. 2835-2842, 2007.
[http://dx.doi.org/10.1021/nl0714637] [PMID: 17718587]

[189]   Z. Yu, H. Li, L.M. Zhang, Z. Zhu, and L. Yang, "Enhancement of phototoxicity against human pancreatic cancer cells with photosensitizer-encapsulated amphiphilic sodium alginate derivative nanoparticles", *Int. J. Pharm.,* vol. 473, no. 1-2, pp. 501-509, 2014.
[http://dx.doi.org/10.1016/j.ijpharm.2014.07.046] [PMID: 25089506]

[190]   Y. Yuan, and B. Liu, "Self-assembled nanoparticles based on PEGylated conjugated polyelectrolyte and drug molecules for image-guided drug delivery and photodynamic therapy", *ACS Appl. Mater. Interfaces,* vol. 6, no. 17, pp. 14903-14910, 2014.
[http://dx.doi.org/10.1021/am5020925] [PMID: 25075548]

[191]   S. Jin, L. Zhou, Z. Gu, G. Tian, L. Yan, W. Ren, W. Yin, X. Liu, X. Zhang, Z. Hu, and Y. Zhao, "A new near infrared photosensitizing nanoplatform containing blue-emitting up-conversion nanoparticles and hypocrellin A for photodynamic therapy of cancer cells", *Nanoscale,* vol. 5, no. 23, pp. 11910-11918, 2013.
[http://dx.doi.org/10.1039/c3nr03515h] [PMID: 24129918]

[192]   T. Youssef, M. Fadel, R. Fahmy, and K. Kassab, "Evaluation of hypericin-loaded solid lipid nanoparticles: Physicochemical properties, photostability and phototoxicity", *Pharm. Dev. Technol.,* vol. 17, no. 2, pp. 177-186, 2012.
[http://dx.doi.org/10.3109/10837450.2010.529148] [PMID: 21047275]

[193]   M. Camerin, M. Magaraggia, M. Soncin, G. Jori, M. Moreno, I. Chambrier, M.J. Cook, and D.A. Russell, "The *in vivo* efficacy of phthalocyanine–nanoparticle conjugates for the photodynamic therapy of amelanotic melanoma", *Eur. J. Cancer,* vol. 46, no. 10, pp. 1910-1918, 2010.
[http://dx.doi.org/10.1016/j.ejca.2010.02.037] [PMID: 20356732]

[194]   S. Ben-Dror, I. Bronshtein, A. Wiehe, B. Röder, M.O. Senge, and B. Ehrenberg, "On the correlation between hydrophobicity, liposome binding and cellular uptake of porphyrin sensitizers", *Photochem. Photobiol.,* vol. 82, no. 3, pp. 695-701, 2006.
[http://dx.doi.org/10.1562/2005-09-01-RA-669] [PMID: 16435882]

[195]   W.G. Love, S. Duk, R. Biolo, G. Jori, and P.W. Taylor, "Liposome-mediated delivery of photosensitizers: localization of zinc (II)-phthalocyanine within implanted tumors after intravenous administration", *Photochem. Photobiol.,* vol. 63, no. 5, pp. 656-661, 1996.
[http://dx.doi.org/10.1111/j.1751-1097.1996.tb05670.x] [PMID: 8628757]

[196]   A. Casas, and A. Batlle, "Aminolevulinic acid derivatives and liposome delivery as strategies for improving 5-aminolevulinic acid-mediated photodynamic therapy", *Curr. Med. Chem.,* vol. 13, no. 10, pp. 1157-1168, 2006.
[http://dx.doi.org/10.2174/092986706776360888] [PMID: 16719777]

[197]   V.P. Torchilin, "Recent advances with liposomes as pharmaceutical carriers", *Nat. Rev. Drug Discov.,* vol. 4, no. 2, pp. 145-160, 2005.
[http://dx.doi.org/10.1038/nrd1632] [PMID: 15688077]

[198]   J.O. Eloy, R. Petrilli, L.N.F. Trevizan, and M. Chorilli, "Immunoliposomes: A review on functionalization strategies and targets for drug delivery", *Colloids Surf. B Biointerfaces,* vol. 159, pp. 454-467, 2017.
[http://dx.doi.org/10.1016/j.colsurfb.2017.07.085] [PMID: 28837895]

[199] E. Paszko, C. Ehrhardt, M.O. Senge, D.P. Kelleher, and J.V. Reynolds, "Nanodrug applications in photodynamic therapy", *Photodiagn. Photodyn. Ther.,* vol. 8, no. 1, pp. 14-29, 2011.
[http://dx.doi.org/10.1016/j.pdpdt.2010.12.001] [PMID: 21333931]

[200] N.A. Peppas, J.Z. Hilt, A. Khademhosseini, and R. Langer, "Hydrogels in biology and medicine: From molecular principles to bionanotechnology", *Adv. Mater.,* vol. 18, no. 11, pp. 1345-1360, 2006.
[http://dx.doi.org/10.1002/adma.200501612]

[201] B.K. Jung, E. Oh, J. Hong, Y. Lee, K.D. Park, and C.O. Yun, "A hydrogel matrix prolongs persistence and promotes specific localization of an oncolytic adenovirus in a tumor by restricting nonspecific shedding and an antiviral immune response", *Biomaterials,* vol. 147, pp. 26-38, 2017.
[http://dx.doi.org/10.1016/j.biomaterials.2017.09.009] [PMID: 28923683]

[202] B. Khurana, P. Gierlich, A. Meindl, L.C. Gomes-da-Silva, and M.O. Senge, "Hydrogels: Soft matters in photomedicine", *Photochem. Photobiol. Sci.,* vol. 18, no. 11, pp. 2613-2656, 2019.
[http://dx.doi.org/10.1039/c9pp00221a] [PMID: 31460568]

[203] F. Bayat, and A.R. Karimi, "Design of photodynamic chitosan hydrogels bearing phthalocyanine-colistin conjugate as an antibacterial agent", *Int. J. Biol. Macromol.,* vol. 129, pp. 927-935, 2019.
[http://dx.doi.org/10.1016/j.ijbiomac.2019.02.081] [PMID: 30772416]

[204] L. Pierau, and D.L. Versace, "Light and hydrogels: A new generation of antimicrobial materials", *Materials,* vol. 14, no. 4, p. 787, 2021.
[http://dx.doi.org/10.3390/ma14040787] [PMID: 33562335]

[205] S. Stolik, J.A. Delgado, A. Pérez, and L. Anasagasti, "Measurement of the penetration depths of red and near infrared light in human "*ex vivo*" tissues", *J. Photochem. Photobiol. B,* vol. 57, no. 2-3, pp. 90-93, 2000.
[http://dx.doi.org/10.1016/S1011-1344(00)00082-8] [PMID: 11154088]

[206] G. Gunaydin, M.E. Gedik, and S. Ayan, "Photodynamic therapy—current limitations and novel approaches", *Front Chem.,* vol. 9, p. 691697, 2021.
[http://dx.doi.org/10.3389/fchem.2021.691697] [PMID: 34178948]

[207] J.R. Mourant, M. Canpolat, C. Brocker, O. Esponda-Ramos, T.M. Johnson, A. Matanock, K. Stetter, and J.P. Freyer, "Light scattering from cells: The contribution of the nucleus and the effects of proliferative status", *J. Biomed. Opt.,* vol. 5, no. 2, pp. 131-137, 2000.
[http://dx.doi.org/10.1117/1.429979] [PMID: 10938776]

[208] D. van Straten, V. Mashayekhi, H. de Bruijn, S. Oliveira, and D. Robinson, "Oncologic photodynamic therapy: Basic principles, current clinical status and future directions", *Cancers,* vol. 9, no. 12, p. 19, 2017.
[http://dx.doi.org/10.3390/cancers9020019] [PMID: 28218708]

[209] K. Wang, B. Yu, and J.L. Pathak, "An update in clinical utilization of photodynamic therapy for lung cancer", *J. Cancer,* vol. 12, no. 4, pp. 1154-1160, 2021.
[http://dx.doi.org/10.7150/jca.51537] [PMID: 33442413]

[210] B.C. Wilson, and M.S. Patterson, "The physics, biophysics and technology of photodynamic therapy", *Phys. Med. Biol.,* vol. 53, no. 9, pp. R61-R109, 2008.
[http://dx.doi.org/10.1088/0031-9155/53/9/R01] [PMID: 18401068]

[211] S.B. Brown, E.A. Brown, and I. Walker, "The present and future role of photodynamic therapy in cancer treatment", *Lancet Oncol.,* vol. 5, no. 8, pp. 497-508, 2004.
[http://dx.doi.org/10.1016/S1470-2045(04)01529-3] [PMID: 15288239]

[212] J. Moan, and K. Berg, "The photodegradation of porphyrins in cells can be used to estimate the lifetime of singlet oxygen", *Photochem. Photobiol.,* vol. 53, no. 4, pp. 549-553, 1991.
[http://dx.doi.org/10.1111/j.1751-1097.1991.tb03669.x] [PMID: 1830395]

[213] V. Bhandari, C. Hoey, L.Y. Liu, E. Lalonde, J. Ray, J. Livingstone, R. Lesurf, Y.J. Shiah, T. Vujcic, X. Huang, S.M.G. Espiritu, L.E. Heisler, F. Yousif, V. Huang, T.N. Yamaguchi, C.Q. Yao, V.Y.

Sabelnykova, M. Fraser, M.L.K. Chua, T. van der Kwast, S.K. Liu, P.C. Boutros, and R.G. Bristow, "Molecular landmarks of tumor hypoxia across cancer types", *Nat. Genet.,* vol. 51, no. 2, pp. 308-318, 2019.
[http://dx.doi.org/10.1038/s41588-018-0318-2] [PMID: 30643250]

[214]   E.F.F. Silva, F.A. Schaberle, C.J.P. Monteiro, J.M. Dąbrowski, and L.G. Arnaut, "The challenging combination of intense fluorescence and high singlet oxygen quantum yield in photostable chlorins : A contribution to theranostics", *Photochem. Photobiol. Sci.,* vol. 12, no. 7, pp. 1187-1192, 2013.
[http://dx.doi.org/10.1039/c3pp25419d] [PMID: 23584281]

[215]   S. Klara, C. Juan, and Z. Gang, "Killer beacons for combined cancer imaging and therapy", *Curr. Med. Chem.,* vol. 14, no. 20, pp. 2110-2125, 2007.
[http://dx.doi.org/10.2174/092986707781389655] [PMID: 17691951]

[216]   S. Yano, S. Hirohara, M. Obata, Y. Hagiya, S. Ogura, A. Ikeda, H. Kataoka, M. Tanaka, and T. Joh, "Current states and future views in photodynamic therapy", *J. Photochem. Photobiol. Photochem. Rev.,* vol. 12, no. 1, pp. 46-67, 2011.
[http://dx.doi.org/10.1016/j.jphotochemrev.2011.06.001]

CHAPTER 5

Biotechnological Importance of Exosomes

Elvan Bakar[1,*], **Zeynep Erim**[2] and **Nebiye Pelin Türker**[3]

[1] *Trakya University, Faculty of Pharmacy, Department of Basic Science, Edirne, Turkey*

[2] *Department of Biotechnology and Genetics, Trakya University, Institute of Natural and Applied Sciences, 22030, Edirne, Türkiye*

[3] *Development Application and Research Center, Trakya University, Technology Research, 22030, Edirne, Türkiye*

Abstract: Extracellular vesicles are molecules secreted by cells, wrapped in phospholipids and carrying some types of RNA, DNA and protein in their inner region. Extracellular vesicles are classified as apoptotic bodies, microvesicles, and exosomes based on their extent and formation process. Exosomes, which have the smallest structure, have received more attention than other extracellular vesicles. Exosomes contain different types of molecules in their structures. Cell membranes comprise a lipid bilayer and contain different cargo molecules and different surface receptors, depending on the cells of origin where biogenesis takes place. The biogenesis of exosomes begins within the endosomal system. Then they mature and are released out of the cell. The biogenesis of exosomes may be associated with the ESCRT complex and may depend on many molecules other than the ESCRT complex. Exosomes excreted by the origin cells are taken up by the target cells in different ways and show their effects. The effects of exosomes on their target cells may vary according to the cargo molecules they carry. They participate in cell-to-cell communication by sending different signals to distant or nearby target cells. Exosomes have a variety of pathological and physiological effects on disease and health. They have different effects on many diseases, especially cancer. They play an active role in cancer development, tumor microenvironment, angiogenesis, drug resistance and immune system. There are many diseases that can be used as a biomarker due to increased secretion from cells of origin in pathological conditions. In addition, exosomes can be utilized as drug transportation systems due to their natural structure. In addition, they are potential candidates as effective vaccines because of their effects on immune system cells or the effects of exosomes secreted from immune system cells.

Keywords: Extracellular vesicles, Exosomes, Biogenesis, Diseases, Cancer, Metastasis, Tumor microenvironment, Biomarker, MiRNAs, Therapy, Immunity, Immune regulation, Clinical applications, Target, Vaccine, Drug resistance, Immune response, Cell metabolism, Intercellular communication, Epithelial-mesenchymal transition,.

---

* **Corresponding author Elvan Bakar:** Trakya University, Faculty of Pharmacy, Department of Basic Science, Edirne, Turkey; Tel: +905327173787; E-mail: elvanbakar@trakya.edu.tr

Habibe Yılmaz (Ed.)

INTRODUCTION

Extracellular vesicles (EVs) are identified by the International Society of Extracellular Vesicles as particles that are bounded by a bilayer lipid membrane that is intrinsically secreted from the cell and that do not proliferate, that is, do not have a nucleus [1].

Wolf discovered EVs in plasma in 1967 and defined them as "platelet dust." [2]. Vesicles are present in all biological fluids tested over time, and they are secreted by cells that have been shown to be of varying sizes. These vesicles have also been given various names over time. Nevertheless, they are now popularly referred to as extracellular vesicles [3]. Because of their potential for use in therapy and diagnosis, they are clinically significant molecules. EVs are crucial molecules in many diseases, such as cancer, neurological diseases, preeclampsia and osteoarthritis. The mechanism of action of these diseases, as well as their potential for use in treatment and diagnosis, have been investigated and are still being investigated.

EVs consist of a lipid bilayer membrane that surrounds the inner molecules. It is well understood that after being released by origin cells, recipient cells can be targeted and bound *via* EV surface proteins and thus mediate communication with different cells [4]. During development, they take an active part in several physiological processes, including morphogen transport, inflammation regulation, coagulation, and sexual behavior in all types of organisms. Furthermore, extracellular vesicles secreted by tumor cells have been demonstrated to function in cell-to-cell communication by promoting angiogenesis, altering the extracellular matrix composition and/or altering the immune response [5].

EVs are heterogeneous nanoparticles formed through various biogenesis pathways. They differ in terms of surface markers and also molecular and genetic content. Their size is highly variable, as are their biogenesis, surface markers, and molecular and genetic contents [6]. They are divided into three classes based on their sizes, biogenesis mechanisms and functions (Fig. 1). Exosomes with sizes ranging from 30nm to 100nm are the smallest of the EVs. Macrovesicles are released from the cell membrane by burgeoning directly outward, and their dimensions range from 100nm to 1μm. Apoptotic bodies are EVs that form from surface bubbles and are especially released by apoptotic cells. Their dimensions range from 50nm to 2μm [7]. Apoptotic bodies differ significantly from exosomes and microvesicles. The primary difference is that apoptotic bodies have organelles, whereas microvesicles and exosomes do not. Therefore, the proteomic properties of apoptotic bodies are extremely similar to the cell's lysate. On the other hand, exosomes have certain differences between their cells of origin. Exosomes are interesting for a variety of reasons [8].

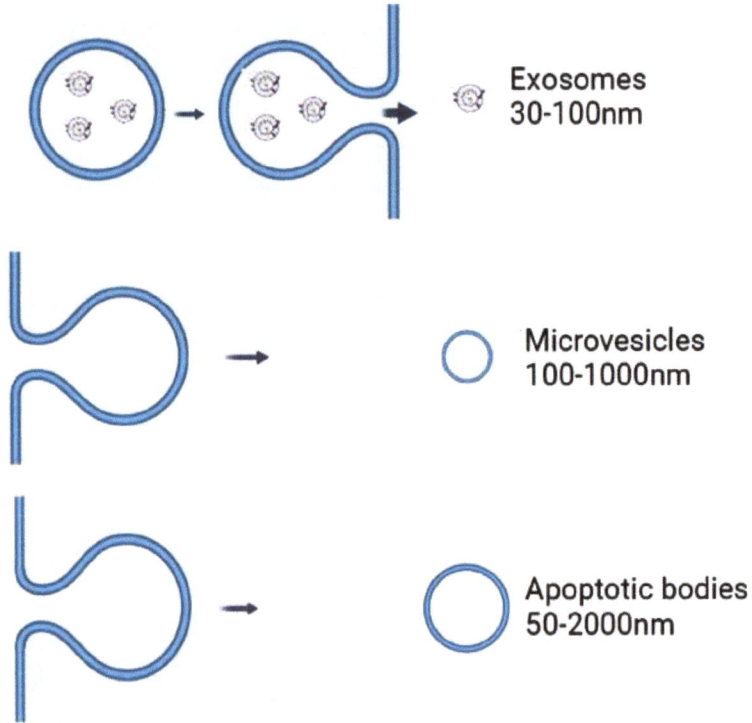

Exosomes
30-100nm

Microvesicles
100-1000nm

Apoptotic bodies
50-2000nm

Fig. (1). Extracellular vesicles of various sizes.

Exosomes are the main topic of this chapter, and in the following topics, their structures, biogenesis, functions, effects on various diseases, particularly cancer, use for treatment and diagnosis, and potential applications will be mentioned in detail.

STRUCTURE AND BIOGENESIS OF EXOSOMES

Exosomes, one type of EV, have received more attention than microvesicles and apoptotic bodies. Therefore, data obtained in order to comprehend their structure, biogenesis, and release from the cell membrane are more abundant than for other types of extracellular vesicles. However, research is ongoing to fully comprehend them.

"Exosome" term was used first time in 1987 by Johnstone in a study of vesicle formation in sheep reticulocytes [9]. In membrane-bound structures, it has been named exo-some because it is a process that generally involves the outward release of internal vesicular contents, as opposed to endocytosis which is the process of incorporating external molecules into the cell [9, 10]. In the years that followed, more detailed information on the formation and release processes of exosomes has been obtained.

Exosomes are composed of a complicated mixture of several macromolecules such as DNA, RNA, lipids, protein, and other metabolites. As previously stated, their sizes range from 30-100 nm [7]. Because of their nanometric size, they are best viewed through an electron microscope. Even though they have a round structure, electron microscope images are classified as cup-plate morphology [3]. The differences in these morphologies' definitions suggest that the effects in the preparation phase for electron microscopy imaging result in different images.

Electron microscopy is the best method used to determine exosomes, and it is critical to see their structures. However, for a more in-depth examination, methods can be used, such as Tunable Resistive Pulse Sensing (TRPS), Nanosight Nanoparticle Tracking Analysis (NNTA) and Flow Cytometry (FC). Furthermore, although not yet commercially developed, exosome-based approaches for surface plasmon resonance (SPR) are seen as potential identification techniques [11].

Exosomes had a hydrophilic inner surface and a hydrophobic crust composed of a lipid bilayer. These amphipathic properties allow them to compartmentalize the molecules they transport and make them suitable for carrying their natural cargo [12]. Exosomes are secreted by various cell types, such as tumor cells, mast cells, dendritic cells, hepatocytes, red blood cells, and epithelial cells [13].

Exosome biogenesis begins in the endosomal system. Cell membrane proteins and extracellular biomolecules form primary endocytic vesicles after endocytosis. The primary endocytic vesicles then merge with each other to create early endosomes (EE). EEs develop into late endosomes (LE). Multivesicular bodies (MVB) are created by LEs. Many membrane invaginations take place in LEs during the formation of MVBs from LEs. Cargo molecules are packaged into intraluminal vesicles (ILV) during this membrane invasion. As a result, MVBs filled with numerous ILVs are produced. MVBs have different intracellular orientations depending on their surface molecules. While some MVBs are transported to the cell membrane by intracellular motor proteins as delivery carriers, others are directed to lysosomes that are responsible for intracellular digestion to remove their contents. ILVs are released from the cell by MVBs and transported to the cell membrane after they fuse with the cell membrane. The ILVs, which remain inside the MVBs, are named exosomes after they are secreted out of the cells [14].

The secretion of exosomes process mentioned in detail above can be considered in summary as four stages: biogenesis, transport of MVBs to the cell membrane, fusion of MVBs with the membrane, and secrete of ILVs out of the cell.

Although some of the molecules involved in these stages have been identified, it is currently not possible to distinguish all of them experimentally due to problems with scientific methodology. Studies have not yet made it crystal clear which phase the under investigation molecule operates in [3]. The molecules mentioned in the following section, which are known to be crucial in the process leading to exosome secretion, may be involved in one or more stages of the process. The molecules involved in these stages cause the development of various MVB and ILV structures, and as a result, the secretion of exosomes with various contents and functions (Fig. **2**).

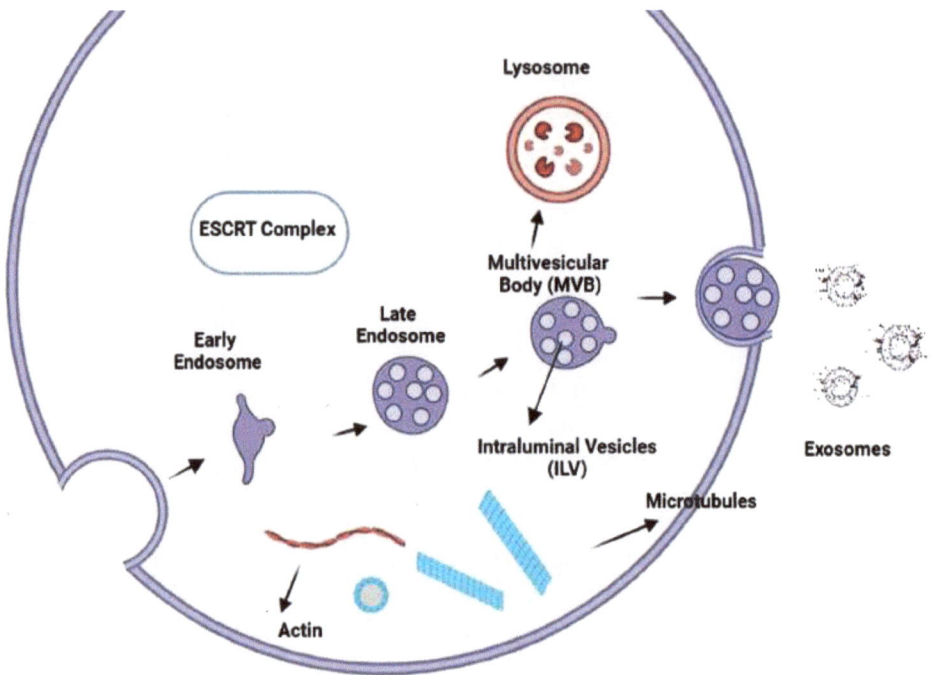

Fig. (2). Exosome Biogenesis.

Endosomal sorting complexes required for transport (ESCRT) machine plays a crucial role in this process. Most descriptions of exosome biogenesis have focused on ESCRT dependence. However, it might also rely on a system independent from ESCRT. These pathways may operate synergistically without being completely separated, or different subtypes of exosomes should be dependent on different machinery. Additionally, cell types and cellular homeostasis play important roles in the mechanism regulating exosome secretion [3, 5, 12].

ESCRT machines are composed of ESCRT-0, ESCRT-1, ESCRT-2, and ESCRT-3 protein complexes containing twenty conserved proteins from yeast to mammals and their associated proteins [ALG-2 interacting protein X (ALIX) and vacuolar protein sorting 4 (VPS4)]. Proteins that have been ubiquitinated in the endosomal membrane are identified and segregated by the ESCRT-0 complex. And it includes hepatocyte growth factor-regulated tyrosine kinase substrate (HRS) proteins in complex with signal transduction adaptor molecule (STAM), epidermal growth factor receptor substrate 15 (EPS 15) and clathrin, which recognize mono-ubiquitinated cargo proteins [15]. It has been shown that the ESCRT-0 protein HRS is necessary for dendritic cells to secrete exosomes [16]. ESCRT-1 and 2 are thought to be in charge of membrane deformation in the process of secreted cargo molecules. ESCRT 2 and 3 protein complexes participate in membrane fusion [17]. The ESCRT-2 complex protein tumor susceptibility gene 101 protein (TSG101) is incorporated by HRS, and then ESCRT-3 is incorporated by ESCRT-1 *via* ESCRT-2 or ALİX. Vesicle division is commanded by ESCRT-3. Finally, interaction with the AAA-ATPase Vps4 is necessary for the ESCRT machine to recycle [15]. Syntenin, ALİX and syndecan proteins are important molecules in exosome biogenesis. The interaction of synthenin with ALIX and syndecan induces ILV formation [16].

The determination that the loss of ESCRT molecules does not decrease the formation of MVBs has shown that ESCRT-independent alternative molecules also be effective in the formation of ILVs and, therefore, exosomes [19]. For example, it has been shown that ESCRT-independent exosome biogenesis is dependent on ceramide formation by sphingomyelinase and tetraspanins [17]. It is also thought to be due to heat shock proteins (HSP), cholesterol, and phosphophatic acids [20]. In addition, as a result of MVB interaction with the microtubule and actin, cell motor proteins are necessary for the carry them to the cell membrane. It has been shown that degradation of cortactin, an actin-binding protein, decreases exosome secretion, while its overexpression increases it [21].

As Hessvik *et al.* have summarized, after exosomes are transported, they are excreted by cell membrane fusion. Meanwhile, it has been shown that macromolecules interactions reduce the energy barrier to overcome energy barriers and facilitate diffusion. It has also been shown that in B-lymphocytes, only MVBs capable of fusing with the cell membrane are those with high cholesterol levels. It has also been stated that the secretion of exosomes can be controlled by calcium, and the increase in intracellular calcium levels increases the release of exosomes. It has been shown that exosome release rise in cases of increased cellular stress, such as cisplatin treatment, hypoxia, and induction of ER stress by tunicamycin [3]. Many molecules, including several small GTPases, have been identified to influence secretion mechanisms. It has been found that the Ral-1 gene is important for the formation of MVBs and the subsequent fusion of MVBs with the cell membrane. Ral-1 function in the secretion process has been shown to be conserved between nematodes and mammals [5]. Exosome secretion is also controlled by ISGylation, a ubiquitin-like posttranslational modification. ISGylation has been shown to inhibit ILV formation without affecting MVB numbers or exosome secretion [22]. Soluble N-ethylmaleimide-sensitive factor binding protein receptors (SNAREs) also mediate the extracellular release of exosomes after the fusion of MVBs to the cell membrane [23].

COMPOSITION OF EXOSOMES

Exosomes are generally composed of nucleic acids, proteins, lipids and other metabolites. Depending on the cells from which they are secreted, exosomes may contain different molecules in addition to the membrane components and metabolites that are present in all exosomes (Fig. **3**).

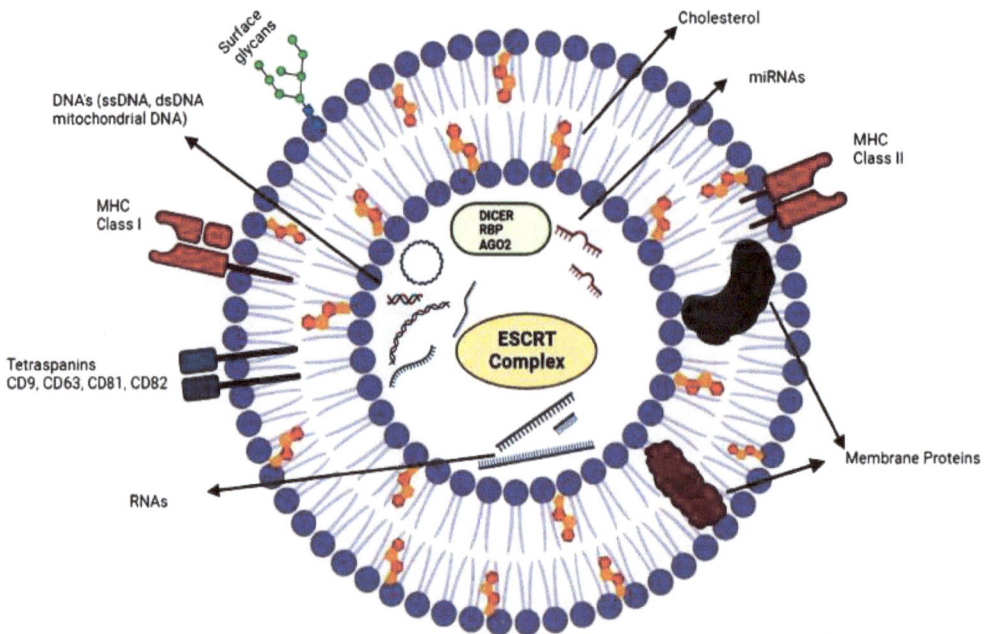

Fig. (3). Exosome Structure.

The membrane components of exosomes are composed of lipids, carbohydrates and proteins, similar to the structure of a typical cell membrane [24].

Depending on the origin cell type, the lipid composition of exosome membranes contains more cholesterol, sphingolipid, phosphatidylserine, sphingomyelin, and glycosphingolipid. It is also richer than the cells of origin in terms of ceramide, cholesterol, long saturated fatty acid chains and phosphoglycerides that can provide structural stability. In contrast, exosomes generally contain less phosphatidylcholine than their origin cells. Only minor variations in phosphatidylethanolamine contents were detected between cells of origin and exosomes [12, 25, 26].

Carbohydrates such as sialoglycoproteins and N-glycans are present on the exosomal membrane lipids' outer side [26]. Glycoproteins in the structure of exosomes not only participate in cell-to-cell interaction outside the cells, but they also serve as cargo molecules. For instance, ATP-dependent P-glycoprotein (P-gp) contributes to drug resistance and is present at high levels in exosomes [17].

The proteome content of exosomes is quite diverse. It includes membrane and cytosolic proteins, as well as types of proteins specific to cells of origin that reflect cell functions and conditions.

Membrane proteins could differ depending on the origin cells and the cell microenvironment. Major histocompatibility complex (MHC) I and II for antigen presentation, integrin proteins for tropism, tetraspanins as marker molecules, biogenesis involved protein molecules such as ALİX, TSG101, endosome-associated membrane proteins (flotilin, annexins, GTPases) are all found in exosome membranes [26, 27]

Exosomes can also contain nucleic acids such as microRNA (miRNA), long non-coding RNA (lncRNA), messenger RNA (mRNA), and DNA [28]. It has been reported that exosomes consist of 10,000 types of mRNAs involved in important cellular activities [29, 30]. Microarray analysis of exosomal RNAs showed that approximately 1300 types of RNAs were found in exosomes secreted by mast cells. Many types of RNA have not even proven to be functional yet, including miRNAs. They are moved to recipient cells and are of broad exosome relevance, which will be discussed in detail in the following topics. The miRNA composition of exosomes depends on the specific binding motifs of the miRNAs, ESCRT required for transport and the surface molecules of the exosomes [13]. Exosomes contain both mature miRNA and pre-miRNA transcripts as well as other essential elements of the miRNA biogenesis process, including DICER, TRBP, and AGO2 [31, 32]. By altering the function of various target mRNAs according to the cell type, miRNAs regulate gene expression in recipient cells [29, 33]. In addition to RNA types, there are also various types of DNA in exosomes, such as double-stranded DNA (dsDNA), single-stranded DNA (ssDNA) and mitochondrial DNA [29].

A list of the molecules discovered in exosomes is accessible in the Carta exosome database [34].

FUNCTION OF EXOSOMES

In studies carried out immediately after exosomes were discovered, it was found that the role of exosomes was to remove cellular waste [9]. The role of exosomes in cellular communication was later discovered, though. Cell communication is traditionally defined as the transfer of signal molecules secreted by one cell to another cell *via* a receptor-mediated synaptic, paracrine, endocrine, or exocrine pathway. However, signal molecules transmit their messages passively. Additionally, exosomes are involved in the transmission of a more complex message [35]. Because of their protein content and genetic characteristics, which reflects the cells from which they originated, exosomes can also serve as carriers to modulate intercellular communication and immune responses [36]. The contents of exosomes, which are wrapped in bilayer lipid membranes, continue to maintain their structure under physiological conditions. Therefore, exosomes are important tools in influencing cells in the surrounding or remote region to target cell responses [37].

The effects of exosomes on target cells begin with the uptake of exosomes by receptor-mediated, membrane fusion or endocytosis [35]. Numerous studies have demonstrated that exosome uptake occurs *via* macropinocytosis, caveolin clathrin dependent endocytosis, or lipid raft endocytosis. It is suggested that exosomes interact with their target cells *via* their surface proteins and thus transmit the materials they contain from cell to cell [13].

Many molecules in the exosome are involved in this process. Caveola-mediated exosome uptake associated with membrane microdomains enriched with heparan sulfate proteoglycans (HSPG) and cholesterol has been described in glioblastoma cell lines. Tetraspanin CD63 has been proposed as a surface marker. Integrin serves as a bridge to the extracellular matrix and is present almost everywhere in cell membranes. It has been stated that tropism for lung, brain or liver metastases depends on the recruitment of exosomes from tumor cells into the healthy host cells. A distinct pattern of integrin expression on the exosomal surface mediates this uptake. For connection with target cells, heparan sulfate (HS) proteoglycans and integrin cooperate to bind the extracellular matrix and cytoskeleton [13].

Exosomes contribute to some pathological and physiological events once they have reached their target cells, especially through the molecules they contain, particularly miRNAs. It has been discovered that exosomes secreted from cardiac fibroblasts are rich in miRNAs. One of the fibroblast exosomal-derived RNA molecule, miR-21, has been analyzed as an effective paracrine mediator in cardiomyocytes. Additionally, miR-208a levels have been found to be elevated in heart failure, both in serum and heart tissue, and this elevation is caused by exosomes released from cardiomyocytes under stress. It has also been demonstrated that both cardiomyocytes treated with high glucose and cardiomyocytes isolated from type 2 diabetic rats secrete exosomes enriched with the antiangiogenic miR-320 [36].

Exosomes from mesenchymal stem cells (MSCs) have been the focus of numerous studies. Exosomes released by MSCs have been demonstrated to have strong anti-liver disease protective effects. Exosomes secreted from MSC have been shown to encourage the regeneration of skeletal muscle, prevent macrophage activation and boost renal tubular epithelial cell proliferation. Furthermore, MSCs that secrete exosomes carrying miRNAs have been found to decrease the growth of breast cancer cells and suppress angiogenesis [13]. The angiogenic and regenerative abilities of exosomes secreted by MSCs have been demonstrated that these exosomes significantly lower myocardial apoptosis and inhibit activation of oxidative stress-induced caspase 3/7 in cardiomyocytes [38].

Exosomes actively contribute to the homeostasis of healthy cells and to several diseases, particularly cancer and exosomes are released in a range of diseases, particularly cancer. This leads to a discussion of their potential for use as a biomarker. Although the specific mechanisms underlying these roles are still being investigated, the following sections discuss how they affect both health and other diseases. Finally, information on the currently used and potential uses of exosomes is given.

THE IMPORTANCE OF EXOSOMES IN HEALTHY PHYSIOLOGICAL PROCESSES AND DISEASES

Exosomes are released into the systemic circulation by many cells, and their importance is related to the surrender of their contents to the target cells and their participation in physiological processes in these cells [39]. Exosomes are receiving increasing amounts of concern due to their effects on possible pathological and physiological situations [40]. Both in the diagnosis and as therapeutic agents, exosomes are becoming more and more significant [41]. Exosomes also contribute to the pathogenesis of diseases. These diseases include cardiovascular diseases, neurodegenerative diseases, and cancer [41, 42]. Furthermore, knowledge of the function of exosomes in diseases is expanding. In the early stages of sickness or in complex situations of joint diseases, exosomes in the blood can be used as a diagnostic tool [42].

EVs carry RNAs and proteins, lipids and other important molecules of biological process [43]. The delivery of the cargo molecules of the exosomes through the circulation to target cells initiates many physiological processes [39].

Exosomes secreted by dendritic cells stimulate T cells with the modulators they contain, such as CD8 and CD4, and provide an adaptive immune response. Those are their key role in immunomodulation [44]. Also, exosomes secreted from antigen-producing cells (APCs) are antigen transporters for T cells [45]. In addition, another effect of exosomes on the immune system is related to the apoptosis of T cells. IL-10 transported by the exosome induces regulatory T cells, stimulating transformation into T cells [46]. Thus, they contribute to the provision of pro-inflammatory and anti-inflammatory balance [39].

Exosomes have effects on implantation during pregnancy. Exosomes synthesized by endometrial cells are absorbed by trophoblasts and increase adhesion [47]. Additionally, it has been demonstrated that trophoblast-derived exosomes activate heat shock protein HSPE1 to promote T cell development into regulatory T (Treg) cells. Moreover, these exosomes contain four MHC class I molecules (HLA-C, HLA-E, HLA-F and HLA-G). HLA-E and HLA-G molecules is essential for a healthy pregnancy [47].

Exosomes contribute to the nutrition and development of the baby after birth. The proteins and RNA found in breast milk take part in important biological processes and contribute to maintaining the health of infants [48, 49]. They are participated in the selective suppression of toll-like receptors 3, 4 and 9 (TLR3, TLR4 and TLR9) through the movement of cystatin-B cathepsin inhibitors and epidermal growth factor where breast milk exosomes increase gingival epithelial cell migration [46]. By suppressing the stimulation of these receptors, the intestinal epithelium is protected against apoptosis and contributes to the development of the microbiota [50].

The content and release of exosomes secreted *via* many cells vary according to the stimulus. The release of harmful molecules from the cell is a significant mechanism for maintaining cellular homeostasis. Exosomes carry cell-specific molecules from which they are secreted. Therefore, they are considered as potential biomarkers in many diseases. Exosomes secreted by stem cells have been considered possible tissue-repair regeneration therapy without the use of cells. Exosomes are easily picked up by other cells. In particular, exosomes secreted by stem cells have been considered a possible cell-free regenerative therapy to repair tissues [40].

Exosomes have a significant place in the healthy functioning of the cardiovascular system. Exosomes can affect cell health or pathogenesis in cardiac cells and other cell types either negatively or positively. It has been demonstrated that some exosomes generated from stem cells can act as a regenerative mediator in heart repair. In order to improve treatment outcomes, it is possible to investigate signaling pathways and potential genetic changes in exosome-releasing cells by having a thorough understanding of the exosomal cargo [40].

It is known that exosomes are involved in paracrine signal transmission in the cardiovascular system [50]. Signal transmission occurs between smooth muscle cells and vascular endothelial cells [51], cardiomyocytes and fibroblasts [52], and smooth muscle cells [53].

In addition, it is stated that exosomes of cardiovascular origin may have a role in endocrine signal transmission. The detection of exosomes in pericardial fluid [54] and blood [55, 56] can be considered as an indicator.

It has been stated that exosomes secreted by cardiomyocytes contain a variety of biomolecules [57]. Some proteins, such as heat shock proteins Hsp20, Hsp60, Hsp70, and TNF-α, are found in exosomes secreted by cardiomyocytes, and these proteins are reported to have beneficial effects on cardiac function [57, 58]. Exosomes have a significant role in traffic between cardiomyocytes and coronary microvascular endothelial cells. Glucose transporters (GLUTs) are transferred to endothelial cells by these vesicles [59]. This transfer ensures that the glucose needs of endothelial cells are met and metabolism is regulated [57].

It is stated that the quantity of circulating exosomes and the cargo they carry can differ in cardiovascular conditions [50, 60]. In cardiac fibrosis, molecular mechanisms are associated with Wnt and IL-11 signaling pathways, TGFβ and nuclear factor-κβ [61]. Therefore, it is thought that any molecule transported in exosomes contributes to the pathogenesis of cardiac fibrosis [50]. Exosomes may play a role in atherosclerosis cell communication that controls tunica intima integrity. It has been showed that miR-205 and miR-712, which are among the proatherogenic vascular miRNAs, reduce the activity of the tissue metalloproteinase inhibitor (TIMP3), increase the activity of the matrix metalloproteinases (MMP), and cause swelling, vascular permeability, and migration of the smooth muscle cell [62]. It has been stated that the other proatherogenic exosomal miRNAs, miRNA-205 and miRNA-92a, increase inflammation, miRNA23b decreases the growth of endothelial cells, miRNA-155 decreases nitric oxide synthesis, and miRNA-222 and miRNA-221 affect calcification of smooth muscle cells [60].

It has been reported that exosomes are secreted by cardiac fibroblasts *via* angiotensin II-stimulation, which are then ingested by cardiomyocytes and can amplify myocardial hypertrophy by changing gene expression [63]. It has been stated that hypertrophic cardiomyocytes secrete more exosomes and IL-6 and IL-8 trigger exosome secretion [64].

Exosomes have a significant role in homeostasis by providing communication between nearby or distant cells in the central nervous system (CNS) [65]. In the CNS, the interaction between different neural cell types promotes exosome-mediated cellular communication, nerve homeostasis, and signal transduction. However, this situation brings with it the transfer of abnormal mediators [66].

Exosomes discharged into the extracellular environment and their effects on the pathogenesis of neurodegenerative disorders have attracted attention recently. The focus of research is on the physiology, etiopathology, diagnosis, biomarkers, and treatment protocols of neurological diseases [67]. Protein aggregation is the primary reason for these diseases. By regulating mRNAs, exosome-derived miRNAs have a significant impact on regulating protein levels [68 - 70].

Exosomes' mechanism of blood-brain barrier (BBB) passage is still unknown. They are thought to cross the BBB *via* endocytosis or transcytosis. Additionally, it is stated that tumor derived exosomes pass through the BBB and pass to the CNS by transferring their contents to neural cells [71]. Oligodendrocyte derived exosomes secrete proteolipids, glycoproteins, cholesterols, sphingolipids, and heat shock proteins [72]. They carry enzymes such as super peroxide dismutase and catalase, which make neurons more resistant to increased oxidative stress and play a role in oxidative stress [73]. In addition, Schwann cells release exosomes that promote healing after nerve injury in the peripheral nervous system [74].

Exosomes have important roles in Alzheimer's and Parkinson's diseases, which are among nervous system diseases [75, 76]. It is predicted that exosomes play a role in the transport of amyloid precursor protein (APP), which plays a role in Alzheimer's disease histopathology, between neurons [77]. It is possible to see exosomal transport of p-tau and A1-42 between cells and bodily fluids in Alzheimer's disease. Early diagnosis of these proteins is crucial for the disease's effective therapy [78]. α-synuclein is an important protein in Parkinson's disease pathology. It is transported by exosomes and leads to cell death [79]. In addition to these diseases, it has been reported that prion protein scrapie (PrPsc) and cellular prion protein (PrPc) are released into the extracellular environment by exosomes and cause the pathological spread of infectious prions [80].

Role of Exosomes in Cancer

Exosomes are secreted by the majority of normal and cancerous cells and have been found to play an important role in every stage of cancer development, including tumor microenvironment, cancer formation, angiogenesis, and metastasis. Additionally, exosomes can modulate the immune system's response to cancer, as well as impact the effectiveness of cancer treatments. These have led to the concept of exosomes being used as cancer biomarkers, making them a significant topic of study in the field of oncology.

Exosomes and their contents are gaining acceptance for their role in cancer biology. Exosomes have been shown to play an important role in cancer progression, and research on this topic is ongoing (Fig. **4**). These investigations may lead to novel methods for detecting and predicting cancer as well as potential therapeutic targets [40].

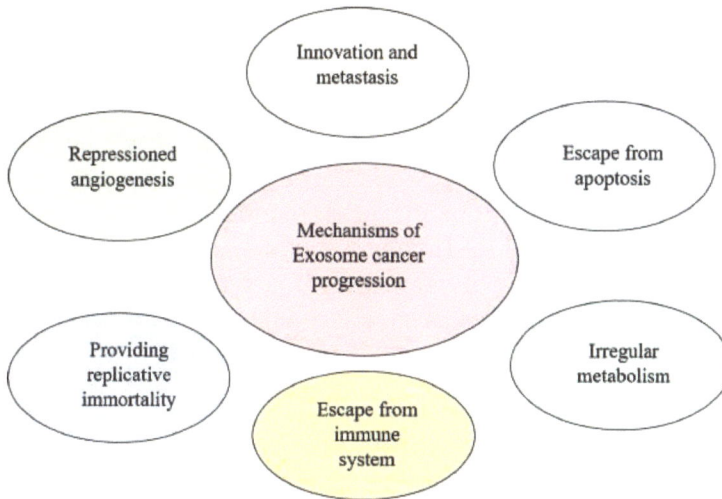

Fig. (**4**). Mechanisms of Exosome-mediated Cancer Progression.

Exosomes released by cancer cells support the extension of cancer by forming a niche for metastasis [81, 82]. They contribute to autocrine and paracrine signal generation to modulate the microenvironment of tumor cells and enhance their effects [83 - 86]. It is generally thought that these vesicles may be effective in creating the appropriate environment for cell proliferation and providing the conditions for the development of tumors [87]. Depending on the types of cells, exosomes have different autocrine actions. For instance, it has been reported that autocrine signals mediated by exosomes promote cellular proliferation [88, 89]. The formation and release of exosomes are crucially regulated by heat shock proteins (HSPs) connected to stressful situations. The levels of HSP70 and HSP90 among these proteins can increase in exosomes because of cellular stress [90 - 92].

During the tumor development, healthy cells transform into cancer cells. When co-injected with non-tumorigenic epithelial cells, exosomes from breast cancer patients' serum have been demonstrated to cause carcinogenesis in mice [93].

The miRNA content in exosomes released from cancer cells may contribute to tumor formation [31]. Exosomes secreted from cancer cells play a critical role in tumor development, metastasis and therapeutic response [94]. It has been stated that exosomes influence cell-independent miRNA biogenesis to promote tumorigenesis [32, 37]. Exosomes secreted from breast cancer cells include miRNAs associated with DICER, RISC, TRBP and AGO2 [32]. These exosomes have the ability to effectively silence mRNA expression when transferred to recipient cells. This results in the Dicer-dependent tumorigenesis of non-tumorigenic epithelial cells [37].

Exosomes can also participate in the processes of senescence and quiescence, which are survival mechanisms brought on by an unfavorable environment during cancer. The cell cycle stops at the G0 phase during quiescence, also known as a dormant state, which is reversible. The continuation of the cell cycle is regulated by gene regulation [95, 96]. On the other hand, senescence is the process of irreversible cell arrest due to Hayflick factors and is defined as a defense mechanism that restricts cell division [95, 97]. Senescence induction facilitates cancer treatment [95, 98]. Senescence's regulatory molecules have also been stated to participate in regulating exosome secretion [95].

Exosomes in the Tumor Microenvironment

The tumor microenvironment (TME) consists of stroma and tumor compartments. The stroma has a heterogeneous structure, consisting of the various cell types and extracellular matrix (ECM). The cellular compartments of the stroma contain vascular cells, fibroblasts, and immune cells, all of which provide functional and structural support to the tumor [17]. Tumor cells act together with TME in tumor development. By contributing to the initiation, progression, and therapeutic response of tumors, the molecular and cellular elements of the TME play a crucial part in the biology of cancer [99].

Exosomes play an important role in the TME. Exosomes that are produced by tumor cells are known as tumor-derived (TD) exosomes [36]. Compared to normal cell derived exosomes, TD exosomes are found in much higher concentrations in the bloodstream of cancer patients. They are crucial for controlling the molecular pathways that lead to cancer. The membranes of TD exosomes contain adhesion receptors and ligands that interact with specific cell types to provide them with a wide range of biomolecules [26]. In addition to TD exosomes, exosomes secreted from other cells of the TME, such as immune cells, fibroblasts, and vascular cells, also play a crucial role in cancer [27].

Through autocrine signals, TD exosomes can influence the growth of cancer locally [88, 89]. Thanks to TD exosomes, tumor cells can form similar cells by causing varied mutations and functional variations in healthy cells [100]. TD exosomes contribute to cancer progression *via* paracrine signaling [101].

Increased collagen production and hyaluronate synthesis during irregular and uncontrolled growth causes fibroblasts to become active as modify in morphology and protein expression progress. Depending on the molecular status of the tumor epithelial cells, activated fibroblasts can either hinder or encourage the growth and progression of tumors [27]. It has been reported that exosomes produced by triple negative breast cancer cells are included in the transformation of fibroblasts into cancer associated fibroblasts (CAF) and the development of the tumor microenvironment [102]. TGFβ/Smad pathway has been revealed to be the mechanism through which TD exosomes cause fibroblast transformation. It has been demonstrated that constitutively active TGFβ cargo from mesothelioma cells and prostate cancer exosomes causes fibroblasts to differentiate into myofibroblasts [103 - 105]. Another showed that exosomal TGFβ mediates the conversion of fibroblasts into CAFs in bladder cancer cells, encouraging the epithelial-mesenchymal transition (EMT), cell growth, invasion and migration [106].

Exosomes released by TME fibroblasts have also been studied for cancer. It has been found that exosomes containing miR -21, -378e and -143 from miRNA types and released from TME fibroblasts support the stemming of cells and epithelial-mesenchymal transition by transfer of human breast cancer cell lines [107].

It has been stated that CAFs cause the development of chemotherapy resistance by secreting exosomes, promoting cell migration and localization. Moreover, it has been reported that cancer stem cells (CSCs) also contribute to EMT initiation and are capable of causing metastasis, recurrence and chemotherapy resistance of cancer cells [85].

TD exosomes participate in the improvement of metastasis by carrying the factors that will facilitate the formation of metastasis into the bloodstream. In addition, there is confirmation to help the role of exosomes in initiating the inflammatory response. It has been stated that some miRNAs and reactive oxygen species (ROS) carried by exosomes can cause breaks in the DNA double helix structure by inducing genotoxic stress [108, 109]. Furthermore, the deregulation of 17 miRNAs that play a role in triggering inflammation in human papillomavirus (HPV) infection and cervical cancer development has been demonstrated *in vitro* [110].

TD exosomes released under hypoxic conditions have been shown to stimulate angiogenesis in a variety of cancers and TME [111]. It has also been stated that mRNAs and miRNAs in exosomes can induce angiogenesis and conduce to promoting angiogenic transition [112].

The most important roles of TD exosomes in cancer therapy stem from their carrying immunomodulatory molecules capable of promoting the getaway of tumor cells in the immune cells [82, 87]. Therefore, exosomes seem to avoid immune monitoring and have the potential to cause the development of drug resistance through the cargo they carry [81, 82, 113]. Tumor cells excrete chemotherapeutic agents out of the cell *via* exosomes [114, 115]. Exosomes play a significant role in tumor pathogenesis and immunosuppression because of their ability to suppress the immune system, hinder the activation of natural killer cells and cause effector T cells to undergo apoptosis [116]. It has been stated that programmed death ligand 1 (PD-L1) released by exosomes participates in the desertion of tumor cells from the immune system [117]. Additionally, the P53 protein affects numerous surveillance pathways and is altered or deleted in the majority of cancers [118]. This protein influences the transcription of numerous genes, such as TSAP6 and CHMP4C, consequently enhancing the synthesis of exosomes [119].

It has been stated that exosomes compose a convenient environment for promote cell proliferation and the accumulation of mutations that eventually result in the growth of a cancerous tumor [50].

Immune cells are a significant component of TME [27]. Tumor associated macrophages support metastasis and tumor progression and are associated with poor prognosis. In addition, the effect of exosomes secreted from other immune system cells causes many pathological responses [17]. Under "Exosome-Mediated Immune Response" topic title, these effects are mentioned.

Exosomes have been shown to control metabolism in prostate cancer. Exosomes released from TME fibroblasts have been found to carry metabolites such as aminoacids, lipids, acetate, lactate and intermediates of the tricarboxylic acid cycle that are used by cancer cells [120].

Almost all cancer cells exhibit the "Warburg effect", which modifies their metabolism to prefer fermentation over aerobic respiration. As a result, compared to non-proliferating normal cells, cancer cells consume significantly more glucose and produce significantly more lactic acid. Increased lactic acid production lowers the pH in the TME. It has been reported that the TME's low pH plays a key role in the increased exosome production and absorption by tumor cells [26].

Exosomes from various sources can have an impact on the TME in four keyways: by increasing metastasis, encouraging angiogenesis, encouraging immune escape, and by increasing drug resistance [121].

Role of Exosomes in Metastasis

In metastasis, cancer cells that spread through cell invasion and migration go to remote areas and hold on to secondary tissues. In metastasis formation, the ECM is disrupted and remodeled. To increase the input of tumor cells into the vessel, tight junctions are opened. The recipient cells' epithelial-mesenchymal transition, a step required for cancer invasion and metastasis, is supported. In addition, premetastatic niche formation is induced [5, 36, 83].

Exosomes actively participate in these processes that necessitate metastasis.

As mentioned in the "Exosomes in the Formation of the Tumor Microenvironment" title, TD exosomes can stimulate the transformation of fibroblasts to myofibroblasts [104]. Then, by increasing the production of pericellular hyaluronic acid, myofibroblasts can modify the TME, disrupting the ECM to promote the spread of cancer cells [36]. Actin rich bulges of the cell membrane, known as invadopodia, are related to the ECM's breakdown in cancer metastasis and invasiveness [122]. It has been documented how TD exosomes affect many phases of the invadopodia life cycle, including invadopodia production and stability. It has also been demonstrated that exosomal proteinases released during invadopodia maturation increase ECM degradation [123].

The capability of tumor cells to enter the vessel during tumor metastasis depends on exosomal miRNAs. It has been published that large levels of exosomal miR-105 are found in metastatic breast cancer cells, which support metastasis by targeting the zonula occludens 1 (ZO-1). ZO-1 organizes endothelial monolayers' and tight junctions' barrier properties [124, 125]. Additionally, by breaking the blood-brain barrier, exosomal miR-181c also encourages brain metastasis of breast cancer cells [71].

EMT was first defined by developmental biologists as a phenotypic change that aids embryonic development [126]. It is a process of transformation in which epithelial cells take on mesenchymal characteristics, lose cell-to-cell contacts, and lose their polarity, giving them the ability to migrate neighboring tissues. EMT is unnaturally activated during the growth and metastasis of tumors [84]. The process of metastasis is started in many cancer types by tumor cells undergoing EMT, which then spreads to the new organ and causes metastasis there [127].

Exosomes have the capacity to carry signals that can activate the EMT process [128]. Exosomes excreted by numerous cells can regulate the processes of angiogenesis, invasion and metastasis [129]. TD exosomes can increase vascular permeability and participate in niche formation [130]. The formation of premetastatic niches, stroma remodeling, and recipient cells' migratory capabilities can all be influenced by TD exosomes. TD exosomes contain a pro-EMT content, which includes β-catenin, TGFβ, HIF1α, and caveolin-1 that promotes proliferation [131]. Studies of the molecules carried by TD exosomes have shown that transcription regulators, which have broad signaling pathway activity, are transported in this manner. These include Annexin A2, Notch-1, HIFα, casein kinase II and matrix metalloproteinases (MMPs), and EMT drivers such as miR-100, LMP1 [85, 86, 132 - 135].

TD exosomes create a favorable environment for EMT by generating paracrine and autocrine signals in the TME to activate its program to neoplastic epithelial cells. These signals result in altered endothelial cell polarity, downregulation of ZO-1, and removal of preexisting metastasis inhibitors, as well as increased mobility in the TME [85, 86, 136 - 138]. With the development of metastatic capacity, exosome composition changes, becoming more enriched with EMT and other proteins that aid in coordinating metastatic activities between the tumor and the microenvironment [137].

It has been reported that TD exosomes secreted under hypoxia, an EMT-related condition with a high risk of metastasis, are more abundant in EMT signaling molecules than exosomes secreted in the normal state [131].

The cancer cell that enters the vein becomes part of the bloodstream and gains the capacity to invade the surrounding primary tissue. The next stage is the development of a premetastatic niche, where organ-targeted exosomes can extravasate, be ingested by organ tissues, and colonize with metastatic cells. In this process, the niche environment is shaped by remodeling of the ECM, angiogenesis, and myofibroblast activation [130, 139, 140].

For a cell to metastasize, it must be able to migrate. However, migration to reach a distant tissue and/or organ is not enough for metastasis formation. Premetastatic niche formation is essential for metastasis. Primary TD exosomes can play a role as mediators for premetastatic niche preparation [27, 31]. Following this, TD exosomes contribute to evading the immune system, help reorganize a premetastatic microenvironment for disease progression and allow unchecked disease development [131]. For instance, in the case of melanoma cancer, premetastatic niche formation is caused by the effect of exosomes secreted by metastatic melanoma cells on bone marrow progenitors *via* the receptor tyrosine kinase MET [139]. In addition, it has been shown that TD exosomes produce unique integrins in the metastasis site to create the premetastatic niche. It has been stated that the arrangement of integrins on exosomes can help predict the locations of metastatic sites where exosomes may be localized in the future. Exosomal integrins α6β1 and α6β4 are linked to lung cancer metastasis, whereas integrin αvβ5 is linked to liver metastasis [141]. Additionally, exosomal miRNA-122 from breast cancer cells has been demonstrated to restrict niche cells' uptake of glucose in order to support metastasis [142].

Role of Exosomes in Angiogenesis

Angiogenesis, which occurs in cancer development, is the process of vascular creation from preexisting vessels. Tumors need to have oxygen, some nutrients, and waste removal to grow up. Tumor cells must reach the vascular system to do this, where they will direct blood flow to the tumor. To establish a tumor blood supply, there must be a sufficient increase in pro-angiogenic factors and a decrease in anti-angiogenic factor activity [27]. TD exosomes contain a variety of pro-angiogenic factors [93]. In a variety of cancers and tumor microenvironments, it has been shown that TD exosomes can stimulate angiogenesis [36].

Cancer cells release exosomes into the TME to start signaling and boost proangiogenic factors to provide sufficient oxygenation when hypoxia is found throughout the tumor [27]. It has been proven that exosomes carrying Rac1/PAK2 proteins, which function as proangiogenic proteins, allow communication between oncogenic cells undergoing EMT and endothelial cells [143]. In addition, TD exosomes containing proangiogenic factors have been demonstrated to induce tubule formation in endothelial cells [30]. Then, as mentioned in the "Role of Exosomes in Metastasis" title, exosomes damage tight junctions and deliver miR-105 to endothelial cells, increasing vascular permeability [124, 125]. Additionally, it has been stated that only exosomes released by CD105+ cancer stem cells (CSC) contain mRNAs that enable the expression of proangiogenic factors such as MMP-2, MMP-9, FGF, VEGF, angiopoietin1, ephrin A3. In addition, exosomes secreted from CD105+ CSC were also discovered to be higher efficient in promoting angiogenesis than exosomes secreted from CD105- tumor cells [95].

Exosome-Mediated Immune Response

In tumor immunity, TD exosomes and immune system cells derived exosomes have various effects. TD exosomes can both enhance and to reduce the actions of immune system cells. Exosomes, which are secreted by immune cells, can also act to stimulate and inhibit the immune response against tumors.

Dendritic cells (DCs) are expert antigen-presenting cells. They have the ability to trigger primary T-cell responses, including distinct antigen presentation and antitumor responses [37]. Myeloid progenitor cells are prevented from differentiating into DCs when they take up TD exosomes. In this instance, there are no longer any DCs that present their antigens to control the immune system's response to the tumor [27]. By altering the expression of pattern recognition receptors (PRRs), TD exosomes can also reduce DCs' ability to recognize antigens. For instance, one study demonstrated that exosomes from pancreatic cancers regulate the expression of toll-like receptor 4 (TLR4), which activates DCs through miRNA-203 [144]. Exosomes secreted from DC have also been demonstrated to promote an antitumor immune response, in addition to TD exosomes [145]. DC exosomes also sensitize other immune cells to tumor antigens [146]. It is still unclear how mast cells affect the TME [147]. However, it has been demonstrated that mast cell exosomes can stimulate B cell and T cell activation and dendritic cell differentiation [148]. It has been stated that exosomes modulate immune responses by supporting the formation of B cells [99].

Exosomes de-regulate T cells by preventing proliferation, increasing regulatory T cell differentiation while decreasing helper T cell differentiation in stimulated T cells and changing the cytokine amounts in these cells. Helper T cells are inactivated by B cells after being stimulated by TD exosomes, which reduces the immune system's reactivity to tumor antigen presentation [27]. Through their miRNAs, TD exosomes have also been shown to reduce T cell differentiation, proliferation and cytokine secretion [149]. Exosomes released by active B cells can also prompt an influential T cell response to cancer cells and induce an immune response that is the antitumor effect [37]. Exosomes secreted by activated CD8$^+$ T cells have been demonstrated to boost tumor immunogenicity by triggering NF-κB and ERK signaling, which can aid tumor cells' capacity for metastasis [150].

Natural killer (NK) cells don't require MHC in order to react to antigens, in contrast to T and B cells [27]. The intricate interaction of inhibited and activated receptors is the primary determinant of the response mechanism of NK cells. It has been stated that TD exosomes interfere with NK cells, thereby suppressing tumor immunity [99]. For example, TD exosomes release exosomes that express the receptor NKG2D recognized by NK cells and direct them to these exosomes to counteract their destruction effect [151]. Exosomes influence the amount of NK cells by regulating the expression of NK cell receptors [99]. Consistent with this, it has been shown that the NKG2D surface expression is decreased in circulating NK cells in cancer [37]. On the other hand, exosomes released from NK cells also showed significant effects. The addition of exosomes enriched with FasL and perforin secreted from NK cells to melanoma cells showed lethal effects for cancer cells [152]. Macrophages, which can be activated by exosomes derived from tumors, are among the most extensively researched inflammatory cell components of the TME [31]. Macrophages are remarkably flexible and can modify their physiological status in response to environmental marks, especially those found in the tumor microenvironment [37]. While M2 macrophages encourage tumor growth and metastasis, M1 macrophages can eradicate tumor cells. Exosomes have been shown to influence macrophage polarization, particularly in M1 and M2, in order to control TME [99]. The M2 anti-inflammatory phenotype of macrophages is the result of factors in the TME [17]. This process also involves TD exosomes. Exosomes from triple negative breast cancer, for instance, encouraged human macrophage M2 differentiation [153]. The polarization of tumor promoting M2 macrophages is also controlled by exosomes that are produced by ovarian tumors [154]. Exosomes from CD163$^+$ M2 macrophages isolated from human colon cancer samples promoted migration and invasion in colorectal cancer cells *in vitro* by delivering miR-155-5p and miR-21-5p [155].

The best cancer vaccines in the future are probably going to be exosomes [156]. The effect of exosomes secreted from TD cells and exosomes released from immune system cells in cancer is quite complex. However, this complexity also provides many alternative avenues for the development of exosome based cancer vaccines, discussed in detail in the following sections.

Exosomes in Drug Resistance

A significant difficulty in the present cancer drug exploration process is tumor drug resistance. Different types of cancer are resistant to different drugs due to several mechanisms [157]. Exosome encapsulation allows for the removal from cells of many anticancer drugs and their metabolites. Target cells can receive proteins and miRNAs associated with multidrug resistance (MDR) from TD exosomes. Additionally, exosomes can block their effects by altering the way that antibody-targeting drugs attach to tumor cells [93]. These briefly discussed topics make it clear that exosomes are crucial molecules in drug resistance.

MDR, which can be inherited or acquired, significantly reduces the effectiveness of chemotherapy [158]. Exosome secretion has been stated to increase after cisplatin therapy of lung cancer cells and platinum-resistant ovarian cancer cells, indicating exosomes as a potential drug resistance mechanism [159]. To avoid the cytotoxic effects of chemotherapy, tumor cells can encapsulate the chemotherapeutic agents in exosomes [27]. It has been discovered that TD exosomes can contain and carry doxorubicin [160]. Furthermore, cisplatin is actively removed by tumor cells *via* exosomes in a pH-dependent, according to a study employing human tumor xenografts [161].

The removal of cytotoxic drugs from cells by ATP-bound P-glycoprotein (P-gp) is a mechanism that contributes to drug resistance (P-gp). It has been stated that P-gp is transferred by exosomes, with the observation of high levels of P-gp in exosomes. Additionally, the exosome-mediated increase in P-gb levels caused drug-resistant characteristics to be passed onto drug-sensitive cells [162]. Cancer cells that resist drugs can affect sensitive cells *via* exosomes. CAF-derived exosomal transport of miRNA-21 has been showed to reason paclitaxel resistance in cancer cells [163].

As mentioned at the beginning of the "Role of Exosomes in Cancer" title, cancer cells are involved in the quiescence or dormant state, one of the survival modes, which is an important process in cancer. For instance, it has been suggested that breast cancer cells may interact with a mesenchymal stem cell during this process in a bone environment, resulting in the secretion of exosomes containing the miR-222 and miR-223, which prevents some cell cycle regulating proteins from being expressed. This can lead to a decrease in sensitivity to chemotherapy by stopping the cell cycle [95, 164, 165].

Additionally, it has been found that tumor exosomes can lessen cytotoxicity by preventing tumor reactive antibodies from adhering to tumor cells [166]. Consistently, exosomes from breast cancer cells that overexpress HER2 have been demonstrated to decrease the activity of the HER2 antibody trastuzumab by binding to it by expressing active HER2 [167]. Additionally, malignant B-cell lymphoma patients are treated with Rituximab, an anti-CD-20 monoclonal antibody. Exosomes are found to have high levels of CD20 expression, bind rituximab, and remove it from the growth medium, according to a study using a lymphoma cell line [17].

In summary, exosomes affect, especially, chemotherapeutic drug resistance by removing drugs and metabolites before they can take effect in the cell. It acts by overexpressing molecules that will recognize drugs, such as monoclonal antibodies, by preventing these drugs from binding to their original targets. In addition, exosomes can carry the drug resistance feature to other cells through miRNAs.

CLINICAL APPLICATIONS OF EXOSOMES

Exosomes are excellent candidates for potential cancer clinical applications because of their association with cancer, as discussed in the previous sections. Exosomes can act as clinical markers to help with disease diagnosis and follow-up. Exosomes can be utilized as drug delivery systems for a variety of molecules and miRNAs, including different anticancer drugs. Exosome secretion can be inhibited, exosomes can be removed from the bloodstream, or exosomes can be prevented from being taken up by target cells to reduce their negative effects. Because exosomes are crucial for drug resistance, metastasis, and growth of cancer, it is possible to combat their negative effects. Finally, cell free vaccines based on exosomes are available for functions exosomes play in the immune system.

EXOSOMES AS BIOMARKERS AND THEIR IMPORTANCE

Exosomes are released by pathological cells and include a range of molecules, such as nucleic acids and proteins which are signs of pathological conditions. This has led to a great deal of research into the use of exosomes as biomarkers. Additionally, the abundance of exosomes in many biological fluids makes their use as a marker more advantageous. As will be discussed in detail in this section, there are many studies relating to the use of exosomes as a biomarker in many diseases, primarily cancer.

Numerous biological fluids, such as amniotic, cerebrospinal fluid, urine, and saliva, have been studied for the determination of exosomes in addition to the potential exosome biomarkers studies carried out in isolation from blood [94].

Many molecular characteristics of the tumors from which they derive are reflected in the molecular characteristics of TD exosomes. As a result, they can be used as distinctive cancer markers for many types of cancer. As a result, it has been stated they can be useful in making clinical decisions related to diagnosis, directing treatment, and determining prognosis [27, 36].

In ovarian cancer, the excess of tumor exosomes and cancer stage has been found to be directly correlated [168]. Further research revealed a significantly higher level of exosomes which contain phosphatidylserine in the circulation of ovarian cancer patients [169]. Furthermore, it has been reported that serum exosomes from ovarian cancer patients contain the same levels of miRNAs that are used as biopsy diagnostic markers for the disease [168]. Thus, as an alternative way to biopsy for diagnosis, a potential biomarker that becomes suitable for miRNA analysis from serum has been identified.

MiR-7641, which has been related to the metastasis and progression of tumor cells in breast cancer, has been found to be significantly more abundant in the plasma of patients with metastatic breast cancer [170].

Exosomes in plasma that contain the proteins CD63+, coveolin-1, and Rab5b have been shown to significantly outnumber those in the plasma of healthy individuals in melanoma patients [171].

Epidermal growth factor receptor v3 (EGFRv3) specific for glioblastoma has been discovered in exosomes isolated from glioblastoma patients' serum sample at a level that can aid in diagnosis [30]. Additionally, EGFRv3 and TGF-β have been found in the exosomes in serum sample of patients with brain tumors [94]. Similarly, it has been demonstrated that the exosomal human epidermal growth factor receptor 2(EGFR2) molecule can be used as a biomarker in lung cancer [172]. Moreover, miR-1246 and miR-96, which are found in high levels in exosomes, have been proposed as possible biomarkers for non-small cell lung cancer (NSCLC), which is a type of lung cancer. Exosomal miR-96 has additionally been linked to vascular invasion and radioresistant NSCLC [173]. Additionally, it has been stated that after resection and irrespective of the histological type of lung cancer, the amounts of two miRNA types (miR-1268b and miR-6075) were significantly reduced [174]. After the cancer cells are attracted, the TD exosomes that normally secrete these RNAs are no longer secreted.

Exosomal miR-21 levels of esophageal squamous cell cancer (ESCC) have been found to be higher in ESCC patient serum sample than in serum sample from patients with benign tumors. Exosomal miR-21 has also been shown to have a positive correlation with tumor development [175].

Exosomes from patients with colorectal cancer's serum samples have been shown to have remarkably higher amounts of the miRNAs [176].

Urine exosome analyses were done in addition to serum exosome analyses for prostate cancer and bladder cancer. Many exosomal proteins have been discovered as potential bladder cancer markers after comparing the protein profile of urinary exosomes between healthy donors and bladder cancer patients [177]. Examples of molecules that may be used as prostate cancer biomarkers include miR-532-5p, miR-486-5p, miR-451a and miR-486-3p, in urinary exosomes [178, 179].

It has been determined that exosomes can be used as a biomarker according to the molecules they carry in diseases other than cancer. In the continuation of this topic, the data obtained by the analysis of exosomes in some diseases are summarized.

In neurological diseases, it has been discovered that some of the molecules carried by exosomes can be used as markers. For instance, it has been noted that amyloid peptides from exosomes are collected in Alzheimer's patients' brains [180]. High concentrations of the phosphorylated tau peptide at Thr-181 were discovered in exosomes that were isolated from spinal fluid samples of Alzheimer's patients [181]. Additionally, some proteins found in the exosomes of Alzheimer's patients have been suggested as early biomarkers of the disease, including autolysosomal proteins, ncRNAs, miRNAs, GAP43, neurogranin, SNAP25, synaptotagmin, and neural origin hemoglobin. In Parkinson's disease, it has been suggested that molecules like synuclein, clusterin and some miRNA types serve as potential biomarkers [50].

In heart diseases, potential exosome markers have also been found. For early detection of acute myocardial infarction, specific plasma exosomal miRNAs and lncRNAs have been identified [182, 183]. Additionally, a lot of studies have demonstrated the possible utility of exosomal mRNAs as biomarkers in clinical diagnosis of heart diseases [28].

Comparing patients with and without acute kidney injury, it has been discovered that patients with acute kidney injury had higher levels of urinary exosome which fetuin-A [184]. Later research has been shown that exosomes from acute kidney injury patients were found to contain activating transcription factor 3, while from controls were not [185].

It has been noted that chronic hepatitis C patients have high serum exosomal CD81 levels, which are related to the degree of fibrosis and inflammation, and it has been discovered that CD81 could perform as a biomarker for the hepatitis C diagnosis [186].

For the early detection or diagnosis of fetal developmental disorders and some pregnancy complications, such as preeclampsia [187] and gestational diabetes [188], analysis of exosomal content has been suggested. In order to diagnose COVID-19 and categorize patients based on the severity of their illnesses, an investigation of exosomal extent has been also done. These analyses consist of RNA analysis, exosome quantification by phenotype, and proteomic profiling [50].

ROLE OF EXOSOMES IN DRUG DELIVERY AND STRATEGIES

A drug delivery vehicle of any kind must be able to encapsulate sufficient drugs to elicit a therapeutic effect, escape macrophages, be non-toxic, biocompatible, and not elicit an immunological response. Additionally, it must possess long-term intrinsic stability to prevent changes in the therapeutic agent's size, structure, or bioactivity during circulation [189].

Exosomes are ideal for delivering drugs or genes because they can pass the blood-brain barrier, are naturally biocompatible, and are stable when circulated in the blood [190]. They also have low toxicity, are stable, and do not trigger an immune reaction because they are biodegradable. Target cells can be absorbed with a lot of uptake mechanisms. They can also be created using different bioengineering techniques [17].

Exosomes are secreted from different cell types and carry various cargo molecules to the target site (Fig. **5**).

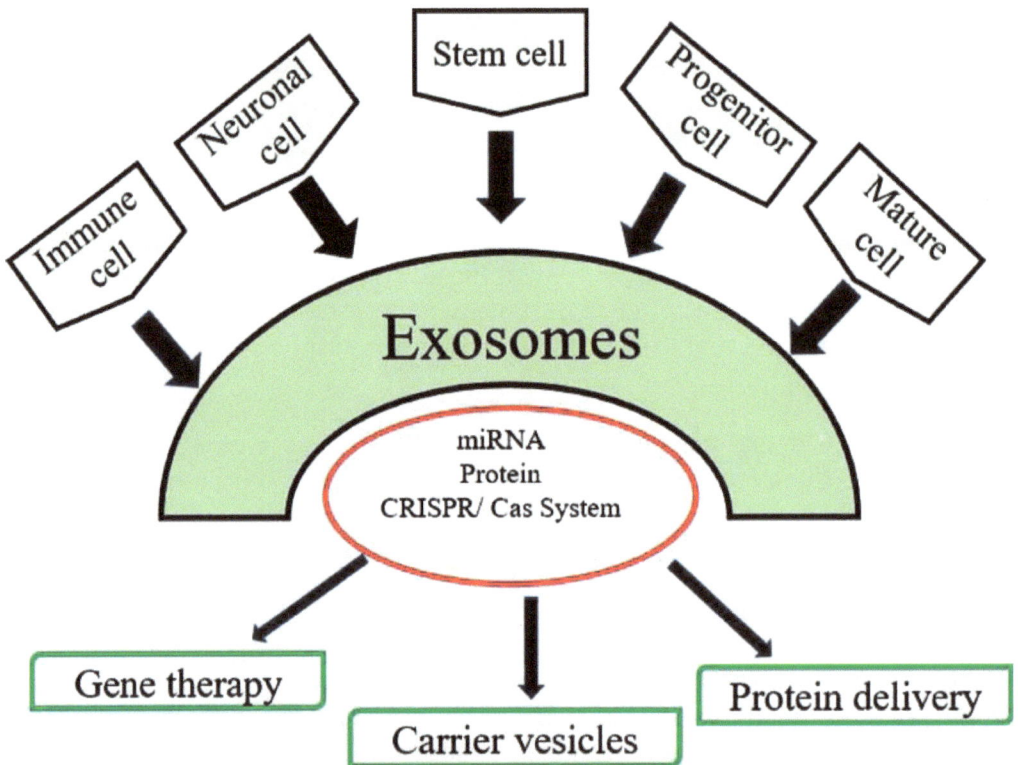

Fig. (5). Exosome-secreting Cells and Their Use as a Transport System.

The development of a cargo loading method is an important step in the use of exosomes as nanocarriers [17]. The creation of an appropriate cargo loading technique is still under development. Exosome loading alternatives include electroporation, chemical-based exosome transfection, and straightforward incubation with target cargo. Sonication is another alternative method for loading exosomes [191]. The choice of cells that produce exosomes is another crucial choice because they influence the function and immune response. The function, distribution, and immunogenicity of these cells should be carefully examined, and they should be chosen from cells that secrete large amounts of exosomes [17]. For instance, exosomes produced by dendritic cells have been used to aim helper T cells in order to promote the improvement of cytotoxic T cells, control T cell differentiation foster an anti-tumor environment [27]. The ability of exosomes secreted from bone marrow MSC to deliver functional anti-miRNA-9 to brain tumor cells has also been investigated. Glioblastoma cells in the brain communicate with MSCs through these exosomes [189]. Chemotherapeutic drugs have been delivered to the target site in cancer using exosomes. Studies have been conducted using the anticancer drugs PTX and DOX to load and deliver them to lung, pancreatic, and prostate cancer cells. In comparison to liposome and free drugs, it has been demonstrated to increase the delivery of PTX (17). Celastrol (CEL) has been reported to have therapeutic value in various cancer types. However, low bioavailability and cytotoxicity are obstacles for this molecule. Transport of CEL by exosomes exhibited enhanced antitumor activity over free administration. Additionally, it did not exhibit any toxicity when delivered by exosomes [192].

Although nucleic acid drugs have therapeutic potential, there is little clinical use for them. This is partially caused by a lack of suitable distribution systems. Exosomes may be the perfect vehicle for carrying nucleic acids. Exosomes were used to deliver siRNA for oncogenes. In addition, it has been stated that exosomes can successfully load and deliver miRNA. According to some reports, tumor-promoting genes are targeted in cancer cells in this way, which suppresses the growth of tumors [17]. The CRISPR-Cas9 delivery vectors, which have recently taken on a significant role in the gene editing system, are also not yet fully functional. Many exosome-based transporter systems have thus been investigated. For instance, SKOV3 ovarian cancer-derived exosomes containing CRISPR/Cas9 have been created. CRISPR-Cas9 exosomes have been used to inactivate poly ADP-ribose polymerase-1 (PARP-1) in mice with SKOV3 tumors [193, 194].

THERAPEUTIC STRATEGIES BASED ON EXOSOMES

Exosomes lead to drug resistance, as was previously mentioned, and to stop these negative effects, treatment options are being investigated. Exosome biogenesis and secretion inhibition, exosome uptake inhibition by target cells, exosome depletion from circulation, and exosomal cargo were the main topics of these studies.

As mentioned earlier, exosome secretion is increased in tumor cells. Pharmacological agents known to inhibit exosome biogenesis and/or secretion have been used to try to make tumors more responsive to therapy. For instance, exosome levels have been shown to be decreased by substances like Cl-amidine and bisindolyllmaleimide-I, which inhibit biogenesis in prostate cancer cells. Additionally, the combination of these inhibitors with the chemotherapy drug 5-FU has elevated apoptosis in prostate cancer cell lines [195]. Exocyst is a protein complex that helps ECVs bind to and be directed toward the plasma membrane before fusing. Exosome secretion has been demonstrated to be constrained by inhibiting exocyst protein expression [196]. It has been reported that exosome secretion is impaired by ISGylation induction, as was mentioned in the "Exosome Structure and Biogenesis" title [22]. In addition, as summarized by Samanta *et al.*, it has been proposed that limiting ceramide production may prevent the formation of exosomes; exosome release can be decreased by preventing the interaction between syndecane proteoglycans, their cytoplasmic adapter synthenin, and the exosomal protein ALIX. GTPase and RAB27A73, involved in exosome release, may serve as therapeutic target to reduce tumor growth. Additionally, blocking specific tetraspanins significant for EV formation may be another possible strategy to prevent the creation of exosomes by reducing tumorigenesis [197].

Additionally, blocking the take-in of exosomes by target cells is a therapeutic strategic approach. As mentioned in the previous sections, it has been shown that the uptake of exosomes by target cells is accomplished in different ways [13]. It has been demonstrated that exosome uptake by cervical cancer cells depends on the presence of exosomal phosphatidylserine [198]. By preventing the interaction between the lipids on the exosome membrane surface and particular membrane regions on the target cells, exosome uptake can be reduced. Exosome uptake was inhibited when treated with Annexin V, which has phosphatidylserine binding properties [199].

The removal of circulating exosomes may be extremely beneficial for cancer patients. A clinical trial that used blood ultrapheresis to remove small particles from cancer patients' blood, possibly including exosomes, led to a reduction in tumor dimensions in a certain percentage of patients [200]. Exosomes isolated from HER2+ breast cancer cells have been demonstrated to inhibit the therapeutic effects of trastuzumab [167]. Exosome removal from patients with advanced breast cancer is likely to enhance trastuzumab therapy response [201]. Aethlon Medical, Inc. is developing a device called ADAPT™ that works similarly to dialysis machines. This device contains the immobilized HER2 antibody and is intended to be used to capture HER2+ breast cancer exosomes and prevent disease progression [17].

Targeting the exosome-carried molecules is another strategy. For instance, Sunitinib, a receptor tyrosine kinase inhibitor, is given to patients with renal cell carcinoma. TD exosomes can also spread sunitinib resistance. It has been demonstrated that the lncRNA lncARSR loaded into TD exosomes increases sunitinib resistance. The response to sunitinib has been restored after the administration of antisense nucleic acid that targets lncARSR [202].

Finally, as discussed previously, dormant or dormant breast cancer cells can interact with a mesenchymal stem cell and release exosomes containing miRNAs 222 and 223, which reduce sensitivity to chemotherapy [164, 165]. Antagonist miR-222/223 treatment has been stated to resensitize these breast cancer cells to chemotherapy [203].

Therapeutic Strategies Based on Exosomes Acting on the Immune System

Exosomes have an impact on the immune system. Strategies to enhance these effects have been developed as a result of this knowledge.

Exosomes released by dendritic cells exposed to diphtheria toxin antigens in infectious diseases have been used to successfully immunize against diphtheria infections [204, 205]. Additionally, it has been discovered that exosomes in human milk can increase the immune response and tip the balance of T cells in favor of a regulatory phenotype [206, 207].

Exosomes are likely to be the most effective cancer vaccines, as was mentioned in the "Exosome-Mediated Immune Response" title [156]. Exosomes secreted by immune system cells and TD exosomes both have complex effects on cancer.

Macrophage derived exosomes have the capability to control the immune response and polarize macrophages [208]. Exosomes have been used by researchers as an adjuvant to anti-cancer vaccines as a result of this. Exosomes released by M1 macrophages have been shown to enhance the effectiveness of cancer vaccines [209].

Exosome co-delivery of adjuvants and tumor antigens is an effective approach for inducing anti-tumor immunity. High concentrations of co-stimulatory molecules, like some tetraspanins, can be found in exosomes released by dendritic cells. These molecules' effects support immune system stimulation [210, 211].

Additionally, the T cell-mediated immune response induced by exosomes released from the dendritic cell with hepatocellular carcinoma antigen added has inhibited tumor growth. Furthermore, clinical trials based on tumor-specific antigen loading on exosomes released from dendritic cells have been conducted in metastatic melanoma and non-small cell lung cancer patients [17, 31].

CONCLUDING REMARKS

Although exosomes are gaining new importance, they continue to be investigated in many areas. While their use as a carrier system ensures that the target gene or molecule reaches its location, exosomes, which are biomarkers of most diseases, can also be detected. Additionally, their effects on the immune system make them excellent candidates for currently popular cancer vaccines. The potential of exosomes will create new areas for many studies on this subject.

REFERENCES

[1]   C. Théry, K.W. Witwer, E. Aikawa, M.J. Alcaraz, J.D. Anderson, R. Andriantsitohaina, A. Antoniou, T. Arab, F. Archer, G.K. Atkin-Smith, D.C. Ayre, J.M. Bach, D. Bachurski, H. Baharvand, L. Balaj, S. Baldacchino, N.N. Bauer, A.A. Baxter, M. Bebawy, C. Beckham, A. Bedina Zavec, A. Benmoussa, A.C. Berardi, P. Bergese, E. Bielska, C. Blenkiron, S. Bobis-Wozowicz, E. Boilard, W. Boireau, A. Bongiovanni, F.E. Borràs, S. Bosch, C.M. Boulanger, X. Breakefield, A.M. Breglio, M.Á. Brennan, D.R. Brigstock, A. Brisson, M.L.D. Broekman, J.F. Bromberg, P. Bryl-Górecka, S. Buch, A.H. Buck, D. Burger, S. Busatto, D. Buschmann, B. Bussolati, E.I. Buzás, J.B. Byrd, G. Camussi, D.R.F. Carter, S. Caruso, L.W. Chamley, Y.T. Chang, C. Chen, S. Chen, L. Cheng, A.R. Chin, A. Clayton, S.P. Clerici, A. Cocks, E. Cocucci, R.J. Coffey, A. Cordeiro-da-Silva, Y. Couch, F.A.W. Coumans, B. Coyle, R. Crescitelli, M.F. Criado, C. D'Souza-Schorey, S. Das, A. Datta Chaudhuri, P. de Candia, E.F. De Santana, O. De Wever, H.A. del Portillo, T. Demaret, S. Deville, A. Devitt, B. Dhondt, D. Di Vizio, L.C. Dieterich, V. Dolo, A.P. Dominguez Rubio, M. Dominici, M.R. Dourado, T.A.P. Driedonks, F.V. Duarte, H.M. Duncan, R.M. Eichenberger, K. Ekström, S. EL Andaloussi, C. Elie-Caille, U. Erdbrügger, J.M. Falcón-Pérez, F. Fatima, J.E. Fish, M. Flores-Bellver, A. Försönits, A. Frelet-Barrand, F. Fricke, G. Fuhrmann, S. Gabrielsson, A. Gámez-Valero, C. Gardiner, K. Gärtner, R. Gaudin, Y.S. Gho, B. Giebel, C. Gilbert, M. Gimona, I. Giusti, D.C.I. Goberdhan, A. Görgens, S.M. Gorski, D.W. Greening, J.C. Gross, A. Gualerzi, G.N. Gupta, D. Gustafson, A. Handberg, R.A. Haraszti, P. Harrison, H. Hegyesi, A. Hendrix, A.F. Hill, F.H. Hochberg, K.F. Hoffmann, B. Holder, H. Holthofer, B. Hosseinkhani, G. Hu, Y. Huang, V. Huber, S. Hunt, A.G.E. Ibrahim, T. Ikezu, J.M. Inal, M. Isin, A. Ivanova, H.K. Jackson, S. Jacobsen, S.M. Jay, M. Jayachandran, G. Jenster, L. Jiang, S.M. Johnson, J.C. Jones, A. Jong, T. Jovanovic-Talisman, S. Jung, R. Kalluri, S. Kano, S. Kaur, Y. Kawamura, E.T. Keller, D. Khamari, E. Khomyakova, A. Khvorova, P. Kierulf, K.P. Kim, T. Kislinger, M. Klingeborn, D.J. Klinke II, M. Kornek, M.M. Kosanović, Á.F. Kovács, E.M. Krämer-Albers, S. Krasemann, M. Krause, I.V. Kurochkin, G.D. Kusuma, S. Kuypers, S. Laitinen, S.M. Langevin, L.R. Languino, J. Lannigan, C. Lässer, L.C. Laurent, G. Lavieu, E. Lázaro-Ibáñez, S. Le Lay, M.S. Lee, Y.X.F. Lee, D.S. Lemos, M. Lenassi, A. Leszczynska, I.T.S. Li, K. Liao, S.F. Libregts, E. Ligeti, R. Lim, S.K. Lim, A. Linē, K. Linnemannstöns, A. Llorente, C.A. Lombard, M.J. Lorenowicz, Á.M. Lörincz, J. Lötvall, J. Lovett, M.C. Lowry, X. Loyer, Q. Lu, B. Lukomska, T.R. Lunavat, S.L.N. Maas, H. Malhi, A. Marcilla, J. Mariani, J. Mariscal, E.S. Martens-Uzunova, L. Martin-Jaular, M.C. Martinez, V.R. Martins, M. Mathieu, S. Mathivanan, M. Maugeri, L.K. McGinnis, M.J. McVey, D.G. Meckes Jr, K.L. Meehan, I. Mertens, V.R. Minciacchi, A. Möller, M. Møller Jørgensen, A. Morales-Kastresana, J. Morhayim, F. Mullier, M. Muraca, L. Musante, V. Mussack, D.C. Muth, K.H. Myburgh, T. Najrana, M. Nawaz, I. Nazarenko, P. Nejsum, C. Neri, T. Neri, R. Nieuwland, L. Nimrichter, J.P. Nolan, E.N.M. Nolte-'t Hoen, N. Noren Hooten, L. O'Driscoll, T. O'Grady, A. O'Loghlen, T. Ochiya, M. Olivier, A. Ortiz, L.A. Ortiz, X. Osteikoetxea, O. Østergaard, M. Ostrowski, J. Park, D.M. Pegtel, H. Peinado, F. Perut, M.W. Pfaffl, D.G. Phinney, B.C.H. Pieters, R.C. Pink, D.S. Pisetsky, E. Pogge von Strandmann, I. Polakovicova, I.K.H. Poon, B.H. Powell, I. Prada, L. Pulliam, P. Quesenberry, A. Radeghieri, R.L. Raffai, S. Raimondo, J. Rak, M.I. Ramirez, G. Raposo, M.S. Rayyan, N. Regev-Rudzki, F.L. Ricklefs, P.D. Robbins, D.D. Roberts, S.C. Rodrigues, E. Rohde, S. Rome, K.M.A. Rouschop, A. Rughetti, A.E. Russell, P. Saá, S. Sahoo, E. Salas-Huenuleo, C. Sánchez, J.A. Saugstad, M.J. Saul, R.M. Schiffelers, R. Schneider, T.H. Schøyen, A. Scott, E. Shahaj, S. Sharma, O. Shatnyeva, F. Shekari, G.V. Shelke, A.K. Shetty, K. Shiba, P.R.M.

Siljander, A.M. Silva, A. Skowronek, O.L. Snyder II, R.P. Soares, B.W. Sódar, C. Soekmadji, J. Sotillo, P.D. Stahl, W. Stoorvogel, S.L. Stott, E.F. Strasser, S. Swift, H. Tahara, M. Tewari, K. Timms, S. Tiwari, R. Tixeira, M. Tkach, W.S. Toh, R. Tomasini, A.C. Torrecilhas, J.P. Tosar, V. Toxavidis, L. Urbanelli, P. Vader, B.W.M. van Balkom, S.G. van der Grein, J. Van Deun, M.J.C. van Herwijnen, K. Van Keuren-Jensen, G. van Niel, M.E. van Royen, A.J. van Wijnen, M.H. Vasconcelos, I.J. Vechetti Jr, T.D. Veit, L.J. Vella, É. Velot, F.J. Verweij, B. Vestad, J.L. Viñas, T. Visnovitz, K.V. Vukman, J. Wahlgren, D.C. Watson, M.H.M. Wauben, A. Weaver, J.P. Webber, V. Weber, A.M. Wehman, D.J. Weiss, J.A. Welsh, S. Wendt, A.M. Wheelock, Z. Wiener, L. Witte, J. Wolfram, A. Xagorari, P. Xander, J. Xu, X. Yan, M. Yáñez-Mó, H. Yin, Y. Yuana, V. Zappulli, J. Zarubova, V. Žėkas, J. Zhang, Z. Zhao, L. Zheng, A.R. Zheutlin, A.M. Zickler, P. Zimmermann, A.M. Zivkovic, D. Zocco, and E.K. Zuba-Surma, "Minimal information for studies of extracellular vesicles 2018 (MISEV2018): A position statement of the international society for extracellular vesicles and update of the MISEV2014 guidelines", *J. Extracell. Vesicles,* vol. 7, no. 1, p. 1535750, 2018.
[http://dx.doi.org/10.1080/20013078.2018.1535750] [PMID: 30637094]

[2] P. Wolf, "The nature and significance of platelet products in human plasma", *Br. J. Haematol.,* vol. 13, no. 3, pp. 269-288, 1967.
[http://dx.doi.org/10.1111/j.1365-2141.1967.tb08741.x] [PMID: 6025241]

[3] N.P. Hessvik, and A. Llorente, "Current knowledge on exosome biogenesis and release", *Cell. Mol. Life Sci.,* vol. 75, no. 2, pp. 193-208, 2018.
[http://dx.doi.org/10.1007/s00018-017-2595-9] [PMID: 28733901]

[4] N.A. Jamaludin, L.M. Thurston, K.J. Witek, A. Meikle, S. Basatvat, S. Elliott, S. Hunt, A. Andronowska, and A. Fazeli, "Efficient isolation, biophysical characterisation and molecular composition of extracellular vesicles secreted by primary and immortalised cells of reproductive origin", *Theriogenology,* vol. 135, pp. 121-137, 2019.
[http://dx.doi.org/10.1016/j.theriogenology.2019.06.002] [PMID: 31207473]

[5] V. Hyenne, M. Labouesse, and J.G. Goetz, "The Small GTPase Ral orchestrates MVB biogenesis and exosome secretion", *Small GTPases,* vol. 9, no. 6, pp. 445-451, 2018.
[http://dx.doi.org/10.1080/21541248.2016.1251378] [PMID: 27875100]

[6] B. Gul, F. Syed, S. Khan, A. Iqbal, and I. Ahmad, "Characterization of extracellular vesicles by flow cytometry: Challenges and promises", *Micron,* vol. 161, p. 103341, 2022.
[http://dx.doi.org/10.1016/j.micron.2022.103341] [PMID: 35985059]

[7] C. Elsner, S. Ergün, and N. Wagner, "Biogenesis and release of endothelial extracellular vesicles: Morphological aspects", *Ann. Anat.,* vol. 245, p. 152006, 2023.
[http://dx.doi.org/10.1016/j.aanat.2022.152006] [PMID: 36183939]

[8] N. Zhang, H. Chen, C. Yang, X. Hu, N. Sun, and C. Deng, "Functionalized nanomaterials in separation and analysis of extracellular vesicles and their contents", *Trends Analyt. Chem.,* vol. 153, p. 116652, 2022.
[http://dx.doi.org/10.1016/j.trac.2022.116652]

[9] R.M. Johnstone, M. Adam, J.R. Hammond, L. Orr, and C. Turbide, "Vesicle formation during reticulocyte maturation. Association of plasma membrane activities with released vesicles (exosomes)", *J. Biol. Chem.,* vol. 262, no. 19, pp. 9412-9420, 1987.
[http://dx.doi.org/10.1016/S0021-9258(18)48095-7] [PMID: 3597417]

[10] R.M. Johnstone, "Revisiting the road to the discovery of exosomes", *Blood Cells Mol. Dis.,* vol. 34, no. 3, pp. 214-219, 2005.
[http://dx.doi.org/10.1016/j.bcmd.2005.03.002] [PMID: 15885604]

[11] E.H. Koritzinsky, J.M. Street, R.A. Star, and P.S.T. Yuen, "Quantification of exosomes", *J. Cell. Physiol.,* vol. 232, no. 7, pp. 1587-1590, 2017.
[http://dx.doi.org/10.1002/jcp.25387] [PMID: 27018079]

[12] R.S. Conlan, S. Pisano, M.I. Oliveira, M. Ferrari, and I. Mendes Pinto, "Exosomes as reconfigurable therapeutic systems", *Trends Mol. Med.,* vol. 23, no. 7, pp. 636-650, 2017.

[http://dx.doi.org/10.1016/j.molmed.2017.05.003] [PMID: 28648185]

[13]   X. Yu, M. Odenthal, and J. Fries, "Exosomes as miRNA carriers: Formation–function–future", *Int. J. Mol. Sci.,* vol. 17, no. 12, p. 2028, 2016.
[http://dx.doi.org/10.3390/ijms17122028] [PMID: 27918449]

[14]   M. Xu, J. Jie, and D. Jin, "The biogenesis and secretion of exosomes and multivesicular bodies (MVBs): Intercellular shuttles and implications in human diseases", *Genes Dis.,* vol. 10, no. 5, pp. 1894-1907, 2022.

[15]   M. Colombo, C. Moita, G. van Niel, J. Kowal, J. Vigneron, P. Benaroch, N. Manel, L.F. Moita, C. Théry, and G. Raposo, "Analysis of ESCRT functions in exosome biogenesis, composition and secretion highlights the heterogeneity of extracellular vesicles", *J. Cell Sci.,* vol. 126, no. Pt 24, p. jcs.128868, 2013.
[http://dx.doi.org/10.1242/jcs.128868] [PMID: 24105262]

[16]   K. Tamai, N. Tanaka, T. Nakano, E. Kakazu, Y. Kondo, J. Inoue, M. Shiina, K. Fukushima, T. Hoshino, K. Sano, Y. Ueno, T. Shimosegawa, and K. Sugamura, "Exosome secretion of dendritic cells is regulated by Hrs, an ESCRT-0 protein", *Biochem. Biophys. Res. Commun.,* vol. 399, no. 3, pp. 384-390, 2010.
[http://dx.doi.org/10.1016/j.bbrc.2010.07.083] [PMID: 20673754]

[17]   N. Milman, L. Ginini, and Z. Gil, "Exosomes and their role in tumorigenesis and anticancer drug resistance", *Drug Resist. Updat.,* vol. 45, pp. 1-12, 2019.
[http://dx.doi.org/10.1016/j.drup.2019.07.003] [PMID: 31369918]

[18]   M.F. Baietti, Z. Zhang, E. Mortier, A. Melchior, G. Degeest, A. Geeraerts, Y. Ivarsson, F. Depoortere, C. Coomans, E. Vermeiren, P. Zimmermann, and G. David, "Syndecan–syntenin–ALIX regulates the biogenesis of exosomes", *Nat. Cell Biol.,* vol. 14, no. 7, pp. 677-685, 2012.
[http://dx.doi.org/10.1038/ncb2502] [PMID: 22660413]

[19]   S. Stuffers, C. Sem Wegner, H. Stenmark, and A. Brech, "Multivesicular endosome biogenesis in the absence of ESCRTs", *Traffic,* vol. 10, no. 7, pp. 925-937, 2009.
[http://dx.doi.org/10.1111/j.1600-0854.2009.00920.x] [PMID: 19490536]

[20]   J. Kowal, M. Tkach, and C. Théry, "Biogenesis and secretion of exosomes", *Curr. Opin. Cell Biol.,* vol. 29, pp. 116-125, 2014.
[http://dx.doi.org/10.1016/j.ceb.2014.05.004] [PMID: 24959705]

[21]   S. Sinha, D. Hoshino, N.H. Hong, K.C. Kirkbride, N.E. Grega-Larson, M. Seiki, M.J. Tyska, and A.M. Weaver, "Cortactin promotes exosome secretion by controlling branched actin dynamics", *J. Cell Biol.,* vol. 214, no. 2, pp. 197-213, 2016.
[http://dx.doi.org/10.1083/jcb.201601025] [PMID: 27402952]

[22]   C. Villarroya-Beltri, F. Baixauli, M. Mittelbrunn, I. Fernández-Delgado, D. Torralba, O. Moreno-Gonzalo, S. Baldanta, C. Enrich, S. Guerra, and F. Sánchez-Madrid, "ISGylation controls exosome secretion by promoting lysosomal degradation of MVB proteins", *Nat. Commun.,* vol. 7, no. 1, p. 13588, 2016.
[http://dx.doi.org/10.1038/ncomms13588] [PMID: 27882925]

[23]   C.M. Fader, D.G. Sánchez, M.B. Mestre, and M.I. Colombo, "TI-VAMP/VAMP7 and VAMP3/cellubrevin: Two v-SNARE proteins involved in specific steps of the autophagy/multivesicular body pathways", *Biochim. Biophys. Acta Mol. Cell Res.,* vol. 1793, no. 12, pp. 1901-1916, 2009.
[http://dx.doi.org/10.1016/j.bbamcr.2009.09.011] [PMID: 19781582]

[24]   A. Llorente, T. Skotland, T. Sylvänne, D. Kauhanen, T. Róg, A. Orłowski, I. Vattulainen, K. Ekroos, and K. Sandvig, "Molecular lipidomics of exosomes released by PC-3 prostate cancer cells", *Biochim. Biophys. Acta Mol. Cell Biol. Lipids,* vol. 1831, no. 7, pp. 1302-1309, 2013.
[http://dx.doi.org/10.1016/j.bbalip.2013.04.011] [PMID: 24046871]

[25]   T. Skotland, K. Sandvig, and A. Llorente, "Lipids in exosomes: Current knowledge and the way forward", *Prog. Lipid Res.,* vol. 66, pp. 30-41, 2017.
[http://dx.doi.org/10.1016/j.plipres.2017.03.001] [PMID: 28342835]

[26]   A. Jafari, A. Babajani, M. Abdollahpour-Alitappeh, N. Ahmadi, and M. Rezaei-Tavirani, "Exosomes and cancer: From molecular mechanisms to clinical applications", *Med. Oncol.,* vol. 38, no. 4, p. 45, 2021.
[http://dx.doi.org/10.1007/s12032-021-01491-0] [PMID: 33743101]

[27]   L.T. Brinton, H.S. Sloane, M. Kester, and K.A. Kelly, "Formation and role of exosomes in cancer", *Cell. Mol. Life Sci.,* vol. 72, no. 4, pp. 659-671, 2015.
[http://dx.doi.org/10.1007/s00018-014-1764-3] [PMID: 25336151]

[28]   H. Valadi, K. Ekström, A. Bossios, M. Sjöstrand, J.J. Lee, and J.O. Lötvall, "Exosome-mediated transfer of mRNAs and microRNAs is a novel mechanism of genetic exchange between cells", *Nat. Cell Biol.,* vol. 9, no. 6, pp. 654-659, 2007.
[http://dx.doi.org/10.1038/ncb1596] [PMID: 17486113]

[29]   Q. Hao, Y. Wu, Y. Wu, P. Wang, and J.V. Vadgama, "Tumor-derived exosomes in tumor-induced immune suppression", *Int. J. Mol. Sci.,* vol. 23, no. 3, p. 1461, 2022.
[http://dx.doi.org/10.3390/ijms23031461] [PMID: 35163380]

[30]   J. Skog, T. Würdinger, S. van Rijn, D.H. Meijer, L. Gainche, W.T. Curry Jr, B.S. Carter, A.M. Krichevsky, X.O. Breakefield, and X.O. Breakefield, "Glioblastoma microvesicles transport RNA and proteins that promote tumour growth and provide diagnostic biomarkers", *Nat. Cell Biol.,* vol. 10, no. 12, pp. 1470-1476, 2008.
[http://dx.doi.org/10.1038/ncb1800] [PMID: 19011622]

[31]   W. Guo, Y. Gao, N. Li, F. Shao, C. Wang, P. Wang, Z. Yang, R. Li, and J. He, "Exosomes: New players in cancer", *Oncol. Rep.,* vol. 38, no. 2, pp. 665-675, 2017.
[http://dx.doi.org/10.3892/or.2017.5714] [PMID: 28627679]

[32]   S.A. Melo, H. Sugimoto, J.T. O'Connell, N. Kato, A. Villanueva, A. Vidal, L. Qiu, E. Vitkin, L.T. Perelman, C.A. Melo, A. Lucci, C. Ivan, G.A. Calin, and R. Kalluri, "Cancer exosomes perform cell-independent microRNA biogenesis and promote tumorigenesis", *Cancer Cell,* vol. 26, no. 5, pp. 707-721, 2014.
[http://dx.doi.org/10.1016/j.ccell.2014.09.005] [PMID: 25446899]

[33]   K. O'Brien, K. Breyne, S. Ughetto, L.C. Laurent, and X.O. Breakefield, "RNA delivery by extracellular vesicles in mammalian cells and its applications", *Nat. Rev. Mol. Cell Biol.,* vol. 21, no. 10, pp. 585-606, 2020.
[http://dx.doi.org/10.1038/s41580-020-0251-y] [PMID: 32457507]

[34]   Exocarta.org, "Exosome protein, RNA and lipid database", Available From: http://www.exocarta.org/ (Accessed on 18th Dec 2022).

[35]   S.L. Liu, P. Sun, Y. Li, S.S. Liu, and Y. Lu, "Exosomes as critical mediators of cell-to-cell communication in cancer pathogenesis and their potential clinical application", *Transl. Cancer Res.,* vol. 8, no. 1, pp. 298-311, 2019.
[http://dx.doi.org/10.21037/tcr.2019.01.03] [PMID: 35116759]

[36]   Z. Wang, J.Q. Chen, J. Liu, and L. Tian, "Exosomes in tumor microenvironment: Novel transporters and biomarkers", *J. Transl. Med.,* vol. 14, no. 1, p. 297, 2016.
[http://dx.doi.org/10.1186/s12967-016-1056-9] [PMID: 27756426]

[37]   Y. Liu, Y. Gu, and X. Cao, "The exosomes in tumor immunity", *OncoImmunology,* vol. 4, no. 9, p. e1027472, 2015.
[http://dx.doi.org/10.1080/2162402X.2015.1027472] [PMID: 26405598]

[38]   Y. Lin, J.D. Anderson, L.M.A. Rahnama, S.V. Gu, and A.A. Knowlton, "Exosomes in disease and regeneration: Biological functions, diagnostics, and beneficial effects", *Am. J. Physiol. Heart Circ.*

Physiol., vol. 319, no. 6, pp. H1162-H1180, 2020.
[http://dx.doi.org/10.1152/ajpheart.00075.2020] [PMID: 32986962]

[39]   C. Corrado, S. Raimondo, A. Chiesi, F. Ciccia, G. De Leo, and R. Alessandro, "Exosomes as intercellular signaling organelles involved in health and disease: Basic science and clinical applications", *Int. J. Mol. Sci.,* vol. 14, no. 3, pp. 5338-5366, 2013.
[http://dx.doi.org/10.3390/ijms14035338] [PMID: 23466882]

[40]   Y. Lin, J.D. Anderson, L.M.A. Rahnama, S.V. Gu, and A.A. Knowlton, "Exosomes in disease and regeneration: Biological functions, diagnostics, and beneficial effects", *Am. J. Physiol. Heart Circ. Physiol.,* vol. 319, no. 6, pp. H1162-H1180, 2020.
[http://dx.doi.org/10.1152/ajpheart.00075.2020] [PMID: 32986962]

[41]   H. Aheget, L. Mazini, F. Martin, B. Belqat, J.A. Marchal, and K. Benabdellah, "Exosomes: Their role in pathogenesis, diagnosis and treatment of diseases", *Cancers,* vol. 13, no. 1, p. 84, 2020.
[http://dx.doi.org/10.3390/cancers13010084] [PMID: 33396739]

[42]   Z. Li, Y. Wang, K. Xiao, S. Xiang, Z. Li, and X. Weng, "Emerging role of exosomes in the joint diseases", *Cell. Physiol. Biochem.,* vol. 47, no. 5, pp. 2008-2017, 2018.
[http://dx.doi.org/10.1159/000491469] [PMID: 29969758]

[43]   S. Keerthikumar, D. Chisanga, D. Ariyaratne, H. Al Saffar, S. Anand, K. Zhao, M. Samuel, M. Pathan, M. Jois, N. Chilamkurti, L. Gangoda, and S. Mathivanan, "ExoCarta: A web-based compendium of exosomal cargo", *J. Mol. Biol.,* vol. 428, no. 4, pp. 688-692, 2016.
[http://dx.doi.org/10.1016/j.jmb.2015.09.019] [PMID: 26434508]

[44]   P.D. Robbins, and A.E. Morelli, "Regulation of immune responses by extracellular vesicles", *Nat. Rev. Immunol.,* vol. 14, no. 3, pp. 195-208, 2014.
[http://dx.doi.org/10.1038/nri3622] [PMID: 24566916]

[45]   E.N.M. Nolte-'t Hoen, S.I. Buschow, S.M. Anderton, W. Stoorvogel, and M.H.M. Wauben, "Activated T cells recruit exosomes secreted by dendritic cells via LFA-1", *Blood,* vol. 113, no. 9, pp. 1977-1981, 2009.
[http://dx.doi.org/10.1182/blood-2008-08-174094] [PMID: 19064723]

[46]   M.I. Zonneveld, M.J.C. Herwijnen, M.M. Fernandez-Gutierrez, A. Giovanazzi, A.M. Groot, M. Kleinjan, T.M.M. Capel, A.J.A.M. Sijts, L.S. Taams, J. Garssen, E.C. Jong, M. Kleerebezem, E.N.M. Nolte-'t Hoen, F.A. Redegeld, and M.H.M. Wauben, "Human milk extracellular vesicles target nodes in interconnected signalling pathways that enhance oral epithelial barrier function and dampen immune responses", *J. Extracell. Vesicles,* vol. 10, no. 5, p. e12071, 2021.
[http://dx.doi.org/10.1002/jev2.12071] [PMID: 33732416]

[47]   V.L. Kaminski, J.H. Ellwanger, and J.A.B. Chies, "Extracellular vesicles in host-pathogen interactions and immune regulation: Exosomes as emerging actors in the immunological theater of pregnancy", *Heliyon,* vol. 5, no. 8, p. e02355, 2019.
[http://dx.doi.org/10.1016/j.heliyon.2019.e02355] [PMID: 31592031]

[48]   Y. Liao, X. Du, J. Li, and B. Lönnerdal, "Human milk exosomes and their microRNAs survive digestion *in vitro* and are taken up by human intestinal cells", *Mol. Nutr. Food Res.,* vol. 61, no. 11, p. 1700082, 2017.
[http://dx.doi.org/10.1002/mnfr.201700082] [PMID: 28688106]

[49]   C. Garcia, R.D. Duan, V. Brévaut-Malaty, C. Gire, V. Millet, U. Simeoni, M. Bernard, and M. Armand, "Bioactive compounds in human milk and intestinal health and maturity in preterm newborn: An overview", *Cell. Mol. Biol.,* vol. 59, no. 1, pp. 108-131, 2013.
[PMID: 25326648]

[50]   M.I. Mosquera-Heredia, L.C. Morales, O.M. Vidal, E. Barceló, C. Silvera-Redondo, J.I. Vélez, and P. Garavito-Galofre, "Exosomes: Potential disease biomarkers and new therapeutic targets", *Biomedicines,* vol. 9, no. 8, p. 1061, 2021.
[http://dx.doi.org/10.3390/biomedicines9081061] [PMID: 34440265]

[51]   E. Hergenreider, S. Heydt, K. Tréguer, T. Boettger, A.J.G. Horrevoets, A.M. Zeiher, M.P. Scheffer, A.S. Frangakis, X. Yin, M. Mayr, T. Braun, C. Urbich, R.A. Boon, and S. Dimmeler, "Atheroprotective communication between endothelial cells and smooth muscle cells through miRNAs", *Nat. Cell Biol.,* vol. 14, no. 3, pp. 249-256, 2012.
[http://dx.doi.org/10.1038/ncb2441] [PMID: 22327366]

[52]   C. Bang, S. Batkai, S. Dangwal, S.K. Gupta, A. Foinquinos, A. Holzmann, A. Just, J. Remke, K. Zimmer, A. Zeug, E. Ponimaskin, A. Schmiedl, X. Yin, M. Mayr, R. Halder, A. Fischer, S. Engelhardt, Y. Wei, A. Schober, J. Fiedler, and T. Thum, "Cardiac fibroblast–derived microRNA passenger strand-enriched exosomes mediate cardiomyocyte hypertrophy", *J. Clin. Invest.,* vol. 124, no. 5, pp. 2136-2146, 2014.
[http://dx.doi.org/10.1172/JCI70577] [PMID: 24743145]

[53]   A.N. Kapustin, M.L.L. Chatrou, I. Drozdov, Y. Zheng, S.M. Davidson, D. Soong, M. Furmanik, P. Sanchis, R.T.M. De Rosales, D. Alvarez-Hernandez, R. Shroff, X. Yin, K. Muller, J.N. Skepper, M. Mayr, C.P. Reutelingsperger, A. Chester, S. Bertazzo, L.J. Schurgers, and C.M. Shanahan, "Vascular smooth muscle cell calcification is mediated by regulated exosome secretion", *Circ. Res.,* vol. 116, no. 8, pp. 1312-1323, 2015.
[http://dx.doi.org/10.1161/CIRCRESAHA.116.305012] [PMID: 25711438]

[54]   C. Beltrami, M. Besnier, S. Shantikumar, A.I.U. Shearn, C. Rajakaruna, A. Laftah, F. Sessa, G. Spinetti, E. Petretto, G.D. Angelini, and C. Emanueli, "Human pericardial fluid contains exosomes enriched with cardiovascular-expressed microRNAs and promotes therapeutic angiogenesis", *Mol. Ther.,* vol. 25, no. 3, pp. 679-693, 2017.
[http://dx.doi.org/10.1016/j.ymthe.2016.12.022] [PMID: 28159509]

[55]   C. Emanueli, A.I.U. Shearn, A. Laftah, F. Fiorentino, B.C. Reeves, C. Beltrami, A. Mumford, A. Clayton, M. Gurney, S. Shantikumar, and G.D. Angelini, "Coronary artery-bypass-graft surgery increases the plasma concentration of exosomes carrying a cargo of cardiac microRNAs: An example of exosome trafficking out of the human heart with potential for cardiac biomarker discovery", *PLoS One,* vol. 11, no. 4, p. e0154274, 2016.
[http://dx.doi.org/10.1371/journal.pone.0154274] [PMID: 27128471]

[56]   G. Pironti, R.T. Strachan, D. Abraham, S. Mon-Wei Yu, M. Chen, W. Chen, K. Hanada, L. Mao, L.J. Watson, and H.A. Rockman, "Circulating exosomes induced by cardiac pressure overload contain functional angiotensin II type 1 receptors", *Circulation,* vol. 131, no. 24, pp. 2120-2130, 2015.
[http://dx.doi.org/10.1161/CIRCULATIONAHA.115.015687] [PMID: 25995315]

[57]   H. Yu, and Z. Wang, "Cardiomyocyte-derived exosomes: Biological functions and potential therapeutic implications", *Front. Physiol.,* vol. 10, p. 1049, 2019.
[http://dx.doi.org/10.3389/fphys.2019.01049] [PMID: 31481897]

[58]   S. Gupta, and A.A. Knowlton, "HSP60 trafficking in adult cardiac myocytes: Role of the exosomal pathway", *Am. J. Physiol. Heart Circ. Physiol.,* vol. 292, no. 6, pp. H3052-H3056, 2007.
[http://dx.doi.org/10.1152/ajpheart.01355.2006] [PMID: 17307989]

[59]   N.A. Garcia, J. Moncayo-Arlandi, P. Sepulveda, and A. Diez-Juan, "Cardiomyocyte exosomes regulate glycolytic flux in endothelium by direct transfer of GLUT transporters and glycolytic enzymes", *Cardiovasc. Res.,* vol. 109, no. 3, pp. 397-408, 2016.
[http://dx.doi.org/10.1093/cvr/cvv260] [PMID: 26609058]

[60]   R.J. Henning, "Cardiovascular exosomes and microRNAs in cardiovascular physiology and pathophysiology", *J. Cardiovasc. Transl. Res.,* vol. 14, no. 2, pp. 195-212, 2021.
[http://dx.doi.org/10.1007/s12265-020-10040-5] [PMID: 32588374]

[61]   P. Kong, P. Christia, and N.G. Frangogiannis, "The pathogenesis of cardiac fibrosis", *Cell. Mol. Life Sci.,* vol. 71, no. 4, pp. 549-574, 2014.
[http://dx.doi.org/10.1007/s00018-013-1349-6] [PMID: 23649149]

[62]   S. Kumar, C.W. Kim, R.D. Simmons, and H. Jo, "Role of flow-sensitive microRNAs in endothelial dysfunction and atherosclerosis: Mechanosensitive athero-miRs", *Arterioscler. Thromb. Vasc. Biol.,* vol. 34, no. 10, pp. 2206-2216, 2014.
[http://dx.doi.org/10.1161/ATVBAHA.114.303425] [PMID: 25012134]

[63]   L. Lyu, H. Wang, B. Li, Q. Qin, L. Qi, M. Nagarkatti, P. Nagarkatti, J.S. Janicki, X.L. Wang, and T. Cui, "A critical role of cardiac fibroblast-derived exosomes in activating renin angiotensin system in cardiomyocytes", *J. Mol. Cell. Cardiol.,* vol. 89, no. Pt B, pp. 268-279, 2015.
[http://dx.doi.org/10.1016/j.yjmcc.2015.10.022] [PMID: 26497614]

[64]   H. Yu, L. Qin, Y. Peng, W. Bai, and Z. Wang, "Exosomes derived from hypertrophic cardiomyocytes induce inflammation in macrophages via miR-155 mediated MAPK pathway", *Front. Immunol.,* vol. 11, p. 606045, 2021.
[http://dx.doi.org/10.3389/fimmu.2020.606045] [PMID: 33613526]

[65]   G. Zhang, and P. Yang, "A novel cell-cell communication mechanism in the nervous system: Exosomes", *J. Neurosci. Res.,* vol. 96, no. 1, pp. 45-52, 2018.
[http://dx.doi.org/10.1002/jnr.24113] [PMID: 28718905]

[66]   C. Frühbeis, D. Fröhlich, and E.M. Krämer-Albers, "Emerging roles of exosomes in neuron-glia communication", *Front. Physiol.,* vol. 3, p. 119, 2012.
[http://dx.doi.org/10.3389/fphys.2012.00119] [PMID: 22557979]

[67]   J. Howitt, and A.F. Hill, "Exosomes in the pathology of neurodegenerative diseases", *J. Biol. Chem.,* vol. 291, no. 52, pp. 26589-26597, 2016.
[http://dx.doi.org/10.1074/jbc.R116.757955] [PMID: 27852825]

[68]   D. Li, Y.P. Li, Y.X. Li, X.H. Zhu, X.G. Du, M. Zhou, W.B. Li, and H.Y. Deng, "Effect of regulatory network of exosomes and microRNAs on neurodegenerative diseases", *Chin. Med. J.,* vol. 131, no. 18, pp. 2216-2225, 2018.
[http://dx.doi.org/10.4103/0366-6999.240817] [PMID: 30203797]

[69]   A. Kalani, A. Tyagi, and N. Tyagi, "Exosomes: Mediators of neurodegeneration, neuroprotection and therapeutics", *Mol. Neurobiol.,* vol. 49, no. 1, pp. 590-600, 2014.
[http://dx.doi.org/10.1007/s12035-013-8544-1] [PMID: 23999871]

[70]   P. Leidinger, C. Backes, S. Deutscher, K. Schmitt, S.C. Mueller, K. Frese, J. Haas, K. Ruprecht, F. Paul, C. Stähler, C.J.G. Lang, B. Meder, T. Bartfai, E. Meese, and A. Keller, "A blood based 12-miRNA signature of Alzheimer disease patients", *Genome Biol.,* vol. 14, no. 7, p. R78, 2013.
[http://dx.doi.org/10.1186/gb-2013-14-7-r78] [PMID: 23895045]

[71]   N. Tominaga, N. Kosaka, M. Ono, T. Katsuda, Y. Yoshioka, K. Tamura, J. Lötvall, H. Nakagama, and T. Ochiya, "Brain metastatic cancer cells release microRNA-181c-containing extracellular vesicles capable of destructing blood–brain barrier", *Nat. Commun.,* vol. 6, no. 1, p. 6716, 2015.
[http://dx.doi.org/10.1038/ncomms7716] [PMID: 25828099]

[72]   H.S. Domingues, A.M. Falcão, I. Mendes-Pinto, A.J. Salgado, and F.G. Teixeira, "Exosome circuitry during (De)(Re) myelination of the central nervous system", *Front. Cell Dev. Biol.,* vol. 8, p. 483, 2020.
[http://dx.doi.org/10.3389/fcell.2020.00483] [PMID: 32612996]

[73]   D. Fröhlich, W.P. Kuo, C. Frühbeis, J.J. Sun, C.M. Zehendner, H.J. Luhmann, S. Pinto, J. Toedling, J. Trotter, and E.M. Krämer-Albers, "Multifaceted effects of oligodendroglial exosomes on neurons: Impact on neuronal firing rate, signal transduction and gene regulation", *Philos. Trans. R. Soc. Lond. B Biol. Sci.,* vol. 369, no. 1652, p. 20130510, 2014.
[http://dx.doi.org/10.1098/rstb.2013.0510] [PMID: 25135971]

[74]   T. Yu, Y. Xu, M.A. Ahmad, R. Javed, H. Hagiwara, and X. Tian, "Exosomes as a promising therapeutic strategy for peripheral nerve injury", *Curr. Neuropharmacol.,* vol. 19, no. 12, pp. 2141-2151, 2021.
[http://dx.doi.org/10.2174/1570159X19666210203161559] [PMID: 33535957]

[75]   S. Rastogi, V. Sharma, P.S. Bharti, K. Rani, G.P. Modi, F. Nikolajeff, and S. Kumar, "The evolving landscape of exosomes in neurodegenerative diseases: Exosomes characteristics and a promising role in early diagnosis", *Int. J. Mol. Sci.,* vol. 22, no. 1, p. 440, 2021.
[http://dx.doi.org/10.3390/ijms22010440] [PMID: 33406804]

[76]   A.F. Hill, "Extracellular vesicles and neurodegenerative diseases", *J. Neurosci.,* vol. 39, no. 47, pp. 9269-9273, 2019.
[http://dx.doi.org/10.1523/JNEUROSCI.0147-18.2019] [PMID: 31748282]

[77]   M. Sardar Sinha, A. Ansell-Schultz, L. Civitelli, C. Hildesjö, M. Larsson, L. Lannfelt, M. Ingelsson, and M. Hallbeck, "Alzheimer's disease pathology propagation by exosomes containing toxic amyloid-beta oligomers", *Acta Neuropathol.,* vol. 136, no. 1, pp. 41-56, 2018.
[http://dx.doi.org/10.1007/s00401-018-1868-1] [PMID: 29934873]

[78]   L. Jiang, H. Dong, H. Cao, X. Ji, S. Luan, and J. Liu, "Exosomes in pathogenesis, diagnosis, and treatment of Alzheimer's disease", *Med. Sci. Monit.,* vol. 25, pp. 3329-3335, 2019.
[http://dx.doi.org/10.12659/MSM.914027] [PMID: 31056537]

[79]   E. Emmanouilidou, K. Melachroinou, T. Roumeliotis, S.D. Garbis, M. Ntzouni, L.H. Margaritis, L. Stefanis, and K. Vekrellis, "Cell-produced α-synuclein is secreted in a calcium-dependent manner by exosomes and impacts neuronal survival", *J. Neurosci.,* vol. 30, no. 20, pp. 6838-6851, 2010.
[http://dx.doi.org/10.1523/JNEUROSCI.5699-09.2010] [PMID: 20484626]

[80]   B. Fevrier, D. Vilette, F. Archer, D. Loew, W. Faigle, M. Vidal, H. Laude, and G. Raposo, "Cells release prions in association with exosomes", *Proc. Natl. Acad. Sci.,* vol. 101, no. 26, pp. 9683-9688, 2004.
[http://dx.doi.org/10.1073/pnas.0308413101] [PMID: 15210972]

[81]   A. Elsherbini, and E. Bieberich, "Ceramide and exosomes: A novel target in cancer biology and therapy", *Adv. Cancer Res.,* vol. 140, pp. 121-154, 2018.
[http://dx.doi.org/10.1016/bs.acr.2018.05.004] [PMID: 30060807]

[82]   L. Mashouri, H. Yousefi, A.R. Aref, A. Ahadi, F. Molaei, and S.K. Alahari, "Exosomes: Composition, biogenesis, and mechanisms in cancer metastasis and drug resistance", *Mol. Cancer,* vol. 18, no. 1, p. 75, 2019.
[http://dx.doi.org/10.1186/s12943-019-0991-5] [PMID: 30940145]

[83]   L. Zhang, and D. Yu, "Exosomes in cancer development, metastasis, and immunity. Biochimica et Biophysica Acta (BBA)", *Rev. Can.,* vol. 1871, no. 2, pp. 455-468, 2019.

[84]   A. Conigliaro, and C. Cicchini, "Exosome-mediated signaling in epithelial to mesenchymal transition and tumor progression", *J. Clin. Med.,* vol. 8, no. 1, p. 26, 2018.
[http://dx.doi.org/10.3390/jcm8010026] [PMID: 30591649]

[85]   M. Aga, G.L. Bentz, S. Raffa, M.R. Torrisi, S. Kondo, N. Wakisaka, T. Yoshizaki, J.S. Pagano, and J. Shackelford, "Exosomal HIF1α supports invasive potential of nasopharyngeal carcinoma-associated LMP1-positive exosomes", *Oncogene,* vol. 33, no. 37, pp. 4613-4622, 2014.
[http://dx.doi.org/10.1038/onc.2014.66] [PMID: 24662828]

[86]   Y. You, Y. Shan, J. Chen, H. Yue, B. You, S. Shi, X. Li, and X. Cao, "Matrix metalloproteinase 13–containing exosomes promote nasopharyngeal carcinoma metastasis", *Cancer Sci.,* vol. 106, no. 12, pp. 1669-1677, 2015.
[http://dx.doi.org/10.1111/cas.12818] [PMID: 26362844]

[87]   K. Stefanius, K. Servage, and K. Orth, "Exosomes in cancer development", *Curr. Opin. Genet. Dev.,* vol. 66, pp. 83-92, 2021.
[http://dx.doi.org/10.1016/j.gde.2020.12.018] [PMID: 33477017]

[88]   A. Khalyfa, I. Almendros, A. Gileles-Hillel, M. Akbarpour, W. Trzepizur, B. Mokhlesi, L. Huang, J. Andrade, R. Farré, and D. Gozal, "Circulating exosomes potentiate tumor malignant properties in a mouse model of chronic sleep fragmentation", *Oncotarget,* vol. 7, no. 34, pp. 54676-54690, 2016.

[http://dx.doi.org/10.18632/oncotarget.10578] [PMID: 27419627]

[89]   H. Gu, R. Ji, X. Zhang, M. Wang, W. Zhu, H. Qian, Y. Chen, P. Jiang, and W. Xu, "Exosomes derived from human mesenchymal stem cells promote gastric cancer cell growth and migration via the activation of the Akt pathway", *Mol. Med. Rep.,* vol. 14, no. 4, pp. 3452-3458, 2016.
[http://dx.doi.org/10.3892/mmr.2016.5625] [PMID: 27513187]

[90]   G.I. Lancaster, and M.A. Febbraio, "Exosome-dependent trafficking of HSP70", *J. Biol. Chem.,* vol. 280, no. 24, pp. 23349-23355, 2005.
[http://dx.doi.org/10.1074/jbc.M502017200] [PMID: 15826944]

[91]   A. Clayton, A. Turkes, H. Navabi, M.D. Mason, and Z. Tabi, "Induction of heat shock proteins in B-cell exosomes", *J. Cell Sci.,* vol. 118, no. 16, pp. 3631-3638, 2005.
[http://dx.doi.org/10.1242/jcs.02494] [PMID: 16046478]

[92]   S. Khan, J.M.S. Jutzy, J.R. Aspe, D.W. McGregor, J.W. Neidigh, and N.R. Wall, "Survivin is released from cancer cells *via* exosomes", *Apoptosis,* vol. 16, no. 1, pp. 1-12, 2011.
[http://dx.doi.org/10.1007/s10495-010-0534-4] [PMID: 20717727]

[93]   X. Zhang, X. Yuan, H. Shi, L. Wu, H. Qian, and W. Xu, "Exosomes in cancer: Small particle, big player", *J. Hematol. Oncol.,* vol. 8, no. 1, p. 83, 2015.
[http://dx.doi.org/10.1186/s13045-015-0181-x] [PMID: 26156517]

[94]   J. Lin, J. Li, B. Huang, J. Liu, X. Chen, X.M. Chen, Y.M. Xu, L.F. Huang, and X.Z. Wang, "Exosomes: Novel biomarkers for clinical diagnosis", *ScientificWorldJournal,* vol. 2015, pp. 1-8, 2015.
[http://dx.doi.org/10.1155/2015/657086] [PMID: 25695100]

[95]   V. Sundararajan, F.H. Sarkar, and T.S. Ramasamy, "Correction to: The versatile role of exosomes in cancer progression: Diagnostic and therapeutic implications", *Cell Oncol.,* vol. 41, no. 4, p. 463, 2018.
[http://dx.doi.org/10.1007/s13402-018-0396-2] [PMID: 30047093]

[96]   T.H. Cheung, and T.A. Rando, "Molecular regulation of stem cell quiescence", *Nat. Rev. Mol. Cell Biol.,* vol. 14, no. 6, pp. 329-340, 2013.
[http://dx.doi.org/10.1038/nrm3591] [PMID: 23698583]

[97]   M. Collado, M.A. Blasco, and M. Serrano, "Cellular senescence in cancer and aging", *Cell,* vol. 130, no. 2, pp. 223-233, 2007.
[http://dx.doi.org/10.1016/j.cell.2007.07.003] [PMID: 17662938]

[98]   P. Kahlem, B. Dörken, and C.A. Schmitt, "Cellular senescence in cancer treatment: Friend or foe?", *J. Clin. Invest.,* vol. 113, no. 2, pp. 169-174, 2004.
[http://dx.doi.org/10.1172/JCI20784] [PMID: 14722606]

[99]   Y. Zhou, Y. Zhang, H. Gong, S. Luo, and Y. Cui, "The role of exosomes and their applications in cancer", *Int. J. Mol. Sci.,* vol. 22, no. 22, p. 12204, 2021.
[http://dx.doi.org/10.3390/ijms222212204] [PMID: 34830085]

[100]   M. Wan, B. Ning, S. Spiegel, C.J. Lyon, and T.Y. Hu, "Tumor–derived exosomes (TDEs): How to avoid the sting in the tail", *Med. Res. Rev.,* vol. 40, no. 1, pp. 385-412, 2020.
[http://dx.doi.org/10.1002/med.21623] [PMID: 31318078]

[101]   L. Fabris, K. Sato, G. Alpini, and M. Strazzabosco, "The tumor microenvironment in cholangiocarcinoma progression", *Hepatology,* vol. 73, no. S1, suppl. 1, pp. 75-85, 2021.
[http://dx.doi.org/10.1002/hep.31410] [PMID: 32500550]

[102]   J.S. Sung, C.W. Kang, S. Kang, Y. Jang, Y.C. Chae, B.G. Kim, and N.H. Cho, "ITGB4-mediated metabolic reprogramming of cancer-associated fibroblasts", *Oncogene,* vol. 39, no. 3, pp. 664-676, 2020.
[http://dx.doi.org/10.1038/s41388-019-1014-0] [PMID: 31534187]

[103]   J. Gu, H. Qian, L. Shen, X. Zhang, W. Zhu, L. Huang, Y. Yan, F. Mao, C. Zhao, Y. Shi, and W. Xu, "Gastric cancer exosomes trigger differentiation of umbilical cord derived mesenchymal stem cells to

carcinoma-associated fibroblasts through TGF-β/Smad pathway", *PLoS One,* vol. 7, no. 12, p. e52465, 2012.
[http://dx.doi.org/10.1371/journal.pone.0052465] [PMID: 23285052]

[104] J. Webber, R. Steadman, M.D. Mason, Z. Tabi, and A. Clayton, "Cancer exosomes trigger fibroblast to myofibroblast differentiation", *Cancer Res.,* vol. 70, no. 23, pp. 9621-9630, 2010.
[http://dx.doi.org/10.1158/0008-5472.CAN-10-1722] [PMID: 21098712]

[105] J.P. Webber, L.K. Spary, A.J. Sanders, R. Chowdhury, W.G. Jiang, R. Steadman, J. Wymant, A.T. Jones, H. Kynaston, M.D. Mason, Z. Tabi, and A. Clayton, "Differentiation of tumour-promoting stromal myofibroblasts by cancer exosomes", *Oncogene,* vol. 34, no. 3, pp. 290-302, 2015.
[http://dx.doi.org/10.1038/onc.2013.560] [PMID: 24441045]

[106] C.R. Goulet, A. Champagne, G. Bernard, D. Vandal, S. Chabaud, F. Pouliot, and S. Bolduc, "Cancer-associated fibroblasts induce epithelial–mesenchymal transition of bladder cancer cells through paracrine IL-6 signalling", *BMC Cancer,* vol. 19, no. 1, p. 137, 2019.
[http://dx.doi.org/10.1186/s12885-019-5353-6] [PMID: 30744595]

[107] E. Donnarumma, D. Fiore, M. Nappa, G. Roscigno, A. Adamo, M. Iaboni, V. Russo, A. Affinito, I. Puoti, C. Quintavalle, A. Rienzo, S. Piscuoglio, R. Thomas, and G. Condorelli, "Cancer-associated fibroblasts release exosomal microRNAs that dictate an aggressive phenotype in breast cancer", *Oncotarget,* vol. 8, no. 12, pp. 19592-19608, 2017.
[http://dx.doi.org/10.18632/oncotarget.14752] [PMID: 28121625]

[108] V. Butin-Israeli, T.M. Bui, H.L. Wiesolek, L. Mascarenhas, J.J. Lee, L.C. Mehl, K.R. Knutson, S.A. Adam, R.D. Goldman, A. Beyder, L. Wiesmuller, S.B. Hanauer, and R. Sumagin, "Neutrophil: Induced genomic instability impedes resolution of inflammation and wound healing", *J. Clin. Invest.,* vol. 129, no. 2, pp. 712-726, 2019.
[http://dx.doi.org/10.1172/JCI122085] [PMID: 30640176]

[109] X. Zhang, S.A. Deeke, Z. Ning, A.E. Starr, J. Butcher, J. Li, J. Mayne, K. Cheng, B. Liao, L. Li, R. Singleton, D. Mack, A. Stintzi, and D. Figeys, "Metaproteomics reveals associations between microbiome and intestinal extracellular vesicle proteins in pediatric inflammatory bowel disease", *Nat. Commun.,* vol. 9, no. 1, p. 2873, 2018.
[http://dx.doi.org/10.1038/s41467-018-05357-4] [PMID: 30030445]

[110] J. Sadri Nahand, M. Moghoofei, A. Salmaninejad, Z. Bahmanpour, M. Karimzadeh, M. Nasiri, H.R. Mirzaei, M.H. Pourhanifeh, F. Bokharaei-Salim, H. Mirzaei, and M.R. Hamblin, "Pathogenic role of exosomes and microRNAs in HPV–mediated inflammation and cervical cancer: A review", *Int. J. Cancer,* vol. 146, no. 2, pp. 305-320, 2020.
[http://dx.doi.org/10.1002/ijc.32688] [PMID: 31566705]

[111] M. Katoh, "Therapeutics targeting angiogenesis: Genetics and epigenetics, extracellular miRNAs and signaling networks (Review)", *Int. J. Mol. Med.,* vol. 32, no. 4, pp. 763-767, 2013.
[http://dx.doi.org/10.3892/ijmm.2013.1444] [PMID: 23863927]

[112] Y. Liu, F. Luo, B. Wang, H. Li, Y. Xu, X. Liu, L. Shi, X. Lu, W. Xu, L. Lu, Y. Qin, Q. Xiang, and Q. Liu, "STAT3-regulated exosomal miR-21 promotes angiogenesis and is involved in neoplastic processes of transformed human bronchial epithelial cells", *Cancer Lett.,* vol. 370, no. 1, pp. 125-135, 2016.
[http://dx.doi.org/10.1016/j.canlet.2015.10.011] [PMID: 26525579]

[113] H. Schwarzenbach, and P.B. Gahan, "Exosomes in immune regulation", *Noncoding RNA,* vol. 7, no. 1, p. 4, 2021.
[http://dx.doi.org/10.3390/ncrna7010004] [PMID: 33435564]

[114] M. Mostafazadeh, N. Samadi, H. Kahroba, B. Baradaran, S. Haiaty, and M. Nouri, "Potential roles and prognostic significance of exosomes in cancer drug resistance", *Cell Biosci.,* vol. 11, no. 1, pp. 1-15, 2021.
[http://dx.doi.org/10.1186/s13578-020-00515-y] [PMID: 33407894]

[115] R. Koch, T. Aung, D. Vogel, B. Chapuy, D. Wenzel, S. Becker, U. Sinzig, V. Venkataramani, T. von Mach, R. Jacob, L. Truemper, and G.G. Wulf, "Nuclear trapping through inhibition of exosomal export by indomethacin increases cytostatic efficacy of doxorubicin and pixantrone", *Clin. Cancer Res.,* vol. 22, no. 2, pp. 395-404, 2016.
[http://dx.doi.org/10.1158/1078-0432.CCR-15-0577] [PMID: 26369630]

[116] K. Li, Y. Chen, A. Li, C. Tan, and X. Liu, "Exosomes play roles in sequential processes of tumor metastasis", *Int. J. Cancer,* vol. 144, no. 7, pp. 1486-1495, 2019.
[http://dx.doi.org/10.1002/ijc.31774] [PMID: 30155891]

[117] Y. Yang, C.W. Li, L.C. Chan, Y. Wei, J.M. Hsu, W. Xia, J.H. Cha, J. Hou, J.L. Hsu, L. Sun, and M.C. Hung, "Exosomal PD-L1 harbors active defense function to suppress T cell killing of breast cancer cells and promote tumor growth", *Cell Res.,* vol. 28, no. 8, pp. 862-864, 2018.
[http://dx.doi.org/10.1038/s41422-018-0060-4] [PMID: 29959401]

[118] P.A.J. Muller, and K.H. Vousden, "p53 mutations in cancer", *Nat. Cell Biol.,* vol. 15, no. 1, pp. 2-8, 2013.
[http://dx.doi.org/10.1038/ncb2641] [PMID: 23263379]

[119] X. Yu, T. Riley, and A.J. Levine, "The regulation of the endosomal compartment by p53 the tumor suppressor gene", *FEBS J.,* vol. 276, no. 8, pp. 2201-2212, 2009.
[http://dx.doi.org/10.1111/j.1742-4658.2009.06949.x] [PMID: 19302216]

[120] H. Zhao, L. Yang, J. Baddour, A. Achreja, V. Bernard, T. Moss, J.C. Marini, T. Tudawe, E.G. Seviour, F.A. San Lucas, H. Alvarez, S. Gupta, S.N. Maiti, L. Cooper, D. Peehl, P.T. Ram, A. Maitra, and D. Nagrath, "Tumor microenvironment derived exosomes pleiotropically modulate cancer cell metabolism", *eLife,* vol. 5, p. e10250, 2016.
[http://dx.doi.org/10.7554/eLife.10250] [PMID: 26920219]

[121] A.E. Massey, S. Malik, M. Sikander, K.A. Doxtater, M.K. Tripathi, S. Khan, M.M. Yallapu, M. Jaggi, S.C. Chauhan, and B.B. Hafeez, "Clinical implications of exosomes: Targeted drug delivery for cancer treatment", *Int. J. Mol. Sci.,* vol. 22, no. 10, p. 5278, 2021.
[http://dx.doi.org/10.3390/ijms22105278] [PMID: 34067896]

[122] D.A. Murphy, and S.A. Courtneidge, "The 'ins' and 'outs' of podosomes and invadopodia: Characteristics, formation and function. nature reviews|", *Mol. Cell. Biol.,* vol. 12, pp. 412-426, 2011.

[123] D. Hoshino, K.C. Kirkbride, K. Costello, E.S. Clark, S. Sinha, N. Grega-Larson, M.J. Tyska, and A.M. Weaver, "Exosome secretion is enhanced by invadopodia and drives invasive behavior", *Cell Rep.,* vol. 5, no. 5, pp. 1159-1168, 2013.
[http://dx.doi.org/10.1016/j.celrep.2013.10.050] [PMID: 24290760]

[124] R. Takahashi, M. Prieto-Vila, A. Hironaka, and T. Ochiya, "The role of extracellular vesicle microRNAs in cancer biology", *Clin. Chem. Lab. Med.,* vol. 55, no. 5, pp. 648-656, 2017.
[http://dx.doi.org/10.1515/cclm-2016-0708] [PMID: 28231055]

[125] W. Zhou, M.Y. Fong, Y. Min, G. Somlo, L. Liu, M.R. Palomares, Y. Yu, A. Chow, S.T.F. O'Connor, A.R. Chin, Y. Yen, Y. Wang, E.G. Marcusson, P. Chu, J. Wu, X. Wu, A.X. Li, Z. Li, H. Gao, X. Ren, M.P. Boldin, P.C. Lin, and S.E. Wang, "Cancer-secreted miR-105 destroys vascular endothelial barriers to promote metastasis", *Cancer Cell,* vol. 25, no. 4, pp. 501-515, 2014.
[http://dx.doi.org/10.1016/j.ccr.2014.03.007] [PMID: 24735924]

[126] M. Saitoh, "JB special review—cellular plasticity in epithelial homeostasis and diseases", *J. Biochem.,* vol. 164, no. 4, pp. 257-264, 2018.
[http://dx.doi.org/10.1093/jb/mvy047] [PMID: 29726955]

[127] I.J. Fidler, "The organ microenvironment and cancer metastasis", *Differentiation,* vol. 70, no. 9-10, pp. 498-505, 2002.
[http://dx.doi.org/10.1046/j.1432-0436.2002.700904.x] [PMID: 12492492]

[128]   L. Li, Z. Liu, and Q. Cheng, "Exosome plays an important role in the development of hepatocellular carcinoma", *Pathol. Res. Pract.*, vol. 215, no. 8, p. 152468, 2019.
[http://dx.doi.org/10.1016/j.prp.2019.152468] [PMID: 31171380]

[129]   K. Al-Nedawi, B. Meehan, and J. Rak, "Microvesicles: Messengers and mediators of tumor progression", *Cell Cycle*, vol. 8, no. 13, pp. 2014-2018, 2009.
[http://dx.doi.org/10.4161/cc.8.13.8988] [PMID: 19535896]

[130]   B. Costa-Silva, N.M. Aiello, A.J. Ocean, S. Singh, H. Zhang, B.K. Thakur, A. Becker, A. Hoshino, M.T. Mark, H. Molina, J. Xiang, T. Zhang, T.M. Theilen, G. García-Santos, C. Williams, Y. Ararso, Y. Huang, G. Rodrigues, T.L. Shen, K.J. Labori, I.M.B. Lothe, E.H. Kure, J. Hernandez, A. Doussot, S.H. Ebbesen, P.M. Grandgenett, M.A. Hollingsworth, M. Jain, K. Mallya, S.K. Batra, W.R. Jarnagin, R.E. Schwartz, I. Matei, H. Peinado, B.Z. Stanger, J. Bromberg, and D. Lyden, "Pancreatic cancer exosomes initiate pre-metastatic niche formation in the liver", *Nat. Cell Biol.*, vol. 17, no. 6, pp. 816-826, 2015.
[http://dx.doi.org/10.1038/ncb3169] [PMID: 25985394]

[131]   N. Syn, L. Wang, G. Sethi, J.P. Thiery, and B.C. Goh, "Exosome-mediated metastasis: From epithelial–mesenchymal transition to escape from immunosurveillance", *Trends Pharmacol. Sci.*, vol. 37, no. 7, pp. 606-617, 2016.
[http://dx.doi.org/10.1016/j.tips.2016.04.006] [PMID: 27157716]

[132]   T.H. Ung, H.J. Madsen, J.E. Hellwinkel, A.M. Lencioni, and M.W. Graner, "Exosome proteomics reveals transcriptional regulator proteins with potential to mediate downstream pathways", *Cancer Sci.*, vol. 105, no. 11, pp. 1384-1392, 2014.
[http://dx.doi.org/10.1111/cas.12534] [PMID: 25220623]

[133]   S. Kruger, Z.Y.A. Elmageed, D.H. Hawke, P.M. Wörner, D.A. Jansen, A.B. Abdel-Mageed, E.U. Alt, and R. Izadpanah, "Molecular characterization of exosome-like vesicles from breast cancer cells", *BMC Cancer*, vol. 14, no. 1, p. 44, 2014.
[http://dx.doi.org/10.1186/1471-2407-14-44] [PMID: 24468161]

[134]   T. Yoshizaki, S. Kondo, N. Wakisaka, S. Murono, K. Endo, H. Sugimoto, S. Nakanishi, A. Tsuji, and M. Ito, "Pathogenic role of Epstein–Barr virus latent membrane protein-1 in the development of nasopharyngeal carcinoma", *Cancer Lett.*, vol. 337, no. 1, pp. 1-7, 2013.
[http://dx.doi.org/10.1016/j.canlet.2013.05.018] [PMID: 23689138]

[135]   D.J. Cha, J.L. Franklin, Y. Dou, Q. Liu, J.N. Higginbotham, M. Demory Beckler, A.M. Weaver, K. Vickers, N. Prasad, S. Levy, B. Zhang, R.J. Coffey, and J.G. Patton, "KRAS-dependent sorting of miRNA to exosomes", *eLife*, vol. 4, p. e07197, 2015.
[http://dx.doi.org/10.7554/eLife.07197] [PMID: 26132860]

[136]   C.A. Franzen, R.H. Blackwell, V. Todorovic, K.A. Greco, K.E. Foreman, R.C. Flanigan, P.C. Kuo, and G.N. Gupta, "Urothelial cells undergo epithelial-to-mesenchymal transition after exposure to muscle invasive bladder cancer exosomes", *Oncogenesis*, vol. 4, no. 8, pp. e163-e163, 2015.
[http://dx.doi.org/10.1038/oncsis.2015.21] [PMID: 26280654]

[137]   D.K. Jeppesen, A. Nawrocki, S.G. Jensen, K. Thorsen, B. Whitehead, K.A. Howard, L. Dyrskjøt, T.F. Ørntoft, M.R. Larsen, and M.S. Ostenfeld, "Quantitative proteomics of fractionated membrane and lumen exosome proteins from isogenic metastatic and nonmetastatic bladder cancer cells reveal differential expression of EMT factors", *Proteomics*, vol. 14, no. 6, pp. 699-712, 2014.
[http://dx.doi.org/10.1002/pmic.201300452] [PMID: 24376083]

[138]   C. Escrevente, S. Keller, P. Altevogt, and J. Costa, "Interaction and uptake of exosomes by ovarian cancer cells", *BMC Cancer*, vol. 11, no. 1, p. 108, 2011.
[http://dx.doi.org/10.1186/1471-2407-11-108] [PMID: 21439085]

[139]   H. Peinado, M. Alečković, S. Lavotshkin, I. Matei, B. Costa-Silva, G. Moreno-Bueno, M. Hergueta-Redondo, C. Williams, G. García-Santos, C.M. Ghajar, A. Nitadori-Hoshino, C. Hoffman, K. Badal, B.A. Garcia, M.K. Callahan, J. Yuan, V.R. Martins, J. Skog, R.N. Kaplan, M.S. Brady, J.D. Wolchok,

P.B. Chapman, Y. Kang, J. Bromberg, and D. Lyden, "Melanoma exosomes educate bone marrow progenitor cells toward a pro-metastatic phenotype through MET", *Nat. Med.,* vol. 18, no. 6, pp. 883-891, 2012.
[http://dx.doi.org/10.1038/nm.2753] [PMID: 22635005]

[140]   J.L. Hood, R.S. San, and S.A. Wickline, "Exosomes released by melanoma cells prepare sentinel lymph nodes for tumor metastasis", *Cancer Res.,* vol. 71, no. 11, pp. 3792-3801, 2011.
[http://dx.doi.org/10.1158/0008-5472.CAN-10-4455] [PMID: 21478294]

[141]   A. Hoshino, B. Costa-Silva, T.L. Shen, G. Rodrigues, A. Hashimoto, M. Tesic Mark, H. Molina, S. Kohsaka, A. Di Giannatale, S. Ceder, S. Singh, C. Williams, N. Soplop, K. Uryu, L. Pharmer, T. King, L. Bojmar, A.E. Davies, Y. Ararso, T. Zhang, H. Zhang, J. Hernandez, J.M. Weiss, V.D. Dumont-Cole, K. Kramer, L.H. Wexler, A. Narendran, G.K. Schwartz, J.H. Healey, P. Sandstrom, K. Jørgen Labori, E.H. Kure, P.M. Grandgenett, M.A. Hollingsworth, M. de Sousa, S. Kaur, M. Jain, K. Mallya, S.K. Batra, W.R. Jarnagin, M.S. Brady, O. Fodstad, V. Muller, K. Pantel, A.J. Minn, M.J. Bissell, B.A. Garcia, Y. Kang, V.K. Rajasekhar, C.M. Ghajar, I. Matei, H. Peinado, J. Bromberg, and D. Lyden, "Tumour exosome integrins determine organotropic metastasis", *Nature,* vol. 527, no. 7578, pp. 329-335, 2015.
[http://dx.doi.org/10.1038/nature15756] [PMID: 26524530]

[142]   M.Y. Fong, W. Zhou, L. Liu, A.Y. Alontaga, M. Chandra, J. Ashby, A. Chow, S.T.F. O'Connor, S. Li, A.R. Chin, G. Somlo, M. Palomares, Z. Li, J.R. Tremblay, A. Tsuyada, G. Sun, M.A. Reid, X. Wu, P. Swiderski, X. Ren, Y. Shi, M. Kong, W. Zhong, Y. Chen, and S.E. Wang, "Breast-cancer-secreted miR-122 reprograms glucose metabolism in premetastatic niche to promote metastasis", *Nat. Cell Biol.,* vol. 17, no. 2, pp. 183-194, 2015.
[http://dx.doi.org/10.1038/ncb3094] [PMID: 25621950]

[143]   S.K. Gopal, D.W. Greening, E.G. Hanssen, H.J. Zhu, R.J. Simpson, and R.A. Mathias, "Oncogenic epithelial cell-derived exosomes containing Rac1 and PAK2 induce angiogenesis in recipient endothelial cells", *Oncotarget,* vol. 7, no. 15, pp. 19709-19722, 2016.
[http://dx.doi.org/10.18632/oncotarget.7573] [PMID: 26919098]

[144]   M. Zhou, J. Chen, L. Zhou, W. Chen, G. Ding, and L. Cao, "Pancreatic cancer derived exosomes regulate the expression of TLR4 in dendritic cells via miR-203", *Cell. Immunol.,* vol. 292, no. 1-2, pp. 65-69, 2014.
[http://dx.doi.org/10.1016/j.cellimm.2014.09.004] [PMID: 25290620]

[145]   L. Zitvogel, A. Regnault, A. Lozier, J. Wolfers, C. Flament, D. Tenza, P. Ricciardi-Castagnoli, G. Raposo, and S. Amigorena, "Eradication of established murine tumors using a novel cell-free vaccine: Dendritic cell derived exosomes", *Nat. Med.,* vol. 4, no. 5, pp. 594-600, 1998.
[http://dx.doi.org/10.1038/nm0598-594] [PMID: 9585234]

[146]   C. Théry, A. Regnault, J. Garin, J. Wolfers, L. Zitvogel, P. Ricciardi-Castagnoli, G. Raposo, and S. Amigorena, "Molecular characterization of dendritic cell-derived exosomes. Selective accumulation of the heat shock protein hsc73", *J. Cell Biol.,* vol. 147, no. 3, pp. 599-610, 1999.
[http://dx.doi.org/10.1083/jcb.147.3.599] [PMID: 10545503]

[147]   A. Aponte-López, E.M. Fuentes-Pananá, D. Cortes-Muñoz, and S. Muñoz-Cruz, "Mast cell, the neglected member of the tumor microenvironment: Role in breast Cancer", *J. Immunol. Res.,* vol. 2018, pp. 1-11, 2018.
[http://dx.doi.org/10.1155/2018/2584243] [PMID: 29651440]

[148]   D. Skokos, H.G. Botros, C. Demeure, J. Morin, R. Peronet, G. Birkenmeier, S. Boudaly, and S. Mécheri, "Mast cell-derived exosomes induce phenotypic and functional maturation of dendritic cells and elicit specific immune responses *in vivo*", *J. Immunol.,* vol. 170, no. 6, pp. 3037-3045, 2003.
[http://dx.doi.org/10.4049/jimmunol.170.6.3037] [PMID: 12626558]

[149]   S. Ye, Z.L. Li, D. Luo, B. Huang, Y.S. Chen, X. Zhang, J. Cui, Y. Zeng, and J. Li, "Tumor-derived exosomes promote tumor progression and T-cell dysfunction through the regulation of enriched exosomal microRNAs in human nasopharyngeal carcinoma", *Oncotarget,* vol. 5, no. 14, pp. 5439-

5452, 2014.
[http://dx.doi.org/10.18632/oncotarget.2118] [PMID: 24978137]

[150] Z. Cai, F. Yang, L. Yu, Z. Yu, L. Jiang, Q. Wang, Y. Yang, L. Wang, X. Cao, and J. Wang, "Activated
      T cell exosomes promote tumor invasion via Fas signaling pathway", *J. Immunol.,* vol. 188, no. 12, pp.
      5954-5961, 2012.
      [http://dx.doi.org/10.4049/jimmunol.1103466] [PMID: 22573809]

[151] L. Mincheva-Nilsson, and V. Baranov, "Cancer exosomes and NKG2D receptor–ligand interactions:
      Impairing NKG2D-mediated cytotoxicity and anti-tumour immune surveillance", *Semin. Cancer Biol.,*
      vol. 28, pp. 24-30, 2014.
      [http://dx.doi.org/10.1016/j.semcancer.2014.02.010] [PMID: 24602822]

[152] L. Zhu, S. Kalimuthu, P. Gangadaran, J.M. Oh, H.W. Lee, S.H. Baek, S.Y. Jeong, S.W. Lee, J. Lee,
      and B.C. Ahn, "Exosomes derived from natural killer cells exert therapeutic effect in melanoma",
      Theranostics, vol. 7, no. 10, pp. 2732-2745, 2017.
      [http://dx.doi.org/10.7150/thno.18752] [PMID: 28819459]

[153] Y.J. Piao, H.S. Kim, E.H. Hwang, J. Woo, M. Zhang, and W.K. Moon, "Breast cancer cell-derived
      exosomes and macrophage polarization are associated with lymph node metastasis", *Oncotarget,* vol.
      9, no. 7, pp. 7398-7410, 2018.
      [http://dx.doi.org/10.18632/oncotarget.23238] [PMID: 29484119]

[154] X. Chen, X. Ying, X. Wang, X. Wu, Q. Zhu, and X. Wang, "Exosomes derived from hypoxic
      epithelial ovarian cancer deliver microRNA-940 to induce macrophage M2 polarization", *Oncol. Rep.,*
      vol. 38, no. 1, pp. 522-528, 2017.
      [http://dx.doi.org/10.3892/or.2017.5697] [PMID: 28586039]

[155] J. Lan, L. Sun, F. Xu, L. Liu, F. Hu, D. Song, Z. Hou, W. Wu, X. Luo, J. Wang, X. Yuan, J. Hu, and
      G. Wang, "M2 macrophage-derived exosomes promote cell migration and invasion in colon cancer",
      Cancer Res., vol. 79, no. 1, pp. 146-158, 2019.
      [http://dx.doi.org/10.1158/0008-5472.CAN-18-0014] [PMID: 30401711]

[156] X. Zhang, Z. Pei, J. Chen, C. Ji, J. Xu, X. Zhang, and J. Wang, "Exosomes for immunoregulation and
      therapeutic intervention in cancer", *J. Cancer,* vol. 7, no. 9, pp. 1081-1087, 2016.
      [http://dx.doi.org/10.7150/jca.14866] [PMID: 27326251]

[157] H. Mirzaei, A. Sahebkar, M.R. Jaafari, M. Goodarzi, and H.R. Mirzaei, "Diagnostic and therapeutic
      potential of exosomes in cancer: The beginning of a new tale?", *J. Cell. Physiol.,* vol. 232, no. 12, pp.
      3251-3260, 2017.
      [http://dx.doi.org/10.1002/jcp.25739] [PMID: 27966794]

[158] K.O. Alfarouk, C.M. Stock, S. Taylor, M. Walsh, A.K. Muddathir, D. Verduzco, A.H.H. Bashir, O.Y.
      Mohammed, G.O. Elhassan, S. Harguindey, S.J. Reshkin, M.E. Ibrahim, and C. Rauch, "Resistance to
      cancer chemotherapy: Failure in drug response from ADME to P-gp", *Cancer Cell Int.,* vol. 15, no. 1,
      p. 71, 2015.
      [http://dx.doi.org/10.1186/s12935-015-0221-1] [PMID: 26180516]

[159] L. Guo, and N. Guo, "Exosomes: Potent regulators of tumor malignancy and potential bio-tools in
      clinical application", *Crit. Rev. Oncol. Hematol.,* vol. 95, no. 3, pp. 346-358, 2015.
      [http://dx.doi.org/10.1016/j.critrevonc.2015.04.002] [PMID: 25982702]

[160] K. Shedden, X.T. Xie, P. Chandaroy, Y.T. Chang, and G.R. Rosania, "Expulsion of small molecules in
      vesicles shed by cancer cells: Association with gene expression and chemosensitivity profiles", *Cancer
      Res.,* vol. 63, no. 15, pp. 4331-4337, 2003.
      [PMID: 12907600]

[161] C. Federici, F. Petrucci, S. Caimi, A. Cesolini, M. Logozzi, M. Borghi, S. D'Ilio, L. Lugini, N.
      Violante, T. Azzarito, C. Majorani, D. Brambilla, and S. Fais, "Exosome release and low pH belong to
      a framework of resistance of human melanoma cells to cisplatin", *PLoS One,* vol. 9, no. 2, p. e88193,
      2014.

[http://dx.doi.org/10.1371/journal.pone.0088193] [PMID: 24516610]

[162] M. Lv, X. Zhu, W. Chen, S. Zhong, Q. Hu, T. Ma, J. Zhang, L. Chen, J. Tang, and J. Zhao, "Exosomes mediate drug resistance transfer in MCF-7 breast cancer cells and a probable mechanism is delivery of P-glycoprotein", *Tumour Biol.,* vol. 35, no. 11, pp. 10773-10779, 2014.
[http://dx.doi.org/10.1007/s13277-014-2377-z] [PMID: 25077924]

[163] C.L. Au Yeung, N.N. Co, T. Tsuruga, T.L. Yeung, S.Y. Kwan, C.S. Leung, Y. Li, E.S. Lu, K. Kwan, K.K. Wong, R. Schmandt, K.H. Lu, and S.C. Mok, "Exosomal transfer of stroma-derived miR21 confers paclitaxel resistance in ovarian cancer cells through targeting APAF1", *Nat. Commun.,* vol. 7, no. 1, p. 11150, 2016.
[http://dx.doi.org/10.1038/ncomms11150] [PMID: 27021436]

[164] P.K. Lim, S.A. Bliss, S.A. Patel, M. Taborga, M.A. Dave, L.A. Gregory, S.J. Greco, M. Bryan, P.S. Patel, and P. Rameshwar, "Gap junction-mediated import of microRNA from bone marrow stromal cells can elicit cell cycle quiescence in breast cancer cells", *Cancer Res.,* vol. 71, no. 5, pp. 1550-1560, 2011.
[http://dx.doi.org/10.1158/0008-5472.CAN-10-2372] [PMID: 21343399]

[165] M. Ono, N. Kosaka, N. Tominaga, Y. Yoshioka, F. Takeshita, R. Takahashi, M. Yoshida, H. Tsuda, K. Tamura, and T. Ochiya, "Exosomes from bone marrow mesenchymal stem cells contain a microRNA that promotes dormancy in metastatic breast cancer cells", *Sci. Signal.,* vol. 7, no. 332, pp. ra63-ra63, 2014.
[http://dx.doi.org/10.1126/scisignal.2005231] [PMID: 24985346]

[166] C. Battke, R. Ruiss, U. Welsch, P. Wimberger, S. Lang, S. Jochum, and R. Zeidler, "Tumour exosomes inhibit binding of tumour-reactive antibodies to tumour cells and reduce ADCC", *Cancer Immunol. Immunother.,* vol. 60, no. 5, pp. 639-648, 2011.
[http://dx.doi.org/10.1007/s00262-011-0979-5] [PMID: 21293856]

[167] V. Ciravolo, V. Huber, G.C. Ghedini, E. Venturelli, F. Bianchi, M. Campiglio, D. Morelli, A. Villa, P.D. Mina, S. Menard, P. Filipazzi, L. Rivoltini, E. Tagliabue, and S.M. Pupa, "Potential role of HER2-overexpressing exosomes in countering trastuzumab-based therapy", *J. Cell. Physiol.,* vol. 227, no. 2, pp. 658-667, 2012.
[http://dx.doi.org/10.1002/jcp.22773] [PMID: 21465472]

[168] D.D. Taylor, and C. Gercel-Taylor, "RETRACTED: MicroRNA signatures of tumor-derived exosomes as diagnostic biomarkers of ovarian cancer", *Gynecol. Oncol.,* vol. 110, no. 1, pp. 13-21, 2008.
[http://dx.doi.org/10.1016/j.ygyno.2008.04.033] [PMID: 18589210]

[169] J. Lea, R. Sharma, F. Yang, H. Zhu, E.S. Ward, and A.J. Schroit, "Detection of phosphatidylserine-positive exosomes as a diagnostic marker for ovarian malignancies: A proof of concept study", *Oncotarget,* vol. 8, no. 9, pp. 14395-14407, 2017.
[http://dx.doi.org/10.18632/oncotarget.14795] [PMID: 28122335]

[170] S. Shen, Y. Song, B. Zhao, Y. Xu, X. Ren, Y. Zhou, and Q. Sun, "Cancer-derived exosomal miR-7641 promotes breast cancer progression and metastasis", *Cell Commun. Signal.,* vol. 19, no. 1, p. 20, 2021.
[http://dx.doi.org/10.1186/s12964-020-00700-z] [PMID: 33618729]

[171] M. Logozzi, A. De Milito, L. Lugini, M. Borghi, L. Calabrò, M. Spada, M. Perdicchio, M.L. Marino, C. Federici, E. Iessi, D. Brambilla, G. Venturi, F. Lozupone, M. Santinami, V. Huber, M. Maio, L. Rivoltini, and S. Fais, "High levels of exosomes expressing CD63 and caveolin-1 in plasma of melanoma patients", *PLoS One,* vol. 4, no. 4, p. e5219, 2009.
[http://dx.doi.org/10.1371/journal.pone.0005219] [PMID: 19381331]

[172] T. Yamashita, H. Kamada, S. Kanasaki, Y. Maeda, K. Nagano, Y. Abe, M. Inoue, Y. Yoshioka, Y. Tsutsumi, S. Katayama, M. Inoue, and S. Tsunoda, "Epidermal growth factor receptor localized to exosome membranes as a possible biomarker for lung cancer diagnosis", *Pharmazie,* vol. 68, no. 12, pp. 969-973, 2013.
[PMID: 24400444]

[173] Q. Zheng, H. Ding, L. Wang, Y. Yan, Y. Wan, Y. Yi, L. Tao, and C. Zhu, "Circulating exosomal miR-96 as a novel biomarker for radioresistant non-small-cell lung cancer", *J. Oncol.,* vol. 2021, pp. 1-11, 2021.
[http://dx.doi.org/10.1155/2021/5893981] [PMID: 33727921]

[174] K. Asakura, T. Kadota, J. Matsuzaki, Y. Yoshida, Y. Yamamoto, K. Nakagawa, S. Takizawa, Y. Aoki, E. Nakamura, J. Miura, H. Sakamoto, K. Kato, S. Watanabe, and T. Ochiya, "A miRNA-based diagnostic model predicts resectable lung cancer in humans with high accuracy", *Commun. Biol.,* vol. 3, no. 1, p. 134, 2020.
[http://dx.doi.org/10.1038/s42003-020-0863-y] [PMID: 32193503]

[175] Y. Tanaka, H. Kamohara, K. Kinoshita, J. Kurashige, T. Ishimoto, M. Iwatsuki, M. Watanabe, and H. Baba, "Clinical impact of serum exosomal microRNA-21 as a clinical biomarker in human esophageal squamous cell carcinoma", *Cancer,* vol. 119, no. 6, pp. 1159-1167, 2013.
[http://dx.doi.org/10.1002/cncr.27895] [PMID: 23224754]

[176] M. Zhu, Z. Huang, D. Zhu, X. Zhou, X. Shan, L. Qi, L. Wu, W. Cheng, J. Zhu, L. Zhang, H. Zhang, Y. Chen, W. Zhu, T. Wang, and P. Liu, "A panel of microRNA signature in serum for colorectal cancer diagnosis", *Oncotarget,* vol. 8, no. 10, pp. 17081-17091, 2017.
[http://dx.doi.org/10.18632/oncotarget.15059] [PMID: 28177881]

[177] D.M. Smalley, N.E. Sheman, K. Nelson, and D. Theodorescu, "Isolation and identification of potential urinary microparticle biomarkers of bladder cancer", *J. Proteome Res.,* vol. 7, no. 5, pp. 2088-2096, 2008.
[http://dx.doi.org/10.1021/pr700775x] [PMID: 18373357]

[178] M.Y. Kim, H. Shin, H.W. Moon, Y.H. Park, J. Park, and J.Y. Lee, "Urinary exosomal microRNA profiling in intermediate-risk prostate cancer", *Sci. Rep.,* vol. 11, no. 1, p. 7355, 2021.
[http://dx.doi.org/10.1038/s41598-021-86785-z] [PMID: 33795765]

[179] Z. Li, L.X. Li, Y.J. Diao, J. Wang, Y. Ye, and X.K. Hao, "Identification of urinary exosomal miRNAs for the non-invasive diagnosis of prostate cancer", *Cancer Manag. Res.,* vol. 13, pp. 25-35, 2021.
[http://dx.doi.org/10.2147/CMAR.S272140] [PMID: 33442291]

[180] L. Rajendran, M. Honsho, T.R. Zahn, P. Keller, K.D. Geiger, P. Verkade, and K. Simons, "Alzheimer's disease β-amyloid peptides are released in association with exosomes", *Proc. Natl. Acad. Sci.,* vol. 103, no. 30, pp. 11172-11177, 2006.
[http://dx.doi.org/10.1073/pnas.0603838103] [PMID: 16837572]

[181] S. Saman, W. Kim, M. Raya, Y. Visnick, S. Miro, S. Saman, B. Jackson, A.C. McKee, V.E. Alvarez, N.C.Y. Lee, and G.F. Hall, "Exosome-associated tau is secreted in tauopathy models and is selectively phosphorylated in cerebrospinal fluid in early Alzheimer disease", *J. Biol. Chem.,* vol. 287, no. 6, pp. 3842-3849, 2012.
[http://dx.doi.org/10.1074/jbc.M111.277061] [PMID: 22057275]

[182] M. Guo, R. Li, L. Yang, Q. Zhu, M. Han, Z. Chen, F. Ruan, Y. Yuan, Z. Liu, B. Huang, M. Bai, H. Wang, C. Zhang, and C. Tang, "Evaluation of exosomal miRNAs as potential diagnostic biomarkers for acute myocardial infarction using next-generation sequencing", *Ann. Transl. Med.,* vol. 9, no. 3, p. 219, 2021.
[http://dx.doi.org/10.21037/atm-20-2337] [PMID: 33708846]

[183] M.L. Zheng, X.Y. Liu, R.J. Han, W. Yuan, K. Sun, J.C. Zhong, and X.C. Yang, "Circulating exosomal long non–coding RNAs in patients with acute myocardial infarction", *J. Cell. Mol. Med.,* vol. 24, no. 16, pp. 9388-9396, 2020.
[http://dx.doi.org/10.1111/jcmm.15589] [PMID: 32649009]

[184] H. Zhou, T. Pisitkun, A. Aponte, P.S.T. Yuen, J.D. Hoffert, H. Yasuda, X. Hu, L. Chawla, R.F. Shen, M.A. Knepper, and R.A. Star, "Exosomal Fetuin-A identified by proteomics: A novel urinary biomarker for detecting acute kidney injury", *Kidney Int.,* vol. 70, no. 10, pp. 1847-1857, 2006.
[http://dx.doi.org/10.1038/sj.ki.5001874] [PMID: 17021608]

[185] H. Zhou, A. Cheruvanky, X. Hu, T. Matsumoto, N. Hiramatsu, M.E. Cho, A. Berger, A. Leelahavanichkul, K. Doi, L.S. Chawla, G.G. Illei, J.B. Kopp, J.E. Balow, H.A. Austin III, P.S.T. Yuen, and R.A. Star, "Urinary exosomal transcription factors, a new class of biomarkers for renal disease", *Kidney Int.,* vol. 74, no. 5, pp. 613-621, 2008.
[http://dx.doi.org/10.1038/ki.2008.206] [PMID: 18509321]

[186] M.W. Welker, D. Reichert, S. Susser, C. Sarrazin, Y. Martinez, E. Herrmann, S. Zeuzem, A. Piiper, and B. Kronenberger, "Soluble serum CD81 is elevated in patients with chronic hepatitis C and correlates with alanine aminotransferase serum activity", *PLoS One,* vol. 7, no. 2, p. e30796, 2012.
[http://dx.doi.org/10.1371/journal.pone.0030796] [PMID: 22355327]

[187] K. Matsubara, Y. Matsubara, Y. Uchikura, and T. Sugiyama, "Pathophysiology of preeclampsia: The role of exosomes", *Int. J. Mol. Sci.,* vol. 22, no. 5, p. 2572, 2021.
[http://dx.doi.org/10.3390/ijms22052572] [PMID: 33806480]

[188] H. Yang, Q. Ma, Y. Wang, and Z. Tang, "Clinical application of exosomes and circulating microRNAs in the diagnosis of pregnancy complications and foetal abnormalities", *J. Transl. Med.,* vol. 18, no. 1, p. 32, 2020.
[http://dx.doi.org/10.1186/s12967-020-02227-w] [PMID: 31969163]

[189] I.M. Chung, G. Rajakumar, B. Venkidasamy, U. Subramanian, and M. Thiruvengadam, "Exosomes: Current use and future applications", *Clin. Chim. Acta,* vol. 500, pp. 226-232, 2020.
[http://dx.doi.org/10.1016/j.cca.2019.10.022] [PMID: 31678573]

[190] T. Yang, P. Martin, B. Fogarty, A. Brown, K. Schurman, R. Phipps, V.P. Yin, P. Lockman, and S. Bai, "Exosome delivered anticancer drugs across the blood-brain barrier for brain cancer therapy in Danio rerio", *Pharm. Res.,* vol. 32, no. 6, pp. 2003-2014, 2015.
[http://dx.doi.org/10.1007/s11095-014-1593-y] [PMID: 25609010]

[191] A. Familtseva, N. Jeremic, and S.C. Tyagi, "Exosomes: Cell-created drug delivery systems", *Mol. Cell. Biochem.,* vol. 459, no. 1-2, pp. 1-6, 2019.
[http://dx.doi.org/10.1007/s11010-019-03545-4] [PMID: 31073888]

[192] F. Aqil, H. Kausar, A.K. Agrawal, J. Jeyabalan, A.H. Kyakulaga, R. Munagala, and R. Gupta, "Exosomal formulation enhances therapeutic response of celastrol against lung cancer", *Exp. Mol. Pathol.,* vol. 101, no. 1, pp. 12-21, 2016.
[http://dx.doi.org/10.1016/j.yexmp.2016.05.013] [PMID: 27235383]

[193] S.M. Kim, Y. Yang, S.J. Oh, Y. Hong, M. Seo, and M. Jang, "Cancer-derived exosomes as a delivery platform of CRISPR/Cas9 confer cancer cell tropism-dependent targeting", *J. Control. Rel.,* vol. 266, pp. 8-16, 2017.
[http://dx.doi.org/10.1016/j.jconrel.2017.09.013] [PMID: 28916446]

[194] Y.K. Kim, Y. Choi, G.H. Nam, and I.S. Kim, "Functionalized exosome harboring bioactive molecules for cancer therapy", *Cancer Lett.,* vol. 489, pp. 155-162, 2020.
[http://dx.doi.org/10.1016/j.canlet.2020.05.036] [PMID: 32623071]

[195] U. Kosgodage, R. Trindade, P. Thompson, J. Inal, and S. Lange, "Chloramidine/bisindolylmaleimide-I-mediated inhibition of exosome and microvesicle release and enhanced efficacy of cancer chemotherapy", *Int. J. Mol. Sci.,* vol. 18, no. 5, p. 1007, 2017.
[http://dx.doi.org/10.3390/ijms18051007] [PMID: 28486412]

[196] S. Bai, W. Hou, Y. Yao, J. Meng, Y. Wei, F. Hu, X. Hu, J. Wu, N. Zhang, R. Xu, F. Tian, B. Wang, H. Liao, Y. Du, H. Fang, W. He, Y. Liu, B. Shen, and J. Du, "Exocyst controls exosome biogenesis via Rab11a", *Mol. Ther. Nucleic Acids,* vol. 27, pp. 535-546, 2022.
[http://dx.doi.org/10.1016/j.omtn.2021.12.023] [PMID: 35036064]

[197] S. Samanta, S. Rajasingh, N. Drosos, Z. Zhou, B. Dawn, and J. Rajasingh, "Exosomes: New molecular targets of diseases", *Acta Pharmacol. Sin.,* vol. 39, no. 4, pp. 501-513, 2018.
[http://dx.doi.org/10.1038/aps.2017.162] [PMID: 29219950]

[198] K. Al-Nedawi, B. Meehan, R.S. Kerbel, A.C. Allison, and J. Rak, "Endothelial expression of autocrine VEGF upon the uptake of tumor-derived microvesicles containing oncogenic EGFR", *Proc. Natl. Acad. Sci.,* vol. 106, no. 10, pp. 3794-3799, 2009.
[http://dx.doi.org/10.1073/pnas.0804543106] [PMID: 19234131]

[199] L.G. Lima, R. Chammas, R.Q. Monteiro, M.E.C. Moreira, and M.A. Barcinski, "Tumor-derived microvesicles modulate the establishment of metastatic melanoma in a phosphatidylserine-dependent manner", *Cancer Lett.,* vol. 283, no. 2, pp. 168-175, 2009.
[http://dx.doi.org/10.1016/j.canlet.2009.03.041] [PMID: 19401262]

[200] M.R. Lentz, "Continuous whole blood ultrapheresis procedure in patients with metastatic cancer", *J. Biol. Response Mod.,* vol. 8, no. 5, pp. 511-527, 1989.
[PMID: 2795094]

[201] A.M. Marleau, C.S. Chen, J.A. Joyce, and R.H. Tullis, "Exosome removal as a therapeutic adjuvant in cancer", *J. Transl. Med.,* vol. 10, no. 1, p. 134, 2012.
[http://dx.doi.org/10.1186/1479-5876-10-134] [PMID: 22738135]

[202] L. Qu, J. Ding, C. Chen, Z.J. Wu, B. Liu, Y. Gao, W. Chen, F. Liu, W. Sun, X.F. Li, X. Wang, Y. Wang, Z.Y. Xu, L. Gao, Q. Yang, B. Xu, Y.M. Li, Z.Y. Fang, Z.P. Xu, Y. Bao, D.S. Wu, X. Miao, H.Y. Sun, Y.H. Sun, H.Y. Wang, and L.H. Wang, "Exosome-transmitted lncARSR promotes sunitinib resistance in renal cancer by acting as a competing endogenous RNA", *Cancer Cell,* vol. 29, no. 5, pp. 653-668, 2016.
[http://dx.doi.org/10.1016/j.ccell.2016.03.004] [PMID: 27117758]

[203] S.A. Bliss, G. Sinha, O.A. Sandiford, L.M. Williams, D.J. Engelberth, K. Guiro, L.L. Isenalumhe, S.J. Greco, S. Ayer, M. Bryan, R. Kumar, N.M. Ponzio, and P. Rameshwar, "Mesenchymal stem cell–derived exosomes stimulate cycling quiescence and early breast cancer dormancy in bone marrow", *Cancer Res.,* vol. 76, no. 19, pp. 5832-5844, 2016.
[http://dx.doi.org/10.1158/0008-5472.CAN-16-1092] [PMID: 27569215]

[204] E.V. Batrakova, and M.S. Kim, "Using exosomes, naturally-equipped nanocarriers, for drug delivery", *J. Control. Release,* vol. 219, pp. 396-405, 2015.
[http://dx.doi.org/10.1016/j.jconrel.2015.07.030] [PMID: 26241750]

[205] J Colino, and CM Snapper, "Dendritic cell-derived exosomes express a streptococcus pneumoniae capsular polysaccharide type 14 cross-reactive antigen that induces protective immunoglobulin responses against pneumococcal infection in mice", *Infect. Immun.,* vol. 75, no. 1, pp. 220-230, 2007.

[206] C. Admyre, S.M. Johansson, K.R. Qazi, J.J. Filén, R. Lahesmaa, M. Norman, E.P.A. Neve, A. Scheynius, and S. Gabrielsson, "Exosomes with immune modulatory features are present in human breast milk", *J. Immunol.,* vol. 179, no. 3, pp. 1969-1978, 2007.
[http://dx.doi.org/10.4049/jimmunol.179.3.1969] [PMID: 17641064]

[207] M.I. Zonneveld, A.R. Brisson, M.J.C. van Herwijnen, S. Tan, C.H.A. van de Lest, F.A. Redegeld, J. Garssen, M.H.M. Wauben, and E.N.M. Nolte-'t Hoen, "Recovery of extracellular vesicles from human breast milk is influenced by sample collection and vesicle isolation procedures", *J. Extracell. Vesicles,* vol. 3, no. 1, p. 24215, 2014.
[http://dx.doi.org/10.3402/jev.v3.24215] [PMID: 25206958]

[208] H. Kim, E.H. Kim, G. Kwak, S.G. Chi, S.H. Kim, and Y. Yang, "Exosomes: Cell-derived nanoplatforms for the delivery of cancer therapeutics", *Int. J. Mol. Sci.,* vol. 22, no. 1, p. 14, 2020.
[http://dx.doi.org/10.3390/ijms22010014] [PMID: 33374978]

[209] L. Cheng, Y. Wang, and L. Huang, "Exosomes from M1-polarized macrophages potentiate the cancer vaccine by creating a pro-inflammatory microenvironment in the lymph node", *Mol. Ther.,* vol. 25, no. 7, pp. 1665-1675, 2017.
[http://dx.doi.org/10.1016/j.ymthe.2017.02.007] [PMID: 28284981]

[210] G. Kibria, E.K. Ramos, Y. Wan, D.R. Gius, and H. Liu, "Exosomes as a drug delivery system in cancer therapy: Potential and challenges", *Mol. Pharm.,* vol. 15, no. 9, pp. 3625-3633, 2018.

[http://dx.doi.org/10.1021/acs.molpharmaceut.8b00277] [PMID: 29771531]

[211] S. Viaud, S. Ploix, V. Lapierre, C. Théry, P.H. Commere, D. Tramalloni, K. Gorrichon, P. Virault-Rocroy, T. Tursz, O. Lantz, L. Zitvogel, and N. Chaput, "Updated technology to produce highly immunogenic dendritic cell-derived exosomes of clinical grade: A critical role of interferon-γ", *J. Immunother.,* vol. 34, no. 1, pp. 65-75, 2011.
[http://dx.doi.org/10.1097/CJI.0b013e3181fe535b] [PMID: 21150714]

CHAPTER 6

Biosynthesis and Function of Glycoconjugates

Elvan Bakar[1,*], **Nebiye Pelin Türker**[2] and **Zeynep Erim**[3]

[1] *Trakya University, Faculty of Pharmacy, Department of Basic Science, Edirne, Turkey*

[2] *Development Application and Research Center, Trakya University, Technology Research, 22030, Edirne, Türkiye*

[3] *Department of Biotechnology and Genetics, Trakya University, Institute of Natural and Applied Sciences, 22030, Edirne, Türkiye*

Abstract: Investigations to ascertain the physiological roles of carbohydrates in biological systems are being given more importance each day. Basically, carbohydrates are biomolecules with a wide range of biological functions, although they represent the primary energy source for metabolic processes. Carbohydrates are found as structural components in connective tissue in animal organisms. They also act as structural elements in both plant and bacterial cell walls. In the cell, they bind to lipids and proteins to form glycoconjugates called glycolipids, glycopeptides, glycoproteins and peptidoglycans. By binding to lipids and proteins on the cell surface, they perform as molecules that support intercellular adhesion and intercellular communication. Glycobiology is the science that investigates the structure, biosynthesis, and impacts of glycans on biological functions. In biology, glycoconjugates serve a variety of key roles. In mammalian cells, the majority of proteins are glycosylated, and this explains how proteins perform their various functions. In the future, these techniques will be crucial for the identification and treatment of specific diseases. The most major area of progress in glycobiology is the development of carbohydrate-based medicines.

Some diseases, including cancer, can be diagnosed *via* altered cell surface glycosylation pathways as a biomarker. Therefore, regulating glycosylation mechanisms and understanding the phenotypic characteristics of glycoconjugates are crucial steps in the design of novel strategies.

This chapter discusses the biosynthesis of glycoconjugates, their wide range of biological functions, and their significance for therapy.

Keywords: Biomarker, Biosynthesis, Carbohydrate, Cancer, Diagnosis, Function, Glycan, Glycosylation, Glycoconjugate, Glycoprotein, Glycolipid, Glycosidases, Membrane, *N*-Glycan, *O*-Glycan, Oligosaccharide, Protein, Proteoglycan, Transferases, Therapeutic Effects.

INTRODUCTION

Glycobiology studies the biosynthesis, structural and biological functions of glycoconjugates. It is an important sub-branch of biology that has been deve-

* **Corresponding author Elvan Bakar**: Trakya University, Faculty of Pharmacy, Department of Basic Science, Edirne, Turkey; Tel: +905327173787; E-mail: elvanbakar@trakya.edu.tr

Habibe Yılmaz (Ed.)

loping rapidly in the last 80 years. Glycans and oligosaccharides, which are covalently bonded to proteins and lipids, surround all cells. In complex multicellular organisms, the biological functions of these glycans are crucial for interactions between cells, tissues, and molecules. Depending on the biomolecules in the cell membrane, there are numerous distinct glycans Fig. (**1**). These glycans regulate cell signaling and cell-cell adhesion [1]. Glycans that are bound to proteins are also prevalent in cells' cytoplasm and nucleus. The sugar components of glycoconjugates also modulate various functions in physiological and pathophysiological processes in addition to forming structural properties [2].

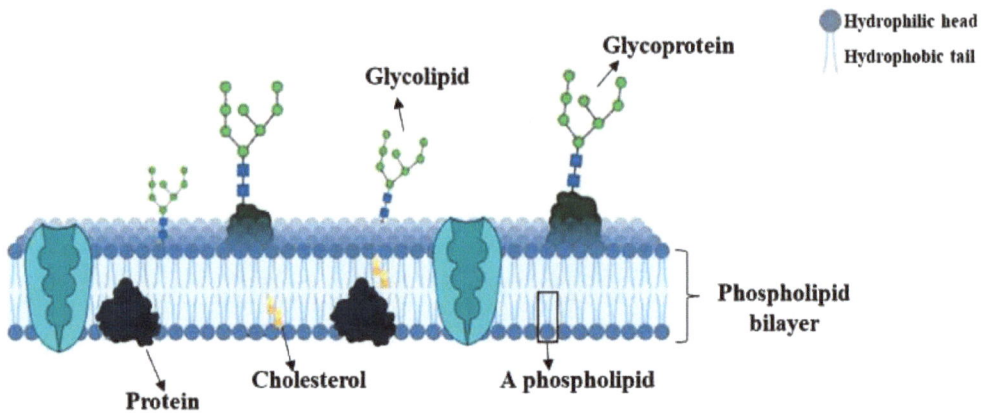

Fig. (**1**). Glycans on the cell surface.

The chemistry and metabolism of carbohydrates were among the most significant issues in the first part of the 20th century. However, it has lagged far behind other molecular biology studies due to the inability to predict the biosynthesis of glycans directly from a DNA template and its structural complexity [1, 3]. Glycobiology has been a broad field of research formed by the merger of many sciences, such as basic research, biotechnology, and biomedicine. This field includes the formation of glycans, the roles of glycans in complex biosystems, their chemistry and metabolism, enzymology and their analysis by various techniques. Therefore glycans glycobiology isn't based only on chemical synthesis, terminology, structure, biosynthesis, and function but also on cell biology, molecular genetics, medicine, physiology, and developmental biology, such as required by interdisciplinary work (3).

Monosaccharides are found in alpha (α) and beta (β) forms. Different connection types, positions, numbers, attachment points, and functional group differences of monosaccharides are important in polymer structures [4]. Each glycoform has a different function as a biological ligand. Many studies conducted in the field of glycobiology are on the determination of glycosylation mechanisms, elucidation of their molecular structure, investigation of biological control mechanisms in different cell types, and their use in therapy [5].

Complex carbohydrate polymers covalently bound to proteins and lipids serve as signals that determine the position and function of hybrid molecules, and these compounds are called glycoconjugates. Glycoconjugates are formed by a biochemical process called glycosylation. Glycosylation starts in the endoplasmic reticulum of eukaryotes and ends in the Golgi cistern. Glycoconjugates are also necessary for long-term immune protection. They participate in detoxification, cell-cell communication, and interactions with the cell matrix because of this [6, 7]. The most varied molecules in nature in terms of structure and function are glycoconjugates. Covalent bonds between carbohydrates and proteins and lipids result in three different forms of glycoconjugates. Proteoglycans function chemically like polysaccharides despite sharing the same types of linkages as glycoproteins. Moreover, mucins composed of glycoproteins and glycos-phingolipids are monosaccharide-modified glycoconjugates [7]. Some of them are structures on a single polypeptide with more than 100 distinct saccharide side chains, which are possibly the most complicated molecules in life [8]. In many studies using cationic dyes such as alcian blue and ruthenium red, it has been shown that almost all cells are surrounded by glycocalyx, known as carbohydrate sheath [6, 9]. Oligosaccharides and polysaccharides that are linked to proteins or lipids in the membrane make up glycocalyx. Even though erythrocytes are small eukaryotic cells, they feature a relatively extensive and complicated glycocalyx [10, 11]. This structure, which is present in all cells, consists of several glycoconjugates, which are quite complex. This structure is critical to cell biology as it specifically mediates and modulates interactions of the cell with small molecules, macromolecules, the extracellular matrix, and other cells. From this point of view, the glycocalyx is much more complex and selective in terms of its molecular interactions, although it has the physical properties of both ion exchange resins and gel filtration. In addition to serving as recognition molecules in multicellular interactions, saccharides of glycocalyx linked to proteins and lipids also act as binding sites for bacterial and viral pathogens [12, 13]. Laminin, collagen, and fibronectin, among other secreted glycoproteins, fill the gaps between eukaryotic cells. Additionally, proteoglycans and glycosaminoglycans contribute substantially to the extracellular matrix's structure. For instance, the three different types of corneal cells secrete proteoglycans and collagens that are highly ordered to preserve the tissue's uniqueness [14 - 18]. Likewise, the flexibility of cartilage is largely provided by the large amount of highly negatively charged proteoglycans and collagen synthesized by chondrocytes [19 - 22]. Extracellular matrix glycoconjugates play an important role not only in development and cellular interaction, but also in determining the physical properties of the matrix [14, 16, 18, 23]. All of these glycoconjugates exhibit developmentally dependent "glyco-types," which are cell-type-specific glycoforms. In addition to having different chain lengths and saccharide bonds, they also have various non-saccharide sub-structural elements. The elucidation of the structure and function of these macromolecules requires separation technologies with high resolution and precision [16, 17, 24].

Glycoconjugates located outside the cell membranes are formed by the covalent attachment of sugar to biomolecules such as lipids or proteins. They participate in specific recognition processes and cell-cell communication.

Together with oligosaccharides, nucleic acids, and proteins, they form the biopolymer. Proteins and nucleic acids are linear polymer structures. But the structures of oligosaccharides are branched [25, 26]. Each monosaccharide contains a number of hydroxyl groups, and they have two distinct stereoisomers (-α and -β). It has also come in two distinct ring configurations, pyranose and furanose. It follows that there are some methods to bind two monosaccharides and a lot of biological information can be contained in a relatively short saccharide Fig. (**2**) [27]. For instance, the three D-hexoses include more than 38,000 different trisaccharides [28]. Oligosaccharides have long been thought to act only as energy storage, as a physical defense, or to also alter the physical properties of lipids and proteins. In this day, it is understood that interactions between carbohydrates and proteins play a critical role in many biological processes, such as protein folding, immunological responses and microbial infections [25, 26, 29 - 31].

The production of nucleic acids and proteins follows defined steps, but the biosynthesis of oligosaccharides is extremely complex and is reliant on the presence of specific glycosylating enzymes in the cell as well as their capacity to glycosylate their substrates [32]. Because these enzyme reactions are not always complete, many oligosaccharides rather than a single isomer are produced. When two identical proteins are synthesized in the same cell, their glycosylation patterns may differ from each other. Cell can contain many glycoforms of a protein [33], and the glycan profile may varies depending on the kind of cell. These proteins are known as two glycoforms [32, 34, 35]. It was discovered in 1968 that the onset of cancer and inflammation causes alterations in the glycosylation pattern of cells [35, 36].

a)

α-D- galactopyranose   α-D- galactofuranose   β-D- galactopyranose

b)

β- Galabiose

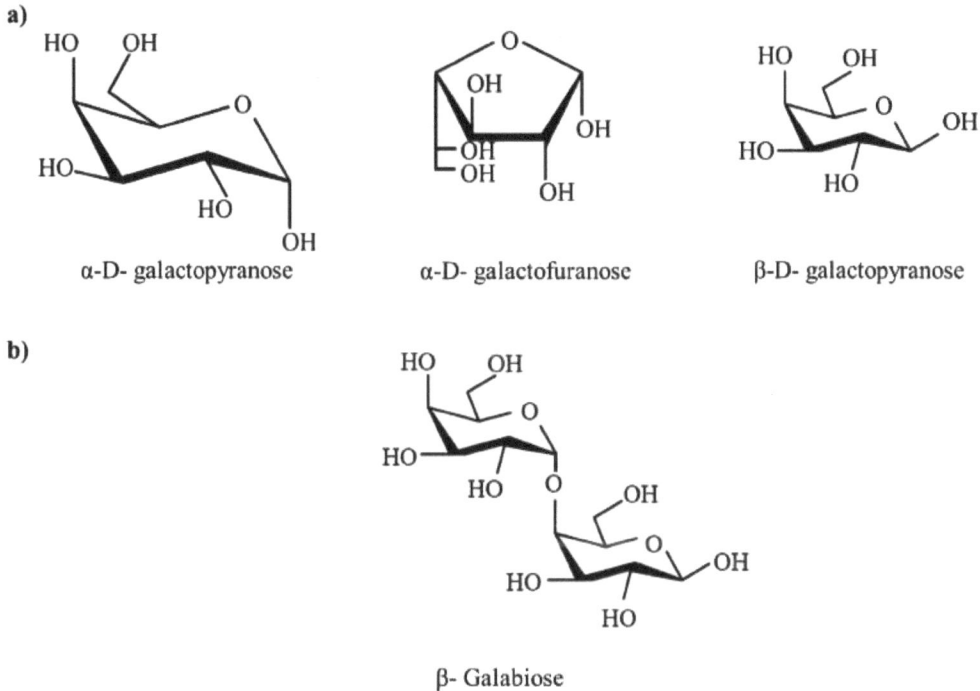

Fig. (2). Saccharides' combinatorial structure **a)** D-galactose isomers. **b)** β-Galabose.

These alterations aid in the diagnosis and treatment of diseases. There are instances of tumor immunotherapy using vaccines with glycoconjugates, as well as phase I and III clinical trials [32]. The biggest obstacle to treatment is that some glycans are endogenous, less frequent in healthy tissues, and present at earlier stages of embryogenesis. However, between cancer cells and healthy cells, specific glycans have different glycosylation patterns [32]. In addition, microbial pathogens have glycan, which is completely different from humans [37]. As a result of this difference, an immune response is formed. The code of information in oligosaccharides is usually decoded by proteins that recognize carbohydrates other than lectins, antibodies, and enzymes [38]. In addition, specific carbohydrate structures are recognized by antibodies [26] and T-cells [39]. The interaction between proteins and carbohydrates is most often a process [40, 41]. This phenomenon, called polyvalence, has several physiological advantages, such as [42] context-dependent interaction [26], fast on/off ratio [43], as well as resilience to shear stress [4, 40].

The types of glycoconjugates, their biosynthesis, biological function, and therapeutic effects of glycoconjugates will all be covered in the following sections.

GLYCOCONJUGATES

Glycoconjugates have a role in a lot of biological processes in living things' synthesis with glycosylation. Carbohydrates are covalently attached to amino acids proteins, lipids and other minor molecules to form glycoconjugates [44]. There are 17 different monosaccharides commonly found in mammals glycoconjugates [2]. Glycoconjugates were thought to be concentrated on the cellular surface, the interior of intracellular organelles, and secretory molecules until the middle of the 1980s. Today, it is understood that glycoconjugates exist throughout the cytoplasm and nucleus of cells in addition to the cell surface. In these regions, glycoconjugates play a role in processes like cytoskeletal structure maintenance [8, 45]. Changes in cellular glycosylation can help viruses escape the immune system, control inflammatory reactions and apoptosis, and encourage the metastasis of cancer cells. New treatments might be created based on the glycom's composition and function. In this way, immunological reactions and the inflammatory process can be altered, therapeutic antibodies' functions can be changed, and the likelihood of boosting immune responses against cancer can be increased [2]. While glycoconjugates such as glycopeptides, glycoproteins, proteoglycan, lipopolysaccarites and glycolipids are well-known examples, a new category called as glycoRNAs has been identified.

GlycoRNAs

Glycans modify lipids and proteins in viable cells to mediate intermolecular and intermolecular interactions. However, it is not believed that the primary target of glycosylation is RNA. It has been discovered that conserved short non-coding RNAs transport sialylated glycans using a variety of chemical and biological methods. These "glycoRNAs" can be found in cultured cells and a wide range of mammalian cell types. The glycoRNA Fig. (3) structure is formed using the standard *N*-glycan biosynthetic process [46]. The process of glycosylation results in structures that are fucose and sialic acid richer. According to an analysis of live cells, the bulk of glycoRNAs are found on the cell surface. GlycoRNAs can interact with anti-dsRNA antibodies and Siglec receptor family members in this area. These results show that RNA biology with glycobiology have a direct contact and that RNA has a broader role in extracellular biology [46 - 48].

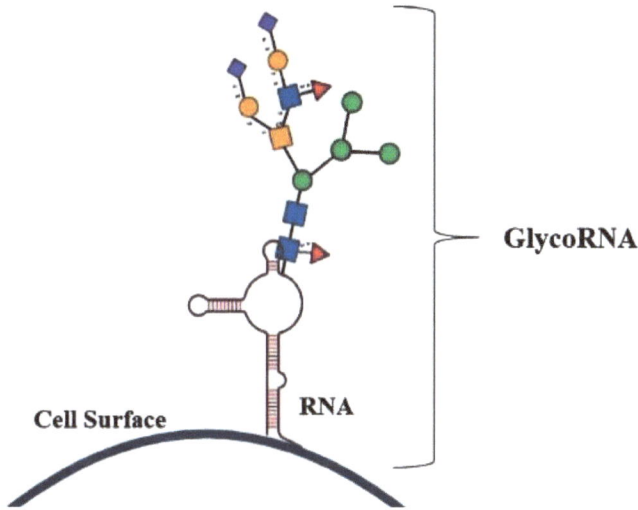

Fig. (3). GlycoRNA structure.

It is thought that glycoRNA found on the external surface of membrane has a role in signaling. It has been proposed that binding to glycolipids or glycoproteins is the basis for all interactions between members of the sialic acid binding-immunoglobulin lectin type (Siglec) receptor family on the cell surface. Siglecs, the largest family of sialoside-binding proteins in humans, are essential for a variety of diseases, including autoimmune disorders, host-pathogen interactions and cancer. According to research by Flynn *et al.,* glycoRNA facilitates contacts between two Siglec family members (-14 and -11) and the cell surface since these associations are sensitive to RNase treatment [46, 49].

GlycoRNA molecules on the cell surface have the potential to be the center of comprehensive studies and provide an opportunity to create various therapeutic approaches. These molecules have properties worth watching that are promising for the future.

Glycolipids

Glycoconjugates, in which the carbohydrate binds to a lipid, are called glycolipids. Glycolipids are glycoconjugates of lipids, which are usually located on the outer face of cell membranes and provide membrane stability and cell-cell interactions Fig. (4).

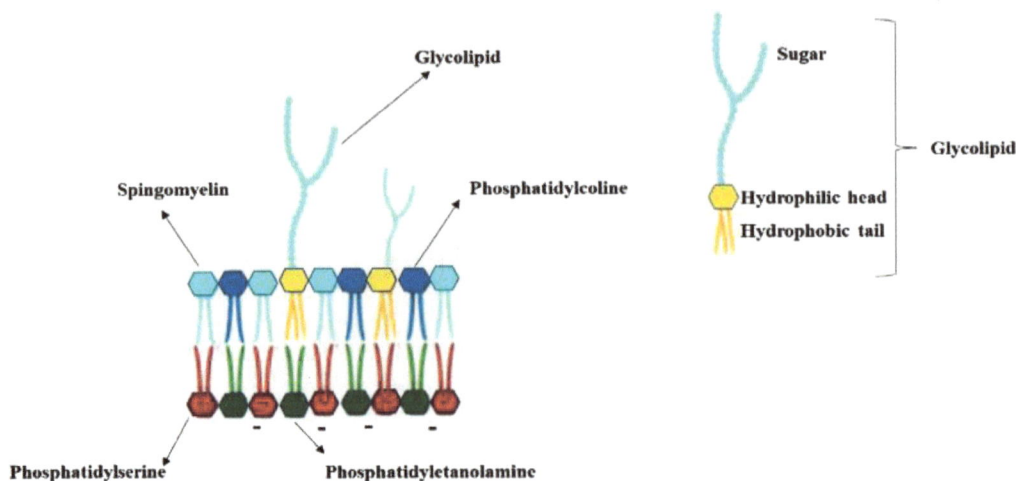

Fig. (4). Glycolipids on cell membrane.

The association of carbohydrates and lipids in the outer surface of the plasma membrane functions as receptors that mediate transmembrane signaling [50]. Despite being limited to the outer bilayer leaflet, membrane glycolipids function as the main receptors for carbohydrates-binding proteins to mediate transmembrane signaling [51]. In addition to acting as receptors, they are also very important for cell aggregation and decomposition. Nonpolar lipid and polar carbohydrate combinations make glycolipids both hydrophobic and hydrophilic [50]. The majority of glycolipids found in animals are glycosphingolipids (GSLs), in which a fatty acid joins a carbohydrate to an acylated sphingoid base Fig. (5) [50]. The extensive heterogeneity of lipid molecules has important, but not yet fully elucidated, roles in glycolipids' function. Glycosphingolipids have a role in managing a diverse range of biological processes [51]. There is little knowledge of the functional of specific GSLs. However, abnormalities in GSL physiology lead to some diseases. Among these diseases is GSL lysosomal storage diseases (LSDs) [52]. These diseases occur because of the accumulation of substrate GSL due to the inability of lysosomal carbohydrate hydrolases to play a role in GSL catabolism. These diseases are given in Table **1**.

a)

b)

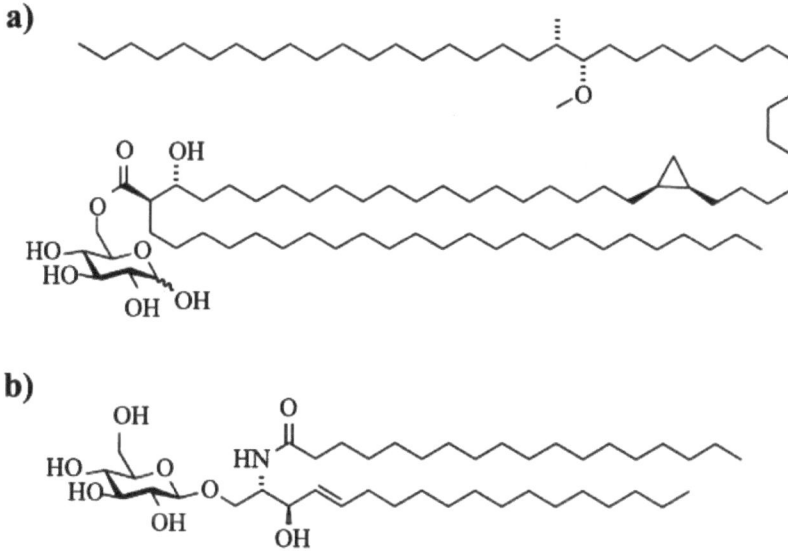

Fig. (5). a) β-Glucosyl-ceramide b) Glucose monomycolate IIa stereoisomer.

Table 1. Diseases Associated with GSL Deposition.

Diseases Associated with GSL Deposition	
Substrate GSL	**Associated Disease**
GM2 gangliosidosis	Sandhoff disease
	Tay-Sachs disease
Globotriaosyl ceramide	Fabry disease
Glucosyl ceramide	Gaucher disease
Galactosyl ceramide	Krabbe disease
Sulfogalactosyl ceramide	Multiple leukodystrophy

Located in the cell membrane, GSLs act as receptors for the GSL binding ligand and interact with other receptors to regulate signal transduction. Examples of fundamental receptor function are receptors to which bacterial toxins that cause infection in humans bind. These receptors include those for cholera toxin, Vero (Shiga) toxin, and *Escherichia coli* toxins [51]. Organisms such as bacteria and protozoa have a heterogeneous composition of both carbohydrate and lipid moieties Fig. (5) [53, 54]. Glycolipids have similar biological functions to other glycoconjugates. Microorganisms including human immunodeficiency virus (HIV) [55] and uropathogenic E.coli typically use carbohydrates as attachment points [56].

Gangliosides, consisting of a glycosphingolipid with one or more sialic acids attached to the sugar chain, are over expressed in neurons and other tissues. Abnormal GSL biosynthesis causes severe neuropathies [57].

Glycolipids and glycoproteins on the surface of cells interact with immune system cells and play a role in the body's immunological response. Glycoconjugates possess the ability to trigger these immunological reactions. The interaction between receptors and ligands underlies the immune system's effectiveness. Recent studies have demonstrated the role of glycolipid antigens in cell-mediated immunity. CD1 proteins, which resemble molecules from the major histocompatibility complex (MHC), are part of the glycolipid pathway and are responsible for supplying glycolipids to a fraction of T lymphocytes that are CD1-restricted.

Glycoproteins

Glycosylation is an extremely important biochemical process for cell physiology. Approximately half of the proteins in cells are in glycosylated form. Furthermore, abnormal glycosylation is associated with more than 50 diseases [58].

There are two ways that proteins can be glycosylated. When carbohydrates bind to a nitrogen atom at the Asn residues, N-linked glycans are created. However, O-linking glycans are formed by the attachment of carbohydrates to Ser or Thr residues of proteins *via* the oxygen atom [59].

In addition to C mannosylation of tryptophan residues, protein glycosylation furthermore involves the addition of N-linked glycans, O-linked glycans, phosphorylated glycans, glycosaminoglycans, and glycosylphosphatidylinositol (GPI) anchors to peptide backbones. N-glycans and O-glycans often have negatively charged sialic acids in the end region [60].

Cell membranes and serum glycoproteins both contain N-glycosidic linked oligosaccharide units. The amido group of the asparagine residues forms a bond with the N-linked chains through N-acetylglucosamine. The polypeptide chain's carbohydrate binding site is made up of the amino acid sequence Asn-X-Thr/Ser. The typical core structure of asparagine-connected carbohydrate chains is $Man_3GlcNAc_2$.

According to their characteristics, the three types of carbohydrates are found in glycoprotein structures. These include complex-type oligosaccharides, hybrid-type carbohydrates, and high-mannose oligosaccharides Fig. (**6**). High mannose oligosaccharides are those that include two to six additional mannose residues connected to the pentasaccharide core. Complex types of oligosaccharide chains two to four, and very rarely five, N-acetylglucosamine residues are formed by the attachment of mannose residues in the pentasaccharide core. These disaccharide units often have short N-acetygalactosamine side branches and are elongated with N-acetygalactosamine residues. In addition, α1→6 fucose is linked to the innermost N–acetylglucosamine residue in complex-type chains. Infrequently, terminal galactosyl groups also bear poly-N-acetylneuraminic acid chains (NeuAca2→8)n, NeuAca2→3 (n = 2-3) or α1→2 fucose, α1→3 galactose, or α2→3 attached N-acetylneuraminic acid residues. The typical substitution for subterminal N-acetylglucosamine is α1→3 fucose [61]. On the other hand, complex type oligosaccharide chains and high mannose oligosaccharide chains are attributes of hybrid carbohydrates.

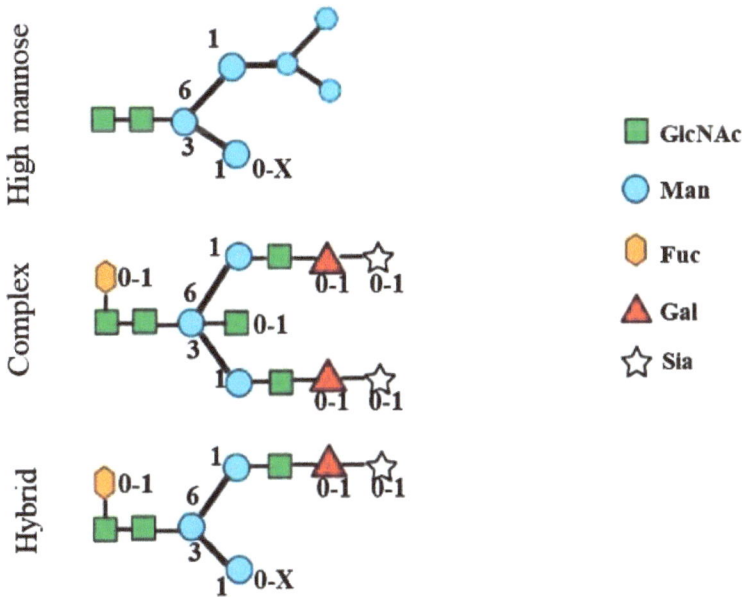

Fig. (6). Structures of the *N*-glycosidic linked oligosaccharide.

Many membrane proteins, mucins, proteoglycans and collagens are mainly *O*-linked glycoproteins. Mucins, which are exocrine gland secretions, contain *O*-glycosidic linked carbohydrate chains in smaller amounts than membrane glycoproteins. They bind to the hydroxyl groups of serine and threonine residues in proteins *via* -N-acetylgalactosamine [59].

NeuAca2→6GalNAc is a disaccharide discovered in submaxillary mucin. In contrast to *N*-linked oligosaccharide chains, *O*-linked units do not contain mannose. The principal carriers of *O*-linked chains are mucins, which are membrane-bound or released into the lumen of the respiratory, reproductive and digestive systems. Mucins are either found connected to the membrane or secreted into the lumen of the gastrointestinal, respiratory, and reproductive tracts and they are known as primary carriers of *O*-linked chains.

In epithelial tissue, there are various types of mucins that are heavily glycosylated. Mucins that are attached to membranes have a hydrophobic domain that covers the membrane. They function as cell adhesion molecules as well. To prevent mechanical harm to tissues, mucous layers' lubricating qualities are crucial [59].

Glycosylphosphatidylinositol (GPI) associated glycoproteins make up a fairly large class of glycoproteins. GPI-bound glycoproteins attach to the outer layer of the plasma membrane [6].

CD1 is a transmembrane glycoprotein found in the membrane of antigen-presenting T cells that contains 2-microglobulin. Antigen-presenting cells (APCs), which are specialized immune cells and play a critical role in the immune system, contain CD1 proteins, similar to MHC. T-cell recognition of glycolipids and lipids was later identified, despite the fact that it is generally known that MHC molecules convey peptide and protein antigens to T cells [62, 63]. CD1 molecules and the class of molecules that present an antigen with structural similarity to MHC class I molecules control this identification [64] hydrophobic groove on CD1 proteins is ideal for attaching lipids [65 - 70]. Depending on the quantity and length of lipid chains, distinct CD1 molecules exhibit varied binding motifs. Fig. (7) displays the four CD1 proteins (a-d) that were examined. Although each of these proteins can be linked by two lipid chains of about 15-30 carbons [71, 72], CD1c also presents single-chain glycolipids [73]. Mycolic acid, which has up to 80 carbons in one chain, is one example of a very long-chain lipid that CD1b can bind. Lipids with more than two chains may also be able to do the same [65]. In addition, recent research has shown that various CD1 isoforms can be used as a tool for determining the lipid contents of intracellular structures Fig. (7).

Proteoglycans

Proteoglycans (PG) are another type of glycoconjugate structure that is created by the association of glycans with proteins. Glycosaminoglycans (GAGs) bind to core proteins to form PGs. The GAGs involved in this structure are a broad family of polysaccharides [74]. GAGs are unbranched, linear, anionic polysaccharides that contain 10–200 repeating disaccharide units [75]. The differentiation and isomerization of sugar residues lead to the great molecular diversity of GAGs [74]. Additionally, selective oxidation, acetylation, phosphorylation, and sulfation can be used to further modify the variety of glycans, adding new information to the receptor [8]. Depending on the disaccharide units they includes, GAGs are divided into four groups: heparin/heparan sulfate (HP/HS), chondroitin/dermatan sulfate (CS/DS), keratan sulfate (KS), and hyaluronan (HA) [76]. While HA is found free in the extracellular matrix, CS/DS, HS, and KS are found bound to core proteins [74].

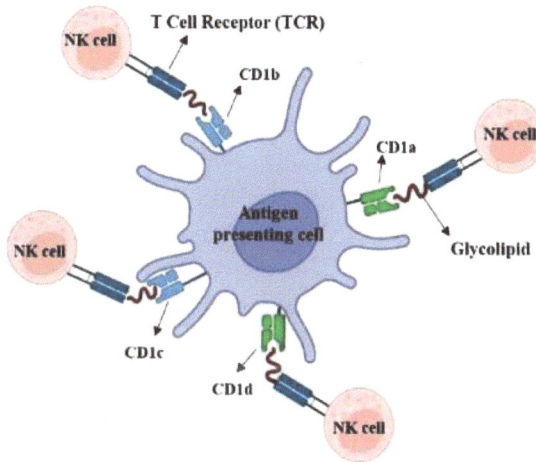

Fig. (7). CD1 protein types.

The various cellular activities of PGs are based on their microheterogeneity [77]. The interactions between PGs and cell surface signaling molecules play a structural role in tissue organization as well as a role in cell signaling [75]. The list of biological processes that PGs are involved in is expanding quickly.

PGs are molecules that are frequently present on cell surfaces, in the extracellular matrix (ECM), and in the basement membrane (BM). Unlike these locations, serglycine is the only proteoglycan species with intracellular localization and heparin side chains. These two factors make it a distinct class unto itself. Serglycine is stored in the granules of mast cells and packaging most intracellular proteases [77].

The diverse types of glycosidic bonds, sulfation patterns, monosaccharide types, differences in polymerization, and core protein variations that make up PGs which are thought to be the most complex and educational molecules in organisms are responsible for these characteristics [78]. Renato *et al.* provided significant information on the classification of proteoglycan gene families, using the general gene/protein homology, cellular and subcellular location, and the use of specific protein modules within the relevant protein core as criteria. These were defined, in order, proteoglycans that are intracellular, cell surface, pericellular, and extracellular [77]. By interacting with numerous extracellular signaling molecules, binding proteins, and enzymes, PGs are able to perform numerous physiological functions in cells, including damage response and development [75, 79, 80]. Due to tissue-specific expression levels and a tissue's enzyme specificities, the organization and distribution of PGs differ between cells and tissues [76, 81, 82]. The physiological and pathological situation of tissues can be determined in part by these differences. These biomolecules have distinguishing characteristics in the diagnosis and prognosis of numerous viral infections, cancer, neurodegenerative diseases, and metabolic diseases like atherosclerosis and diabetes.

Important data regarding the categorization of proteoglycan gene families was presented by Renato *et al.,* who used the cellular and subcellular location, general gene/protein homology, and the use of particular protein modules within the pertinent protein core as criteria. These are, in order, proteoglycans that are intracellular, cell surface, pericellular, and extracellular. PG's in these territory are presented in the format discussed by the researchers in Table **2** [77].

Table 2. PG's position in the organism.

Intracellular Proteoglycans	**1. Serglycin**
Cell Surface Proteoglycans	1. Syndecans 2. Chondroitin sulfate proteoglycan 4/nerve glial antigen 2 (CSPG4/NG2) 3. Betaglycan/TGFβ type III receptor 4. Phosphacan/receptor-type protein tyrosine phosphatase β 5. Glypicans/GPI-anchored proteoglycans
Pericellular and Basement Membrane Zone Proteoglycans	1. Perlecan 2. Agrin 3. Collagens XVIII and XV
Extracellular Proteoglycans	1. Hyaluronan and lectin binding proteoglycans (hyalectans) 2. Aggrecan 3. Versican 4. Neurocan and brevican 5. Small leucine-rich proteoglycans (SLRPs) 6. Testican/SPOCK family

Heparan sulfate proteoglycans (HSPGs) are commonly related to the cell surface or intercellular matrix. HSPGs interact with the cell membrane of cells either by binding directly to a protein core or by a glycosyl-phosphatidyl-inositol (GPI) anchor and act as biological modifiers for growth factors such as PDGF, VEGF and FGF. They perform analogous functions in the basement membrane region of the cell and are in contact with ECM members such as HSPGs, collagen type IV and various laminins in this region. Although the main function of pericellular HSPGs is to provide growth factors and receptor interaction, they also perform different functions in embryogenesis and regenerative processes.

Small leucine-rich proteoglycans (SLRPs) have both structural and signaling roles in tissue's remodeling in diseases like cancer, diabetes, inflammation, and atherosclerosis. They are found in large amounts in the extracellular matrix. They control crucial processes such as proliferation, innate immunity, migration, angiogenesis, autophagy and apoptosis, by interacting with receptor tyrosine kinases (RTKs) and Toll-like receptors [77].

A glycoconjugate, or proteoglycan, is a biomolecule having a central protein covalently joined to chains of glycosaminoglycan (GAG). Long, unbranched polysaccharides comprised of repeated disaccharide units are known as mucopolysaccharides or glycosaminoglycans. Proteoglycans are highly glycosylated glycoconjugates since they are made of polysaccharides. Using a trisaccharide bridge, the protein's serine (Ser) residues connect to the glycosaminoglycan chains as places of attachment. Under physiological settings, the GAG chains are negatively charged due to the sulfate and uronic acid groups on them. One of the primary elements of the extracellular matrix (ECM), which lines the area around cells, are proteoglycans (Fig. **8**).

Biosynthesis of Glycoconjugates

Glycoconjugate biosynthesis is a dynamic process that is influenced by the cell type, the amount of available enzymes in the cell, sugar precursors, organelle structures, and cellular signals [60].

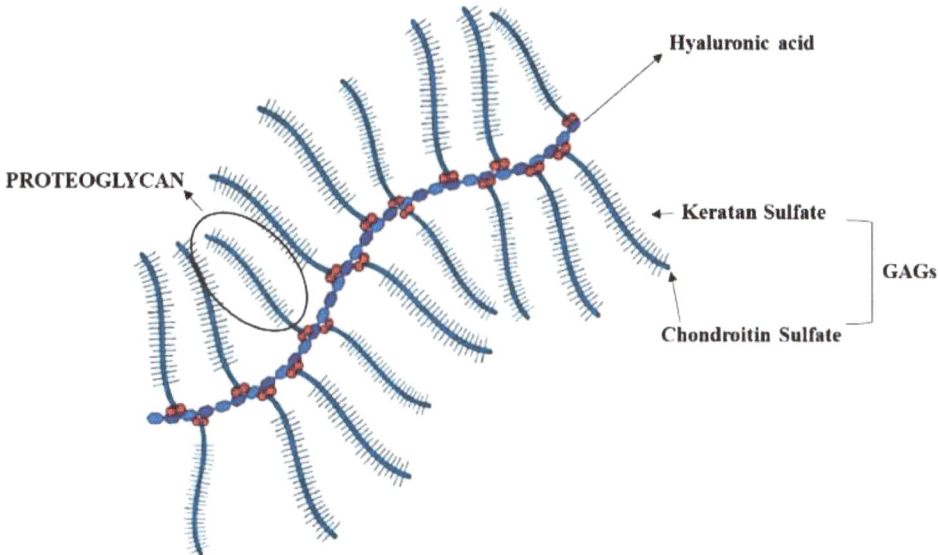

Fig. (8). Structure of a proteoglycan.

Glycoproteins, glycolipids, and proteoglycans are complex molecules synthesized by glycosylation. Glycosylation of proteins is only one of the post-translational modifications, and proteins thus acquire a wide variety of critical functions. The synthesis of *N*- and *O*-linked glycoproteins takes place in the lumen of the Endoplasmic reticulum (ER) and in the Golgi apperatus.The ER is the starting point for protein glycosylation processes, which continue in the cis-, medial-, and trans-Golgi compartments of the Golgi apparatus [83]. Regulation of glycosylation enzymes, gene transcription levels and the presence of substrates and enzymes in organelles play a role in the formation of carbohydrate structures in these organelles [60]. Lipid carriers are not used in the synthesis of *O*-linked glycoproteins; however, Dolicol, a lipid structure, and its phosphorylated derivative, Dolicol Pyrophosphate, are necessary for the synthesis of *N*-linked glycoproteins. The two stages of the *N*-glycosylation process are the assembly and transfer of the oligosaccharide P-P dolicol and the subsequent processing of the oligosaccharide chain. The synthesis of *O*-linked oligosaccharides takes place in the Golgi by the cascading addition of sugars from sugar nucleotides. Lipid carriers are not involved. In conclusion, the glycome is a reflection of the gene expression model that regulates the activity of the glycoconjugation enzymes, and unlike the proteome, the glycome is produced without the use of templates [60].

Glycoproteins are molecules made up of one or more carbohydrate chains that are covalently linked to polypeptide chains because of enzymatic activities. Galactose, Glucose, Mannose, N-Acetyl Neuraminic acid, Fucose, N-Acetyl galactosamine, N-Acetyl glucosamine, Xylose are carbohydrates added to the glycoprotein structure. N-Acetyl Neuraminic acid is attached to Galactose/N-Acetylgalactosamine at the end of the oligosaccharide chains, while the others are in the inner position [84].

The carbohydrates are attached to a polypeptide backbone. Carbohydrates are attached to proteins *via N*-glycosidic or *O*-glycosidic bonds Fig. (**9**).

An oligosaccharide, a carbohydrate made up of multiple sugar molecules, is attached to the amide nitrogen of an asparagine (Asn) residue of a protein to produce *N*-linked glycans [85]. *O*-linked glycans are created when a sugar molecule binds to a protein's serine or threonine residue's oxygen atom. The transfer of N-acetylgalactosamine onto serine/threonine in the cis Golgi cistern is the initial step in *O*-glycosylation [86]. When sugar molecule binds to the oxygen atom of a serine or threonine residue in protein, *O*-linked glycans are produced. The first step in *O*-glycosylation is the transfer of N-acetylgalactosamine onto serine/threonine in the cis Golgi cistern [86].

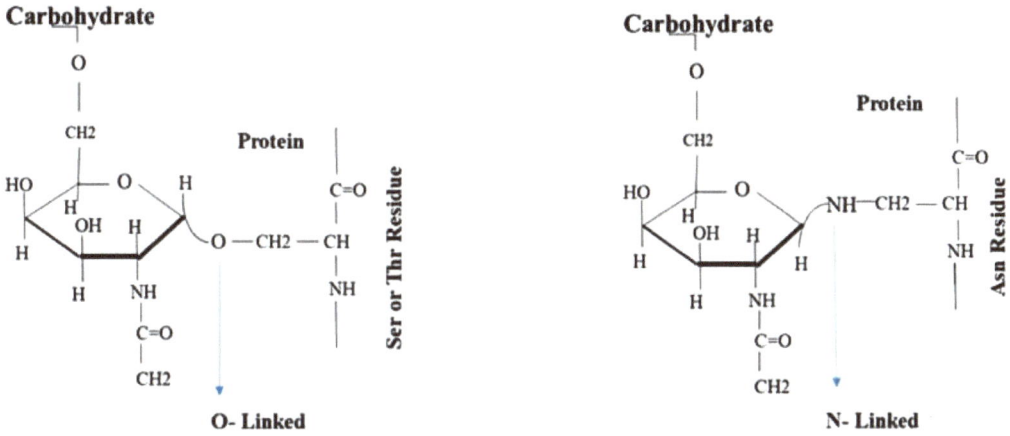

Fig. (9). Types of glycoprotein linkages.

In two distinct phases, the protein backbone's amino nitrogen and asparagine residues are used to bind the carbohydrate chains of glycoproteins Fig. (**10**).

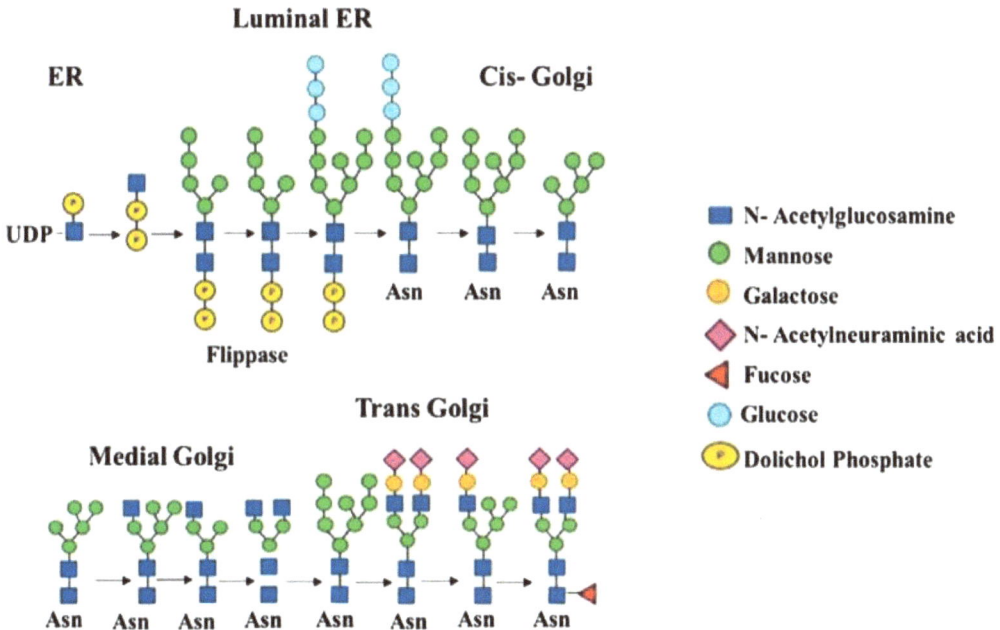

Fig. (10). Oligosaccharide processing steps in the Endoplasmic reticulum and Golgi apparatus.

In the first phase, N-acetylglucosamine, glucose, and mannose residues are transferred onto dolichol pyrophosphate as an intermediate carrier molecule, resulting in the formation of a complex branching oligosaccharide unit in the endoplasmic reticulum membrane. The oligosaccharide that results from this process has Glc3Man5GlcNAc2 on average Fig. (**11**). During the second stage, the synthesized oligosaccharide unit is transferred from the lipid molecule to the protein. The usual amino acid sequences for the polypeptide chain's glycosyl acceptor site are Asn-X-Thr and Asn-X-Ser [85, 87].

N-linked glycoprotein biosynthesis follows the following steps, respectively;

a) Creation of a mannose- and glucose-rich oligosaccharide on a phosphoryl dolichol intermediate,

b) The oligosaccharide is transferred from the phosphoryl dolichol to the protein acceptor,

c) The removal of particular, terminal glucose and mannose residues by glycosidases from protein-bound oligosaccharide,

d) The oligosaccharide is lengthened by the sequential actions of particular glycosyltransferases.

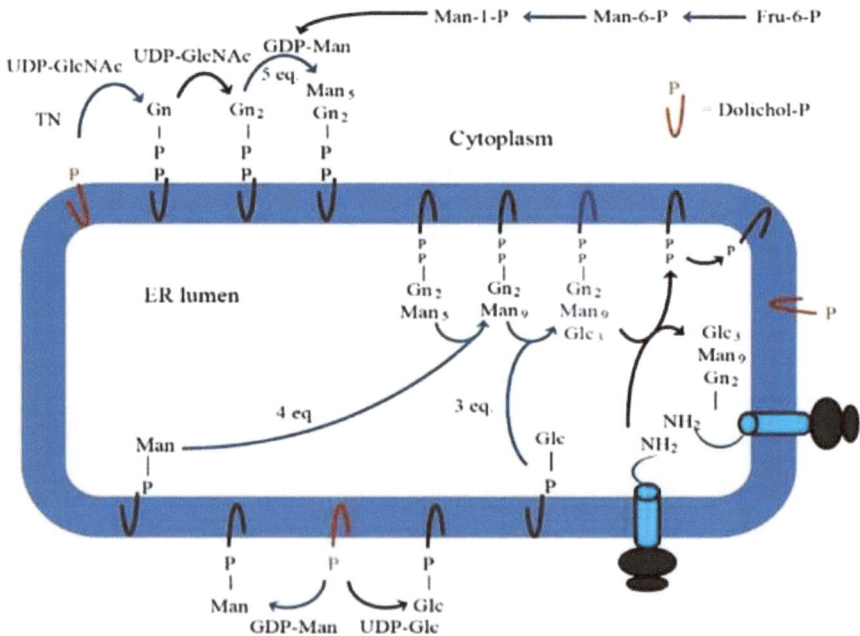

Fig. (11). Dolicol and precursor oligosaccharides based on dolichol pyrophosphate.

The order of the reactions in steps a, b, and c has been established, despite the fact that little is understood about the involved glycosyltransferases. While many of the transferases involved in elongation step d have been isolated and thoroughly characterized, it is unclear exactly how they work to create the final non-reducing terminal sequences. For a complete understanding of oligosaccharide biosynthesis, knowledge of the reactions leading to the synthesis of terminal sites in step d is crucial because the majority of *N*-linked oligosaccharides differ primarily in structure from one glycoprotein to another in their non-reducing terminal sequences [85].

Elongation

Common complex-type oligosaccharides have two to four branches made up of the Galβ1-4GlcNAc sequence linked to the core oligosaccharide's mannose residues through α-link. At various links, they are then replaced by sialic acid or fucose. As a result, many potential reaction sequences can result in the development of the same finished oligosaccharide chain. In theory, the formation of each of these branches can be completely independent of the elongation of the other branches. However, several of these theoretically viable pathways appear to be effectively prohibited by the relevant glycosyltransferases, and in other cases, they may restrict the synthesis of a certain construct to a specific reaction sequence [88, 89].

Transferring N-acetylglucosamine from the β1-2 link to the α-linked mannose residues in the core oligosaccharide is the first step specifically required for the formation of complex-type chains [90, 91].

Chain Termination

The unbinding of one or more -linked sugar residues, typically sialic acid, fucose, and more rarely galactose, N-acetylgalactosamine, N-acetylglucosamine, or mannose, is the final step in the synthesis of complex-type oligosaccharide chains.

This process is called chain termination. Because the binding of these residues generally prevents further elongation of the oligosaccharide structure. It is not clear why the addition of specific sugars results in chain termination or why the addition of different sugars is necessary for the termination in different glycoproteins, except perhaps for lysosomal hydrolases containing phospho-mannose residues [12, 13]. However, the additional order of terminal residues cannot simply be deduced from the structure of the product, as often multiple residues are added to a single chain, producing a branched structure. However, certain aspects of the biosynthesis of some of these constructs are predicted from the acceptor substrate specificities of the glycosyltransferases forming these linkages. In recent years, several glycosyltransferases involved in chain termination of complex-type oligosaccharides have been obtained in pure or highly purified form. Among these enzymes are β-galactoside α2→6 sialyltransferase, β-galactoside α1→2 fucosyltransferase, β-N-acetylglucosamine α1→3 fucosyltransferase fucosyl αl→2galactoside αl→3N-acetylgalactosaminyl-transferase [92 - 96].

For the insertion of *O*-linked N-acetylgalactosamine residues into proteins, there is no standard sequence [97]. Glycoproteins with high *O*-glycosylation have 2, 3, or 4 consecutive serine and threonine residues. Typical Thr/Ser-linked oligosaccharide chains of epithelial mucins exhibit much greater heterogeneity than *N*-linked oligosaccharides, ranging from 1 to 20 monosaccharide residues in size.

This is partly because these chains do not originate from a common lipid-bound precursor but instead are synthesized one residue at a time directly on the protein. These oligosaccharides fall into two main classes: those containing only a core of the Galβ1→3GalNAc disaccharide sequence and those that additionally have one or more branches consisting of a repeating galactose-N-acetylglucosamine sequence. Much of our knowledge of the biosynthesis of these structures has been obtained from studies of glycosyltransferases involved in the formation of sheep and pig submaxillary mucins.

All the transferases necessary for the synthesis of the most complex oligosaccharide chains on these proteins were first described by Roseman *et al.* Then, four of these enzymes were homogeneously purified and their acceptor substrate properties were extensively characterized. The antifreeze glycoprotein, a model containing oligosaccharides with the Galβl → 3GalNAc structure, was used to investigate all possible sequential reactions of the four enzymes Fig. (**12**).

Antifreeze glycoprotein (Core 1 structure), α-N-acetylgalactosaminide α2→6 sialyltransferase (Core 2 structure), β galactoside α2→3 sialyltransferase (Core 3 structure), and β-galactoside α1→2 fucosyl-fully glycosylated transferase (Core 4), are desialylated product (Core 5 construct) found in many serum and membrane glycoproteins are synthesized *via* the core1→ core2→ core5 or core1→ core3→ core5 pathway. Of these reactions, the latter is preferred. Because the core 2 structure is a relatively weak acceptor for α2→3 sialyltransferase. Similarly, the core 7 structure are formed as one of the oligosaccharides on porcine submaxillary mucin, the receptor for the sialyltransferase, are probably the former route is preferred, since structure 4 is weak. The other monosialyl, monofucosyl product, structure 6, cannot be generated because structure 3 is not an acceptor for fucosyltransferase, and structure 4 is not an acceptor for α2→3 sialyltransferase. Both structures 4 and 7 are acceptors for blood group A N-acetylgalactosaminyltransferase and structure 8 yields oligosaccharides found on A+ porcine submaxillary mucin. However, the 8 structure is not an acceptor for the α→6 sialyltransferase [97 - 100].

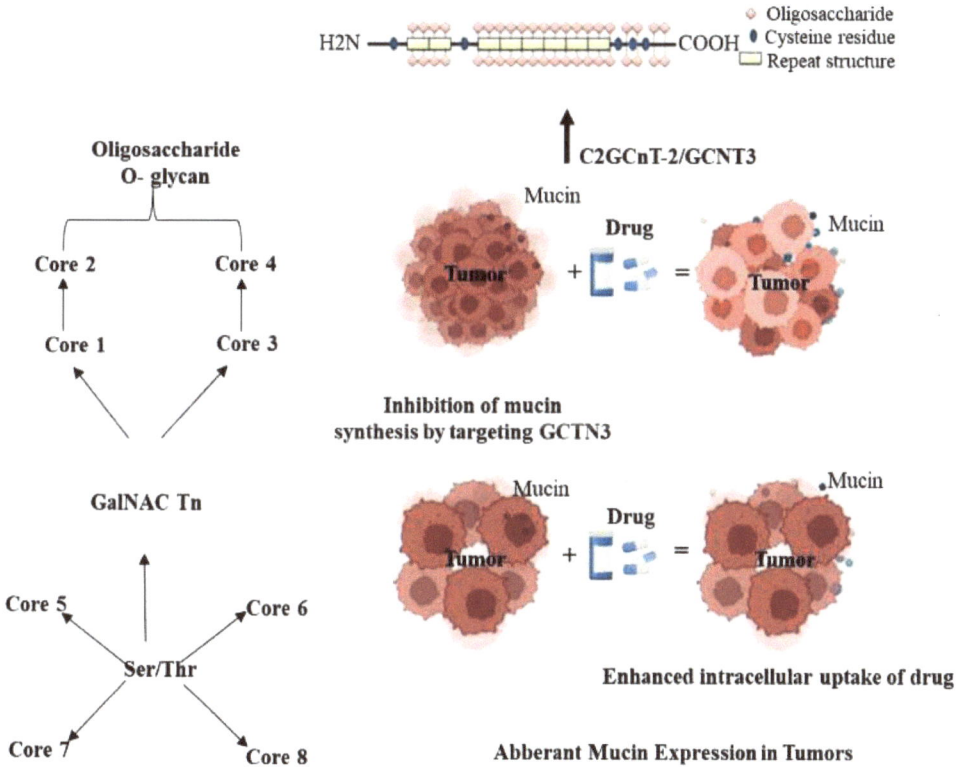

Fig. (12). Biosynthetic pathways of mucin-type oligosaccharide chains.

GLYCOSYLATION RELATED ENZYMES

The carbohydrate-active enzymes (CAZymes) include numerous enzymes involved in carbohydrate synthesis and degradation [101].

In glycosylation, two families of enzymes are in effect. Glycosyltransferases and glycosidases are the two classes of enzymes served in the formation of glycoconjugates. Glycosyltransferases are the enzymes that regulate this process and contribute to the transfer of sugar residues. However, glycosidases are responsible for separating monosaccharides, forming glycans. Depending on where they function, glycosidases are classified exoglycosidases and endoglycosidases. Inner glycosidic linkages are degraded by enzymes called endo-glycosidases. On the other hand, exo-glycosidases, catalyze reactions that allow cascading by breaking off sugars from the non-reducing end of oligosaccharides [102].

In conclusion, glycosyltransferases attach glycans to proteins, and glycosidases remove them away [103]. Table **3** displays the monosaccharide linkages that transferases catalyze.

Table 3. Monosaccharide linkages catalyzed by transferases.

Sialic Acid Link Types	Galactose Link Types	Fucosyltranferase Link Types
Sialic acid-galactose	**Galactose-N-acetylglucosamine**	**Fucose-galactose**
✓ Siaα2,6Galβ1,4GlcNAc-[a]		
✓ Siaα2,3Galβ1,4GlcNAc-[b]	✓ Galβ1,4GlcNAc-[a]	✓ Fucα,2Galβ-[a]
✓ Siaα2,3Galβ1,3GalNAc-[a]	✓ Galβ1,3GlcNAc-	✓ Fucα,6Galβ-
✓ Siaα2,4Galβ1,4GlcNAc-	✓ Galβ1,6GlcNAc-	✓ Fucα,3Galβ
✓ Siaα2,4Galβ1,3GlcNAc-		
Sialic acid-N-acetylgalactosamine	**Galactose-N-acetylgalactosamine**	**Fucose-N-acetylglucosamine**
✓ Siaα2,6GalNAccx-[a]	✓ Galβ1,3GalNAc-[b]	✓ Fucα 1,4(Galβ1,3)GlcNAcβ-[a]
✓ Siaα2,4GlcNAc-	✓ Galβ1,6GalNAc-	✓ Fucα 1,3(Galβ1,4)GlcNAcβ-
✓ Siaα2,6GlcNAc-	✓ Galβ1,3GalNAc-	✓ Fucα 1,6(R-GlcNAcβ1,4)GlcNAβ-[b]
	Galactose-galactose	
Sialic acid-sialic acid	✓ Galα1,3Gal-[a]	**Fucose-fucose**
✓ Siaα2,8Siaα2[b]	✓ Galα1,6Gal-	✓ Fucα,3Fuc
	✓ Galα1,3Gal-	

[a] It has been purified. [b] Defined in tissues.

Glycosyltransferases

Oligosaccharide groups in glycoproteins are synthesized by several glycosyltransferases; each transferase extends the length of the oligosaccharide by one monosaccharide residue. Glycosyltransferases are biosynthetic enzymes that produce interglycosidic connections. A specific transferase is thought to be required for the synthesis of each of the approximately 100 different disaccharide sequences known to be found in glycoconjugates [12, 35], but less than half of the predicted transferase activities have been identified *in vitro* systems, of which only a dozen have been purified and enzymatically characterized. Therefore, with the increasing number of reported new oligosaccharide structures [104, 105], so does the number of glycosyltransferases that need to be identified that are important for oligosaccharide biosynthesis. Most of the glycosyltransferases currently purified and characterized are those associated with the synthesis of non-reducing terminal sequences in glycoproteins. Monosaccharides such as fucose, sialic acid, mannose, galactose, N-acetylglucosamine and N-acetylgalactosamine are found in non-reducing terminal positions in the oligosaccharides of glycoproteins. More than one monosaccharide may be in the terminal position because most oligosaccharides have branched structures. Also, due to structural microheterogeneity, the oligosaccharide groups in pure glycoproteins may differ slightly from molecule to molecule. Oligosaccharides attached to a specific amino acid side chain may have the same carbohydrate-protein linkage and the same core structures but differ in the number of branches or the sequence of terminal residues in a given branch [106].

Controlling the selectivity of glycosyl reactions during the synthesis of carbohydrates is exceedingly challenging. This carefully controlled synthesis of carbohydrate linkages has no known method. However, different chemical processes are employed to create glycans [65]. In recent years, studies on the molecular properties, analysis and purification of these transferases have been at the forefront [99, 106, 107]. Only nine sugar nucleotide donors are utilized by mammals' glycosyltransferases. These include CMP-sialic acid, UDP-GlcNAc, GDP-mannose, UDP-glucose, UDP-GalNAc, UDP-galactose, GDP-fucose, UDP-glucuronic acid, and UDP-xylose [13].

Glycosyltransferases are a large family of enzymes with many members. Sialyltransferases, Galactosyltransferases, Fucosyltransferases, N-Acetylglucosaminetransferases, N-Acetylgalactosaminyltransferases, and Mannosyltransferases are members of the Glycosyltransferases family.

We will discuss respectively the glycosyl transferases that catalyze significant processes.

Sialyltransferases

Sialyltransferases, important enzymes in the formation of glycoconjugates and oligosaccharides containing sialic acid, are involved in the control of the immune system as well as cellular communication and recognition [108]. Also, this enzyme crucial for drug design. Each of the known sialyltransferase linkages is thought to be synthesized by a different transferase, although only five of the nine predicted transferase activities have been demonstrated to exist *in vitro* systems, and only three of the five have undergone significant purification. They are crucial for drug design as well. Each of the listed linkages is thought to be synthesized by a different transferase, but of the nine expected transferase activities, only five have been shown to exist *in vitro* systems and only three of the five have been extensively purified. Each of the known transferases catalyzes a reaction as follows, where 'n' denotes the substituted hydroxyl group at the acceptor.

The most common sequences containing sialic acid are Siaα2-6Gal and Siaα2-3Gal, which are found not only in various soluble glycoproteins but also in blood plasma and membrane-associated glycoproteins on cell surfaces or intracellular membranes [109].

Sialyltransferases facilitate the post-translational addition of sialic acid, and three enzymes catalyze polysialylation, which is the addition of α-2,8-linked sialic acid residues. These enzymes are ST8SIA2 (STX), ST8SIA3 and ST8SIA4 (PST), known as polysialyltransferases [110]. Polysialic acid, which is involved in cellular adhesion, and ST8SIA2, which regulates its production, are linked to the development of cancer and its prognosis. Polysialic acid is a significant polymer for cellular adhesion, neuronal migration, and tumor metastasis. In a lot of tumors, it is strongly expressed [111].

Galactosyltransferases

Galactosyltransferase is an adhesion receptor as well as a protein involved in the Golgi apparatus [112]. The most common sequence seen in *O*- and *N*-linked glycoprotein oligosaccharides, proteoglycans and glycolipids is Galβ1→ 4GlcNAc, formed by N-acetylglucosaminide β1→4 galactosyltransferase. The enzyme has been isolated in homogeneous form from extramammary tissues and fluids that appear to be involved in glycoprotein biosynthesis, such as bovine and human milk, where it interacts with -lactalbumin to form lactose synthase [113, 114].

Fucosyltransferases

Fucosyltransferases are enzymes that transfer L-fucose sugar from a GDP-fucose sugar substrate to an acceptor substrate [115]. Fucosylated glycoconjugates are essential for biological sensing processes. Four of the seven linkage types are synthesized by glycosyltransferase activities *in vitro* systems, and three of the enzymes were highly purified. The transferase, Fucα1→2Gal, forming the H blood group structure, was homogeneously purified from porcine submaxillary glands and enzymatically characterized. Unlike many transferases, the receptor specificity is quite broad [115, 116].

N-Acetylglucosaminetransferases and N-Acetylgalactosaminyltransferases

N-Acetylglucosamine has been found in significantly higher concentrations than other monosaccharides and in 14 different types of glycosidic linkages in glycoproteins from higher organisms. But most of them are not in terminal non-reducing sequences. The most common terminal N-acetylglucosamine residues are those that form branches with mannose in di-, tri-, and tetraantenary *N*-linked oligosaccharides, as shown in the hypothetical sequence below. However, only two of the N-acetylglucosaminyltransferases that make up such sequences have been purified and characterized. One catalyzes the formation of the GlcNAcβ1→2Manαl→3Man sequence, and the other catalyzes the formation of the GlcNAcβ1→2Manαl→6Man sequence. The only N-acetylgalactosaminy-ltransferase, which forms terminal sequences in the A rh (+) blood group structure, catalyzes the reaction. Thise enzyme was purified from human serum and porcine submaxillary glands [117, 118].

Mannosyltransferases

Enzymes called mannosyltransferases allow the transfer of mannose to the glycosylphosphatidyl group [119, 120]. Most of the known mannosyltransferases are involved in the biosynthesis of "high mannose" dolicylphosphoryl oligosaccharide intermediates in *N*-linked oligosaccharides. They use GDP-Man or dolicylphosphoryl-mannose as donor substrates in reactions of the following form, where R is any of a set of receivers. Currently, many of these transferases have not been purified and characterized. Therefore, little is known about their enzymatic properties. However, it is possible to make inferences about them based on their substrates and products [121, 122].

Glycosidases

Glycosidases (EC3.2.1) are imperative enzymes that cleave glycoside bonds. Because of their function in the biological process, glycosyl hydrolases (GHs), one of the targets of treatment strategies, needs to be studied in many ways. Members of this class regulate biological processes by catalytically breaking down the glycosidic bond of glycans, which reveals a range of physiological effects [123, 124]. Based on the specificity of the substrate, GH has been divided into two classes. The former has members of the EC 3.2.1 class that break down *O*- or *S* glycosides, and the other has members of the EC 3.2.2 class, which includes hydrolases of N glycosides. However, classification according to amino acid sequence similarities GH families has become possible in line with genomic information [101]. Glycosidases are used to identify specific diseases, label cells and tissues, and recognize pathogens. Sialidases (Neuraminidase), Fucosidases, galactosidases, glucosidases, hexosaminidases,, and glucuronidase are all members of this family.

Sialidase

Sialidase is a hydrolytic enzyme that separates non-reducing terminal sialic acid residues from sialoglycans Fig. (**13**). Sialidase is associated with Voltage-gated sodium and calcium channels. In addition, it has essential roles whit axonal growth, neurotransmission and memory processes [125, 126]. Sialic acids are linear homopolymers. Polysialic acid (PSA), is found in a large number of eukaryotes, including prokaryotes, vertebrates, and invertebrates [127]. In the mammalian brain, PSAs function to trap dopamine and brain-derived neurotrophic factor (BDNF) and to prevent synaptogenesis *via* neural cell adhesion molecule (NCAM)-mediated mechanisms [128, 129]. For the brain's synaptic plasticity, learning, memory, and synaptic transmission, removing sialic acid *via* sialidase is crucial. As a result, sialidase's role in the desorption of sialic acid is crucial in the control of brain activity. Based on data from imaging of sialidase activity in the pancreas, it has been determined that the sialidase inhibitor can be used as a diabetic drug to prevent hypoglycemia, a severe side effect of insulin secretagogues [129, 130].

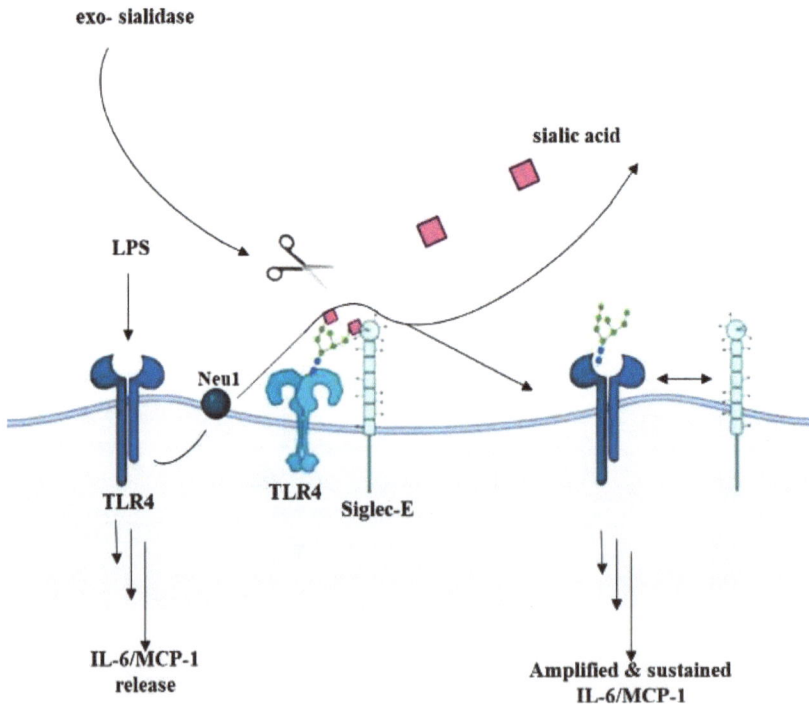

Fig. (**13**). Function of sialidase.

In mammals, including humans, SAs and their ST counterparts have conflicting roles in conditions like diabetes, cancer, and atherosclerosis. Despite having low main structure sequence identity (26%), all SAs have a strikingly comparable -propeller architecture made up of antiparallel -sheets with about six-fold symmetry. However, the catalytic tyrosine residue and arginine triad are retained in all SAs. Rossman fold conservation is seen in STs, which is common among many glycosyltransferases [131].

The creation of effective antibiotics, antivirals, and antiparasitic medications, as well as innovative treatments, depends on the discovery of inhibitors of these enzymes. It is believed that sialidase contributes to the adhesion and invasion of microorganisms into the host cell. As long as the amino sugar metabolism is functioning properly, sialyltransferase and sialidase regulate the amount of sialic acid in mammalian cells. Similar to sialyltransferases, sialidases exhibit consistent expression variations during cell differentiation, growth, and malignant transformation and they are implicated in a variety of biological processes [132].

Fucosidases

Fucosidases remove fucose from the glycan chain. Few natural fucosidases have been characterized [133]. The widespread availability of fucosylated oligosaccharides in nature suggests that the fucose hydrolyzing enzyme may also be abundant. Most of the fucose hydrolases (EC3.2.1) currently processed in the Carbohydrate Active Enzymes (CAZy) database are classified into two main families. These are the GH29 and GH9 [101]. In addition to these, GH141, which is involved in the release of fucose from plant cell wall polysaccharides, and GH139, which has -2-O-methyl-L-fucosidase activity, are also present [134]. The GH1 and GH30 families have also been reported to have fucosidase activity [101].

In the CAZy database, GH95, which offers hydrolysis of fucose, has 2400 entries. To date, no GH95 enzyme in animals has been discovered [135]. In contrast, the CAZy database has more than 4700 entries for the largest fucosidase family, GH29, which consists members from all spheres of life [136, 137]. Even though the CAZy family contains a large number of GH29 sequences, only a small number of these enzymes have been functionally characterized [138].

For example, the enzyme α-L-fucosidase 1 (EC 3.2.1.51) is found in lysosomes and is involved in the degradation of *N*-linked glycans. Baudot A. *et al.* declared that, defucosylation affects macroautophagy, loss of FUCA1 expression results in an accumulation of autophagosomes, and FUCA1 modulates induced autophagy [103]. The fucosylation of oligosaccharides is important in the processing of signals, embryogenesis, regulation of development and growth processes, apoptosis, and immune response [139].

Galactosidases

The hydrolysis of β-galactosidic bonds in oligosaccharides and polysaccharides is catalyzed by a significant class of glycosidases known as β-galactosidases. β-Galactosidases are widely distributed in microorganisms, animals and plants and are classified as GH families 1, 2, 35, 42, 59 and 147 in the CAZy database (http://www.cazy.org/). These enzymes are primarily found in the GH2, GH35, and GH 42 families, based on the number of enzyme sequences that have been stored in the database [140].

Lactose hydrolysis by β-galactosidases yields galactose and glucose. β-Galactosidases (EC3.2.1.23) are promising glycosidases because they have hydrolytic and transglycosylation activities [141].

Glucosidases

β-Glucosidases (β-D-glucoside glucohydrolase, EC 3.2.1.21) are a diverse group of enzymes found in Archaea, Eubacteria, and Eukaryotes. In general, it hydrolyzes the glycosidic bonds of cyanogenic glycosides, aryl-β-D-glucosides, disaccharides, and short oligosaccharides [142].

Hexosaminidases

β-N-Acetyl-D-hexosaminidases (EC 3.2.1.52) catalyze the removal reactions of N-acetyl-D-galactosamine or b-linked N-acetyl-D-glucosamine from glycoconjugates [143]. It was discovered that GH84 β-N-acetyl-D-hexosaminidase was connected to the O-GlcNAc cycle in metazoan cells. Abnormalities in this cycle may lead to diseases such as type II diabetes, Alzheimer's and cancer [144].

Glucuronidase

The enzyme β-glucuronidase (EC 3.2.1.31) is a member of the GH families 1, 2, 30, 79, and 154. β-glucuronidase (βGLU) is a lysosomal hydrolase that is widely found in many mammalian tissues, including body fluids and microbiota, and is particularly associated with the endoplasmic reticulum [145]. Studies have shown that the enzyme is associated with some diseases. For instance arherosclerosis [146] and the lysosomal storage disease mucopolysaccharidosis type VII [147] are related to mutations in the GUS gene, which codes for βGLU. An increase in the expression of βGLU was observed in the tissues of patients with some cancer types, and therefore the enzyme's. It was thought that it could be a biomarker for tumor diagnosis. At the same time, increased expression has been detected in diabetes, neuropathy, rheumatoid arthritis, and urinary tract infections, and it has been reported that it may have the same potential in these diseases [145].

BIOLOGICAL FUNCTIONS OF GLYCOCONJUGATES

There are numerous methods for explaining the biological functions of glycoconjugates. These consist of structural and regulatory functions, internal and extrinsic recognition, roles in recognition, and molecular mimicry of host glycans. Glycans in cells, the extracellular matrix, and body fluids have important biological effects because of their structural characteristics and capacity to alter the functions of the structures to which they are attached. These effects are mediated by their structural characteristics and the functions of the structures to which they are connected. The study of the genetic alterations that disease-related glycosylation causes is one of the most significant of these. Genetic variations in the expression of enzymes involved in biosynthesis processes have been found to cause abnormalities in the biosynthesis pathways of glycans in multicellular organisms. Glycoconjugates have biological functions such as differentiation, metastasis, cell-cell adhesion, cell growth and cytokinesis, viral attachment, bacterial infection, signal transduction, ligand receptor binding and protein regulation, as well as identification of blood group antigens, tissue transplantation and targeting of lysosomal enzymes to the lysosome. Biological functions of glycoconjugates are summarized in Fig. (14) [2, 13, 148].

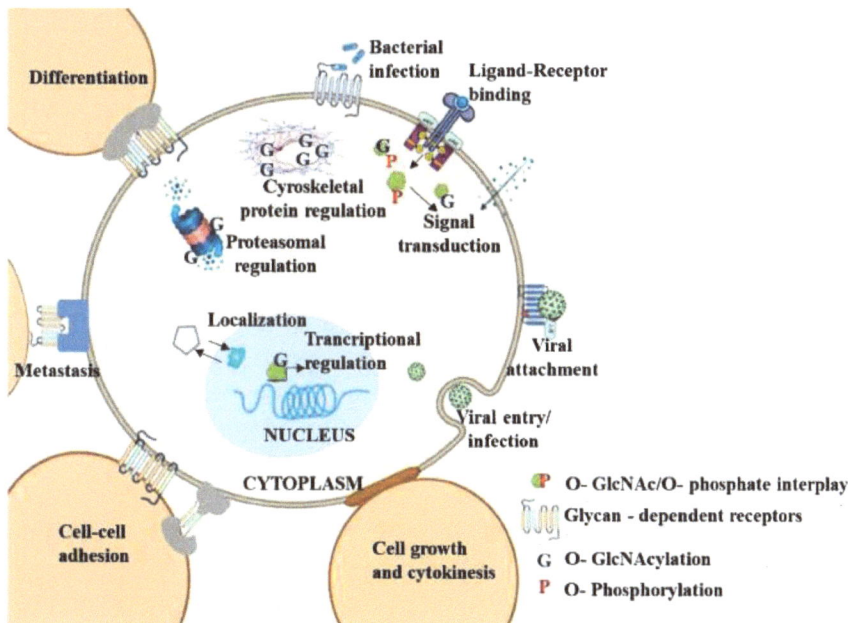

Fig. (14). The biological functions of glycoconjugates.

Association of Glycoconjugates with Diseases and Therapeutic Properties

Glycosylation is important for both physiological and pathological processes. Through oxidation, acetylation, phosphorylation, and sulfation, glycans further alter proteins. The receptors receive a variety of information as a result of this modification, which also controls biological processes [149, 150]. Congenital defects in glycosylation have provided significant insights into the underlying mechanisms involved in the disease-related association of specific glycoconjugates [60]. Immune cell interactions, which are linked to cell surface molecules and result in cellular activation, control the types of glycosylation of glycoconjugates and their interactions with receptors. Glycosylation, affects apoptosis, metastasis, and chemotherapy resistance in canser. Furthermore, cells display oncofetal phenotypes. Abnormal glycosylation of glycoproteins plays a role in the pathogenesis of many autoimmune diseases [151]. Glycosylation-related genetic diseases are frequently congenital. Congenital glycosylation disorders (CDG) were first described by Jaeken *et al.* in 1980 [152]. Congenital glycosylation disorders of glycoconjugates due to a decrease or increase in glycosylation It is a large group of genetic diseases. There are many genetic diseases in this group. These diseases are disunited into four groups: disorders of protein *N*-glycosylation, protein *O*-glycosylation, combined *N*- and *O*-glycosylation, and lipid glycosylation. Congenital glycosylation disorders are divided into 2 main groups as Type I and II. In CDG-I, *N*-glycan formation and transport to proteins in the cytosol and endoplasmic reticulum (ER) are impaired, and in CDG-II, there is a defect in *N*-glycan processing in the ER and Golgi [153]. Type I CDGs disrupt the normal generation of the oligosaccharide structure on the glycolipid precursor before binding to the Asn residue of a protein. On the other hand, Type II CDGs have flaws in the regulation of the *N*-linked branching structure of the developing glycoprotein [60]. Similar to all other cells, immune system cells express glycan-binding proteins, glycolipids, and other molecules that sense signals on their cell surfaces. Numerous immune receptors, also referred to as pathogen-associated molecular models, are expressed in innate and adaptive immune cells and recognize glycans found on the surface of microorganisms. Bacterial lipopolysaccharides, peptidoglycans, teichoic acids, and capsular polysaccharides are examples of this type of glycan [151]. The development of vaccines has made use of the immune system's ability to recognize these glycosylated microbial models [154, 155]. For instance, a combination of capsular polysaccharides is used in pneumococcal vaccines [151]. Understanding the effects of the HIV-1 envelope (Env) glycoprotein and the glycans that make up it on immune responses and immune evasion has also aided recent advances in HIV-1 vaccine development. Carbohydrates are becoming increasingly important in the life sciences due to their numerous biological functions. Changes in the composition of glycans on the surface of the cell determine the prognosis of the illness. Therefore, carbohydrates that induce interactions *via* glycans are used as therapeutic agents to treat a variety of diseases [60]. Many scientists are researching carbohydrate-based targeted drug design. Carbohydrates were glycoconjugated with cytotoxic agents such as busulfan, gluphosphamide, paclitaxel, chlorambucil, and docetaxel, and they were found to be less toxic in healthy cells than parent aglycones. These intracellular glycosidases are thought to degrade these sugar drugs. The majority of prodrugs based on carbohydrates are used to enhance pharmacokinetics. The glycosidase site of action of these drugs is typically extracellular and allows the release of active drugs. A known distinguishing feature of cancer progression is the biosynthesis of certain glycans, such as *N*-glycans, through altered glycosylation. Overexpression of various glycosyltransferases, such as N-acetylglucosami-nyltransferase-V, causes abnormal *N*-glycan glycosylation in tumor cells [156, 157]. In conclusion, intensive studies are underway to develop suitable polyvalent carbohydrate sequences for drug design. Anticoagulants such as heparin, polysaccharide vaccines, and aminoglycoside antibiotics are a few examples of carbohydrate-based medications [158]. Numerous glycan-based therapeutics have been developed as a result of research into glycobiology. For instance, altering the glycosylation of the envelope glycoprotein gp120 expressed in HIV-1 facilitates the removal of the virus by the immune system [159, 160]. Monoclonal antibodies have been modified with new glycan-linked structures to enhance their neutralizing activity. Thus, it is ensured that the recombinant protein recognizes HIV-1 better [161, 162]. Between receptor-glycan interactions, glycosylation is essential for controlling the immune response. It has also been attempted to lessen angiogenesis in cancer patients by targeting the glycosylation pathways of Galectin 1, and a decrease in angiogenesis has been seen. An increase in *N*-glycan levels was seen in a study that involved adding GlcNAc to T cells obtained from patients with ulcerative colitis. However, thioglycosides block SGLT1 and SGLT2, preventing renal glucose reabsorption in diabetic patients.

Saccharide-receptor interactions are widely used by bacteria and viruses to mediate cell attachment, colonization, or invasion. In order to recognize the oligosaccharide structures on the cell surfaces of the organs they invade, microorganisms produce molecules called adhesins that resemble lectins. Also, many parasites use lectin-saccharide interactions to spread their infection [163]. It is possible to inhibit bacterial binding by competitive inhibition using soluble oligosaccharides that are identical or similar to the ligands recognized by the adhesins. However, it has proven difficult to isolate large quantities of glycoconjugates. Cloning of glycosyltransferases into bacterial cells is a promising technique to facilitate the synthesis of very large quantities of related oligosaccharides for testing anti-adhesive pharmaceutical agents. Therefore, there is much interest in the development of oligodendromers [164, 165]. The most important of the functions of glycans on viral proteins are the changes that occur in the immune system of the host.

The addition of new glycosylation sites has a compensatory effect on receptor binding in addition to increasing compliance [166]. The identification of conserved glycosylation patterns among rapidly mutating viruses is important for the development of effective vaccines. For instance, by identifying conserved glycosides in the root region of influenza strain H5, new vaccine candidates have been created. The development of a vaccine that induces a B-cell response was made possible by the mutation of some conserved glycosides and increased neutralization of heterosubtypic, heterologous, and homologous viruses.

Modulating the glycan density on immunogens is one strategy utilized in the design of vaccines [167, 168]. The hepatitis virus's surface antigen become more immunogenic as the *N*-glycan quantity increase [169]. Numerous broadly neutralizing HIV-1 antibodies that target glycan-dependent epitopes have been discovered [170]. Synthetic carbohydrate antigens were sought for the induction of such antibodies, but unfortunately, vaccination strategies failed to achieve neutralization, and poor reactivity to gp120 was observed [171, 172]. Although there is a lot of information on the HIV envelo-1 envelope glycoprotein's natural glycosylation, experiments on vaccination with native trimers are few and far between [173, 174].

New and unexplored approaches is necessary to control atypical pneumonia and find a solution to the global pandemic. For this reason, recently studies have increased to develop potential antivirals and new candidate vaccines.

Cleanliness, patient isolation, distance, travel restrictions, and quarantine have all been suggested as potential strategic control measures against COVID-19, but they weren't enough to stop the infection from spreading. as some of the possibilities, but they were insufficient to stop the spread of this infection. As a result, it became urgent to enhance and reuse current medications [174]. The spike-shaped (S) glycoproteins that are present on the surface of these viruses are encoded by the SARS-CoV-2 genome. The S glycoprotein is essential for the virus's adhesion, fusion, and entry into the host cell. The focus has been on antibodies that neutralize the S glycoprotein since its location on the surface makes it a direct target for host immune responses. The S protein is the main target of the majority of vaccination methods as well as therapeutic therapies because of its crucial involvement in viral infection and adaptive immunity [175].

Studies have revealed that lectin-like non-peptidic mimic Pradimycin-A (PRM-A), a carbohydrate-binding agent (CBA), and plant lectins have antiviral efficacy against enveloped viruses, including coronavirus [176 - 178]. The antiviral action of lectins isolated from diverse sources, including cyanobacteria, algae, and plants, is well recognized. Mannose (Man) binding lectins have also been shown to have antiviral action against COVID-19, including *Allium porrum* agglutinin (APA), *Galanthus nivalis* aglutinin (GNA), *Hippeastrum hybrid* lectin (Amaryllis) (HHA), and *Narcissus pseudonarcissus* agglutinin (NPA). In addition, specific lectins from *Nicotiana tabacum* (Nictaba) and *Urtica dioica* agglutinin are known to have potent antiviral activity [177]. Because viral glycoproteins contain mannose-linked *N*-linked glycans, it is known that CBAs have high antiviral activity [174, 179, 180].Human influenza viruses A and B produce polyvalent sialylated glycoconjugates, including Sia (α2-6) Gal-binding liposomes, neoglycoproteins, and hemagglutinins, which allow them to attach to the epithelium of the human respiratory tract. There have been developed polyvalent sialylated glycoconjugates, such as polymers, liposomes, and neoglycoproteins, that can prevent viral binding to epithelial tissue cells. The creation of polymer libraries decorated with sialic acid residues in addition to significantly hydrophobic, non-carbohydrate structures is another strategy that can be used to combat other infection processes with a similar structure. Investigating the virus's binding sites revealed that sialic acids and other non-carbohydrate substances were very effective at preventing influenza virus binding. As they reduce viral replication and lessen the severity and duration of infection, sialidase inhibitors are effective treatments for influenza virus infection. By attaching to sialidase, GlaxoSmithKline's RelenzaTM, an analogue of sialic acid, renders influenza inactive when used as an inhalation agent [163, 181]. *Helicobacter pylori* infections are the root cause of chronic gastritis and gastric ulcers. Binding is *via* the Lewis b (Leb) antigen on epithelial cells of the gastrointestinal tract, where *H. pylori* bacterial adhesion is recognized. Breast milk, rich in various free sugars and sialylation and fucosylation glycoconjugates, may play a role in protecting infants from infection. The idea of developing carbohydrate-based prophylactics for the prevention and treatment of *H. pylori* infection is a possibility, but clinical trials have failed to yield convincingly beneficial results. In the past, *H. pylori* infection has been difficult to study as it is only a human pathogen and no animal models exist. Recently, transgenic mice have been generated that express human Lewis α1,3/4 Fuc-T in the gastric epithelium. H. pylori binds to the mucosa of these animals in a Leb-dependent manner. Equal numbers of *H. pylori* bacteria were found in the stomachs of transgenic and non-transgenic mice. *H. pylori* was discovered in mucus and the gastric pit that is connected to the surface of mucosal cells in transgenic animals, but it was only found in mucus in normal mice [163, 182, 183]. Human immunodeficiency virus-1 (HIV-1) recognizes CD4 in most cells. Some CD4-negative cells express a saccharide ligand on, for example, neurons and colon epithelium. In an experimental system, these saccharide structural analogues can block HIV infection. This is considered a promising area for developing new approaches to prevent or control HIV infection [174, 182].

PRP-D (polyribosyl ribitol phosphate diphtheria toxoid (DT) conjugate), the first CPS-conjugate vaccine against Hib, was created in the US in 1987 [184, 185]. In the development of vaccine formulations for the treatment of various infectious diseases, naturally isolated capsular polysaccharides and their conjugates containing carrier proteins have been utilized Today, gluconjugate vaccines developed against various microorganisms are licensed [83].

Oligo- and polysaccharide vaccines against bacterial infections have great potential for stimulating immunity against disease-causing microorganisms. Some of them have been used successfully for a long time. These include vaccines against *Neisseria meningitides, Salmonella typhi,* and *Streptococcus pneumonia.* With the increasing and worrying emergence of more antibiotic-resistant bacterial strains, this approach will be of great clinical importance in the future. Carbohydrate-based vaccines are generally ineffective in very young children and in patients with compromised immunity. This can be circumvented by conjugating the glycan to an immunogenic polypeptide carrier. Successful applications of this type of approach include developing a vaccine against *Haemophilus influenzae type* b (Hib) and *Streptococci pneumonia.* In addition, a fully synthetic glycan-peptide vaccine has been developed to prevent *Shigella sp.* infection that causes dysentery [163, 164, 174, 181].

Carbohydrate-based therapeutics are also used in the treatment of some neurological diseases. Topiramate is used in combination with traditional anti-convulsant drugs and is a derivative of d-fructose that is effective in the treatment of epilepsy. It increases the inhibitory selection of gamma aminobutyrate (GABA) on neurons and inhibits the excitatory effect of the α-amino-3-hydroxy-methyl isoxazole-4-propionic acid type glutamate receptor. The drug SygenTM is another example. A drug formulation called SygenTM is based on GM1 ganglioside. It has been used to treat acute trauma such as stroke, spinal cord injury, and chronic neurological diseases such as Parkinson's disease, although its efficacy is not fully known [186, 187]. Hyaluronan (HA) is a naturally occurring polysaccharide that is a component of vertebrate glycosaminoglycans and has elasto-viscous properties. HA is a biomaterial with appealing properties, is extensively researched for use in biomedical and pharmaceutical applications. HA has been successfully used in surgery, where the elastic and lubricating properties of the polymer are beneficial. Hyaluronic acid products are used in eye surgery because they prevent dehydration and, therefore, tissue damage. Loss of HA and other extracellular matrix (ECM) components, such as cholesterol sulfate, keratan sulfate, and dermatan sulfate, leads to many chronic conditions in the eye, such as osteoarthritis and macular degeneration. In such cases, carbohydrate-based substances can be used, which are aimed at replacing components or stimulating their regeneration HA, a crucial component of synovial fluid in healthy knee joints, that gives synovial fluid viscoelastic characteristics. The viscoelastic and lubricating character traits of the synovial fluid in the joint areas are lost in osteoarthritis. Therefore, it is thought that intra-articular injection of HA will give the synovial fluid its former viscoelasticity and lubricating properties [174, 188].

Cancer is an extremely common disease with a different pathogenesis and is difficult to treat. Therefore, there is a need for new techniques for the early diagnosis and treatment of cancer. Glycans are currently promising biomolecules for the creation of new biomarkers [189, 190]. Modifications in glycosylation patterns were identified as a biomarker for the development of human cancers. Due to its excellent target specificity and minimal side effects, cancer immunotherapy has recently gained popularity as a potential treatment [191]. Programmed death ligand-1 (PDL-1) and cytotoxic T lymphocyte-associated protein 4 are two examples of tumor-specific surface proteins that are the focus of the majority of immunotherapeutic drugs (CTLA-4). An alternative strategy for creating immunotherapeutic drugs and therapeutic vaccines is to target the tumor-associated carbohydrate antigens (TACAs) (191). TACAs are divided into four subcategories based on their structural characteristics: globo series glycolipids (Globo H, SSEA4, and SSEA3), the gangliosides (GD2, GD3, GM2, GM3, and fucosyl GM1), the blood group antigens (Lewisx, Lewisy, sialyl Lewisx, and sialyl Lewisy) and the glycoproteins (Thomsennouveau (Tn), Thomsen–Friendreich (TF), and sialyl-Tn (STn)) [163, 192 - 194]. TACAs are clearly overexpressed in many tumors, including breast, prostate, colon, ovary, and lung cancer, melanoma, neuroblastoma, B-cell lymphoma, and others, but they aren't all like that [163, 192, 193]. The immunogenic TACAs on tumor cells make them a potential target for vaccine development [195].

In addition to the serological markers used clinically for the diagnosis and monitoring of cancer diagnosis and prognosis, some of the prognostic biomarkers used to follow the development of the disease are also in the glycoprotein structure [189, 190]. It is known that almost all of these markers arise from abnormal glycosylation in cancer cells. However, due to their low specificity, they have limited application. The AFP-L3 fraction (AFP), a glycobiomarker for the detection of liver disorders, is a recognized protein for the diagnosis of HCC [190]. A glycosylated form of AFP that shows significantly increased levels of fucosylation in HCC patients compared to chronic liver patients is used as a tumor biomarker [190, 196]. Liver-secreted proteins such as haptoglobin, kininogen, and GP73 are promising biomarkers for the early diagnosis of HCC and the development of the illness. With the development of new glycan research techniques, numerous instances of aberrant glycans linked to cancer have been discovered [197].

The ability of cancer cells to get beyond cellular division checkpoints, avoid immune surveillance and death signals, and move to metastatic areas is essential for tumor growth. Glycosylation plays a part in each of these processes. For instance, a crucial step in the formation of cancer is aberrant growth factor signaling, which can be inhibited by the right ligands, receptors, and glycosylation patterns on the molecular scaffold [198]. One of the earliest biomarkers of cancer is the pattern of glycosylation. It is still used to identify stem cells in both healthy and cancerous tissue [198, 199]. Examples of glycosylation changes frequently found in cancer cells include increased sialyl Lewis structures, aberrant nuclear fucosylation, increased *N*-glycan branching, or the synthesis of mucin-type *O*-glycan, Tn antigen. Table **4** lists the modifications to glycosylation that take place in cancer cells. Because they reflect patterns frequently seen in early development, several of these distinct glycosylation patterns found in cancer have been dubbed "oncofetal" [200]. The makeup of the glycans in developing cancer cells may alter in accordance with modifications in cellular metabolism [201, 202].

Table 4. Glycosylation and cancer development.

Cancer Associated Glycosylation Changes	Mechanism Leading to Altered Glycan Structure	Mechanism of Regulation of Glycogenes
β-1,6 branching	Altered glycosidase expression	Regulation by glycogenes by oncogenes and tumor supressor genes
Sialyl Lewis antigens	Masking of sugar structure by substituent groups	Glycosylation changes
α-2,6 sialyl-Tn antigens	Altered expression of sugar and sugar	Hypoxia
Gangliosides	Competition between normal and cancer associated carbohydrate structure	Epgenetic regulation

Glycoconjugates in Treatment Strategies

It is well known that the pharmacokinetics, pharmacodynamics, immunogenicity, biological activity, and production yield of therapeutic proteins are all significantly influenced by their glycosylation profiles. The development of glycoprotein therapeutics, such as glycoconjugate vaccines, antibody-drug conjugates (ADCs), glycoengineered monoclonal antibodies, and other recombinant proteins for the treatment of many diseases, such as autoimmune disorders and cancer, has been accelerated by advances in glycobiology [203].

Recombinant therapeutic proteins have been investigated for the treatment of diseases for a very long time. According to their uses and functions, therapeutic recombinant proteins are divided into four groups. These include protein vaccines, protein diagnostics, proteins with enzymatic activity, and proteins with targeting activity. Nearly 70% of all therapeutic recombinant proteins are glycosylated, making it the most common modification of recombinant proteins [204, 205].

In nature, glycoproteins are quite diverse and they have a glycoform structure. Glycoproteins have a glycoform structure and are quite diverse in nature. To properly understand the characteristics that carbohydrates add to this structure, numerous techniques have been developed. The advancements in this field were increased by the study of precisely defined glycan, polypeptide, and lipid structures. In order to create *O*- and *N*-glycoproteins, glycan and peptide synthesis was combined with either the ligation of expressed proteins or synthetic glycopeptides [206].

During infection, a host's immune system is stimulated by a pathogen-associated carbohydrate, which results in the production of cytokines and antiglycan antibodies [207]. Carbohydrate-based vaccines using keyhole limpet hemocyanin (KLH), tetanus toxoid carrier protein, or its non-toxic variant, CRM, are being developed for both cancer and infectious diseases in order to facilitate glycan delivery for glycan-specific immune responses [181, 195].

In addition to carrier proteins, a number of other platforms, including ferritin, dendrimers, polymers, nanoparticles, and carbon nanotubes, have been investigated for the synthesis of glycoconjugates [208].

Glycans become highly immunogenic when they are conjugated to lipids, changing them from being weak immune activators. Immuno-active lipids have been conjugated to carbohydrate antigens in synthetic vaccines to demonstrate the adjuvant activity of glycolipids [209]. For the creation of vaccines against a number of infectious diseases, including meningitis and pneumonia, the capsular polysaccharides (CPS) on the surface of pathogens serves as an immunogen [210]. There are glycoconjugate-based synthetic vaccines developed against some pathogens.*Streptococcus pneumoniae, Shigella dysenteriae, Shigella flexneri, Bacillus antracis, and Clostridium difficile* are some of these pathogens [83].

The biocompatibility and safety of carbohydrate-containing adjuvants in human vaccines have been thoroughly studied in both preclinical and clinical studies. These include -Galactosyl Ceramide-Derived Adjuvants, Saponin-Based Adjuvant QS-21, and Monophosphoryl Lipid A (MPLA) [83].

It has become clear that cell surface carbohydrates play a crucial role in the creation of vaccines. A T-cell-dependent immunological response to the glycosyl moiety is induced by polysaccharide conjugation to the carrier protein. To prevent meningitis brought on by *Pneumococci, Meningococci* genus, and *Haemophilus influenzae type b*, known glycoconjugate vaccines are created by chemically conjugating polysaccharides. However, biomolecules such as O-antigens, exopolysaccharides, and teichoic acid in the cell wall structure are seen as important targets for the development of vaccines. Vaccination also helps reduce the use of broad-spectrum antibiotics [211]. Many antibiotics today are glycoconjugates. These include aminoglycoside-derived antibiotics such as streptomycin, gentamicin, tobramycin, netilmicin, and abecacin, which are effective against aerobic Gram (-) bacilli and Gram (+) bacteria such as *Staphylococcus aureus* and *Enterococcus spp*. Altering the glycosylation of molecules in antibiotics can lead to more beneficial or harmful effects. This method has proven to be a useful method for the development of new and active compounds in the face of emerging antibiotic resistance. For example, altering the glycosylation of vancomycin has produced a product that is effective against previously resistant *Enterococcus* strains.

Everninomycins are new glycoconjugate antibiotics belonging to the orthosomycin family used to treat methicillin-resistant *Staphylococcus* genus members and vancomycin-resistant *Enterococcus* genus members.

Some glycoside antibiotics are also used as antineoplastic agents in the treatment of cancer. Daunorubicin and doxorubicin contain tetracycline ring structures covalently linked to daunosamine, an unusual amino sugar that binds to tetracycline *via* an *O*-glycosidic bond. Daunosamine plays an important role in DNA binding. Daunorubicin is mainly effective in hematological neoplasms.

In addition to being effective against acute leukemias and lymphomas, doxorubicin is also active against a number of solid neoplasms, including lung cancer, breast cancer, sarcomas, and neuroblastomas. Bleomycin and its derivatives are also antibiotics used as anti-cancer agents. Bleomycin is a water-soluble glycopeptide. Different bleomycin types carry alternative terminal amine residues. They block the cell cycle and are effective against some difficult-to-treat tumors; for example, Jung is squamous cell carcinoma and head and neck cancers. As a result, many antibiotics are glycoconjugates; altering the glycosylation of these antibiotics could be a way to produce new antibiotics to overcome antibiotic resistance developed by pathogenic bacteria [164, 174].

Erythropoietin, which is used to treat anemia brought on by the inhibition of erythroid progenitor cells in the bone marrow, is a well-known example of a recombinant glycoprotein. Erythropoietin is a naturally occurring 4-sialyl complex *N*-linked oligosaccharide. In recombinant erythropoietin, correct glycosylation is essential for biological activity, and incorrectly glycosylated molecules have only 10% of full biological activity. This is because improperly glycosylated molecules remove fermentation tropoietin from circulation quickly *via* Gal/GlcNAc/Man receptors in hepatocytes and macrophages [174, 212, 213].

Specific consideration must be given to inhibitory molecules when formulating treatment plans for diseases associated with glycosylation.

Glycosyltransferase and glycosidase inhibitors are crucial for glycosylation processes in all organisms. Glycosylation inhibitors occur naturally in bacteria, plants, and fungi as part of their chemical defense. On the other hand, synthetic inhibitors have been synthesized using known substrates for a particular enzyme [13]. Research on this aims to create antiviral, anticancer, and anti-diabetic therapeutic agents. Genetic disorders are a specific area of research. As a result, extensive research is still being done on synthetic inhibitor molecules that focus on inhibiting enzymes [123, 124]. Inhibitors generally act as precursors with altered specificity and increased activity. Several inhibitors of glycosylation exist that inhibit the metabolism or intracellular transport activities of common precursors [13]. Some of these affect proteins in the region between the ER, Golgi, and trans-Golgi apparatus, which causes indirect action. On the other hand, some inhibitors work on important stages of the intermediate metabolism, such as glycosylation. Some of the inhibitor molecules block glycosyltransferases (for example, substrate analogs), while others act on glycosidases (for example, alkaloids). They are indirect inhibitors, tunicamycin, plant alkaloids, substrate analogs, glycoside primers, glycolipid inhibitors, and neuraminidase inhibitors. Glycosylation inhibitors have an impact on various glycan synthesis pathways. Some inhibitor molecules affect central reactions, but others are highly specific and affect only one specific enzyme reaction (*e.g.*, neurominidase inhibitors) [86, 214, 215].

In diabetes, it is aimed at reducing the flow of glucose absorbed through the intestinal wall after eating. Reduction of glucose absorption can be accomplished with glucose hydrolase inhibitors, such as azasugars, similar to acarbose, which have proven successful in diabetic treatment by slowing the conversion of dietary carbohydrates to glucose. Miglitol is a derivative of 1-deuximmojirimycin, which is an initiator of maltase and sucrase and is also effective in the treatment of diabetes [163, 174, 181].

CONCLUDING REMARKS

Glycobiology's significance is growing in the fields of medicine, basic research, and biotechnology. Various disorders can originate and evolve as a result of modifications to the glycosylation system. Changes in the function, structure, and quantity of the substrate molecules used in glycosylation as well as the end products are crucial to the emergence of various disorders. Glycoprotein synthesis has been found to rise in several cancer kinds and illnesses, according to numerous research. Serum sialic acid measures in conjunction with other clinical and biochemical criteria are used to diagnose and treat cancer patients who have particularly high sialic acid levels. Glycan biology has taken on a significant role in research as a key element in comprehending various disease mechanisms, diagnoses, and therapy approaches. Additionally, glycans serve as the primary regulatory component of a number of pathological processes. Glycans are a possible source for the creation of novel biomarkers, which are desperately needed today for the early diagnosis, risk assessment, and treatment of diseases, including cancer. The most often utilized serological indicators for clinical illness diagnosis, tracking disease progression, and prognosis of disease relapse are glycoproteins. These biomarkers have been demonstrated by abnormal glycosylation, particularly in cancer cells. Due to their low specificity, these biomarkers can only be used in limited screening and diagnostic procedures. In order to diagnose and detect diseases early, new research and high specificity biomarker methodologies are needed. The quick development of glycoengineering and model platforms will be made possible by new information and understanding in glycobiology. Growing data combined with current developments in glycomics, glycoproteomics, genomes, metabolomics, and proteomics will offer new targets and effective approaches for early disease diagnosis, prognosis, and enhanced treatments. In short, it's crucial for the design and synthesis of carbohydrate-based medications to have a better understanding of aberrant glycan processes.

The goal of current glycoengineering research, which is promising, is to offer options for treating existing diseases by creating humanized or human glycoconjugates from a variety of organisms and discovering natural or synthetic inhibitor molecules. It is crucial that the chemical synthesis of glycoconjugates be sustainable due to the importance of these compounds as potential therapeutics. Glycoconjugates must be thoroughly characterized, biocompatible, and scaleable for this objective to be accomplished.

REFERENCES

[1]   A.W.T. Chiang, H.M. Baghdassarian, B.P. Kellman, B. Bao, J.T. Sorrentino, C. Liang, C.C. Kuo, H.O. Masson, and N.E. Lewis, "Systems glycobiology for discovering drug targets, biomarkers, and rational designs for glyco-immunotherapy", *J. Biomed. Sci.,* vol. 28, no. 1, p. 50, 2021. [http://dx.doi.org/10.1186/s12929-021-00746-2] [PMID: 34158025]

[2]   A. Varki, "Biological roles of glycans", *Glycobiology,* vol. 27, no. 1, pp. 3-49, 2017.
        [http://dx.doi.org/10.1093/glycob/cww086] [PMID: 27558841]

[3]   G.Y. Wiederschain, "Glycobiology: Progress, problems, and perspectives", *Biochemistry,* vol. 78, no. 7, pp. 679-696, 2013.
        [http://dx.doi.org/10.1134/S0006297913070018] [PMID: 24010832]

[4]   J. Yin, "Chemical glycobiology drives the discovery of carbohydrate-based drugs", *Chin. J. Nat. Med.,* vol. 18, no. 10, pp. 721-722, 2020.
        [http://dx.doi.org/10.1016/S1875-5364(20)60011-5] [PMID: 33039050]

[5]   N. Taniguchi, "Integrative glycobiology and future perspectives", *Biochem. Biophys. Res. Commun.,* vol. 453, no. 2, pp. 199-200, 2014.
        [http://dx.doi.org/10.1016/j.bbrc.2014.10.034] [PMID: 25438779]

[6]   A. Frey, K.T. Giannasca, R. Weltzin, P.J. Giannasca, H. Reggio, W.I. Lencer, and M.R. Neutra, "Role of the glycocalyx in regulating access of microparticles to apical plasma membranes of intestinal epithelial cells: Implications for microbial attachment and oral vaccine targeting", *J. Exp. Med.,* vol. 184, no. 3, pp. 1045-1059, 1996.
        [http://dx.doi.org/10.1084/jem.184.3.1045] [PMID: 9064322]

[7]   Z. Ma, F. Yang, J. Fan, X. Li, Y. Liu, W. Chen, H. Sun, T. Ma, Q. Wang, Y. Maihaiti, and X. Ren, "Identification and immune characteristics of molecular subtypes related to protein glycosylation in Alzheimer's disease", *Front. Aging Neurosci.,* vol. 14, p. 968190, 2022.
        [http://dx.doi.org/10.3389/fnagi.2022.968190] [PMID: 36408104]

[8]   G.W. Hart, R.S. Haltiwanger, G.D. Holt, and W.G. Kelly, "Glycosylation in the nucleus and cytoplasm", *Annu. Rev. Biochem.,* vol. 58, no. 1, pp. 841-874, 1989.
        [http://dx.doi.org/10.1146/annurev.bi.58.070189.004205] [PMID: 2673024]

[9]   P.L. Debbage, "A systematic histochemical investigation in mammals of the dense glycocalyx glycosylations common to all cells bordering the interstitial fluid compartment of the brain", *Acta Histochem.,* vol. 98, no. 1, pp. 9-28, 1996.
        [http://dx.doi.org/10.1016/S0065-1281(96)80046-8] [PMID: 9054194]

[10]   R. Huisjes, T.J. Satchwell, L.P. Verhagen, R.M. Schiffelers, W.W. van Solinge, A.M. Toye, and R. van Wijk, "Quantitative measurement of red cell surface protein expression reveals new biomarkers for hereditary spherocytosis", *Int. J. Lab. Hematol.,* vol. 40, no. 4, pp. e74-e77, 2018.
        [http://dx.doi.org/10.1111/ijlh.12841] [PMID: 29746727]

[11]   K.J. Kang, M.J. Choi, T.J. Min, T.M. You, G. Lee, S.Y. Ko, and Y.J. Jang, "Cell surface accumulation of intracellular leucine proline-enriched proteoglycan 1 enhances odontogenic potential of human dental pulp stem cells", *Stem Cells Dev.,* vol. 31, no. 21-22, pp. 684-695, 2022.
        [http://dx.doi.org/10.1089/scd.2022.0174] [PMID: 35859453]

[12]   A. Varki, "Biological roles of oligosaccharides: All of the theories are correct", *Glycobiology,* vol. 3, no. 2, pp. 97-130, 1993.
        [http://dx.doi.org/10.1093/glycob/3.2.97] [PMID: 8490246]

[13]   A Varki, RD Cummings, JD Esko, P Stanley, GW Hart, and M Aebi, *Essentials of glycobiology* 4th. Cold Spring Harbor Laboratory Press: Cold Spring Harbor (NY), 2015.

[14]   F. Leisico, J. Omeiri, C. Le Narvor, J. Beaudouin, M. Hons, D. Fenel, G. Schoehn, Y. Couté, D. Bonnaffé, R. Sadir, H. Lortat-Jacob, and R. Wild, "Structure of the human heparan sulfate polymerase complex EXT1-EXT2", *Nat. Commun.,* vol. 13, no. 1, p. 7110, 2022.
        [http://dx.doi.org/10.1038/s41467-022-34882-6] [PMID: 36402845]

[15]   M. Salmivirta, K. Lidholt, and U. Lindahl, "Heparan sulfate: A piece of information", *FASEB J.,* vol. 10, no. 11, pp. 1270-1279, 1996.
        [http://dx.doi.org/10.1096/fasebj.10.11.8836040] [PMID: 8836040]

[16]   I. Alcalde, C. Sánchez-Fernández, S. Del Olmo-Aguado, C. Martín, C. Olmiere, E. Artime, L.M.

Quirós, and J. Merayo-Lloves, "Synthetic heparan sulfate mimetic polymer enhances corneal nerve regeneration and wound healing after experimental laser ablation injury in mice", *Polymers,* vol. 14, no. 22, p. 4921, 2022.
[http://dx.doi.org/10.3390/polym14224921] [PMID: 36433048]

[17]   J.M. Keller, "Specificity in heparin/heparan sulphate-protein interactions", *Glycobiology,* vol. 4, no. 1, pp. 1-2, 1994.
[http://dx.doi.org/10.1093/glycob/4.1.1-a] [PMID: 8186545]

[18]   O.M. Saad, H. Ebel, K. Uchimura, S.D. Rosen, C.R. Bertozzi, and J.A. Leary, "Compositional profiling of heparin/heparan sulfate using mass spectrometry: assay for specificity of a novel extracellular human endosulfatase", *Glycobiology,* vol. 15, no. 8, pp. 818-826, 2005.
[http://dx.doi.org/10.1093/glycob/cwi064] [PMID: 15843596]

[19]   Y.J. Kim, A.J. Grodzinsky, and A.H.K. Plaas, "Compression of cartilage results in differential effects on biosynthetic pathways for aggrecan, link protein, and hyaluronan", *Arch. Biochem. Biophys.,* vol. 328, no. 2, pp. 331-340, 1996.
[http://dx.doi.org/10.1006/abbi.1996.0181] [PMID: 8645012]

[20]   M.I. Wiweger, C.M. Avramut, C.E. de Andrea, F.A. Prins, A.J. Koster, R.B.G. Ravelli, and P.C.W. Hogendoorn, "Cartilage ultrastructure in proteoglycan-deficient zebrafish mutants brings to light new candidate genes for human skeletal disorders", *J. Pathol.,* vol. 223, no. 4, pp. 531-542, 2011.
[http://dx.doi.org/10.1002/path.2824] [PMID: 21294126]

[21]   J. Frisch, and M. Cucchiarini, "Gene- and stem cell-based approaches to regulate hypertrophic differentiation in articular cartilage disorders", *Stem Cells Dev.,* vol. 25, no. 20, pp. 1495-1512, 2016.
[http://dx.doi.org/10.1089/scd.2016.0106] [PMID: 27269415]

[22]   I. Alho, L. Costa, M. Bicho, and C. Coelho, "Characterization of low molecular weight protein tyrosine phosphatase isoforms in human breast cancer epithelial cell lines", *Anticancer Res.,* vol. 33, no. 5, pp. 1983-1987, 2013.
[PMID: 23645747]

[23]   C.P. Dietrich, I.L. Tersariol, L. Toma, C.T. Moraes, M.A. Porcionatto, F.W. Oliveira, and H.B. Nader, "Structure of heparan sulfate: Identification of variable and constant oligosaccharide domains in eight heparan sulfates of different origins", *Cell. Mol. Biol.,* vol. 44, no. 3, pp. 417-429, 1998.
[PMID: 9620437]

[24]   S. Faham, R.E. Hileman, J.R. Fromm, R.J. Linhardt, and D.C. Rees, "Heparin structure and interactions with basic fibroblast growth factor", *Science,* vol. 271, no. 5252, pp. 1116-1120, 1996.
[http://dx.doi.org/10.1126/science.271.5252.1116] [PMID: 8599088]

[25]   A. Baim-Lance, M. Angulo, M.A. Chiasson, H.M. Lekas, R. Schenkel, J. Villarreal, A. Cantos, C. Kerr, A. Nagaraja, M.T. Yin, and P. Gordon, "Challenges and opportunities of telehealth digital equity to manage HIV and comorbidities for older persons living with HIV in New York State", *BMC Health Serv. Res.,* vol. 22, no. 1, p. 609, 2022.
[http://dx.doi.org/10.1186/s12913-022-08010-5] [PMID: 35524251]

[26]   L. Zhu, X. Wei, J. Cong, J. Zou, L. Wan, and S. Xu, "Structural insights into mechanism and specificity of the plant protein O-fucosyltransferase SPINDLY", *Nat. Commun.,* vol. 13, no. 1, p. 7424, 2022.
[http://dx.doi.org/10.1038/s41467-022-35234-0] [PMID: 36456586]

[27]   A. Imberty, and S. Pérez, "Structure, conformation, and dynamics of bioactive oligosaccharides: theoretical approaches and experimental validations", *Chem. Rev.,* vol. 100, no. 12, pp. 4567-4588, 2000.
[http://dx.doi.org/10.1021/cr990343j] [PMID: 11749358]

[28]   R.A. Laine, "A calculation of all possible oligosaccharide isomers both branched and linear yields 1.05 x 10(12) structures for a reducing hexasaccharide: The Isomer Barrier to development of single-

method saccharide sequencing or synthesis systems", *Glycobiology,* vol. 4, no. 6, pp. 759-767, 1994.
[http://dx.doi.org/10.1093/glycob/4.6.759] [PMID: 7734838]

[29]   P.H. Seeberger, and W.C. Haase, "Solid-phase oligosaccharide synthesis and combinatorial carbohydrate libraries", *Chem. Rev.,* vol. 100, no. 12, pp. 4349-4394, 2000.
[http://dx.doi.org/10.1021/cr9903104] [PMID: 11749351]

[30]   K.A. Karlsson, "Microbial recognition of target-cell glycoconjugates", *Curr. Opin. Struct. Biol.,* vol. 5, no. 5, pp. 622-635, 1995.
[http://dx.doi.org/10.1016/0959-440X(95)80054-9] [PMID: 8574698]

[31]   T. Matsuzaki, D. Terutsuki, S. Sato, K. Ikarashi, K. Sato, H. Mitsuno, R. Okumura, Y. Yoshimura, S. Usami, Y. Mori, M. Fujii, S. Takemi, S. Nakabayashi, H.Y. Yoshikawa, and R. Kanzaki, "Low surface potential with glycoconjugates determines insect cell adhesion at room temperature", *J. Phys. Chem. Lett.,* vol. 13, no. 40, pp. 9494-9500, 2022.
[http://dx.doi.org/10.1021/acs.jpclett.2c01673] [PMID: 36201238]

[32]   D.H. Dube, and C.R. Bertozzi, "Glycans in cancer and inflammation: Potential for therapeutics and diagnostics", *Nat. Rev. Drug Discov.,* vol. 4, no. 6, pp. 477-488, 2005.
[http://dx.doi.org/10.1038/nrd1751] [PMID: 15931257]

[33]   R.A. Dwek, "Glycobiology: Toward understanding the function of sugars", *Chem. Rev.,* vol. 96, no. 2, pp. 683-720, 1996.
[http://dx.doi.org/10.1021/cr940283b] [PMID: 11848770]

[34]   "Society for glycobiology awards-2022", *Glycobiology,* vol. 32, no. 11, pp. 912-916, 2022.
[PMID: 36315011]

[35]   J.S. dos Reis, M.A. Rodrigues da Costa Santos, D.P. Mendonça, S.I. Martins do Nascimento, P.M. Barcelos, R.G. Correia de Lima, K.M. da Costa, C.G. Freire-de-Lima, A. Morrot, J.O. Previato, L. Mendonça Previato, L.M. da Fonseca, and L. Freire-de-Lima, "Glycobiology of cancer: Sugar drives the show", *Medicines,* vol. 9, no. 6, p. 34, 2022.
[http://dx.doi.org/10.3390/medicines9060034] [PMID: 35736247]

[36]   S.I. Hakomori, and W.T. Murakami, "Glycolipids of hamster fibroblasts and derived malignant-transformed cell lines", *Proc. Natl. Acad. Sci.,* vol. 59, no. 1, pp. 254-261, 1968.
[http://dx.doi.org/10.1073/pnas.59.1.254] [PMID: 4298334]

[37]   M.T.C. Walvoort, A.G. Volbeda, N.R.M. Reintjens, H. van den Elst, O.J. Plante, H.S. Overkleeft, G.A. van der Marel, and J.D.C. Codée, "Automated solid-phase synthesis of hyaluronan oligosaccharides", *Org. Lett.,* vol. 14, no. 14, pp. 3776-3779, 2012.
[http://dx.doi.org/10.1021/ol301666n] [PMID: 22780913]

[38]   M. Ambrosi, N.R. Cameron, and B.G. Davis, "Lectins: Tools for the molecular understanding of the glycocode", *Org. Biomol. Chem.,* vol. 3, no. 9, pp. 1593-1608, 2005.
[http://dx.doi.org/10.1039/b414350g] [PMID: 15858635]

[39]   B. Holm, J. Bäcklund, M.A.F. Recio, R. Holmdahl, and J. Kihlberg, "Glycopeptide specificity of helper T cells obtained in mouse models for rheumatoid arthritis", *ChemBioChem,* vol. 3, no. 12, pp. 1209-1222, 2002.
[http://dx.doi.org/10.1002/1439-7633(20021202)3:12<1209::AID-CBIC1209>3.0.CO;2-0]   [PMID: 12465029]

[40]   C.R. Bertozzi, and L.L. Kiessling, "Chemical glycobiology", *Science,* vol. 291, no. 5512, pp. 2357-2364, 2001.
[http://dx.doi.org/10.1126/science.1059820] [PMID: 11269316]

[41]   P.H. Seeberger, "Chemical glycobiology: Why now?", *Nat. Chem. Biol.,* vol. 5, no. 6, pp. 368-372, 2009.
[http://dx.doi.org/10.1038/nchembio0609-368] [PMID: 19448600]

[42]   M. Mammen, S.K. Choi, and G.M. Whitesides, "Polyvalent interactions in biological systems:

Implications for design and use of multivalent ligands and inhibitors", *Angew. Chem. Int. Ed.,* vol. 37, no. 20, pp. 2754-2794, 1998.
[http://dx.doi.org/10.1002/(SICI)1521-3773(19981102)37:20<2754::AID-ANIE2754>3.0.CO;2-3]
[PMID: 29711117]

[43]   J. Holgersson, A. Gustafsson, and M.E. Breimer, "Characteristics of protein–carbohydrate interactions as a basis for developing novel carbohydrate–based antirejection therapies", *Immunol. Cell Biol.,* vol. 83, no. 6, pp. 694-708, 2005.
[http://dx.doi.org/10.1111/j.1440-1711.2005.01373.x] [PMID: 16266322]

[44]   S.L. Hart, A.M. Knight, R.P. Harbottle, A. Mistry, H.D. Hunger, D.F. Cutler, R. Williamson, and C. Coutelle, "Cell binding and internalization by filamentous phage displaying a cyclic Arg-Gly-Ap-containing peptide", *J. Biol. Chem.,* vol. 269, no. 17, pp. 12468-12474, 1994.
[http://dx.doi.org/10.1016/S0021-9258(18)99898-4] [PMID: 8175653]

[45]   M.O. Kanev, and E. Bakar, "Glycoconjgates in cancer", *J. Heal. Sci.Koca. Uni.,* vol. 2, no. 1, pp. 1-5, 2016.

[46]   R.A. Flynn, K. Pedram, S.A. Malaker, P.J. Batista, B.A.H. Smith, A.G. Johnson, B.M. George, K. Majzoub, P.W. Villalta, J.E. Carette, and C.R. Bertozzi, "Small RNAs are modified with N-glycans and displayed on the surface of living cells", *Cell,* vol. 184, no. 12, pp. 3109-3124.e22, 2021.
[http://dx.doi.org/10.1016/j.cell.2021.04.023] [PMID: 34004145]

[47]   M.D. Disney, "A glimpse at the glycoRNA world", *Cell,* vol. 184, no. 12, pp. 3080-3081, 2021.
[http://dx.doi.org/10.1016/j.cell.2021.05.025] [PMID: 34115968]

[48]   S. Nachtergaele, and C. He, "Chemical modifications in the life of an mRNA transcript", *Annu. Rev. Genet.,* vol. 52, no. 1, pp. 349-372, 2018.
[http://dx.doi.org/10.1146/annurev-genet-120417-031522] [PMID: 30230927]

[49]   P.R. Crocker, J.C. Paulson, and A. Varki, "Siglecs and their roles in the immune system", *Nat. Rev. Immunol.,* vol. 7, no. 4, pp. 255-266, 2007.
[http://dx.doi.org/10.1038/nri2056] [PMID: 17380156]

[50]   R. Zeng, F. Vingopoulos, M. Wang, A. Bannerman, H.E. Wescott, and G. Baldwin, "Structure-function association between contrast sensitivity and retinal thickness (total, regional, and individual retinal layer) in patients with idiopathic epiretinal membrane", *Graefes Arch. Clin. Exp. Ophthalmol.,* vol. 261, no. 3, pp. 631-639, 2022.
[PMID: 36149494]

[51]   C.A. Lingwood, "Glycosphingolipid functions", *Cold Spring Harb. Perspect. Biol.,* vol. 3, no. 7, p. a004788, 2011.
[http://dx.doi.org/10.1101/cshperspect.a004788] [PMID: 21555406]

[52]   H. Schulze, and K. Sandhoff, "Lysosomal lipid storage diseases", *Cold Spring Harb. Perspect. Biol.,* vol. 3, no. 6, p. a004804, 2011.
[http://dx.doi.org/10.1101/cshperspect.a004804] [PMID: 21502308]

[53]   Y. Kinjo, and K. Ueno, "iNKT cells in microbial immunity: Recognition of microbial glycolipids", *Microbiol. Immunol.,* vol. 55, no. 7, pp. 472-482, 2011.
[http://dx.doi.org/10.1111/j.1348-0421.2011.00338.x] [PMID: 21434991]

[54]   J. Chan, T. Fujiwara, P. Brennan, M. McNeil, S.J. Turco, J.C. Sibille, M. Snapper, P. Aisen, and B.R. Bloom, "Microbial glycolipids: Possible virulence factors that scavenge oxygen radicals", *Proc. Natl. Acad. Sci.,* vol. 86, no. 7, pp. 2453-2457, 1989.
[http://dx.doi.org/10.1073/pnas.86.7.2453] [PMID: 2538841]

[55]   A. Magérus-Chatinet, H. Yu, S. Garcia, E. Ducloux, B. Terris, and M. Bomsel, "Galactosyl ceramide expressed on dendritic cells can mediate HIV-1 transfer from monocyte derived dendritic cells to autologous T cells", *Virology,* vol. 362, no. 1, pp. 67-74, 2007.
[http://dx.doi.org/10.1016/j.virol.2006.11.035] [PMID: 17234232]

[56]   B.B. Finlay, and S. Falkow, "Common themes in microbial pathogenicity revisited", *Microbiol. Mol. Biol. Rev.,* vol. 61, no. 2, pp. 136-169, 1997.
[PMID: 9184008]

[57]   A. Varki, and T. Angata, "Siglecs—the major subfamily of I-type lectins", *Glycobiology,* vol. 16, no. 1, pp. 1R-27R, 2006.
[http://dx.doi.org/10.1093/glycob/cwj008] [PMID: 16014749]

[58]   H.H. Freeze, "Understanding human glycosylation disorders: biochemistry leads the charge", *J. Biol. Chem.,* vol. 288, no. 10, pp. 6936-6945, 2013.
[http://dx.doi.org/10.1074/jbc.R112.429274] [PMID: 23329837]

[59]   "The nomenclature of lipids (Recommendations 1976) IUPAC-IUB commission on biochemical nomenclature", *Biochem. J.,* vol. 171, no. 1, pp. 21-35, 1978.
[http://dx.doi.org/10.1042/bj1710021] [PMID: 646817]

[60]   C. Reily, T.J. Stewart, M.B. Renfrow, and J. Novak, "Glycosylation in health and disease", *Nat. Rev. Nephrol.,* vol. 15, no. 6, pp. 346-366, 2019.
[http://dx.doi.org/10.1038/s41581-019-0129-4] [PMID: 30858582]

[61]   G.C. Hansson, K.A. Karlsson, G. Larson, N. Strömberg, and J. Thurin, "Carbohydrate-specific adhesion of bacteria to thin-layer chromatograms: A rationalized approach to the study of host cell glycolipid receptors", *Anal. Biochem.,* vol. 146, no. 1, pp. 158-163, 1985.
[http://dx.doi.org/10.1016/0003-2697(85)90410-5] [PMID: 3993927]

[62]   Y. Dutronc, and S.A. Porcelli, "The CD1 family and T cell recognition of lipid antigens", *Tissue Antigens,* vol. 60, no. 5, pp. 337-353, 2002.
[http://dx.doi.org/10.1034/j.1399-0039.2002.600501.x] [PMID: 12492810]

[63]   S. Joyce, and L. Van Kaer, "CD1-restricted antigen presentation: An oily matter", *Curr. Opin. Immunol.,* vol. 15, no. 1, pp. 95-104, 2003.
[http://dx.doi.org/10.1016/S0952-7915(02)00012-2] [PMID: 12495740]

[64]   M. Brigl, and M.B. Brenner, "CD1: Antigen presentation and T cell function", *Annu. Rev. Immunol.,* vol. 22, no. 1, pp. 817-890, 2004.
[http://dx.doi.org/10.1146/annurev.immunol.22.012703.104608] [PMID: 15032598]

[65]   S.D. Gadola, N.R. Zaccai, K. Harlos, D. Shepherd, J.C. Castro-Palomino, G. Ritter, R.R. Schmidt, E.Y. Jones, and V. Cerundolo, "Structure of human CD1b with bound ligands at 2.3 Å, a maze for alkyl chains", *Nat. Immunol.,* vol. 3, no. 8, pp. 721-726, 2002.
[http://dx.doi.org/10.1038/ni821] [PMID: 12118248]

[66]   D.M. Zajonc, C. Cantu III, J. Mattner, D. Zhou, P.B. Savage, A. Bendelac, I.A. Wilson, and L. Teyton, "Structure and function of a potent agonist for the semi-invariant natural killer T cell receptor", *Nat. Immunol.,* vol. 6, no. 8, pp. 810-818, 2005.
[http://dx.doi.org/10.1038/ni1224] [PMID: 16007091]

[67]   T. Batuwangala, D. Shepherd, S.D. Gadola, K.J.C. Gibson, N.R. Zaccai, A.R. Fersht, G.S. Besra, V. Cerundolo, and E.Y. Jones, "The crystal structure of human CD1b with a bound bacterial glycolipid", *J. Immunol.,* vol. 172, no. 4, pp. 2382-2388, 2004.
[http://dx.doi.org/10.4049/jimmunol.172.4.2382] [PMID: 14764708]

[68]   D.M. Zajonc, M.A. Elsliger, L. Teyton, and I.A. Wilson, "Crystal structure of CD1a in complex with a sulfatide self antigen at a resolution of 2.15 Å", *Nat. Immunol.,* vol. 4, no. 8, pp. 808-815, 2003.
[http://dx.doi.org/10.1038/ni948] [PMID: 12833155]

[69]   B. Giabbai, S. Sidobre, M.D.M. Crispin, Y. Sanchez-Ruìz, A. Bachi, M. Kronenberg, I.A. Wilson, and M. Degano, "Crystal structure of mouse CD1d bound to the self ligand phosphatidylcholine: A molecular basis for NKT cell activation", *J. Immunol.,* vol. 175, no. 2, pp. 977-984, 2005.
[http://dx.doi.org/10.4049/jimmunol.175.2.977] [PMID: 16002697]

[70]   A.M. Luoma, C.D. Castro, T. Mayassi, L.A. Bembinster, L. Bai, D. Picard, B. Anderson, L. Scharf, J.E. Kung, L.V. Sibener, P.B. Savage, B. Jabri, A. Bendelac, and E.J. Adams, "Crystal structure of Vδ1 T cell receptor in complex with CD1d-sulfatide shows MHC-like recognition of a self-lipid by human γδ T cells", *Immunity,* vol. 39, no. 6, pp. 1032-1042, 2013.
[http://dx.doi.org/10.1016/j.immuni.2013.11.001] [PMID: 24239091]

[71]   A. Shamshiev, H.J. Gober, A. Donda, Z. Mazorra, L. Mori, and G. De Libero, "Presentation of the same glycolipid by different CD1 molecules", *J. Exp. Med.,* vol. 195, no. 8, pp. 1013-1021, 2002.
[http://dx.doi.org/10.1084/jem.20011963] [PMID: 11956292]

[72]   A. Jahng, I. Maricic, C. Aguilera, S. Cardell, R.C. Halder, and V. Kumar, "Prevention of autoimmunity by targeting a distinct, noninvariant CD1d-reactive T cell population reactive to sulfatide", *J. Exp. Med.,* vol. 199, no. 7, pp. 947-957, 2004.
[http://dx.doi.org/10.1084/jem.20031389] [PMID: 15051763]

[73]   D.B. Moody, T. Ulrichs, W. Mühlecker, D.C. Young, S.S. Gurcha, E. Grant, J.P. Rosat, M.B. Brenner, C.E. Costello, G.S. Besra, and S.A. Porcelli, "CD1c-mediated T-cell recognition of isoprenoid glycolipids in Mycobacterium tuberculosis infection", *Nature,* vol. 404, no. 6780, pp. 884-888, 2000.
[http://dx.doi.org/10.1038/35009119] [PMID: 10786796]

[74]   Q. Wang, and L. Chi, "The alterations and roles of glycosaminoglycans in human diseases", *Polymers,* vol. 14, no. 22, p. 5014, 2022.
[http://dx.doi.org/10.3390/polym14225014] [PMID: 36433141]

[75]   J.R. Couchman, "Transmembrane signaling proteoglycans", *Annu. Rev. Cell Dev. Biol.,* vol. 26, no. 1, pp. 89-114, 2010.
[http://dx.doi.org/10.1146/annurev-cellbio-100109-104126] [PMID: 20565253]

[76]   H. Nakato, and K. Kimata, "Heparan sulfate fine structure and specificity of proteoglycan functions", *Biochim. Biophys. Acta, Gen. Subj.,* vol. 1573, no. 3, pp. 312-318, 2002.
[http://dx.doi.org/10.1016/S0304-4165(02)00398-7] [PMID: 12417413]

[77]   R.V. Iozzo, and L. Schaefer, "Proteoglycan form and function: A comprehensive nomenclature of proteoglycans", *Matrix Biol.,* vol. 42, pp. 11-55, 2015.
[http://dx.doi.org/10.1016/j.matbio.2015.02.003] [PMID: 25701227]

[78]   H.E. Bülow, and O. Hobert, "The molecular diversity of glycosaminoglycans shapes animal development", *Annu. Rev. Cell Dev. Biol.,* vol. 22, no. 1, pp. 375-407, 2006.
[http://dx.doi.org/10.1146/annurev.cellbio.22.010605.093433] [PMID: 16805665]

[79]   B. Gesslbauer, A. Rek, F. Falsone, E. Rajkovic, and A.J. Kungl, "Proteoglycanomics: Tools to unravel the biological function of glycosaminoglycans", *Proteomics,* vol. 7, no. 16, pp. 2870-2880, 2007.
[http://dx.doi.org/10.1002/pmic.200700176] [PMID: 17654462]

[80]   J.D. Esko, and S.B. Selleck, "Order out of chaos: Assembly of ligand binding sites in heparan sulfate", *Annu. Rev. Biochem.,* vol. 71, no. 1, pp. 435-471, 2002.
[http://dx.doi.org/10.1146/annurev.biochem.71.110601.135458] [PMID: 12045103]

[81]   J.T. Gallagher, "Multiprotein signalling complexes: Regional assembly on heparan sulphate", *Biochem. Soc. Trans.,* vol. 34, no. 3, pp. 438-441, 2006.
[http://dx.doi.org/10.1042/BST0340438] [PMID: 16709181]

[82]   H. Habuchi, G. Miyake, K. Nogami, A. Kuroiwa, Y. Matsuda, M. Kusche-Gullberg, O. Habuchi, M. Tanaka, and K. Kimata, "Biosynthesis of heparan sulphate with diverse structures and functions: two alternatively spliced forms of human heparan sulphate 6-O-sulphotransferase-2 having different expression patterns and properties", *Biochem. J.,* vol. 371, no. 1, pp. 131-142, 2003.
[http://dx.doi.org/10.1042/bj20021259] [PMID: 12492399]

[83]   S.S. Shivatare, V.S. Shivatare, and C.H. Wong, "Glycoconjugates: Synthesis, functional studies, and therapeutic developments", *Chem. Rev.,* vol. 122, no. 20, pp. 15603-15671, 2022.
[http://dx.doi.org/10.1021/acs.chemrev.1c01032] [PMID: 36174107]

[84]   A. Kobata, "Structures and functions of the sugar chains of glycoproteins", *Eur. J. Biochem.,* vol. 209, no. 2, pp. 483-501, 1992.
[http://dx.doi.org/10.1111/j.1432-1033.1992.tb17313.x] [PMID: 1358608]

[85]   T. Higashiyama, M. Umekawa, M. Nagao, T. Katoh, H. Ashida, and K. Yamamoto, "Chemo-enzymatic synthesis of the glucagon containing N-linked oligosaccharide and its characterization", *Carbohydr. Res.,* vol. 455, pp. 92-96, 2018.
[http://dx.doi.org/10.1016/j.carres.2017.11.007] [PMID: 29175660]

[86]   Y. Akimoto, I. Brockhausen, K.J. Colley, P.R. Crocker, T.L. Doering, and A.D. Elbein, *Essentials of Glycobiology* 2nd. Cold Spring Harbor Laboratory Press: Cold Spring Harbor (NY), 2009. Available From: https://www.ncbi.nlm.nih.gov/books/NBK1908/

[87]   D.R. Beriault, V.T. Dang, L.H. Zhong, C.I. Petlura, C.S. McAlpine, Y. Shi, and G.H. Werstuck, "Glucosamine induces ER stress by disrupting lipid-linked oligosaccharide biosynthesis and N -linked protein glycosylation", *Am. J. Physiol. Endocrinol. Metab.,* vol. 312, no. 1, pp. E48-E57, 2017.
[http://dx.doi.org/10.1152/ajpendo.00275.2016] [PMID: 27879249]

[88]   H. Yoshitake, N. Hashii, N. Kawasaki, S. Endo, K. Takamori, A. Hasegawa, H. Fujiwara, and Y. Araki, "Chemical characterization of N-Linked oligosaccharide as the antigen epitope recognized by an anti-sperm auto-monoclonal antibody, Ts4", *PLoS One,* vol. 10, no. 7, p. e0133784, 2015.
[http://dx.doi.org/10.1371/journal.pone.0133784] [PMID: 26222427]

[89]   H. Abe, K. Tomimoto, Y. Fujita, T. Iwaki, Y. Chiba, K.I. Nakayama, and Y. Nakajima, "Development of N- and O-linked oligosaccharide engineered Saccharomyces cerevisiae strain", *Glycobiology,* vol. 26, no. 11, pp. 1248-1256, 2016.
[PMID: 27496768]

[90]   A. Maddi, and S.J. Free, "α-1,6-Mannosylation of n-linked oligosaccharide present on cell wall proteins is required for their incorporation into the cell wall in the filamentous fungus neurospora crassa", *Eukaryot. Cell,* vol. 9, no. 11, pp. 1766-1775, 2010.
[http://dx.doi.org/10.1128/EC.00134-10] [PMID: 20870880]

[91]   Y. Kanoh, S. Egawa, S. Baba, and T. Akahoshi, "Associations of IgG N -linked oligosaccharide chains and proteases in sera of prostate cancer patients with and without α2-macroglobulin deficiency", *J. Clin. Lab. Anal.,* vol. 23, no. 2, pp. 125-131, 2009.
[http://dx.doi.org/10.1002/jcla.20302] [PMID: 19288446]

[92]   H. Nothaft, and C.M. Szymanski, "Bacterial protein N-glycosylation: New perspectives and applications", *J. Biol. Chem.,* vol. 288, no. 10, pp. 6912-6920, 2013.
[http://dx.doi.org/10.1074/jbc.R112.417857] [PMID: 23329827]

[93]   D. Calo, L. Kaminski, and J. Eichler, "Protein glycosylation in archaea: Sweet and extreme", *Glycobiology,* vol. 20, no. 9, pp. 1065-1076, 2010.
[http://dx.doi.org/10.1093/glycob/cwq055] [PMID: 20371512]

[94]   O. Haji-Ghassemi, M. Gilbert, J. Spence, M.J. Schur, M.J. Parker, M.L. Jenkins, J.E. Burke, H. van Faassen, N.M. Young, and S.V. Evans, "Molecular basis for recognition of the cancer glycobiomarker, LacdiNAc (GalNAc[β1→4]GlcNAc), by Wisteria floribunda Agglutinin", *J. Biol. Chem.,* vol. 291, no. 46, pp. 24085-24095, 2016.
[http://dx.doi.org/10.1074/jbc.M116.750463] [PMID: 27601469]

[95]   I.C. Schoenhofen, N.M. Young, and M. Gilbert, "Biosynthesis of legionaminic acid and its incorporation into glycoconjugates", *Methods Enzymol.,* vol. 597, pp. 187-207, 2017.
[http://dx.doi.org/10.1016/bs.mie.2017.06.042] [PMID: 28935102]

[96]   T.R. Tivey, J.E. Parkinson, P.E. Mandelare, D.A. Adpressa, W. Peng, X. Dong, Y. Mechref, V.M. Weis, and S. Loesgen, "N-Linked surface glycan biosynthesis, composition, inhibition, and function in cnidarian-dinoflagellate symbiosis", *Microb. Ecol.,* vol. 80, no. 1, pp. 223-236, 2020.
[http://dx.doi.org/10.1007/s00248-020-01487-9] [PMID: 31982929]

[97]   D.K. Hansen, J.R. Winther, and M. Willemoës, "Steady state kinetic analysis of O-linked GalNAc glycan release catalyzed by endo-α-N-acetylgalactosaminidase", *Carbohydr. Res.,* vol. 480, pp. 54-60, 2019.
[http://dx.doi.org/10.1016/j.carres.2019.05.009] [PMID: 31176190]

[98]   T. Kouka, S. Akase, I. Sogabe, C. Jin, N.G. Karlsson, and K.F. Aoki-Kinoshita, "Computational modeling of O-Linked glycan biosynthesis in CHO Cells", *Molecules,* vol. 27, no. 6, p. 1766, 2022.
[http://dx.doi.org/10.3390/molecules27061766] [PMID: 35335136]

[99]   J. Li, S. Mu, J. Yang, C. Liu, Y. Zhang, P. Chen, Y. Zeng, Y. Zhu, and Y. Sun, "Glycosyltransferase engineering and multi-glycosylation routes development facilitating synthesis of high-intensity sweetener mogrosides", *iScience,* vol. 25, no. 10, p. 105222, 2022.
[http://dx.doi.org/10.1016/j.isci.2022.105222] [PMID: 36248741]

[100]   W. Li, K. De Schutter, E.J.M. Van Damme, and G. Smagghe, "Developmental O –glycan profile analysis shows pentasaccharide mucin–type O –glycans are linked with pupation of Tribolium castaneum", *Arch. Insect Biochem. Physiol.,* vol. 109, no. 1, p. e21852, 2022.
[http://dx.doi.org/10.1002/arch.21852] [PMID: 34796531]

[101]   V. Lombard, H. Golaconda Ramulu, E. Drula, P.M. Coutinho, and B. Henrissat, "The carbohydrate-active enzymes database (CAZy) in 2013", *Nucleic Acids Res.,* vol. 42, no. D1, pp. D490-D495, 2014.
[http://dx.doi.org/10.1093/nar/gkt1178] [PMID: 24270786]

[102]   P. Goettig, "Effects of glycosylation on the enzymatic activity and mechanisms of proteases", *Int. J. Mol. Sci.,* vol. 17, no. 12, p. 1969, 2016.
[http://dx.doi.org/10.3390/ijms17121969] [PMID: 27898009]

[103]   A.D. Baudot, V.M.Y. Wang, J.D. Leach, J. O'Prey, J.S. Long, V. Paulus-Hock, S. Lilla, D.M. Thomson, J. Greenhorn, F. Ghaffar, C. Nixon, M.H. Helfrich, D. Strathdee, J. Pratt, F. Marchesi, S. Zanivan, and K.M. Ryan, "Glycan degradation promotes macroautophagy", *Proc. Natl. Acad. Sci.,* vol. 119, no. 26, p. e2111506119, 2022.
[http://dx.doi.org/10.1073/pnas.2111506119] [PMID: 35737835]

[104]   A.J. Parodi, E.W. Blank, J.A. Peterson, and R.L. Ceriani, "Dolichol-bound oligosaccharides and the transfer of distal monosaccharides in the synthesis of glycoproteins by normal and tumor mammary epithelial cells", *Breast Cancer Res. Treat.,* vol. 2, no. 3, pp. 227-237, 1982.
[http://dx.doi.org/10.1007/BF01806935] [PMID: 6817834]

[105]   J.W. Dennis, S. Laferté, C. Waghorne, M.L. Breitman, and R.S. Kerbel, "Beta 1-6 branching of Asn-linked oligosaccharides is directly associated with metastasis", *Science,* vol. 236, no. 4801, pp. 582-585, 1987.
[http://dx.doi.org/10.1126/science.2953071] [PMID: 2953071]

[106]   Y. Audet-Delage, M. Rouleau, L. Villeneuve, and C. Guillemette, "The glycosyltransferase pathway: An integrated analysis of the cell metabolome", *Metabolites,* vol. 12, no. 10, p. 1006, 2022.
[http://dx.doi.org/10.3390/metabo12101006] [PMID: 36295907]

[107]   C. Liu, Q. Zhou, Y. Li, L.V. Garner, S.P. Watkins, L.J. Carter, J. Smoot, A.C. Gregg, A.D. Daniels, S. Jervey, and D. Albaiu, "Research and development on therapeutic agents and vaccines for COVID-19 and related human coronavirus diseases", *ACS Cent. Sci.,* vol. 6, no. 3, pp. 315-331, 2020.
[http://dx.doi.org/10.1021/acscentsci.0c00272] [PMID: 32226821]

[108]   C.W. Fu, H.E. Tsai, W.S. Chen, T.T. Chang, C.L. Chen, P.W. Hsiao, and W.S. Li, "Sialyltransferase inhibitors suppress breast cancer metastasis", *J. Med. Chem.,* vol. 64, no. 1, pp. 527-542, 2021.
[http://dx.doi.org/10.1021/acs.jmedchem.0c01477] [PMID: 33371679]

[109]   N.C. Hait, A. Maiti, R. Wu, V.L. Andersen, C.C. Hsu, Y. Wu, D.G. Chapla, K. Takabe, M.E. Rusiniak, W. Bshara, J. Zhang, K.W. Moremen, and J.T.Y. Lau, "Extracellular sialyltransferase st6gal1 in breast tumor cell growth and invasiveness", *Cancer Gene Ther.,* vol. 29, no. 11, pp. 1662-1675, 2022.
[http://dx.doi.org/10.1038/s41417-022-00485-y] [PMID: 35676533]

[110] J.M. Fullerton, P. Klauser, R.K. Lenroot, A.D. Shaw, B. Overs, A. Heath, M.J. Cairns, J. Atkins, R. Scott, P.R. Schofield, C.S. Weickert, C. Pantelis, A. Fornito, T.J. Whitford, T.W. Weickert, and A. Zalesky, "Differential effect of disease-associated ST8SIA2 haplotype on cerebral white matter diffusion properties in schizophrenia and healthy controls", *Transl. Psychiatry,* vol. 8, no. 1, p. 21, 2018.
[http://dx.doi.org/10.1038/s41398-017-0052-z] [PMID: 29353880]

[111] N. Berois, and E. Osinaga, "Glycobiology of neuroblastoma: Impact on tumor behavior, prognosis, and therapeutic strategies", *Front. Oncol.,* vol. 4, p. 114, 2014.
[PMID: 24904828]

[112] D. Echeverri, and J. Orozco, "β-1,4-Galactosyltransferase-V colorectal cancer biomarker immunosensor with label-free electrochemical detection", *Talanta,* vol. 243, p. 123337, 2022.
[http://dx.doi.org/10.1016/j.talanta.2022.123337] [PMID: 35255430]

[113] C. Ren, Y. Guo, L. Xie, Z. Zhao, M. Xing, Y. Cao, Y. Liu, J. Lin, D. Grierson, B. Zhang, C. Xu, K. Chen, and X. Li, "Identification of UDP-rhamnosyltransferases and UDP-galactosyltransferase involved in flavonol glycosylation in Morella rubra", *Hortic. Res.,* vol. 9, p. uhac138, 2022.
[http://dx.doi.org/10.1093/hr/uhac138] [PMID: 36072838]

[114] S.R. Sharma, G. Crispell, A. Mohamed, C. Cox, J. Lange, S. Choudhary, S.P. Commins, and S. Karim, "Alpha-Gal Syndrome: Involvement of Amblyomma americanum α-D-Galactosidase and β-1,4 galactosyltransferase enzymes in α-Gal metabolism", *Front. Cell. Infect. Microbiol.,* vol. 11, p. 775371, 2021.
[http://dx.doi.org/10.3389/fcimb.2021.775371] [PMID: 34926322]

[115] Z. Wu, T. Lin, P. Kang, Z. Zhuang, H. Wang, W. He, Q. Wei, and Z. Li, "Overexpression of fucosyltransferase 8 reverses the inhibitory effect of high-dose dexamethasone on osteogenic response of MC3T3-E1 preosteoblasts", *PeerJ,* vol. 9, p. e12380, 2021.
[http://dx.doi.org/10.7717/peerj.12380] [PMID: 34966572]

[116] R. Battat, A. Qatomah, U. Kopylov, J. Wyse, A. Cohen, W. Afif, P.L. Lakatos, E. Seidman, A. Bitton, and T. Bessissow, "Fucosyltransferase 2 mutations are associated with a favorable clinical course in Crohn's Disease", *J. Clin. Gastroenterol.,* vol. 56, no. 3, pp. e166-e170, 2022.
[http://dx.doi.org/10.1097/MCG.0000000000001626] [PMID: 34739405]

[117] H. Ise, Y. Araki, I. Song, and G. Akatsuka, "N-acetylglucosamine-bearing polymers mimicking O-GlcNAc-modified proteins elicit anti-fibrotic activities in myofibroblasts and activated stellate cells", *Glycobiology,* vol. 33, no. 1, pp. 17-37, 2022.
[PMID: 36190502]

[118] Y.S. Hsu, P.J. Wu, Y.M. Jeng, C.M. Hu, and W.H. Lee, "Differential effects of glucose and N-acetylglucosamine on genome instability", *Am. J. Cancer Res.,* vol. 12, no. 4, pp. 1556-1576, 2022.
[PMID: 35530290]

[119] C. Kadooka, D. Hira, Y. Tanaka, K. Miyazawa, M. Bise, S. Takatsuka, and T. Oka, "Identification of an α-(1 → 6)-Mannosyltransferase contributing to biosynthesis of the fungal-type galactomannan α-core-mannan structure in aspergillus fumigatus", *MSphere,* vol. 7, no. 6, p. e00484-22, 2022.
[http://dx.doi.org/10.1128/msphere.00484-22] [PMID: 36445154]

[120] Z. Wen, H. Tian, Y. Xia, and K. Jin, "O-mannosyltransferase MaPmt2 contributes to stress tolerance, cell wall integrity and virulence in Metarhizium acridum", *J. Invertebr. Pathol.,* vol. 184, p. 107649, 2021.
[http://dx.doi.org/10.1016/j.jip.2021.107649] [PMID: 34343571]

[121] L. Bai, A. Kovach, Q. You, A. Kenny, and H. Li, "Structure of the eukaryotic protein O-mannosyltransferase Pmt1−Pmt2 complex", *Nat. Struct. Mol. Biol.,* vol. 26, no. 8, pp. 704-711, 2019.
[http://dx.doi.org/10.1038/s41594-019-0262-6] [PMID: 31285605]

[122] J. Zhang, Y. Dai, Y. Fan, N. Jiang, Y. Zhou, L. Zeng, and Y. Li, "Glycosylphosphatidylinositol

mannosyltransferase I protects chinese giant salamander, Andrias davidianus, against Iridovirus", *Int. J. Mol. Sci.,* vol. 23, no. 16, p. 9009, 2022.
[http://dx.doi.org/10.3390/ijms23169009] [PMID: 36012277]

[123] N. Asano, "Glycosidase inhibitors: Update and perspectives on practical use", *Glycobiology,* vol. 13, no. 10, pp. 93R-104, 2003.
[http://dx.doi.org/10.1093/glycob/cwg090] [PMID: 12851286]

[124] S.A.W. Gruner, E. Locardi, E. Lohof, and H. Kessler, "Carbohydrate-based mimetics in drug design: Sugar amino acids and carbohydrate scaffolds", *Chem. Rev.,* vol. 102, no. 2, pp. 491-514, 2002.
[http://dx.doi.org/10.1021/cr0004409] [PMID: 11841252]

[125] D. Isaev, E. Isaeva, T. Shatskih, Q. Zhao, N.C. Smits, N.W. Shworak, R. Khazipov, and G.L. Holmes, "Role of extracellular sialic acid in regulation of neuronal and network excitability in the rat hippocampus", *J. Neurosci.,* vol. 27, no. 43, pp. 11587-11594, 2007.
[http://dx.doi.org/10.1523/JNEUROSCI.2033-07.2007] [PMID: 17959801]

[126] M.S.L. Hammond, C. Sims, K. Parameshwaran, V. Suppiramaniam, M. Schachner, and A. Dityatev, "Neural cell adhesion molecule-associated polysialic acid inhibits NR2B-containing N-methyl-D-aspartate receptors and prevents glutamate-induced cell death", *J. Biol. Chem.,* vol. 281, no. 46, pp. 34859-34869, 2006.
[http://dx.doi.org/10.1074/jbc.M602568200] [PMID: 16987814]

[127] Z. Yang, L.E. Harris, D.E. Palmer-Toy, and W.S. Hancock, "Multilectin affinity chromatography for characterization of multiple glycoprotein biomarker candidates in serum from breast cancer patients", *Clin. Chem.,* vol. 52, no. 10, pp. 1897-1905, 2006.
[http://dx.doi.org/10.1373/clinchem.2005.065862] [PMID: 16916992]

[128] G. Di Cristo, B. Chattopadhyaya, S.J. Kuhlman, Y. Fu, M.C. Bélanger, C.Z. Wu, U. Rutishauser, L. Maffei, and Z.J. Huang, "Activity-dependent PSA expression regulates inhibitory maturation and onset of critical period plasticity", *Nat. Neurosci.,* vol. 10, no. 12, pp. 1569-1577, 2007.
[http://dx.doi.org/10.1038/nn2008] [PMID: 18026099]

[129] A. Minami, Y. Kurebayashi, T. Takahashi, T. Otsubo, K. Ikeda, and T. Suzuki, "The function of sialidase revealed by sialidase activity imaging probe", *Int. J. Mol. Sci.,* vol. 22, no. 6, p. 3187, 2021.
[http://dx.doi.org/10.3390/ijms22063187] [PMID: 33804798]

[130] H. Varbanov, and A. Dityatev, "Regulation of extrasynaptic signaling by polysialylated NCAM: Impact for synaptic plasticity and cognitive functions", *Mol. Cell. Neurosci.,* vol. 81, pp. 12-21, 2017.
[http://dx.doi.org/10.1016/j.mcn.2016.11.005] [PMID: 27865768]

[131] W.H.D. Bowles, and T.M. Gloster, "Sialidase and sialyltransferase inhibitors: Targeting pathogenicity and disease", *Front. Mol. Biosci.,* vol. 8, p. 705133, 2021.
[http://dx.doi.org/10.3389/fmolb.2021.705133] [PMID: 34395532]

[132] T. Miyagi, "Aberrant expression of sialidase and cancer progression", *Proc. Jpn. Acad., Ser. B, Phys. Biol. Sci.,* vol. 84, no. 10, pp. 407-418, 2008.
[http://dx.doi.org/10.2183/pjab.84.407] [PMID: 19075514]

[133] H. Grootaert, L. Van Landuyt, P. Hulpiau, and N. Callewaert, "Functional exploration of the GH29 fucosidase family", *Glycobiology,* vol. 30, no. 9, pp. 735-745, 2020.
[http://dx.doi.org/10.1093/glycob/cwaa023] [PMID: 32149359]

[134] D. Ndeh, A. Rogowski, A. Cartmell, A.S. Luis, A. Baslé, J. Gray, I. Venditto, J. Briggs, X. Zhang, A. Labourel, N. Terrapon, F. Buffetto, S. Nepogodiev, Y. Xiao, R.A. Field, Y. Zhu, M.A. O'Neill, B.R. Urbanowicz, W.S. York, G.J. Davies, D.W. Abbott, M.C. Ralet, E.C. Martens, B. Henrissat, and H.J. Gilbert, "Complex pectin metabolism by gut bacteria reveals novel catalytic functions", *Nature,* vol. 544, no. 7648, pp. 65-70, 2017.
[http://dx.doi.org/10.1038/nature21725] [PMID: 28329766]

[135] A. Rogowski, J.A. Briggs, J.C. Mortimer, T. Tryfona, N. Terrapon, E.C. Lowe, A. Baslé, C. Morland, A.M. Day, H. Zheng, T.E. Rogers, P. Thompson, A.R. Hawkins, M.P. Yadav, B. Henrissat, E.C.

Martens, P. Dupree, H.J. Gilbert, and D.N. Bolam, "Glycan complexity dictates microbial resource allocation in the large intestine", *Nat. Commun.,* vol. 6, no. 1, p. 7481, 2015.
[http://dx.doi.org/10.1038/ncomms8481] [PMID: 26112186]

[136] T. Katayama, A. Sakuma, T. Kimura, Y. Makimura, J. Hiratake, K. Sakata, T. Yamanoi, H. Kumagai, and K. Yamamoto, "Molecular cloning and characterization of Bifidobacterium bifidum 1,2-alpha-L-fucosidase (AfcA), a novel inverting glycosidase (glycoside hydrolase family 95)", *J. Bacteriol.,* vol. 186, no. 15, pp. 4885-4893, 2004.
[http://dx.doi.org/10.1128/JB.186.15.4885-4893.2004] [PMID: 15262925]

[137] M. Nagae, A. Tsuchiya, T. Katayama, K. Yamamoto, S. Wakatsuki, and R. Kato, "Structural basis of the catalytic reaction mechanism of novel 1,2-alpha-L-fucosidase from Bifidobacterium bifidum", *J. Biol. Chem.,* vol. 282, no. 25, pp. 18497-18509, 2007.
[http://dx.doi.org/10.1074/jbc.M702246200] [PMID: 17459873]

[138] B.L. Cantarel, P.M. Coutinho, C. Rancurel, T. Bernard, V. Lombard, and B. Henrissat, "The Carbohydrate-Active EnZymes database (CAZy): An expert resource for Glycogenomics", *Nucleic Acids Res.,* vol. 37, no. Database, pp. D233-D238, 2009.
[http://dx.doi.org/10.1093/nar/gkn663] [PMID: 18838391]

[139] N.P. Türker, and E. Bakar, "The importance of glycans in cancer metabolism and metastasis", *J. Cumhu. Uni. Heal. Sci. Inst.,* vol. 6, no. 2, pp. 112-119, 2021.
[http://dx.doi.org/10.51754/cusbed.867416]

[140] N.F. Brás, P.A. Fernandes, and M.J. Ramos, "QM/MM Studies on the β-Galactosidase catalytic mechanism: Hydrolysis and transglycosylation reactions", *J. Chem. Theory Comput.,* vol. 6, no. 2, pp. 421-433, 2010.
[http://dx.doi.org/10.1021/ct900530f] [PMID: 26617299]

[141] L. Lu, M. Xiao, X. Xu, Z. Li, and Y. Li, "A novel β-galactosidase capable of glycosyl transfer from Enterobacter agglomerans B1", *Biochem. Biophys. Res. Commun.,* vol. 356, no. 1, pp. 78-84, 2007.
[http://dx.doi.org/10.1016/j.bbrc.2007.02.106] [PMID: 17336932]

[142] J.R. Ketudat Cairns, and A. Esen, "β-Glucosidases", *Cell. Mol. Life Sci.,* vol. 67, no. 20, pp. 3389-3405, 2010.
[http://dx.doi.org/10.1007/s00018-010-0399-2] [PMID: 20490603]

[143] B. Henrissat, and G. Davies, "Structural and sequence-based classification of glycoside hydrolases", *Curr. Opin. Struct. Biol.,* vol. 7, no. 5, pp. 637-644, 1997.
[http://dx.doi.org/10.1016/S0959-440X(97)80072-3] [PMID: 9345621]

[144] G.W. Hart, M.P. Housley, and C. Slawson, "Cycling of O-linked β-N-acetylglucosamine on nucleocytoplasmic proteins", *Nature,* vol. 446, no. 7139, pp. 1017-1022, 2007.
[http://dx.doi.org/10.1038/nature05815] [PMID: 17460662]

[145] P. Awolade, N. Cele, N. Kerru, L. Gummidi, E. Oluwakemi, and P. Singh, "Therapeutic significance of β-glucuronidase activity and its inhibitors: A review", *Eur. J. Med. Chem.,* vol. 187, p. 111921, 2020.
[http://dx.doi.org/10.1016/j.ejmech.2019.111921] [PMID: 31835168]

[146] S. Kayahan, "Atherosclerosis and beta-glucuronidase", *Lancet,* vol. 276, no. 7152, pp. 667-669, 1960.
[http://dx.doi.org/10.1016/S0140-6736(60)91744-X] [PMID: 13752036]

[147] F.I. Khan, M. Shahbaaz, K. Bisetty, A. Waheed, W.S. Sly, F. Ahmad, and M.I. Hassan, "Large scale analysis of the mutational landscape in β-glucuronidase: A major player of mucopolysaccharidosis type VII", *Gene,* vol. 576, no. 1, pp. 36-44, 2016.
[http://dx.doi.org/10.1016/j.gene.2015.09.062] [PMID: 26415878]

[148] K. Fiedler, and K. Simons, "The role of n-glycans in the secretory pathway", *Cell,* vol. 81, no. 3, pp. 309-312, 1995.
[http://dx.doi.org/10.1016/0092-8674(95)90380-1] [PMID: 7736583]

[149] D. Macmillan, and A. Daines, "Recent developments in the synthesis and discovery of oligosaccharides and glycoconjugates for the treatment of disease", *Curr. Med. Chem.,* vol. 10, no. 24, pp. 2733-2773, 2003.
[http://dx.doi.org/10.2174/0929867033456413] [PMID: 14529463]

[150] D.P. Gamblin, E.M. Scanlan, and B.G. Davis, "Glycoprotein synthesis: An update", *Chem. Rev.,* vol. 109, no. 1, pp. 131-163, 2009.
[http://dx.doi.org/10.1021/cr078291i] [PMID: 19093879]

[151] Z. Polonskaya, S. Deng, A. Sarkar, L. Kain, M. Comellas-Aragones, C.S. McKay, K. Kaczanowska, M. Holt, R. McBride, V. Palomo, K.M. Self, S. Taylor, A. Irimia, S.R. Mehta, J.M. Dan, M. Brigger, S. Crotty, S.P. Schoenberger, J.C. Paulson, I.A. Wilson, P.B. Savage, M.G. Finn, and L. Teyton, "T cells control the generation of nanomolar-affinity anti-glycan antibodies", *J. Clin. Invest.,* vol. 127, no. 4, pp. 1491-1504, 2017.
[http://dx.doi.org/10.1172/JCI91192] [PMID: 28287405]

[152] J. Jaeken, M. Vanderschueren-Lodeweyckx, P. Casaer, L. Snoeck, L. Corbeel, E. Eggermont, and R. Eeckels, "Familial psychomotor retardation with markedly fluctuating serum prolactin, FSH and GH levels, partial TBG-deficiency, increased serum arylsulphatase A and increased CSF protein: A new syndrome?: 90", *Pediatr. Res.,* vol. 14, no. 2, p. 179, 1980.
[http://dx.doi.org/10.1203/00006450-198002000-00117]

[153] E. Özaydın, F. Yalçın, M. Gündüz, and G. Köse, "Congenital glycosylation disorder type II", *Turk. J. Pedia.,* vol. 6, no. 1, pp. 47-53, 2012.

[154] N. Matsubara, A. Imamura, T. Yonemizu, C. Akatsu, H. Yang, A. Ueki, N. Watanabe, H. Abdu-Allah, N. Numoto, H. Takematsu, S. Kitazume, T.F. Tedder, J.D. Marth, N. Ito, H. Ando, H. Ishida, M. Kiso, and T. Tsubata, "CD22-Binding synthetic sialosides regulate B lymphocyte proliferation through CD22 ligand-dependent and independent pathways, and enhance antibody production in mice", *Front. Immunol.,* vol. 9, p. 820, 2018.
[http://dx.doi.org/10.3389/fimmu.2018.00820] [PMID: 29725338]

[155] R. Kuai, X. Sun, W. Yuan, L.J. Ochyl, Y. Xu, A. Hassani Najafabadi, L. Scheetz, M.Z. Yu, I. Balwani, A. Schwendeman, and J.J. Moon, "Dual TLR agonist nanodiscs as a strong adjuvant system for vaccines and immunotherapy", *J. Control. Release,* vol. 282, pp. 131-139, 2018.
[http://dx.doi.org/10.1016/j.jconrel.2018.04.041] [PMID: 29702142]

[156] D. Vasudevan, and R.S. Haltiwanger, "Novel roles for O-linked glycans in protein folding", *Glycoconj. J.,* vol. 31, no. 6-7, pp. 417-426, 2014.
[http://dx.doi.org/10.1007/s10719-014-9556-4] [PMID: 25186198]

[157] S.A.M. Laarse, A.C. Leney, and A.J.R. Heck, "Crosstalk between phosphorylation and O–Glc NA cylation: Friend or foe", *FEBS J.,* vol. 285, no. 17, pp. 3152-3167, 2018.
[http://dx.doi.org/10.1111/febs.14491] [PMID: 29717537]

[158] F. Chiaradonna, F. Ricciardiello, and R. Palorini, "The nutrient-sensing hexosamine biosynthetic pathway as the hub of cancer metabolic rewiring", *Cells,* vol. 7, no. 6, p. 53, 2018.
[http://dx.doi.org/10.3390/cells7060053] [PMID: 29865240]

[159] M. Raska, L. Czernekova, Z. Moldoveanu, K. Zachova, M.C. Elliott, Z. Novak, S. Hall, M. Hoelscher, L. Maboko, R. Brown, P.D. Smith, J. Mestecky, and J. Novak, "Differential glycosylation of envelope gp120 is associated with differential recognition of HIV-1 by virus-specific antibodies and cell infection", *AIDS Res. Ther.,* vol. 11, no. 1, p. 23, 2014.
[http://dx.doi.org/10.1186/1742-6405-11-23] [PMID: 25120578]

[160] J. Mikulak, C. Di Vito, E. Zaghi, and D. Mavilio, "Host immune responses in HIV-1 Infection: The emerging pathogenic role of siglecs and their clinical correlates", *Front. Immunol.,* vol. 8, p. 314, 2017.
[http://dx.doi.org/10.3389/fimmu.2017.00314] [PMID: 28386256]

[161] R.J. O'Connell, J.H. Kim, and J.L. Excler, "The HIV-1 gp120 V1V2 loop: Structure, function and

importance for vaccine development", *Expert Rev. Vaccines,* vol. 13, no. 12, pp. 1489-1500, 2014.
[http://dx.doi.org/10.1586/14760584.2014.951335] [PMID: 25163695]

[162] R.C. Doran, G.P. Tatsuno, S.M. O'Rourke, B. Yu, D.L. Alexander, K.A. Mesa, and P.W. Berman, "Glycan modifications to the gp120 immunogens used in the RV144 vaccine trial improve binding to broadly neutralizing antibodies", *PLoS One,* vol. 13, no. 4, p. e0196370, 2018.
[http://dx.doi.org/10.1371/journal.pone.0196370] [PMID: 29689099]

[163] S. Hakomori, "New directions in cancer therapy based on aberrant expression of glycosphingolipids: Anti-adhesion and ortho-signaling therapy", *Cancer Cells,* vol. 3, no. 12, pp. 461-470, 1991.
[PMID: 1820092]

[164] F. Hossain, and P.R. Andreana, "Developments in carbohydrate-based cancer therapeutics", *Pharmaceuticals,* vol. 12, no. 2, p. 84, 2019.
[http://dx.doi.org/10.3390/ph12020084] [PMID: 31167407]

[165] D.H. Joziasse, and R. Oriol, "Xenotransplantation: The importance of the Galα1,3Gal epitope in hyperacute vascular rejection", *Biochim. Biophys. Acta Mol. Basis Dis.,* vol. 1455, no. 2-3, pp. 403-418, 1999.
[http://dx.doi.org/10.1016/S0925-4439(99)00056-3] [PMID: 10571028]

[166] I. Kosik, W.L. Ince, L.E. Gentles, A.J. Oler, M. Kosikova, M. Angel, J.G. Magadán, H. Xie, C.B. Brooke, and J.W. Yewdell, "Correction: Influenza a virus hemagglutinin glycosylation compensates for antibody escape fitness costs", *PLoS Pathog.,* vol. 14, no. 6, p. e1007141, 2018.
[http://dx.doi.org/10.1371/journal.ppat.1007141] [PMID: 29924863]

[167] A.J. Behrens, A. Kumar, M. Medina-Ramirez, A. Cupo, K. Marshall, V.M. Cruz Portillo, D.J. Harvey, G. Ozorowski, N. Zitzmann, I.A. Wilson, A.B. Ward, W.B. Struwe, J.P. Moore, R.W. Sanders, and M. Crispin, "Integrity of glycosylation processing of a Glycan-Depleted Trimeric HIV-1 Immunogen Targeting Key B-Cell Lineages", *J. Proteome Res.,* vol. 17, no. 3, pp. 987-999, 2018.
[http://dx.doi.org/10.1021/acs.jproteome.7b00639] [PMID: 29420040]

[168] J. Ingale, K. Tran, L. Kong, B. Dey, K. McKee, W. Schief, P.D. Kwong, J.R. Mascola, and R.T. Wyatt, "Hyperglycosylated stable core immunogens designed to present the CD4 binding site are preferentially recognized by broadly neutralizing antibodies", *J. Virol.,* vol. 88, no. 24, pp. 14002-14016, 2014.
[http://dx.doi.org/10.1128/JVI.02614-14] [PMID: 25253346]

[169] M. Hyakumura, R. Walsh, M. Thaysen-Andersen, N.J. Kingston, M. La, L. Lu, G. Lovrecz, N.H. Packer, S. Locarnini, and H.J. Netter, "Modification of asparagine-linked glycan density for the design of hepatitis B virus virus-like particles with enhanced immunogenicity", *J. Virol.,* vol. 89, no. 22, pp. 11312-11322, 2015.
[http://dx.doi.org/10.1128/JVI.01123-15] [PMID: 26339047]

[170] L.E. McCoy, and D.R. Burton, "Identification and specificity of broadly neutralizing antibodies against HIV", *Immunol. Rev.,* vol. 275, no. 1, pp. 11-20, 2017.
[http://dx.doi.org/10.1111/imr.12484] [PMID: 28133814]

[171] R.D. Astronomo, H.K. Lee, C.N. Scanlan, R. Pantophlet, C.Y. Huang, I.A. Wilson, O. Blixt, R.A. Dwek, C.H. Wong, and D.R. Burton, "A glycoconjugate antigen based on the recognition motif of a broadly neutralizing human immunodeficiency virus antibody, 2G12, is immunogenic but elicits antibodies unable to bind to the self glycans of gp120", *J. Virol.,* vol. 82, no. 13, pp. 6359-6368, 2008.
[http://dx.doi.org/10.1128/JVI.00293-08] [PMID: 18434393]

[172] Z. Wang, C. Qin, J. Hu, X. Guo, and J. Yin, "Recent advances in synthetic carbohydrate-based human immunodeficiency virus vaccines", *Virol. Sin.,* vol. 31, no. 2, pp. 110-117, 2016.
[http://dx.doi.org/10.1007/s12250-015-3691-3] [PMID: 26992403]

[173] J.L. Excler, M.L. Robb, and J.H. Kim, "Prospects for a Globally Effective HIV-1 Vaccine", *Am. J. Prev. Med.,* vol. 49, no. 6, suppl. 4, pp. S307-S318, 2015.
[http://dx.doi.org/10.1016/j.amepre.2015.09.004] [PMID: 26590431]

[174] N.P. Türker, and E. Bakar, "Carbohydrate based therapeutics", *Int. j. life sci. biotech.,* vol. 4, no. 3, pp. 581-607, 2021.
[http://dx.doi.org/10.38001/ijlsb.875364]

[175] L. Duan, Q. Zheng, H. Zhang, Y. Niu, Y. Lou, and H. Wang, "The SARS-CoV-2 Spike glycoprotein biosynthesis, structure, function, and antigenicity: Implications for the design of spike-based vaccine immunogens", *Front. Immunol.,* vol. 11, p. 576622, 2020.
[http://dx.doi.org/10.3389/fimmu.2020.576622] [PMID: 33117378]

[176] K.B. Lokhande, G.R. Apte, A. Shrivastava, A. Singh, J.K. Pal, and K.V. Swamy, "Sensing the interactions between carbohydrate-binding agents and N-linked glycans of SARS-CoV-2 spike glycoprotein using molecular docking and simulation studies", *J. Biomol. Struct. Dyn.,* vol. 40, no. 9, pp. 3880-3898, 2020.
[PMID: 33292056]

[177] E. Keyaerts, L. Vijgen, C. Pannecouque, E. Van Damme, W. Peumans, H. Egberink, J. Balzarini, and M. Van Ranst, "Plant lectins are potent inhibitors of coronaviruses by interfering with two targets in the viral replication cycle", *Antiviral Res.,* vol. 75, no. 3, pp. 179-187, 2007.
[http://dx.doi.org/10.1016/j.antiviral.2007.03.003] [PMID: 17428553]

[178] C.A. Mitchell, K. Ramessar, and B.R. O'Keefe, "Antiviral lectins: Selective inhibitors of viral entry", *Antiviral Res.,* vol. 142, pp. 37-54, 2017.
[http://dx.doi.org/10.1016/j.antiviral.2017.03.007] [PMID: 28322922]

[179] F.J.U.M. van der Meer, C.A.M. de Haan, N.M.P. Schuurman, B.J. Haijema, M.H. Verheije, B.J. Bosch, J. Balzarini, and H.F. Egberink, "The carbohydrate-binding plant lectins and the non-peptidic antibiotic pradimicin A target the glycans of the coronavirus envelope glycoproteins", *J. Antimicrob. Chemother.,* vol. 60, no. 4, pp. 741-749, 2007.
[http://dx.doi.org/10.1093/jac/dkm301] [PMID: 17704516]

[180] R.K. Gupta, G.R. Apte, K.B. Lokhande, S. Mishra, and J.K. Pal, "Carbohydrate-binding agents: Potential of repurposing for COVID-19 therapy", *Curr. Protein Pept. Sci.,* vol. 21, no. 11, pp. 1085-1096, 2020.
[http://dx.doi.org/10.2174/1389203721666200918153717] [PMID: 32951577]

[181] R. Mettu, C.Y. Chen, and C.Y. Wu, "Synthetic carbohydrate-based vaccines: Challenges and opportunities", *J. Biomed. Sci.,* vol. 27, no. 1, p. 9, 2020.
[http://dx.doi.org/10.1186/s12929-019-0591-0] [PMID: 31900143]

[182] Y. Lyu, H. Kaddour, S. Kopcho, T.D. Panzner, N. Shouman, E.Y. Kim, J. Martinson, H. McKay, O. Martinez-Maza, J.B. Margolick, J.T. Stapleton, and C.M. Okeoma, "Human immunodeficiency virus (HIV) infection and use of Illicit substances promote secretion of semen exosomes that enhance monocyte adhesion and induce actin reorganization and chemotactic migration", *Cells,* vol. 8, no. 9, p. 1027, 2019.
[http://dx.doi.org/10.3390/cells8091027] [PMID: 31484431]

[183] CW Cheng, Y Zhou, WH Pan, S Dey, CY Wu, and WL Hsu, "Hierarchical and programmable one-pot oligosaccharide synthesis", *J Vis Exp.,* no. 151, 2019.

[184] C.G. Vinuesa, and P.P. Chang, "Innate B cell helpers reveal novel types of antibody responses", *Nat. Immunol.,* vol. 14, no. 2, pp. 119-126, 2013.
[http://dx.doi.org/10.1038/ni.2511] [PMID: 23334833]

[185] K. Pobre, M. Tashani, I. Ridda, H. Rashid, M. Wong, and R. Booy, "Carrier priming or suppression: Understanding carrier priming enhancement of anti-polysaccharide antibody response to conjugate vaccines", *Vaccine,* vol. 32, no. 13, pp. 1423-1430, 2014.
[http://dx.doi.org/10.1016/j.vaccine.2014.01.047] [PMID: 24492014]

[186] C.L. Schengrund, "Glycoconjugates: Roles in neural diseases caused by exogenous pathogens", *CNS Neurol. Disord. Drug Targets,* vol. 5, no. 4, pp. 381-389, 2006.
[http://dx.doi.org/10.2174/187152706777950701] [PMID: 16918390]

[187]   H.H. Freeze, E.A. Eklund, B.G. Ng, and M.C. Patterson, "Neurology of inherited glycosylation disorders", *Lancet Neurol.,* vol. 11, no. 5, pp. 453-466, 2012.
[http://dx.doi.org/10.1016/S1474-4422(12)70040-6] [PMID: 22516080]

[188]   Z. Cai, H. Zhang, Y. Wei, M. Wu, and A. Fu, "Shear-thinning hyaluronan-based fluid hydrogels to modulate viscoelastic properties of osteoarthritis synovial fluids", *Biomater. Sci.,* vol. 7, no. 8, pp. 3143-3157, 2019.
[http://dx.doi.org/10.1039/C9BM00298G] [PMID: 31168540]

[189]   E.J. Kumpulainen, R.J. Keskikuru, and R.T. Johansson, "Serum tumor marker CA 15.3 and stage are the two most powerful predictors of survival in primary breast cancer", *Breast Cancer Res. Treat.,* vol. 76, no. 2, pp. 95-102, 2002.
[http://dx.doi.org/10.1023/A:1020514925143] [PMID: 12452445]

[190]   G.Y. Locker, S. Hamilton, J. Harris, J.M. Jessup, N. Kemeny, J.S. Macdonald, M.R. Somerfield, D.F. Hayes, and R.C. Bast Jr, "ASCO 2006 update of recommendations for the use of tumor markers in gastrointestinal cancer", *J. Clin. Oncol.,* vol. 24, no. 33, pp. 5313-5327, 2006.
[http://dx.doi.org/10.1200/JCO.2006.08.2644] [PMID: 17060676]

[191]   A. Schietinger, M. Philip, and H. Schreiber, "Specificity in cancer immunotherapy", *Semin. Immunol.,* vol. 20, no. 5, pp. 276-285, 2008.
[http://dx.doi.org/10.1016/j.smim.2008.07.001] [PMID: 18684640]

[192]   W.B. Hamilton, F. Helling, K.O. Lloyd, and P.O. Livingston, "Ganglioside expression on human malignant melanoma assessed by quantitative immune thin-layer chromatography", *Int. J. Cancer,* vol. 53, no. 4, pp. 566-573, 1993.
[http://dx.doi.org/10.1002/ijc.2910530407] [PMID: 8436430]

[193]   S. Hakomori, "Antigen structure and genetic basis of histo-blood groups A, B and O: Their changes associated with human cancer", *Biochim. Biophys. Acta, Gen. Subj.,* vol. 1473, no. 1, pp. 247-266, 1999.
[http://dx.doi.org/10.1016/S0304-4165(99)00183-X] [PMID: 10580143]

[194]   K. Sivasubramaniyan, A. Harichandan, K. Schilbach, A.F. Mack, J. Bedke, A. Stenzl, L. Kanz, G. Niederfellner, and H.J. Bühring, "Expression of stage-specific embryonic antigen-4 (SSEA-4) defines spontaneous loss of epithelial phenotype in human solid tumor cells", *Glycobiology,* vol. 25, no. 8, pp. 902-917, 2015.
[http://dx.doi.org/10.1093/glycob/cwv032] [PMID: 25978997]

[195]   S. Lang, and X. Huang, "Carbohydrate conjugates in vaccine developments", *Front Chem.,* vol. 8, p. 284, 2020.
[http://dx.doi.org/10.3389/fchem.2020.00284] [PMID: 32351942]

[196]   Y. Sato, K. Nakata, Y. Kato, M. Shima, N. Ishii, T. Koji, K. Taketa, Y. Endo, and S. Nagataki, "Early recognition of hepatocellular carcinoma based on altered profiles of alpha-fetoprotein", *N. Engl. J. Med.,* vol. 328, no. 25, pp. 1802-1806, 1993.
[http://dx.doi.org/10.1056/NEJM199306243282502] [PMID: 7684823]

[197]   B. Adamczyk, T. Tharmalingam, and P.M. Rudd, "Glycans as cancer biomarkers", *Biochim. Biophys. Acta, Gen. Subj.,* vol. 1820, no. 9, pp. 1347-1353, 2012.
[http://dx.doi.org/10.1016/j.bbagen.2011.12.001] [PMID: 22178561]

[198]   S.S. Pinho, and C.A. Reis, "Glycosylation in cancer: Mechanisms and clinical implications", *Nat. Rev. Cancer,* vol. 15, no. 9, pp. 540-555, 2015.
[http://dx.doi.org/10.1038/nrc3982] [PMID: 26289314]

[199]   L. Oliveira-Ferrer, K. Legler, and K. Milde-Langosch, "Role of protein glycosylation in cancer metastasis", *Semin. Cancer Biol.,* vol. 44, pp. 141-152, 2017.
[http://dx.doi.org/10.1016/j.semcancer.2017.03.002] [PMID: 28315783]

[200]   R. Isomura, K. Kitajima, and C. Sato, "Structural and functional impairments of polysialic acid by a

mutated polysialyltransferase found in schizophrenia", *J. Biol. Chem.,* vol. 286, no. 24, pp. 21535-21545, 2011.
[http://dx.doi.org/10.1074/jbc.M111.221143] [PMID: 21464126]

[201]   A. Vojta, I. Samaržija, L. Bočkor, and V. Zoldoš, "Glyco-genes change expression in cancer through aberrant methylation", *Biochim. Biophys. Acta, Gen. Subj.,* vol. 1860, no. 8, pp. 1776-1785, 2016.
[http://dx.doi.org/10.1016/j.bbagen.2016.01.002] [PMID: 26794090]

[202]   S. Chugh, J. Meza, Y.M. Sheinin, M.P. Ponnusamy, and S.K. Batra, "Loss of N-acetylgalactosaminyltransferase 3 in poorly differentiated pancreatic cancer: Augmented aggressiveness and aberrant ErbB family glycosylation", *Br. J. Cancer,* vol. 114, no. 12, pp. 1376-1386, 2016.
[http://dx.doi.org/10.1038/bjc.2016.116] [PMID: 27187683]

[203]   R.J. Solá, and K. Griebenow, "Glycosylation of therapeutic proteins: An effective strategy to optimize efficacy", *BioDrugs,* vol. 24, no. 1, pp. 9-21, 2010.
[http://dx.doi.org/10.2165/11530550-000000000-00000] [PMID: 20055529]

[204]   D. Ghaderi, M. Zhang, N. Hurtado-Ziola, and A. Varki, "Production platforms for biotherapeutic glycoproteins. occurrence, impact, and challenges of non-human sialylation", *Biotechnol. Genet. Eng. Rev.,* vol. 28, no. 1, pp. 147-176, 2012.
[http://dx.doi.org/10.5661/bger-28-147] [PMID: 22616486]

[205]   B. Leader, Q.J. Baca, and D.E. Golan, "Protein therapeutics: A summary and pharmacological classification", *Nat. Rev. Drug Discov.,* vol. 7, no. 1, pp. 21-39, 2008.
[http://dx.doi.org/10.1038/nrd2399] [PMID: 18097458]

[206]   L.X. Wang, and M.N. Amin, "Chemical and chemoenzymatic synthesis of glycoproteins for deciphering functions", *Chem. Biol.,* vol. 21, no. 1, pp. 51-66, 2014.
[http://dx.doi.org/10.1016/j.chembiol.2014.01.001] [PMID: 24439206]

[207]   Y. van Kooyk, and G.A. Rabinovich, "Protein-glycan interactions in the control of innate and adaptive immune responses", *Nat. Immunol.,* vol. 9, no. 6, pp. 593-601, 2008.
[http://dx.doi.org/10.1038/ni.f.203] [PMID: 18490910]

[208]   C. Müller, G. Despras, and T.K. Lindhorst, "Organizing multivalency in carbohydrate recognition", *Chem. Soc. Rev.,* vol. 45, no. 11, pp. 3275-3302, 2016.
[http://dx.doi.org/10.1039/C6CS00165C] [PMID: 27146554]

[209]   Q. Wang, Z. Zhou, S. Tang, and Z. Guo, "Carbohydrate-monophosphoryl lipid a conjugates are fully synthetic self-adjuvanting cancer vaccines eliciting robust immune responses in the mouse", *ACS Chem. Biol.,* vol. 7, no. 1, pp. 235-240, 2012.
[http://dx.doi.org/10.1021/cb200358r] [PMID: 22013921]

[210]   M. Heidelberger, and O.T. Avery, "The soluble specific substance of pneumococcus", *J. Exp. Med.,* vol. 38, no. 1, pp. 73-79, 1923.
[http://dx.doi.org/10.1084/jem.38.1.73] [PMID: 19868772]

[211]   F. Micoli, P. Costantino, and R. Adamo, "Potential targets for next generation antimicrobial glycoconjugate vaccines", *FEMS Microbiol. Rev.,* vol. 42, no. 3, pp. 388-423, 2018.
[http://dx.doi.org/10.1093/femsre/fuy011] [PMID: 29547971]

[212]   N. Inoue, M. Takeuchi, H. Ohashi, and T. Suzuki, "The production of recombinant human erythropoietin", *Biotechnol. Annu. Rev.,* vol. 1, pp. 297-313, 1995.
[http://dx.doi.org/10.1016/S1387-2656(08)70055-3] [PMID: 9704092]

[213]   Y. Zhang, L. Wang, S. Dey, M. Alnaeeli, S. Suresh, H. Rogers, R. Teng, and C. Noguchi, "Erythropoietin action in stress response, tissue maintenance and metabolism", *Int. J. Mol. Sci.,* vol. 15, no. 6, pp. 10296-10333, 2014.
[http://dx.doi.org/10.3390/ijms150610296] [PMID: 24918289]

[214]   A.D. Elbein, "Inhibitors of the biosynthesis and processing of N-linked oligosaccharide chains", *Annu. Rev. Biochem.,* vol. 56, no. 1, pp. 497-534, 1987.

[http://dx.doi.org/10.1146/annurev.bi.56.070187.002433] [PMID: 3304143]

[215] M. von Itzstein, W.Y. Wu, G.B. Kok, M.S. Pegg, J.C. Dyason, B. Jin, T. Van Phan, M.L. Smythe, H.F. White, S.W. Oliver, P.M. Colman, J.N. Varghese, D.M. Ryan, J.M. Woods, R.C. Bethell, V.J. Hotham, J.M. Cameron, and C.R. Penn, "Rational design of potent sialidase-based inhibitors of influenza virus replication", *Nature,* vol. 363, no. 6428, pp. 418-423, 1993. [http://dx.doi.org/10.1038/363418a0] [PMID: 8502295]

CHAPTER 7

Nanoparticle Targeting Strategies In Cancer Therapy

Hande Balyapan[1] and Güliz Ak[1,*]

[1] *Department of Biochemistry, Faculty of Science, Ege University, 35100, İzmir, Türkiye*

Abstract: This review outlines major cancer targeting strategies for nanoparticle systems. Targeted therapies have superiority over conventional chemotherapy or radiotherapy methods. Nanoparticles as drug nanocarriers enable drug delivery to the tumoral regions. For targeted drug delivery, nanoparticles are designed and tailored depending on the cancer and the purpose of the targeting mechanism. In this review, nanoparticle targeting for cancer therapy was summarized into three sections: passive, active, and physical targeting. Each issue was described and discussed with recent nanoparticular studies and their findings. In addition, a combination of targeting with diagnostics and theranostics was also presented.

Keywords: Active targeting, Antibody targeted delivery, Aptamer targeted delivery, Cancer, Cancer electrotherapy mediated drug delivery, Drug delivery systems, Epr effect, Lectin targeted delivery, Magnetic targeted delivery, Nanomedicine, Nanoparticle, Nir-triggered drug delivery, Passive targeting, Physical targeting, Receptor targeted delivery, Stimuli responsive nanoparticle, Targeted nanoparticle, Theranostic, Thermoliposome, Ultrasound sensitive/ mediated drug delivery.

INTRODUCTION

One of the main challenges of cancer treatment is to treat the disease site without damaging healthy tissues. Drug targeting basically refers to the delivery of a drug to a specific disease site in order to achieve safer and more efficient therapeutic results. A smart practice for drug transportation can be enabled using nanotechnology. Thanks to this technology, nanocarriers can escape from the reticuloendothelial framework, prevent the drug from degradation despite biological obstructions, and deliver the drug to the target cancerous area [1 - 3]. There are several nanoparticle targeting strategies for cancer treatment.

* **Corresponding author Güliz Ak:** Department of Biochemistry, Faculty of Science, Ege University, 35100, İzmir, Türkiye; Tel: +902323111711; E-mail: guliz.ak@ege.edu.tr

Habibe Yılmaz (Ed.)

Types of Targeting

Passive Targeting

With passive targeting, the behavior of drug delivery systems in the body is controlled by using body defense mechanisms such as metabolism, excretion, opsonization, and phagocytosis. The high permeability of the vasculature and the immaturity of the lymphatic drainage system allow the cytotoxic agent to accumulate in the tumor mass. For targeting, the modification and design of the surface charge of the system are carried out by taking into account the molecular weight, size, and surface hydrophobicity of the system. Thus, by allowing the long-term circulation of drugs in the blood, it is ensured that the desired areas are targeted. Uncontrolled drug release with passive targeting can be met with negative consequences such as off-target drug delivery and, as a result, multidrug resistance. Furthermore, the permeability of vessels can be heterogeneous throughout a single tumor. These two situations limit passive targeting [4].

Increased Permeability and Retention Effect (EPR)

When solid tumors reach a certain size, the vascular system surrounding them is not sufficient to provide the oxygen supply necessary for the tumor to proliferate. Cells start to die from a lack of oxygen. As a result, they release growth factors from the surrounding capillaries that trigger the budding of new blood vessels. In this process, called angiogenesis, due to rapid growth, blood vessels are irregular with discontinuous epithelium and have no basement membrane, forming a leaky vasculature with a fenestration of 200 to 2000 nm. This allows better penetration of blood components as it reduces resistance to extravasation into the tumor interstitium. This indicates the improved permeability portion of the EPR effect [5].

The extracellular fluid of healthy tissues is continuously emptied into the lymph vessels at an average flow rate of 0.1-2 μm/s. This allows for the continuous replenishment of interstitial fluid as well as the return of extravasated solutes and colloids to the circulation. Because the lymphatic functions are defective in tumor cells, the outflow of interstitial fluid is blocked. Molecules smaller than 4 nm diffuse back into the bloodstream and are reabsorbed, while the diffusion of macromolecules and nanoparticles are hindered by their large hydrodynamic size. Therefore, nanoparticles reaching the perivascular space are not cleaned sufficiently and concentrate in the tumor interstitium. This aspect represents the enhanced retention part of the EPR effect [6].

Due to the abnormal and leaky vasculature and EPR effect in tumor tissues, nanoparticles accumulate more in this region than in other tissues. Moreover, the nanoparticle system increases the half-life of the drug owing to the escape of the drug from renal clearance. Factors such as protein binding and nanoparticle aggregation have an impact on the EPR effect by increasing the size of the nanoparticle system based on complex formation. It is known that the reticuloendothelial system in the liver and spleen reduces the effect of EPR [7]. Wang *et al.* developed ultrafine iron oxide nanoparticles and studied a breast tumor model. Nanoparticles were accumulated in the tumor interstitial space due to their size, which was formed by self-assembling in the acidic area. Thus, tumor targeting occurs with EPR driven passive targeting [8].

Localized Delivery

The physiological barriers faced by passively and actively targeted nanoparticles need to be overcome. Administering drugs directly to the disease area helps to overcome such obstacles as it does not interfere with the systemic circulation barrier. This method is very effective because it is easy to administer drugs to certain parts of the body, such as the lungs, bladder, brain, peritoneum, and eyes. Topical application, which is one of the local delivery methods, can significantly increase the pharmacological action in the disease area and reduce systemic toxicity since it does not enter the circulation and can take effect in a short time. A variety of drug carrier systems such as liposomes, microparticles, polymeric films, and hydrogels have been prepared for localized delivery. However, while *in vitro* and preclinical studies have been reported for a number of nanoparticles for local delivery in cancer, just a few have reached the clinical stage.

This targeting method is particularly effective when localized chemotherapy of non-metastatic primary tumors and surgical resection are contraindicated. Moreover, for debulking surgeries requiring adjuvant or neoadjuvant chemotherapy to minimize local regional recurrence, local delivery of chemotherapeutics may lead to better therapeutic results with lower toxicity. However, the use of a local delivery strategies is limited in cancer types that are difficult to reach [9].

Tumor Microenvironment

There are some parameters that distinguish the tumor microenvironment from healthy cells. Compared to blood vessels in healthy tissues, the vasculature at the tumor site exhibits varying structural and functional properties. The tumor microenvironment is highly acidic as a result of anaerobic glycolysis and lactic acid production of cancer cells, and this microenvironment is characterized by hypoxia and restricted nutrient supply. As a result of the hypoxia state the gene expression of tumor cells can be altered, thereby cell survival can be increased. The variations in vascular networks generate a distinct tumor microenvironment and ultimately affect therapeutic efficacy. By targeting these physical variables, drugs can be distributed specifically to these regions [10 - 12]. Ling *et al.* prepared pH-sensitive multifunctional iron oxide nanoparticles to target the tumor microenvironment. They have added a ligand, including the imidazole group, which is easy to ionize, to the nanoparticles to increase pH sensitivity and have found that MR imaging and the photoactivity of nanoparticles enhanced with the response to tumor acidic sites [12].

Active Targeting

Active targeting is a method that involves modifying the delivery system using chemical, biological, and physical properties to direct drugs to the desired site. Ligands such as antibodies, peptides, and sugar chains are used as targeting agents or necrotic cells excluding normal tissues. Thanks to these targeting tools, molecules that can recognize specific molecules in the target tissue are designed. Nanoparticles are taken into the target cell, organelle or microenvironment through the target molecule and deposited there. Since therapeutics accumulate in the tumor area, it allows longer and more continuous treatment. In addition, with the delivery of nanoparticulate drug delivery systems to tumor tissues through active targeting, drug delivery to healthy tissues is reduced, and as a result, the damage to healthy tissues is minimized. Thus, the therapeutic efficacy in the desired tumor region is increased [13].

Tumor Cells

The goal of active targeting of cell surface receptors overexpressed by cancer cells is to enhance the cellular uptake of nanocarriers. In particular, active targeting comes to the fore in the intracellular delivery of macromolecular drugs like DNA, siRNA, and proteins. Active targeting is responsible for the transport of the nanoparticular system to the targeted tumor site, its accumulation in that area, and its effect. The critical point in targeting tumor cells is the selection of the appropriate targeting ligand to the tumor site. Thanks to the ligand, it binds to the targeted cell surface receptor and internalizes the cell. Here, delivery systems targeting endocytosis-prone surface receptors come to the fore. In this strategy, ligand-targeted systems cause direct death of cells at the tumor periphery [14].

In a study, Ruan *et al.* synthesized gold nanoparticles targeting glioma cells and penetrating through the blood brain barrier (BBB) for glioma treatment. Nanoparticles were modified *via* angiopep2, a ligand of low-density lipoprotein receptor-related protein-1 (LRP1), which is highly expressed both on BBB and glioma cells and loaded with doxorubicin (Dox). As a result of the study, it was determined that the targeted nanoparticles specifically accumulated in glioma cells [15].

Receptor- Ligand

Some specific receptors are highly expressed by malignant cells but not by normal cells. Various drug nanocarrier systems are used to specifically target the tumor site before they reach healthy tissues. Anticancer drug carrying systems usually bind to bioactive vector molecules such as specific small molecule ligands, peptides, or antibodies, and the receptors on the surface of targeted cancer cells. So they can be directed against specific malignant cells. They are then internalized into cells with high specificity and efficiency. This way, increased selectivity can be achieved and multidrug resistance can be reduced. Therefore, targeting specific receptors overexpressed on the surface of cancer cells aids in appropriate patient selection for personalized therapy. A variety of ligands specific to different receptors are available for use as a drug delivery system.

Transferrin receptor is a membrane glycoprotein and a carrier for transferrin. Transferrin carries the import of iron into the cell from the blood. It binds to transferrin receptors and then internalizes into cells *via* receptor-mediated endocytosis. Folate receptor is a receptor that binds with high affinity to vitamins, folic acid, folate drug conjugates, and folate-containing nanocarriers. It carries out the transport of folate-linked molecules into cells *via* receptor-mediated endocytosis. Epidermal growth factor (EGF) receptor (EGFR) is a member of the ErbB/HER family of tyrosine kinase (TK) receptors. EGFR is overexpressed in most cancers, especially breast cancer. It stimulates tumor growth and progression such as proliferation, angiogenesis, invasion and metastasis [14, 16, 17].

It is known that the folate receptor is overexpressed in many cancers such as ovarian, mammary, lung, prostate, colon, and *etc.* cancer [18]. In a study, Ak and Sanlier synthesized and radiolabelled a folate-poly(ethylene glycol)-doxorubicin conjugate system. 99mTc-Radiolabelled compounds indicated significant uptake in prostate tissue whereas the conjugate without folate ligand demonstrated lower uptake [19]. In another study, Ak and Sanlier developed a magnetically sensitive and folate receptor targeted biomimetic doxorubicin carrying system for use in ovarian cancer therapy. The magnetic nanoparticles were developed and coated with erythrocyte membrane vesicles to provide biomimetic properties. The folate molecule was added to the vesicle to target the folate receptor. As a result of *in vitro* studies on SKOV3, an ovarian cancer cell line, it was concluded that the IC_{50} value of the folate receptor targeted biomimetic delivery system was lower than the delivery system without folate ligand, meaning the targeting ability of the folate molecule [20].

Carbohydrate- Lectin

Carbohydrates in the structure of tumor cells differ from those of healthy cells, and carbohydrates are cell-specific molecules that can bind. Lectin is a non-immunological protein that has the ability to bind and recognize glycoproteins found on the cell surface. The affinity between lectins and carbohydrates can be used as a targeting vehicle in drug delivery systems. Lectin-carbohydrate interactions can be achieved in two ways. The first way is adding lectins to nanoparticles for targeting carbohydrates on the cell surface and the second is adding carbohydrates to the nanoparticles for targeting the lectins on the cell surface. The selection of these two types is made according to the cancer type and the targeted cell [21, 22].

Galectin-1 (Gal-1) is a β-galactoside-binding lectin that causes the progression of cancer by regulating the interactions of cancer with immune cells. Prostate cancer cells result in upregulation of Gal-1. Besford *et al.* reported the development of β-galactoside-functionalized glyconanoparticles for prostate cancer through Gal-1 targeting. It was found that these nanoparticles were successfully interacted with peanut agglutinin, β-galactoside-specific lectins. It was concluded that β-galactoside-functionalized glyconanoparticles could have targeted binding to prostate cancer cells [23].

Antigen- Antibody

Antibodies have high affinity for specific antigens. Although there are many types of antibodies, most antibodies used for therapeutics are IgG antibodies. They are Y-shaped proteins in the IgG structure, consisting of an invariant constant (Fc) region and a variable (Fab) region specific to each antibody, with two heavy and two light chain units linked by disulfide bonds. Antigens are specifically expressed on cancer cells. Targeting tumor regions can be achieved by utilizing the antibody-antigen relationship. Antibodies are positioned on the surface of nanoparticles from the Fc region and the Fab region is released to bind antigens. Thanks to this finding, nanoparticles can be transported directly to the target tissue, so the selectivity to the target increases and the toxic effects on other cells are reduced [13, 24].

Dancy and co-workers prepared a targeted drug-carrying system for breast cancer. It is known that the fibroblast growth factor-inducible 14 (Fn14) receptor, which is one of the tumor necrosis factor receptor family, is synthesized at a low level in healthy cells, however a higher levels in breast cancer cells. Decreased non-specific adhesivity receptor-targeted (DART) nanoparticles were loaded with paclitaxel and linked with Fn14 monoclonal antibody. Targeted DART nanoparticles had minimal nonspecific binding and had significant therapeutic efficacy for breast cancer [25].

Aptamer-Ligand

Aptamers are short single-stranded RNA or DNA oligonucleotides with three-dimensional structures that can bind to antigens with high affinity and specificity like antibodies. Aptamers are attached to nanoparticles and mediate targeting of tumor sites. Aptamers have a very broad spectrum of target recognition and binding. They show almost no immunological properties. They can easily be attached to the ends with a chemical group to assemble the nanoparticles. They are small and do not cause a significant increase in nanoparticle size. They can easily enter the cell at the target site and release the drug. Peptide aptamers that bind to peptide-binding and ATP-binding domains increase the susceptibility of cells to apoptosis induced by anticancer drugs. However, aptamers are still expensive to produce in large quantities. Major structural modifications to protect against nuclease degradation and improve pharmacokinetic properties result in increased cost. This limits the use of aptamers [13, 26].

In 2017, Taghavi and co-workers synthesized epirubicin loaded poly (lactic-co-glycolic acid) modified chitosan nanoparticles. For MUC1 receptor targeting 5TR1, DNA aptamer was conjugated to the nanoparticles. These targeted nanoparticles indicated specificity for cancer cells expressing MUC1 receptors, and it was confirmed that 5TR1 DNA aptamer had an important role for targeting based on the results of *in vitro* and *in vivo* experiments on MCF7 breast cancer and C26 colon carcinoma cells (MUC1 positive cells) [27].

Vascular Endothelium

The vascular endothelial growth factor receptor (VEGFR) is considered the most suitable inducer of tumor angiogenesis *via* VEGF receptor signaling mechanism. VEGFR-1 and VEGFR-2 have roles in tumor angiogenesis and neovascularization. Tumor hypoxia and oncogenes upregulate VEGF levels in tumor cells, which lead to upregulation of VEGF receptors in tumor endothelial cells. The VEGFR of interest for the most active targeting systems is VEGFR-2. Apart from angiogenesis, it was found that few melanoma and leukemia cell lines express VEGFR-2 in non-endothelial tumors. VEGFR-2 phosphorylates tyrosine upon VEGF binding and starts the angiogenesis mechanism. Therefore, VEGF and VEGFR-2 appear to be promising targets for nanocarrier systems. There could be two main approaches for targeting angiogenesis through VEGF and VEGFR-2. The first is to target VEGFR-2 to decrease VEGF binding and the second is to target VEGF to inhibit ligand binding to VEGFR-2 [14, 28].

In a study, Ruiz *et al.* synthesized polyurethane-urea nanoparticles and coated them with streptavidin. The VEGFR2 blocker, CBO-P11, was incorporated into nanoparticles. This inhibitor allowed a targeted pharmacological action *via* reducing the proliferation just in inflamed endothelial cells (HUVEC, human umbilical vein endothelial cells) [29].

Physical Targeting

Physical targeting includes a variety of agents that affect activity in the target tissue, either stimulus-sensitive materials or environmentally sensitive materials. Materials sensitive to stimuli undergo physical or chemical changes in response to a particular stimulus. These materials exhibit environmentally sensitive behavior and respond to external stimuli due to their biomimetic nature. The therapy, which responds to many physical stimuli, focuses the external stimuli specifically on the tumor with minimized damage to the normal tissues and increased therapeutic effect to the diseased area. Some physical stimuli, such as a magnetic field, provide enrichment of therapeutic agents at the tumor site to increase therapeutic efficacy. Some physical stimuli control the internalization of the drug into the cells. It also greatly increases the effectiveness of drug release and chemotherapy drugs in the tumor and reduces toxicity. Many physical stimulant responsive therapies are not only used to directly kill cancer cells, but are also useful for triggering or improving different cancer treatments to obtain desired anticancer effects by different mechanisms [30, 31].

Magnetic Field

A magnetic field targeting strategy can be used to specifically concentrate drugs in a particular region. The drugs are encapsulated with crystalline magnetic structures and taken into the cell. Materials such as iron, nickel, and cobalt are generally used to create magnetics. Magnetic nanoparticle-coated drug complexes are injected intravenously, intraarterially, or intraperitoneally, and a magnet is placed near the cancerous region to generate a localized magnetic field. The magnetically coated drug complex is directed to the magnetic field created by the magnet during circulation and concentrates in that area. The particles must have sufficient strength to be attracted by the magnetic field and be of a size to allow them to enter the tumor or the surrounding vascular system. Since magnetic nanoparticles can cross the blood-brain barrier, their use, especially in brain cancer, comes to the fore [32, 33].

Huang *et al.* prepared magnetic graphene oxide nanoparticles and loaded with Dox active substance. In studies with glioblastoma cells, they placed magnets on plates containing the cell line. It has been determined that the cells attract magnetic nanoparticles and take them into the cell. Thus, magnetic nanoparticles killed cancer cells at a high level. Compared to cells without magnetic targeting, the effect was found to be quite high in cells with targeting [34]. In a study described by Wu and co-workers, poly(L-glutamic acid) was conjugated to SN-38 and was complexed with chitosan. Citrate coated superparamagnetic iron oxide nanoparticles were encapsulated into the complex. These magnetic nanocomplexes exhibited targeted accumulation and enhanced colorectal tumor growth inhibition with the help of external magnetic application as a result of *in vivo* studies on HCT-116 tumor-bearing mice [35].

Ultrasound

Ultrasound is mechanical waves that periodically oscillate at a frequency higher than 20 kHz, a frequency higher than the sounds that the human ear can hear. These waves are usually produced by a mechanical converter of an electrical signal. Ultrasound's real-time, portable, non-ionizing, and deep-tissue penetrating capabilities have made it attractive in tumor imaging and treatment applications. High intensity focused ultrasound could cause coagulative tumor cell death by specifically targeting the tumor site and rapidly reaching a thermal ablation temperature in this area. In addition, nanosystems used as sonosensitizers or contrast agents enable drugs to target the tumor site. With the sonosensitizer, ROS production occurs in the tumor area. It has been shown that especially polymer-based nanoparticles and hydrogels can be released from microbubbles by ultrasonic targeting, causing significant drug accumulation and cell death in tumor tissues [36, 37].

Other perspectives are also possible about ultrasound targeting. Xing *et al.* reported paclitaxol encapsulated CA19-9 antibody-modified nanoparticles and combined with ultrasound-mediated microbubble destruction (UMMD) for targeted pancreatic cancer drug delivery. UMMD method contributes to increased cellular uptake of drugs *via* creating transient and repairable pores on the cell membrane that would enhance the permeability with the aid of gas-filled microbubbles under ultrasound pressure. As a result of the study, it was determined that cellular uptake of paclitaxel increased with UMMD, and also higher pancreatic tumor inhibition was confirmed with the UMMD combination [38]. In another approach, a study on non-small cell lung cancer, Şanlıer and colleagues developed magnetic-targeted and ultrasound-sensitive dual drugs (pemetrexed and pazopanib) carrying a nanobubble system. Drugs were linked to magnetic nanoparticles, which were loaded into liposomes, then nanobubbles formed from these liposomes using argon gas. It has been thought that nanobubbles accumulated in the tumor area based on magnetic field application and were disrupted by ultrasound energy resulting drug delivery in the target sites [39].

Heat

Although studies on the manipulation of nanoparticles by temperature are limited, it is considered a promising targeting strategy. The experiments indicated various applications of achieving and monitoring heat for drug release except for direct treatment of high intensity focussed ultrasound or radiofrequency ablation to kill tumor cells [40]. In order to achieve a rise in temperature of cancerous tissue for investigation of heat-responsive nanoplatforms, high intensity focussed ultrasound (HIFU), radiofrequency ablation (RFA), microwave, near infrared (NIR), magnetic resonance imaging (MRI) and alternating magnetic field applications could have been used [41, 42].

Metallic nanoparticles strongly absorb and scatter light near localized surface plasmon resonances. It allows selective targeting of tumors without damaging the surrounding tissues with increased temperature in a particular area. Otherwise, thermosensitive liposomes (TSLs) come to the fore with heat-sensitive nanocarriers [40]. TSLs could destabilize at their melting transition temperature (T_m) (for instance 40-44.5°C) as transformed from gel to the liquid-crystalline phase mediates drug release [42]. In 2022, doxorubicin loaded thermosensitive liposomes were prepared for hyperthermia induced drug delivery. During the experiment with healthy pigs, mild hyperthermia (10 min to 60 min) was applied using MR guided HIFU at a particular muscle tissue while infusion of TSLs. As a comparison with untreated muscle tissues, doxorubicin uptake was clearly higher in the heated target tissue. The authors thought that this treatment could be used for sarcoma therapy [43].

Light

There are two types of light-triggered therapy: photothermal therapy and photodynamic therapy. Photothermal therapy converts light energy into heat energy, causing a rise in temperature in the cancerous area. Photodynamic therapy, on the other hand, absorbs optical energy and transfers it to oxygen molecules, thus triggering the production of ROS, resulting in the death of cancer cells. By coating the tumor area with suitable photothermal agents, other inorganic nanoparticles, especially nanoparticles containing transition metals, are directed to that area [30]. Lei *et al.* prepared mesoporous silica nanoparticles loaded with indocyanine green (ICG) and doxorubicin for NIR-triggered drug release and chemo/photothermal therapy. Nanoparticles were conjugated with β-cyclodextrin with thermal-labile azo linker. ICG was used as the photothermal agent to trigger the release of ICG and doxorubicin and heat formation resulting in the breakdown of thermal-labile links which also leads to drug release [44].

Electric

The electrical properties of the cells determine the basic functions of the cells. The electrical properties of a cancerous cell and a healthy cell, such as membrane potential, cellular dielectric constant, or electrical capacitance, are different from each other. It is known that cancerous cells are more electronegative than healthy cells. Thanks to this feature that distinguishes cancerous cells from others, electrotherapy comes to the fore in cancer treatment. With the presence of ion channels, the cell membrane is known to be an electrically polarizable environment and can be significantly affected by the electric field. Cancer electrotherapy focuses specifically on stimulating the cell by giving short electrical pulses to change the permeability of the cell membrane. As a result of the stimulation of the targeted area, enlargements occur in the cell membrane pores. Thus, the transported drug molecules are intensely transported to the stimulated region, and the opening of the pores facilitates the entry of drug molecules into the cell [45, 46].

Rodzinski and colleagues synthesized paclitaxel loaded and targeted magnetoelectric nanoparticles capable of distinguishing cancer cells from normal cells based on the membrane's electrical properties. Thanks to the attraction force between magnetoelectric nanoparticles and the intense electric charges of cancer cells in addition to magnetic field application, it was determined that the nanoparticles were penetrating the cancer cells efficiently [47].

Targeting and Diagnostic

Theranostics is a couple of terms therapeutics and diagnostics. Theranostic represents an integrated nanotherapeutic system that can diagnose, deliver targeted therapy, and monitor response to therapy. The location, stage, and progression of cancer can be easily monitored with theranostic. Theranostic nanoparticles have the ability to selectively accumulate in cancerous tissue and provide a therapeutic effect. It has a long half-life in circulation, so it can act on the tumor site for a long time and is easily biodegraded. In order to create a theranostic agent, it is necessary to have a good understanding of cancer at the molecular level and to have extensive knowledge about the diagnosis and treatment mechanisms. The dual-purpose nanoparticle system accelerates drug development, improves the treatment process, reduces potential risks, and lowers costs.

Care is taken to ensure that the components used in the preparation of the combined nanoparticle system are compatible with each other. The diagnostic aspect of theranostics mostly aims to take images by using different contrast agents. Magnetic resonance imaging is the most preferred imaging mechanism.

Especially magnetic particles are used as contrast material. Gadolinium, iron oxide, gold, and silver are low toxic metals used for this purpose. By using these particles, the diseased area is visualized as opposed to healthy tissues. In addition to imaging and diagnosis, magnetic nanoparticles have the feature of targeting the tumor site and allowing drugs to accumulate in that area. The drugs increase the effect of the treatment by performing a slow and long-term release in the cancerous tissue [48, 49].

Satpathy *et al.* produced HER2-targeted magnetic iron oxide nanoparticles for image-guided treatment and targeted drug delivery. Cisplatin carrying magnetic nanoparticles were conjugated with a near infrared dye labeled HER2 affibody. These nanocarriers suppressed tumor growth both in primary tumor and lung metastases in the ovarian cancer model. In addition, it has been shown that therapy responses were detectable by magnetic resonance imaging [50]. In another study reported by Ren *et al.*, fluorescence carbon dots were linked with epidermal growth factor (EGF) (targeting agent), chlorin e6 (photosensitizer, generating reactive oxygen species) was incorporated and Pt(IV) (chemotherapy agent) was grafted on carbon dots for image guided therapy of EGFR+ esophageal cancer. Based on the *in vitro* and *in vivo* studies, the theranostic nanoplatform provided greater tumor imaging and therapy [51].

CONCLUDING REMARKS

Radiotherapy and chemotherapy in traditional cancer treatment are known to damage healthy tissues as well as cancerous tissues. For this reason, the side effects are felt quite a lot. The potential of nanoparticles and their derivatives as therapeutic agents has led to many studies in recent years. But translating scientific research into marketable products is still a major challenge, only a few therapeutic nanoparticles have been approved by regulatory agencies.

In the future, nanotechnology may advance in personalized medicine. Considering that each cancer differs from person to person, its effect and treatment process vary, and it is thought that personalized treatment will yield positive results. Moreover, in the future, nanodiagnostics may come to the fore in diagnostic tests for improved sensitivity and specificity [52].

REFERENCES

[1]   A. Shah, S. Aftab, J. Nisar, M.N. Ashiq, and F.J. Iftikhar, "Nanocarriers for targeted drug delivery", *J. Drug Deliv. Sci. Technol.,* vol. 62, p. 102426, 2021.
[http://dx.doi.org/10.1016/j.jddst.2021.102426]

[2]   M.T. Manzari, Y. Shamay, H. Kiguchi, N. Rosen, M. Scaltriti, and D.A. Heller, "Targeted drug delivery strategies for precision medicines", *Nat. Rev. Mater.,* vol. 6, no. 4, pp. 351-370, 2021.
[http://dx.doi.org/10.1038/s41578-020-00269-6] [PMID: 34950512]

[3]   M.F. Attia, N. Anton, J. Wallyn, Z. Omran, and T.F. Vandamme, "An overview of active and passive targeting strategies to improve the nanocarriers efficiency to tumour sites", *J. Pharm. Pharmacol.,* vol. 71, no. 8, pp. 1185-1198, 2019.
[http://dx.doi.org/10.1111/jphp.13098] [PMID: 31049986]

[4]   T.D. Clemons, R. Singh, A. Sorolla, N. Chaudhari, A. Hubbard, and K.S. Iyer, "Distinction between active and passive targeting of nanoparticles dictate their overall therapeutic efficacy", *Langmuir,* vol. 34, no. 50, pp. 15343-15349, 2018.
[http://dx.doi.org/10.1021/acs.langmuir.8b02946] [PMID: 30441895]

[5]   N. Bertrand, J. Wu, X. Xu, N. Kamaly, and O.C. Farokhzad, "Cancer nanotechnology: The impact of passive and active targeting in the era of modern cancer biology", *Adv. Drug Deliv. Rev.,* vol. 66, pp. 2-25, 2014.
[http://dx.doi.org/10.1016/j.addr.2013.11.009] [PMID: 24270007]

[6]   M.A. Subhan, S.S.K. Yalamarty, N. Filipczak, F. Parveen, and V.P. Torchilin, "Recent advances in tumor targeting via epr effect for cancer treatment", *J. Pers. Med.,* vol. 11, no. 6, p. 571, 2021.
[http://dx.doi.org/10.3390/jpm11060571] [PMID: 34207137]

[7]   D. Kalyane, N. Raval, R. Maheshwari, V. Tambe, K. Kalia, and R.K. Tekade, "Employment of enhanced permeability and retention effect (EPR): Nanoparticle-based precision tools for targeting of therapeutic and diagnostic agent in cancer", *Mater. Sci. Eng. C,* vol. 98, pp. 1252-1276, 2019.
[http://dx.doi.org/10.1016/j.msec.2019.01.066] [PMID: 30813007]

[8]   L. Wang, J. Huang, H. Chen, H. Wu, Y. Xu, Y. Li, H. Yi, Y.A. Wang, L. Yang, and H. Mao, "Exerting enhanced permeability and retention effect driven delivery by ultrafine iron oxide nanoparticles with T1 – T2 switchable magnetic resonance imaging contrast", *ACS Nano,* vol. 11, no. 5, pp. 4582-4592, 2017.
[http://dx.doi.org/10.1021/acsnano.7b00038] [PMID: 28426929]

[9]   D. Rosenblum, N. Joshi, W. Tao, J.M. Karp, and D. Peer, "Progress and challenges towards targeted delivery of cancer therapeutics", *Nat. Commun.,* vol. 9, no. 1, p. 1410, 2018.
[http://dx.doi.org/10.1038/s41467-018-03705-y] [PMID: 29650952]

[10]   T. Wu, and Y. Dai, "Tumor microenvironment and therapeutic response", *Cancer Lett.,* vol. 387, pp. 61-68, 2017.
[http://dx.doi.org/10.1016/j.canlet.2016.01.043] [PMID: 26845449]

[11]   R. Wakaskar, "Passive and active targeting in tumor microenvironment", *Int. J. Drug. Dev. Res.,* vol. 9, no. 2, pp. 37-41, 2017.

[12]   D. Ling, W. Park, S. Park, Y. Lu, K.S. Kim, M.J. Hackett, B.H. Kim, H. Yim, Y.S. Jeon, K. Na, and T. Hyeon, "Multifunctional tumor pH-sensitive self-assembled nanoparticles for bimodal imaging and treatment of resistant heterogeneous tumors", *J. Am. Chem. Soc.,* vol. 136, no. 15, pp. 5647-5655, 2014.
[http://dx.doi.org/10.1021/ja4108287] [PMID: 24689550]

[13]   S. Sultana, M.R. Khan, M. Kumar, S. Kumar, and M. Ali, "Nanoparticles-mediated drug delivery approaches for cancer targeting: A review", *J. Drug Target.,* vol. 21, no. 2, pp. 107-125, 2013.
[http://dx.doi.org/10.3109/1061186X.2012.712130] [PMID: 22873288]

[14]   F. Danhier, O. Feron, and V. Préat, "To exploit the tumor microenvironment: Passive and active tumor targeting of nanocarriers for anti-cancer drug delivery", *J. Control. Release.,* vol. 148, no. 2, pp. 135-146, 2010.
[http://dx.doi.org/10.1016/j.jconrel.2010.08.027] [PMID: 20797419]

[15]   S. Ruan, M. Yuan, L. Zhang, G. Hu, J. Chen, X. Cun, Q. Zhang, Y. Yang, Q. He, and H. Gao, "Tumor microenvironment sensitive doxorubicin delivery and release to glioma using angiopep-2 decorated gold nanoparticles", *Biomaterials,* vol. 37, pp. 425-435, 2015.
[http://dx.doi.org/10.1016/j.biomaterials.2014.10.007] [PMID: 25453970]

[16] G. Gocheva, and A. Ivanova, "A look at receptor–ligand pairs for active-targeting drug delivery from crystallographic and molecular dynamics perspectives", *Mol. Pharm.*, vol. 16, no. 8, pp. 3293-3321, 2019.
[http://dx.doi.org/10.1021/acs.molpharmaceut.9b00250] [PMID: 31274322]

[17] A. Ahmad, F. Khan, R.K. Mishra, and R. Khan, "Precision cancer nanotherapy: Evolving role of multifunctional nanoparticles for cancer active targeting", *J. Med. Chem.*, vol. 62, no. 23, pp. 10475-10496, 2019.
[http://dx.doi.org/10.1021/acs.jmedchem.9b00511] [PMID: 31339714]

[18] A.R. Hilgenbrink, and P.S. Low, "Folate receptor-mediated drug targeting: From therapeutics to diagnostics", *J. Pharm. Sci.*, vol. 94, no. 10, pp. 2135-2146, 2005.
[http://dx.doi.org/10.1002/jps.20457] [PMID: 16136558]

[19] G. Ak, F. Yurt Lambrecht, and S.H. Sanlier, "Radiolabeling of folate targeted multifunctional conjugate with technetium-99m and biodistribution studies in rats", *J. Drug. Target.*, vol. 20, no. 6, pp. 509-514, 2012.
[http://dx.doi.org/10.3109/1061186X.2012.686038] [PMID: 22643314]

[20] G. Ak, and Ş. Hamarat Şanlıer, "Erythrocyte membrane vesicles coated biomimetic and targeted doxorubicin nanocarrier: Development, characterization and *in vitro* studies", *J. Mol. Struct.*, vol. 1205, p. 127664, 2020.
[http://dx.doi.org/10.1016/j.molstruc.2019.127664]

[21] A. Sharma, N. Jain, and R. Sareen, "Nanocarriers for diagnosis and targeting of breast cancer", *BioMed Res. Int.*, vol. 2013, pp. 1-10, 2013.
[http://dx.doi.org/10.1155/2013/960821] [PMID: 23865076]

[22] Z.R. Goddard, M.J. Marín, D.A. Russell, and M. Searcey, "Active targeting of gold nanoparticles as cancer therapeutics", *Chem. Soc. Rev.*, vol. 49, no. 23, pp. 8774-8789, 2020.
[http://dx.doi.org/10.1039/D0CS01121E] [PMID: 33089858]

[23] Q.A. Besford, M. Wojnilowicz, T. Suma, N. Bertleff-Zieschang, F. Caruso, and F. Cavalieri, "Lactosylated glycogen nanoparticles for targeting prostate cancer cells", *ACS Appl. Mater. Interfaces*, vol. 9, no. 20, pp. 16869-16879, 2017.
[http://dx.doi.org/10.1021/acsami.7b02676] [PMID: 28362077]

[24] R. Bazak, M. Houri, S. El Achy, S. Kamel, and T. Refaat, "Cancer active targeting by nanoparticles: A comprehensive review of literature", *J. Cancer Res. Clin. Oncol.*, vol. 141, no. 5, pp. 769-784, 2015.
[http://dx.doi.org/10.1007/s00432-014-1767-3] [PMID: 25005786]

[25] J.G. Dancy, A.S. Wadajkar, N.P. Connolly, R. Galisteo, H.M. Ames, S. Peng, N.L. Tran, O.G. Goloubeva, G.F. Woodworth, J.A. Winkles, and A.J. Kim, "Decreased nonspecific adhesivity, receptor-targeted therapeutic nanoparticles for primary and metastatic breast cancer", *Sci. Adv.*, vol. 6, no. 3, p. eaax3931, 2020.
[http://dx.doi.org/10.1126/sciadv.aax3931] [PMID: 31998833]

[26] K. Alt, F. Carraro, E. Jap, M. Linares-Moreau, R. Riccò, M. Righetto, M. Bogar, H. Amenitsch, R.A. Hashad, C. Doonan, C.E. Hagemeyer, and P. Falcaro, "Self–assembly of oriented antibody–decorated metal–organic framework nanocrystals for active–targeting applications", *Adv. Mater.*, vol. 34, no. 21, p. 2106607, 2022.
[http://dx.doi.org/10.1002/adma.202106607] [PMID: 34866253]

[27] S. Taghavi, M. Ramezani, M. Alibolandi, K. Abnous, and S.M. Taghdisi, "Chitosan-modified PLGA nanoparticles tagged with 5TR1 aptamer for *in vivo* tumor-targeted drug delivery", *Cancer. Lett.*, vol. 400, pp. 1-8, 2017.
[http://dx.doi.org/10.1016/j.canlet.2017.04.008] [PMID: 28412238]

[28] J.D. Byrne, T. Betancourt, and L. Brannon-Peppas, "Active targeting schemes for nanoparticle systems in cancer therapeutics", *Adv. Drug. Deliv. Rev.*, vol. 60, no. 15, pp. 1615-1626, 2008.
[http://dx.doi.org/10.1016/j.addr.2008.08.005] [PMID: 18840489]

[29] G. Morral-Ruíz, P. Melgar-Lesmes, C. Solans, and M.J. García-Celma, "Multifunctional polyurethane–urea nanoparticles to target and arrest inflamed vascular environment: A potential tool for cancer therapy and diagnosis", *J. Control. Release,* vol. 171, no. 2, pp. 163-171, 2013.
[http://dx.doi.org/10.1016/j.jconrel.2013.06.027] [PMID: 23831054]

[30] Q Chen, H Ke, Z Dai, and Z Liu, "Nanoscale theranostics for physical stimulus-responsive cancer therapies", *Biomaterials,* vol. 73, no. 2, pp. 14-30, 2015.
[http://dx.doi.org/10.1016/j.biomaterials.2015.09.018]

[31] M.A. Rahim, N. Jan, S. Khan, H. Shah, A. Madni, A. Khan, A. Jabar, S. Khan, A. Elhissi, Z. Hussain, H.C. Aziz, M. Sohail, M. Khan, and H.E. Thu, "Recent advancements in stimuli responsive drug delivery platforms for active and passive cancer targeting", *Cancers,* vol. 13, no. 4, p. 670, 2021.
[http://dx.doi.org/10.3390/cancers13040670] [PMID: 33562376]

[32] Y. Shen, C. Wu, T.Q.P. Uyeda, G.R. Plaza, B. Liu, Y. Han, M.S. Lesniak, and Y. Cheng, "Elongated nanoparticle aggregates in cancer cells for mechanical destruction with low frequency rotating magnetic field", *Theranostics,* vol. 7, no. 6, pp. 1735-1748, 2017.
[http://dx.doi.org/10.7150/thno.18352] [PMID: 28529648]

[33] C. Li, L. Li, and A.C. Keates, "Targeting cancer gene therapy with magnetic nanoparticles", *Oncotarget,* vol. 3, no. 4, pp. 365-370, 2012.
[http://dx.doi.org/10.18632/oncotarget.490] [PMID: 22562943]

[34] Y.S. Huang, Y.J. Lu, and J.P. Chen, "Magnetic graphene oxide as a carrier for targeted delivery of chemotherapy drugs in cancer therapy", *J. Magn. Magn. Mater.,* vol. 427, pp. 34-40, 2017.
[http://dx.doi.org/10.1016/j.jmmm.2016.10.042]

[35] D. Wu, Y. Li, L. Zhu, W. Zhang, S. Xu, Y. Yang, Q. Yan, and G. Yang, "A biocompatible superparamagnetic chitosan-based nanoplatform enabling targeted SN-38 delivery for colorectal cancer therapy", *Carbohydr. Polym.,* vol. 274, p. 118641, 2021.
[http://dx.doi.org/10.1016/j.carbpol.2021.118641] [PMID: 34702462]

[36] E. Alphandéry, "Ultrasound and nanomaterial: An efficient pair to fight cancer", *J. Nanobiotech.,* vol. 20, no. 1, p. 139, 2022.
[http://dx.doi.org/10.1186/s12951-022-01243-w] [PMID: 35300712]

[37] L.Q. Zhou, P. Li, X.W. Cui, and C.F. Dietrich, "Ultrasound nanotheranostics in fighting cancer: Advances and prospects", *Cancer Lett.,* vol. 470, no. 1095, pp. 204-219, 2020.
[http://dx.doi.org/10.1016/j.canlet.2019.11.034] [PMID: 31790760]

[38] L. Xing, Q. Shi, K. Zheng, M. Shen, J. Ma, F. Li, Y. Liu, L. Lin, W. Tu, Y. Duan, and L. Du, "Ultrasound-mediated microbubble destruction (UMMD) facilitates the delivery of CA19-9 targeted and paclitaxel loaded mPEG-PLGA-PLL nanoparticles in pancreatic cancer", *Theranostics,* vol. 6, no. 10, pp. 1573-1587, 2016.
[http://dx.doi.org/10.7150/thno.15164] [PMID: 27446491]

[39] Ş. Hamarat Şanlıer, G. Ak, H. Yılmaz, A. Ünal, Ü.F. Bozkaya, G. Tanıyan, Y. Yıldırım, and G. Yıldız Türkyılmaz, "Development of ultrasound-triggered and magnetic-targeted nanobubble system for dual-drug delivery", *J. Pharm. Sci.,* vol. 108, no. 3, pp. 1272-1283, 2019.
[http://dx.doi.org/10.1016/j.xphs.2018.10.030] [PMID: 30773203]

[40] N.M. Dimitriou, G. Tsekenis, E.C. Balanikas, A. Pavlopoulou, M. Mitsiogianni, T. Mantso, G. Pashos, A.G. Boudouvis, I.N. Lykakis, G. Tsigaridas, M.I. Panayiotidis, V. Yannopapas, and A.G. Georgakilas, "Gold nanoparticles, radiations and the immune system: Current insights into the physical mechanisms and the biological interactions of this new alliance towards cancer therapy", *Pharmacol. Ther.,* vol. 178, pp. 1-17, 2017.
[http://dx.doi.org/10.1016/j.pharmthera.2017.03.006] [PMID: 28322970]

[41] M. Chaudhry, P. Lyon, C. Coussios, and R. Carlisle, "Thermosensitive liposomes: A promising step toward localised chemotherapy", *Expert Opin. Drug. Deliv.,* vol. 19, no. 8, pp. 899-912, 2022.
[http://dx.doi.org/10.1080/17425247.2022.2099834] [PMID: 35830722]

[42]   H. Bi, J. Xue, H. Jiang, S. Gao, D. Yang, Y. Fang, and K. Shi, "Current developments in drug delivery with thermosensitive liposomes", *Asian J. Pharm. Sci,* vol. 14, no. 4, pp. 365-379, 2019.
[http://dx.doi.org/10.1016/j.ajps.2018.07.006] [PMID: 32104466]

[43]   L.C. Sebeke, J.D. Castillo Gómez, E. Heijman, P. Rademann, A.C. Simon, S. Ekdawi, S. Vlachakis, D. Toker, B.L. Mink, C. Schubert-Quecke, S.Y. Yeo, P. Schmidt, C. Lucas, S. Brodesser, M. Hossann, L.H. Lindner, and H. Grüll, "Hyperthermia-induced doxorubicin delivery from thermosensitive liposomes *via* MR-HIFU in a pig model", *J. Control. Release,* vol. 343, pp. 798-812, 2022.
[http://dx.doi.org/10.1016/j.jconrel.2022.02.003] [PMID: 35134460]

[44]   Q Lei, W Qiu, J Hu, P Cao, C Zhu, and H. Cheng, "Multifunctional mesoporous silica nanoparticles with thermal-responsive gatekeeper for nir light-triggered chemo / photothermal-therapy", *Small,* vol. 12, no. 31, pp. 4286-4298, 2016.

[45]   E. Hondroulis, R. Zhang, C. Zhang, C. Chen, K. Ino, T. Matsue, and C.Z. Li, "Immuno nanoparticles integrated electrical control of targeted cancer cell development using whole cell bioelectronic device", *Theranostics,* vol. 4, no. 9, pp. 919-930, 2014.
[http://dx.doi.org/10.7150/thno.8575] [PMID: 25057316]

[46]   R. Guduru, P. Liang, C. Runowicz, M. Nair, V. Atluri, and S. Khizroev, "Magneto-electric nanoparticles to enable field-controlled high-specificity drug delivery to eradicate ovarian cancer cells", *Sci. Rep.,* vol. 3, no. 1, p. 2953, 2013.
[http://dx.doi.org/10.1038/srep02953] [PMID: 24129652]

[47]   A. Rodzinski, R. Guduru, P. Liang, A. Hadjikhani, T. Stewart, E. Stimphil, C. Runowicz, R. Cote, N. Altman, R. Datar, and S. Khizroev, "Targeted and controlled anticancer drug delivery and release with magnetoelectric nanoparticles", *Sci. Rep.,* vol. 6, no. 1, p. 20867, 2016.
[http://dx.doi.org/10.1038/srep20867] [PMID: 26875783]

[48]   N. Ahmed, H. Fessi, and A. Elaissari, "Theranostic applications of nanoparticles in cancer", *Drug. Discov. Today.,* vol. 17, no. 17-18, pp. 928-934, 2012.
[http://dx.doi.org/10.1016/j.drudis.2012.03.010] [PMID: 22484464]

[49]   M. Mendes, J.J. Sousa, A. Pais, and C. Vitorino, "Targeted theranostic nanoparticles for brain tumor treatment", *Pharmaceutics,* vol. 10, no. 4, p. 181, 2018.
[http://dx.doi.org/10.3390/pharmaceutics10040181] [PMID: 30304861]

[50]   M. Satpathy, L. Wang, R.J. Zielinski, W. Qian, Y.A. Wang, A.M. Mohs, B.A. Kairdolf, X. Ji, J. Capala, M. Lipowska, S. Nie, H. Mao, and L. Yang, "Targeted drug delivery and image-guided therapy of heterogeneous ovarian cancer using HER2-targeted theranostic nanoparticles", *Theranostics,* vol. 9, no. 3, pp. 778-795, 2019.
[http://dx.doi.org/10.7150/thno.29964] [PMID: 30809308]

[51]   G. Ren, Z. Wang, Y. Tian, J. Li, Y. Ma, L. Zhou, C. Zhang, L. Guo, H. Diao, L. Li, L. Lu, S. Ma, Z. Wu, L. Yan, and W. Liu, "Targeted chemo-photodynamic therapy toward esophageal cancer by GSH-sensitive theranostic nanoplatform", *Biomed. Pharmacother.,* vol. 153, p. 113506, 2022.
[http://dx.doi.org/10.1016/j.biopha.2022.113506] [PMID: 36076595]

[52]   E.S. Ali, S.M. Sharker, M.T. Islam, I.N. Khan, S. Shaw, M.A. Rahman, S.J. Uddin, M.C. Shill, S. Rehman, N. Das, S. Ahmad, J.A. Shilpi, S. Tripathi, S.K. Mishra, and M.S. Mubarak, "Targeting cancer cells with nanotherapeutics and nanodiagnostics: Current status and future perspectives", *Semin. Cancer. Biol.,* vol. 69, pp. 52-68, 2021.
[http://dx.doi.org/10.1016/j.semcancer.2020.01.011] [PMID: 32014609]

CHAPTER 8

Nanomedicine Based Therapies Against Cancer Stem Cells

Aslı Sade Memişoğlu[1,*] and **Zehra Tavşan**[1]

¹ Faculty of Education, Department of Science Education, Dokuz Eylul University, Izmir, Turkey

Abstract: A tumor consists of not only cancer cells but also an ecosystem including different subpopulations. Cancer stem cells (CSCs) are a rare subpopulation in the tumor cell population. Traditional therapies, such as chemotherapy and radiotherapy target cancer cells except for CSCs. Therefore, the self-renewal and colony formation capacity of CSCs provides the recurrence of tumors as well as drug resistance. Different strategies are used to eradicate CSCs with the knowledge of CSC properties. The recent technologic revolution gives a chance to design nanoscale medicines for the effective treatment of CSCs. Nanoparticle-based delivery systems improve the transport of traditional therapeutic drugs across biological barriers with maximum bioavailability, less toxicity, and side effects, and take advantage in combination with specific CSC targets, controlled and site-specific release. This chapter summarizes the current models of CSCs, the molecular mechanisms leading to metastases and drug resistance of CSCs, strategies to target CSCs, examples of currently approved nanomedicine drugs and future perspectives.

Keywords: Biodistribution, Cancer stem cell, Chemosensitivity, Chemotherapy, Differentiation, Drug delivery system, Drug resistance, Epithelial-to-mesen-chymal transition, Metastasis, Nanomedicine, Nanoparticle, Nanotechnology, Radiotherapy, Regenerative medicine, Self-renewal, Stability, Solubility, Traditional therapy, Tumor-initiating cell, Tumor relapse.

INTRODUCTION

Although there are improvements in cancer therapy, many patients continue to have therapy failure, which causes the illness to grow, return, and lower overall survival. Recent advances in screening tumors revealed that a tumor is not only a collection of uniform cancerous cells, instead, a tumor is an ecosystem that includes heterogeneous tumor cells as well as a microenvironment that can affect how the tumor functions as a whole [1]. Individual tumor cells within a tumor also differ from each other through both genetic and non-genetic mechanisms, result-

---

* **Corresponding author Aslı Sade Memişoğlu**: Faculty of Education, Department of Science Education, Dokuz Eylul University, Izmir, Turkey; Tel: +90542374464; E-mail: asli.memisoglu@deu.edu.tr

Habibe Yılmaz (Ed.)

ing in variance in the "hallmarks of cancer" and the formation of different tumor cell populations. In addition, the heterogeneity within a tumor seems not constant but plastic in nature, which adds another level of complexity to the process of tumorigenesis [2].

Cancer stem cells (CSCs) are defined as a separate population of cells in the tumor and have been demonstrated to exhibit long-term clonal repopulation and self-renewal capacity in many cancers [3, 4]. The defining features of this stemness have also been shown to drive therapy resistance. Evidence from both clinical studies and experimental models suggests that CSCs endure many of the cancer therapies that are frequently used and contribute to metastasis and recurrence of the disease. Additionally, CSC-specific characteristics and transcriptional patterns are significantly prognostic of overall patient survival, demonstrating their therapeutic importance [5, 6].

Traditional therapies, such as chemotherapy, radiotherapy, and drugs that target tumors, frequently lead to a selection of the CSC population, increasing their likelihood of survival and dissemination. To specifically determine CSC response to treatment, additional preclinical and clinical research is required. Additionally, effective therapeutic techniques against CSCs must be developed in order to boost the efficacy of cancer therapy. Over the past few years, a growing number of therapeutic drugs that can destroy CSCs have been evaluated or developed [5]. Unfortunately, most such agents share traits with other anticancer medications, such as small peptides and molecule drugs, that restrict their clinical applications. These include side effects, inadequate biodistribution, low solubility, insufficient circulation time, instability and low therapeutic effects.

The ultimate goal of nanomedicine is to enhance the quality of life by using nanoscale instruments for disease detection, prevention, and therapy as well as to gain insights into the intricate pathophysiology of disease. The limitations of traditional pharmaceutical delivery methods make them less suitable for delivering bioactive compounds into distant tissues [7]. The use of biosensors for diagnostic purposes and biocompatible nanomaterials as delivery systems for therapy, such as nanocapsules for the treatment of cancer, are among the main research areas of nanomedicine. One of the main goals of nanomedicine is to use nanotechnology to find treatment for diseases, such as cancer, and to use targeted drug delivery for more efficient treatment with fewer adverse effects. Targeted delivery methods include liposomes, polymers, micelles, conjugates, nanoparticles, and conjugates of this nanopharmaceuticals [8].

WHAT ARE CANCER STEM CELLS?

As early as the 19th century, some striking parallels between embryonic development and tumor growth were recognized. Malignant tumors were shown to contain rare cell populations that are capable of self-renewal and proliferation. Several decades ago, the first evidence supporting the CSC hypothesis was produced [9]. This study revealed that a limited number of cells harbored the potential to initiate leukemia in mice. These cells also exhibited high similarities with hematopoietic stem cells (HSCs). In addition to expressing typical HSC markers (CD34+/CD38−), these cells could also renew themselves and differentiate. Notably, tumors that developed in mice after *in vivo* transplantation of CSCs, formed a very comparable cell population to the primary tumor. Therefore, the same process may be repeated at every stage of the development of cancer: CSCs produce cells that are the same as the transplanted ones, as well as additional differentiated cells undergoing a process of differentiation and losing their tumorigenic potential. CSCs have now been seen in a variety of malignancies, including those that affect solid tissues. For example, very few CSCs (*e.g.*, 100 cells) could form breast cancer tumors in mice, in contrast to thousands of different phenotype cells that could not start tumors [10]. Later research on colon, brain, head and neck, and prostate cancers has determined other small populations of uncommon cells that can grow *in vivo* tumors [6].

The existence of a distinct subset of tumor cells with specific proliferation, self-renewal, and differentiation abilities - often referred to as cancer stem (-like) cells or tumor-initiating cells (TICs) - has been confirmed by a growing body of research on the heterogeneity of tumors and relevant mechanisms [4, 11]. Later studies focused on the origin of CSCs as well as their methods of eradication. CSCs are thought to have self-renewal and differentiation capacity. They can produce differentiated cells, which make up the majority of tumor tissue. CSCs within solid tumors have an unknown origin; nevertheless, some studies suggest they might come from normal stem cells and others argue that they may also originate from differentiated cells [12]. The critical role of epithelial-to-mesenchymal transition (EMT) programs in the development of CSC-like cells in a variety of cancers has also been pointed out in several studies [13]. Therefore, the CSC theory can be summarized in three different models according to the proposed origin of how they emerge.

Hierarchical Model

The hierarchical model was the original CSC model, in which the rare stem cells residing in a tumor give rise to a stem cell and a differentiated cancer cell (DCC), providing the continuity of the stem cell population in a tumor mass. Accordingly, these stem-like cells were proposed to easily be distinguished from DCCs by their ability to initiate new tumors *in vivo* [14]. The model suggests a constant CSC population that is responsible for tumor relapse and therapy resistance. Therefore, the best treatment strategy would be to identify and target these cells in order to eradicate the tumor [15]. However, attempts to define markers for separating CSCs from DCCs were often unsuccessful. Markers were identified according to the tumor-initiating capacity in some cell lines, which did not confer the same ability in others, even in tumors of the same origin. The discovery that tumor growth and dissemination were sometimes seen even after CSCs were successfully destroyed [16] revealed further flaws in the hierarchical paradigm. Studies showing that more differentiated cell types exhibiting lineage markers can act as leukemia-initiating cells in some leukemias have led the initial CSC theory to be reconsidered [17, 18].

Stochastic Model

Based on the accumulating evidence of the tumor-forming ability of differentiated cancer cells, a stochastic model has been postulated [19]. According to this model, every cancer cell possesses an intrinsic capacity to initiate new tumors, provided that the conditions are favorable. This introduced the concept of plasticity, the ability of cells to go back and forth through transitional states and even not commit a permanent phenotypic state [20, 21]. The model could better explain why CSCs could not be separated from non-CSCs in certain types of cancers and most of the cells have tumor-initiating potential in these cancers. Even when DCCs are isolated and grown under non-permissive (low-attachment) conditions in cell culture, they exhibit a dedifferentiated stem-like phenotype. The CSC phenotype reverts to a differentiated phenotype when the conditions are permissive (high-attachment), and the original CSC/non-CSC ratios are established, which is also relevant in *in vivo* models [15, 16]. The plastic feature of cancer cells is now referred to as epithelial-mesenchymal plasticity (EMP), which became a hot topic in cancer research in the last decade. The concept of EMP has changed the original view of CSCs to interchangeable phenotypes rather than being permanently committed to one [22]. In the meanwhile, an update to the stochastic model also has emerged with the recent understanding of the relationship of the tumor with its surrounding environment.

Dynamic Tumor Initiation Model

The non-tumor cell components present in tumors are generally referred to as the tumor microenvironment (TME) [23]. It involves cellular components other than cancer cells (fibroblasts, immune cells, *etc.*), cell-secreted molecules, extracellular vesicles, and the extracellular matrix. Therefore, a dynamical tumor initiation model for a mixed population has been suggested in order to recognize all of these significant players (Fig. **1**). A tumor cell is affected by its TME, leading to a substantial difference in its function. Interaction between tumor cells and the TME increases the complexity of the disease. Because cells with the same genetic makeup might be sensitive or resistant to medications depending on their environment, TME contributes to adaptive drug resistance. Recent research suggests that TME may start stem cell-like processes in cancer cells [24, 25]. The complexity of tumor growth regulation altered the perceptions of how well it can be eliminated. According to the model, the idea of targeting the promoter of the CSC phenotype rather than CSC itself as a more effective strategy, has gained attention.

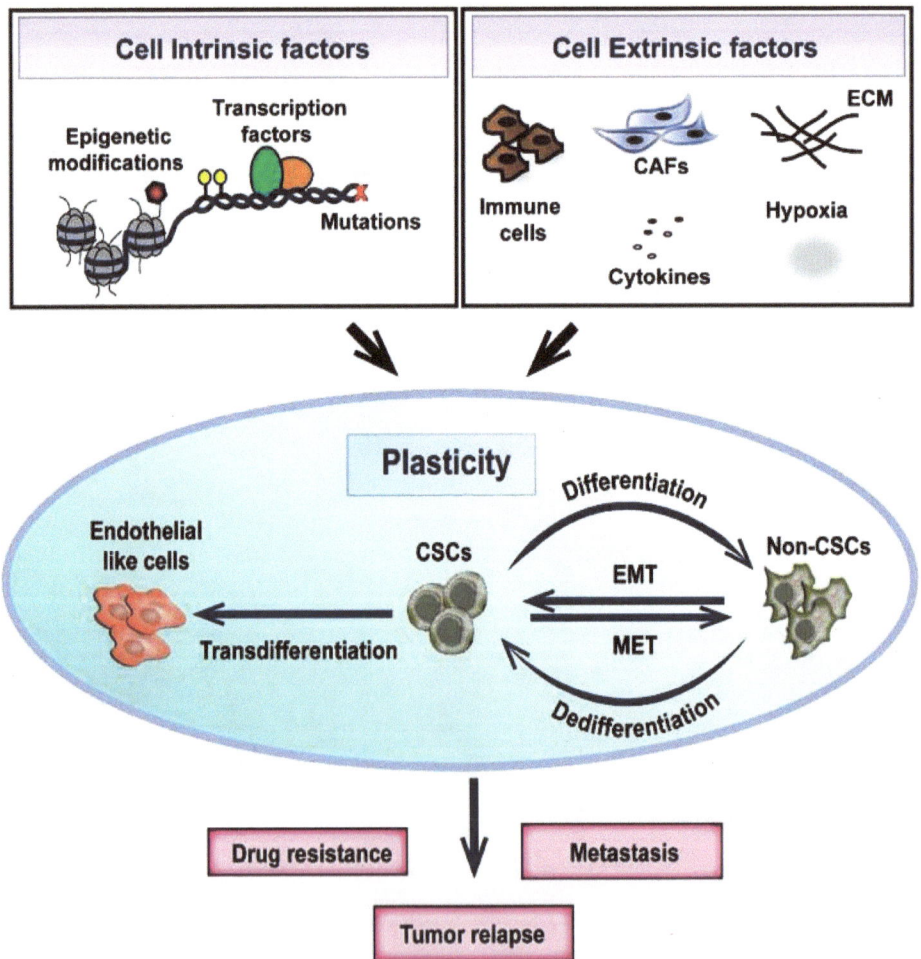

Fig. (1). Dynamic tumor initiation model of CSCs. (Figure used with permission from [26]).

STRATEGIES AGAINST CSCS AND THEIR MICROENVIRONMENT

Research on CSCs provides evidence for their similarity to multidrug resistant cells, such as increased levels of efflux transporters, resistance to apoptosis, more efficient repair of DNA, and altered metabolism [27]. Therefore, targeting CSC surface indicators, blocking drug transporters, inhibiting self-renewal signaling, inducing apoptosis, or eliminating/inhibiting tumor microenvironment signaling, are some potential methods to target CSCs [28, 29]. In order to develop therapeutics that can successfully eliminate CSCs, the use of contemporary drug delivery technologies and a thorough knowledge of the properties of CSCs are fundamental. The following sections describe the discriminating properties of CSCs which form the basis of drug development.

Efflux Transporters

The majority of ATP-binding cassette transporters (ABC transporters) use ATP hydrolysis to pump a wide variety of substrates across biological membranes. Seven subfamilies (A - G) of ABC transporters are encoded by 48 genes in the human genome [30]. ABCG2 and/or ABCB1 transporters were reported to be highly expressed in hematopoietic stem cells (HSCs) [31]. Mice that lack ABCB1, ABCG2, or ABCC1 were shown to be particularly sensitive to several chemicals, suggesting that these transporters may help shield stem cells from harmful toxins [32]. Mesenchymal stem cells (MSCs), multipotent adult progenitor cells (MAPCs), and unrestricted somatic stem cells (USSCs) also show high expression of ABC transporters [33]. According to research, ABC transporters are also involved in drug resistance of cancer stem cells [34]. Since they give cells a special defense mechanism by reducing the accumulation of various therapeutic drugs, ABC transporters are frequently linked to multidrug resistance (MDR). The drugs that can be pumped out by ABCG2 involve tyrosine kinase inhibitors, doxorubicin, topotecan, mitoxantrone, and methotrexate. Fumitremorgin C [35] and tryprostatin A [36] are two examples of low molecular weight ABCG2 transporter inhibitors that have been studied as one of promising methods for sensitizing and eliminating CSCs. Over 50% of all cancers resistant to treatment have been shown to express ABCB1, which helps cells to pump out many hydrophobic substances. Leukemia and solid tumor cancer therapy are currently thought to be adversely affected mostly by high ABCB1 expression [37, 38]. Another ABC transporter, ABCB5, is related to CSC drug resistance in many tumor types. It was shown to serve as an MDR transporter and induce chemotherapy resistance in malignant melanomas [39]. It was discovered that a monoclonal antibody (mAb) could be used to sensitize melanoma cells to doxorubicin, underlining the function of efflux pumps in drug resistance [40]. Additionally, the acquired or intrinsic nature of drug resistance is also important.

Even before receiving chemotherapy, certain tumor cells have increased expression of efflux transporters, which is known as intrinsic resistance [41]. The increase of efflux transporters after treatment is known as acquired resistance and is observed in several *in vitro* and clinical studies [42 - 44]. Nevertheless, ABC transporters may be targeted directly by mAb or drugs or indirectly by inhibiting signaling [45]. Although several strategies have been developed against ABC transporters, clinical trials have been hampered by limited inhibitory efficacy and toxicity to normal cells. In this aspect, nanocarriers may provide significant promise for avoiding drug resistance brought on by ABC transporters.

Pathways Dysregulated in CSCs

Overexpression of efflux pumps is usually not an independent event in CSCs. Instead, it is mostly a result of the dysregulation of some important signaling pathways regulating survival and cell fate. Some of these major pathways are Wnt/β-catenin, Hedgehog, Notch, PI3K/Akt, Bcl-2, and NF-kB, and understanding the molecular mechanisms leading to aberrant signaling is essential for cancer therapy (Fig. **2**).

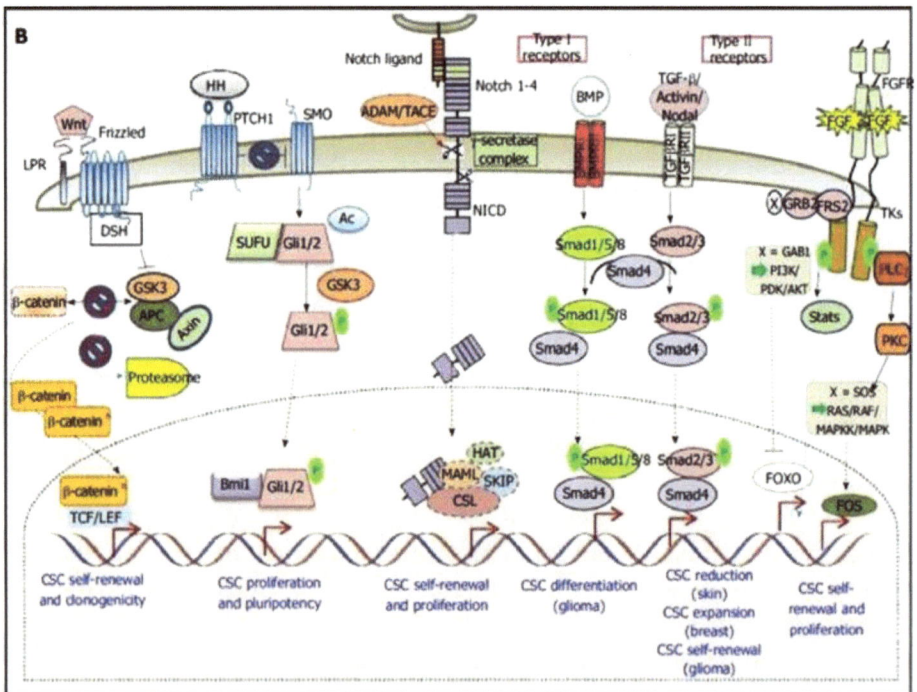

Fig. (2). Major signaling pathways shared between embryonic stem cells and cancer stem cells (Used with permission from [46]).

The modification or reactivation of signaling pathways is closely related to stem cell signaling pathways. CSC self-renewal, proliferation, dedifferentiation, metastasis, and escape from apoptosis are shown to be related to amplified Wnt, Hedgehog and Notch signaling [47]. In CSCs and developing tumors, these pathways frequently interact. Early clinical trials for inhibiting Notch and Hedgehog pathways have seen significant advancements, however it has proved challenging to target the Wnt pathway [48]. Clinical targets for downstream transcription factors, including β-catenin, STAT3, and Nanog have also been discovered [49]. The efficacy and clinical impact of these approaches may be significantly limited by the shared expression of numerous genes and signaling pathways between CSCs and normal stem cells, as well as by the overlapping regulatory circuits. Recently, a wide range of biological and small chemical Wnt signaling inhibitors have been developed [50]. However, none of the inhibitors have been authorized for use in clinical settings as of yet. The majority of therapeutics that target the Notch pathway are γ-secretase inhibitors and antibodies against DLL4. Inhibiting the Notch pathway was shown to impede the activation of self-renewal target genes.

One of the key signaling pathways in CSCs that contributes to stemness maintenance, proliferation, differentiation, EMT, migration, and autophagy is the PI3K/Akt/mTOR pathway [51]. Therefore, inhibiting the PI3K/Akt/mTOR pathway may be a promising approach for specific cancer therapy. There are several inhibitors of this pathway, some of which are approved by the Food and Drug Administration (FDA). Our knowledge of the precise processes, pathway regulators, and the activities of the pathway in CSCs is currently restricted. Therefore, additional studies will be necessary to ascertain the role and constituents of the pathway that enables the development of more effective anticancer drugs for usage in the clinical context.

Escaping apoptosis is a well-known hallmark of carcinogenesis, which is also common in CSCs [52]. Increased production of antiapoptotic proteins, such as Bcl-XL and Bcl-2, or inactivation of proapoptotic proteins, such as p53 and caspases, in CSCs were shown in several studies [53 - 55]. Additionally, resistance to apoptosis may be caused by other signaling pathways like Notch and Hedgehog, which combinatorially induce the development of drug resistance [56]. A Bcl-2 inhibitor, Venetoclax, was approved by FDA for the treatment of several hematological malignancies as single or in combination with other drugs [47, 57].

Inflammation is another well-known hallmark of cancer [1]. Nuclear factor-κB (NF-κB), a transcription factor with a quick response time, is crucial for controlling immunological and inflammatory responses [58]. The NF-κB pathway also contributes to cellular proliferation, differentiation, and survival [59].

Important links between the NF-κB pathway and inflammation, self-renewal, maintenance, and metastasis of CSCs were established [60]. Drugs like sulforaphane and curcumin were found to inhibit CSC self-renewal, metastasis, and proliferation, acting through the NF-κB pathway [61, 62].

Many pathways don't function as autonomous entities but rather frequently collaborate with one another to form a physiological network. The interplay between pathways contributes to the preservation of the CSC population. The crosstalk between signaling pathways adds another level of complexity for drug discovery in cancer research. Significant work has been done in recent years to create combination medicines that target different pathways in cancer treatment. For example, a recent study found that blocking Notch and Hedgehog signaling together reduced the number of CSC subpopulations in prostate cancer [56].

Tumor Microenvironment

Increasing data indicate that a tumor environment may play a role in the "stemness" trait. The potential to form tumors has been associated with a more dedifferentiated state, which can be induced and maintained by the microenvironment (TME) or sometimes called "the tumor niche" [1]. The surrounding tissue cells, microvessels, immune cells, and secreted molecules and extracellular matrix (ECM) make up the majority of the TME. In addition to adapting to environmental changes, CSCs also have an impact on the TME. The microenvironment simultaneously maintains CSC's self-renewal, stimulates angiogenesis, attracts immune and stromal cells, retains phenotypic plasticity, guards against drug-induced apoptosis, and encourages tumor invasion and metastasis [47]. Additionally, environmental modifications like pH and hypoxia can influence the CSC niche. Therefore, targeting the TME and contributing factors may also be a potential therapeutic approach and preventive measure for the growth of tumors.

Targeting specific niches has already produced some encouraging outcomes. CSCs respond to TME through secreted factors and cell-cell/cell-matrix attachment molecules. For example, antagonists for CXCR4, a chemokine receptor and CSC marker, are undergoing clinical trials [63, 64]. The restoration of tumor vasculature, breakdown of the CSC niche, and reduction of tumor development can all be achieved by inhibiting VEGF [25]. Another approach to control the niche of dormant, drug-resistant cells involves focusing on tumor hypoxia. For patients with gliomas, HIF-1α and HIF-2α, constitute a prospective therapeutic target [65]. Tumor-associated stromal cells constitute important regulators of CSC homeostasis and targeting these cells would be promising.

Surface Markers

Several biomarkers have been used to identify CSCs in human malignancies. Combining particular biomarkers, primarily found on the cell membrane, allows for the separation of CSCs. As a recent form of cancer therapy, monoclonal antibodies (mAbs) or inhibitors that target these unique surface indicators are being used. Surface marker inhibitors provide high specificity so that normal cells of the tissues are minimally affected. mAbs against CD20 (rituximab), CD52 (alemtuzumab), and CD44v6 (bivatuzumab) are some examples of FDA-approved drugs [47, 66, 67]. The production of antibodies that target CSCs has advanced significantly owing to a better understanding of the surface indicators of CSCs. However, CSC phenotypes can differ significantly between individuals or cancer types, and CSC populations with multiple phenotypes may coexist. Upon recurrence, CSCs also evolve and obtain unique characteristics. Accordingly, these differences should be kept in mind when determining the therapy approach. A perfect therapeutic outcome may be achieved by the combination of surface antibodies with existing chemotherapeutics.

Metabolism

CSCs can adapt to a novel way of metabolizing cell energy and prevent hypoxia-induced apoptosis. Numerous studies have demonstrated that CSCs perform predominantly aerobic glycolysis, that is, even in the presence of oxygen, they depend on glycolysis as the major form of energy metabolism. Major transcription factors involved in hypoxia response are known as hypoxia-inducible factors (HIFs) and their high expression is correlated with tumor malignancy [68]. In addition to CSC survival and tumorigenesis, HIFs regulate the expression of phosphoinositide-dependent protein kinase-1 (PDK1), lactate dehydrogenase A (LDHA), glucose transporter 1 (GLUT1) and glucose transporter 3 (GLUT3) which are all involved in cellular metabolism [69]. Among these, GLUT3 is shown to be particularly expressed in brain CSCs, and GLUT3 knockdown significantly decreased the amount of tumor stem cell population and inhibited the progression of glioblastoma [70]. Another key molecule in metabolism is adenosine 5′-monophosphate protein kinase (AMPK). AMPK is mostly investigated in diabetes and other metabolic illnesses, but is also activated and highly expressed in several CSCs [47]. The diabetic drug metformin was shown to affect AMPK signaling and reduce cellular growth and metabolism in colon cancer and hepatocellular cancer stem cells [71, 72].

ESSENTIALS OF NANOMEDICINE

In recent years, nanotechnology as a revolution in technology has been carried out in a variety of fields in which several applications and redesigned nanoscale products have become available. This revolution also occurred in the field of pharmaceutical research. Nanomedicine has been described as the application of nanotechnology and nanomaterials that have medical purposes in different areas: disease diagnosis, regenerative medicine, and drug delivery systems.

Disease Diagnosis

With the technologic revolution, the definitive objective of nanotechnology is the empowerment of clinical diagnostics to detect diseases as early as possible. Efficient and precise diagnosis leads to improved health outcomes with the assistance of regenerative medicine and drug delivery systems.

One of the most urgent diseases in need of developments in nanotechnology is cancer. Molecular imaging with the assistance of nanotechnology helps early detection of cancer and precise diagnosis distinguishing harmful cells in biological samples. These advanced molecular imaging systems use different nanomaterial-based molecules in two specific fields: probes and contrast agents. Nanomaterial-based probes are convenient contrasting materials, including fluorescent, radioactive, paramagnetic, and super paramagnetic or electron-dense [73]. Nanomaterial-based contrast agents can be classified as micelles, liposomes, polymersomes, dendrimers, carbon nanoparticles, and magnetic nanoparticles (iron oxides, metal alloys) [74].

Magnetic Resonance Imaging (MRI), Positron Emission Tomography (PET), and Computed Tomography (CT) are imaging procedures providing scalability to investigate biological processes on both the cellular and molecular levels. The contrast agents in MRI are referred to as T1 and T2, which are nanomaterial-based agents with paramagnetic compounds, micelles, liposomes, polymersomes, dendrimers, and carbon nanoparticles and super-paramagnetic agents, such as iron oxides and metal alloys, respectively [75]. PET imaging is especially important for the early detection of cancer cells and imaging probes used in this technique are metal oxide nanomaterial. In recent years, liposomes with positron-emitting radionuclides, polymeric micelles and hydrogels are also used [76]. Similar to MRI, the development of nanomaterial-based contrast agents for CT imaging is the focus of extensive research. These agents amplify the contrast while reducing high radiation exposure. Moreover, targeting with cancer-specific antibodies in combination with contrast agents enhances the visibility of tumors and allows for distinguishing cancerous and healthy cells more precisely [77, 78].

Regenerative Medicine and Drug Delivery Systems

In addition to nanomaterial-guided imaging, nanotechnology holds promising challenges in effective treatment by 1) maximizing bioavailability and reducing the dose and toxicity, 2) improving transport across biological barriers and favorable distribution, and 3) providing a combination of specific targets, controlled and site-specific release.

Novel properties of nanomaterials, such as smaller size, specific nanoformulation, *etc.*, provide several advantages. The smaller size than conventional chemical equivalents increases bioavailability and reduces the doses needed. The specific nanoformulation allows penetration of biological membranes or barriers more easily, neutralization of toxic properties, and boosts their persistence in the biological environment. Consequently, this property improves the pharmacokinetics of nanomaterials including absorption, distribution, metabolism, and elimination. Nowadays and in the future, there is a good chance of success with the use of nanoparticles as a personalized and efficient therapy with fewer side effects.

Nanoparticles

Various types of nanoscale materials are produced by nanotechnology. Nanoparticles (NPs) are a broad class of nanoscale materials, ranging from 1 to 100 nm. They are classified based on their properties and dimensions. The variations in size and shape give them characteristic colors and properties which can be utilized in imaging applications [79] and therapeutic proposes.

Three-layer compositions of NPs possess unique physical and chemical properties. The surface layer consists of several small molecules, metal ions, surfactants, and polymers, which make NPs functional. The core layer is an essential central part of NPs and the shell layer is completely different from the core layer. These exceptional properties got the attention in multidisciplinary fields.

NPs can be classified into three categories based on physical and chemical characteristics.

1- Carbon-based NPs contain carbon and are represented by fullerenes, carbon nanotubes, carbon nanofibers, and graphene. These NPs are made of tubular tubes, ellipsoids, and spheres. Their unique physical, chemical and mechanical characteristics such as electrical conductivity and high strength [80] are important properties for commercial applications [81 - 84].

2- Inorganic-based NPs are divided into three subclasses, metallic, non-metallic, and composite.

Metallic NPs contain alkali and noble metals, such as Cu, Ag, and Au. Non-metallic NPs are solids and found in different forms, porous, hollow, and amorphous. Non-metallic NPs are used mostly in catalysis, photocatalysis applications as well as imaging systems [85], while broad absorption bands and advanced optical properties of metallic NPs give them opportunity in many research areas, such as sampling of SEM to obtain high-quality images by enhancing the electronic stream. There are also composite NPs that possess properties of both metal and non-metal NPs. These composite NPs are semiconductors and have wide bandgap tuning. Therefore, they are efficient in water-splitting systems [86].

3- Organic-based NPs are special polymeric nanoparticles or contain lipid moieties. Their shapes are nanospheroidal or nanocapsular [87]. Matrix particles of polymeric NPs are solids in overall mass, adsorbed at the outside of the spheroidal surface and encapsulated within the particle [88]. Contrary to polymeric NPs, lipid NPs have a solid lipid core and a matrix of soluble lipophilic molecules. The surfactants or emulsifiers stabilize the core of lipid NPs which are used as drug carriers of the delivery systems in cancer therapy [89, 90].

NPs are also classified depending on the electron movement along the dimensions, such as 0D, 1D, 2D, and 3D after 2007. The electrons in 0D NPs are entrapped in a dimensionless space, whereas the electrons of 1D NPs move along the x-axis. In a similar nature, 2D and 3D NPs have electrons that move along the x, y-axis and x, y, and z-axis, respectively. The characteristics of NPs affect the ability and are attributed to the dimensionality and classical inner side effects. Therefore, the focus on the arrangement of dimensionality along with other properties extends the applications of nanoparticles.

NANOPARTICLES AS THERAPEUTICS

Cancer is one of the leading causes of death worldwide. Chemotherapy and radiation therapy are used as traditional cancer therapies. In recent years, targeted therapy and immunotherapy have been used for the treatment. However, the complex pathophysiology of cancer, cytotoxicity and multidrug resistance hold a major challenge for efficient cancer therapy. NPs offer a new class of therapeutics that can perform in ways that the therapeutic entities they contain cannot. NPs as therapeutics comprise entities including small-molecule drugs, proteins or peptides, nucleic acids or other components. The specific advantages achieved by nanoparticles through nanotechnology give enhanced anticancer effects owing to biocompatibility, more specific targeting, more stability, reduced toxicity, and enhanced permeability.

Examples of Nanomedicines against CSCs

Therapeutic Candidates

The significant problem in cancer treatment is resistant tumor cells to conventional chemotherapeutics. Since these therapeutics are not targeting CSCs, residual CSCs in resistant cancer cells cause a relapse of disease. The delivery of conventional chemotherapeutics by specifically designed NPs may contribute to improved anticancer efficacy. One of the conventional chemotherapeutics, antitumorigenic polyphenolics or an inhibitor of signalin,g pathways can be a drug candidate targeting CSCs. Several drug delivery platforms were explored for the delivery of these CSCs-targeting therapeutic candidates.

Sun *et al.* showed that doxorubicin-loaded gold nanoparticles transferred doxorubicin to breast CSCs more efficiently [91]. These NPs decreased the mammosphere formation capacity and cancer initiation feature of enriched CSCs from breast cancer cells. In addition, the delivery system overcame the drug resistance mechanism and inhibited tumor growth *in vivo*. Similarly, when oxaliplatin (OXA) was loaded into stearic acid-g-chitosan oligosaccharide polymeric micelles, the uptake ratio and cytotoxicity of OXA increased [92] in colorectal cancer cells. Moreover, the studies indicated that the delivery of OXA by polymeric micelles reversed the chemoresistance of colon CSCs *in vitro* and *in vivo*. Besides improving the cytotoxicity of conventional therapeutics or reversing the resistance, there are several identified agents targeting CSCs with the same undesired effects as conventional therapeutics, such as poor solubility in the aqueous solution and toxicity. High-throughput screening (HTS) for agents identified salinomycin (SAL) among a collection of ~16,000 compounds with epithelial CSC-specific toxicity [93]. SAL treatment inhibited tumor growth in mammary tissue in mice and resulted in the loss of CSC gene expression in the breast tumor tissues of cancer patients. However, poor solubility and toxicity retained SAL for clinical applications. Liu *et al.* overcame this problem by loading SAL into chitosan-coated carbon nanotubes functionalized with hyaluronic acid (HA) [94]. This drug delivery system enhanced the bioavailability and increased the cytotoxic effects of SAL in gastric cancer. Moreover, it was selectively uptaken by gastric CSCs, and revealed CD44 receptor-mediated endocytosis, showing HA-functionalization. After all modifications, the delivery of SAL with this system decreased spheroid and colony formation and induced apoptosis. Zhao *et al.* designed a different delivery system with the conjugation of SAL to hydrophilic, immune-tolerant, elastin-like polypeptide (iTEP) and encapsulation in NPs with N,N-dimethylhexylamine (DMHA) and α-tocopherol termed as iTEP–Sali NP3s [95]. This formulation improved the pharmacokinetics and SAL has accumulated in the tumor by boosting CSC elimination, resulting in the delay of tumor formation. Another anticancer therapeutic candidate, curcumin, is the primary bioactive substance in turmeric, but poor solubility and bioavailability hinder clinical applications. Similar to OXA, loading of curcumin into stearic acid-g-chitosan oligosaccharide polymeric micelles, increased its internalization and solubility, resulting in the accumulation inside tumor cells [96]. Additionally, the CSC subpopulation in the colon cancer cell population has markedly decreased. Wei *et al.* used the same delivery system for etoposide, SAL and curcumin [97]. The loading of these CSC-targeting drugs into a nanogel based on membranotropic cholesteryl conjugated with HA, demonstrated about 2 to 7 times higher cytotoxicity in human breast and pancreatic adenocarcinoma cells. The NPs were more effective against CD44-expressed drug-resistant cells. Cholesterol moiety and CD44 conjugation to nanogels ensured the anchorage to the cellular membrane and internalization of NPs *via* CD44 receptor-mediated endocytosis, respectively. These NPs couldalso penetrate the cancer spheroids. Recently, anti-psychotic, anti-diabetic, and anti-helminthic drugs have been shown to be effective against CSCs. The self-assembled diblock copolymer of poly(ethylene glycol) (PEG) was functionalized with urea and loaded with an antidiabetic drug, phenformin by Krishnamurthy *et al.* [98]. These micelles were stable in the cell growth medium with no cytotoxic effect on non-cancerous cells. The released phenformin from phenformin-loaded micelles inhibited the growth of both stem cell and non-stem cell subpopulations in lung cancer. These effects were also confirmed in *in vivo* studies with greater anti-tumorigenic activity and reduction in the stem cell population in tumor tissues. Kopecek *et al.* designed N-(2-hydroxypropyl)methacrylamide copolymer-cyclopamine conjugate as HPMA, for the delivery of Hedgehog signaling inhibitor, cyclopamine [99]. This delivery system has improved drug solubility, resulting in decreased systemic toxicity. Efficient cyclopamine delivery resulted in decreased expression of CSC markers, possibly targeting CSCs in the human prostate cancer epithelial cell population. Shen *et al.* studied the amphiphilic copolymer PEG-block-poly(D, L-lactide) (PEG-b-PLA) loaded with Bortezomib, which is a proteasome inhibitor in breast cancer [100]. Bortezomib encapsulated nanoparticles delivered bortezomib into CSCs as well as non-CSCs, and then inhibited proliferation and activated apoptosis. The nanodiamonds loaded with an ATP binding cassette (ABC) transporter substrate, epirubicin, were passively transported and impaired the growth of tumors that arose from chemoresistant cancer stem cells, as a result of enhanced penetration of epirubicin in the murine model for MYC-driven tumor [94].

Nucleic Acid Drugs

MicroRNAs (miRNAs) are small, single-stranded, non-coding RNA molecules which regulate genes post-transcriptionally. The regulated genes are involved in cell growth, differentiation, development, and apoptosis. In recent years, several studies uncovered the role of miRNAs that regulate the functions of CSCs *via* various oncogenic signaling pathways [101]. Nevertheless, the therapeutic effects of miRNAs are limited by tissue specificity, cellular uptake and systemic toxicity [102]. NPs have been shown to increase the stability, bioavailability, delivery, and targeted cytotoxicity of miRNAs [103].

Yang *et al.* explored a tumor suppressor miRNA in gastric cancer cells by miRNA microarray. miR145 was found to negatively regulate stem cell markers, Oct4 and Sox2 in CD133+ stem cell subpopulations [104]. Loading miR145 into a polyurethane-short branch polyethylenimine (PU-PEI) vehicle and delivery to CD133+ CSCs, resulted in the growth inhibition of the tumor and the differentiation of CD133+ CSCs into CD133⁻ non-CSCs. Another PU-PE--mediated miR145 delivery has also improved chemoradioresistance in lung adenocarcinoma CSCs and reduced CSC percentage [105]. Moreover, these NPs directly targeted embryonic stem cell markers, Oct4/Sox2/Fascin1. Cui *et al.* prepared a delivery system consisting of gelatinase cleavage peptide with a poly(ethylene glycol) (PEG) and poly (ε-caprolactone) (PCL)-based structure [106]. miR-200c-loaded gelatinase-stimuli PEG-Pep-PCL NPs reduced CD44 expression resulting in a decrease of $CD44^+$ gastric CSCs. Upregulation of miR-200c levels by delivery with miR-200c-loaded NPs, enhanced radiotherapy efficiency while inducing little radiosensitization in normal cells. In head and neck squamous cell carcinoma (HNSCC), a cationic lipid nanoparticle delivery system was employed to deliver miR107, which is normally reduced in HNSCC [107]. Following the treatment of NP-miR107, miR107 levels were increased in the cells and miR107 targets, protein kinase Cε (PKCε), cyclin-dependent kinase 6 (CDK6) and hypoxia-inducible factor 1-β (HIF1-β) levels were decreased. Along with the reduction in stem cell transcription factors, Nanog, Oct3/4, and Sox2, CSC or cancer-initiating cell (CIC) numbers were declined and tumor spheroid formation efficiency was diminished. In another study, the delivery of miR-34a within solid lipid NPs has directly suppressed CSC marker CD44 and inhibited CSC differentiation and metastasis of $CD44^+$ lung CSCs [108].

Small or short interfering RNAs (siRNA) are double-stranded RNAs. Similar to miRNAs, they operate in post-transcriptional gene silencing. siRNAs require loading into an NP for protection and efficient delivery into target cells, in order to overcome limiting obstacles such as low cellular uptake and nuclease-mediated degradation [109]. Singh *et al.* formulated a short hairpin RNA targeting annexin A2, AnxA2 (shAnxA2), in a liposomal (cationic ligand-guided, CLG) carrier [110]. These shAnxA2-loaded liposomes were incorporated into lung CSCs resistant to chemotherapy for 2 hours and were shown to reduce Anx2 both at the protein and mRNA levels. *in vivo* studies confirmed mass reduction in the orthotopic lung tumors of lung CSCs as a result of decreases in the AnxA2 as well as SOX2, total β-catenin and S100A10. Lo *et al.* studied the effects of PU-PE--based administration of double-stranded DNA (dsDNA) encoding small interfering RNA (siRNA) against EZH2 and Oct4 that are upregulated in HNSCC-derived ALDH1$^+$/CD44$^+$ CSCs [111]. Treatment with NPs resulted in partial suppression of cancer capacity and CSC properties. siRNA expressing dsDNA, conjugated with nuclear localization signal (NLS) enhanced nuclear delivery of siRNAs and significantly suppressed epithelial-mesenchymal transition and improved radiosensitization in ALDH1+/CD44+ CSCs. Gul-Uludağ *et al.* reported CD44 siRNA delivery with lipid-substituted PEI/siRNA complexes into difficult-to-transfect acute myeloid leukemia (AML) cell lines [112]. NPs were shown to reduce CD44 protein and induce apoptosis in CD34+ leukemic stem/progenitor cells (LSPC).

In addition to targeting signaling pathways, cancer metabolism is another strategic candidate for cancer therapy. Xu *et al.* chose glucose transporter 3 (GLUT3) and loaded siGLUT3 into PEG-PLA NPs [113]. These NPs reduced CSC percentage and inhibited the self-renewal of glioblastoma CSCs by arresting cell metabolism and proliferation. Another promising strategy is to silence genes encoding drug efflux transporters that are predominantly responsible for drug resistance. Delivery of multidrug resistance gene (MDR1) siRNA with a carrier composed of cationic oligomer (PEI1200), a hydrophilic polymer (PEG) and biodegradable lipid-based crosslinking moiety effectively reduced MDR1 expression in human colon CD133$^+$ CSCs and sensitized to paclitaxel [114].

Application of Combinational Delivery Systems

The single-target drug delivery may not be sufficient to destroy tumors because of the heterogeneity and plasticity of tumor cells. The incorporation of multiple high capacity therapeutic agents within NPs may bear more therapeutic promise to tackle both CSCs and non-CSCs. In this sense, NPs developed by Liu *et al.* played a synergistic role in improving paclitaxel chemosensitivity in colon CD133+ CSCs [114]. In breast cancer, the combination of octreotide-modified paclitaxel-loaded active targeting PEG-b-PCL polymeric micelles and salinomycin-loaded PEG-b-PCL polymeric passive targeting micelles showed a more effective antitumor response in suppressing breast CD44+/CD24-CSCs [115]. Kopecek *et al.* studied the effects of cyclopamine-loaded N-(2-hydroxypropyl)methacryl-amide (HPMA) and docetaxel-loaded HPMA combination [116]. The combined therapy resulted in the growth inhibition of the tumor by targeting CSCs with cyclopamine and non-CSCs in the tumor mass with docetaxel. Yang *et al.* designed a combined therapy using CSC-targeted thioridazine-loaded acid-functionalized poly(carbonate) (PAC) and PEG diblock copolymer, and doxorubicin-loaded urea-functionalized poly(carbonate) (PUC) and PEG diblock copolymer [117]. The co-delivery of thioridazine and doxorubicin eradicated both breast CD44+/CD24- CSCs and non-CSCs *in vitro* and *in vivo*. Wang *et al.* developed a combination therapy with a preferential CSC-targeted organic small molecule, 8-Hydroxyquinoline (8-HQ)- and DOX-loaded hyaluronan-modified mesoporous silica NPs [118]. DOX-loaded NPs showed much cytotoxicity in breast cancer cells, whereas 8-HQ-loaded NPs demonstrated enhanced cytotoxicity against breast cancer mammospheres. In a xenograft breast cancer model, combination therapy also generated more antitumor efficiency. Another liposomal formulation by the encapsulation of a new semisynthetic vinca alkaloid, Vinorelbine, and a sesquiterpene lactone in the herbal medicine, parthenolide was developed [119]. The application of vinorelbine-loaded liposomes in combination with parthenolide-loaded liposomes produced a higher inhibitory effect on breast CSCs. In addition, combination therapy was shown to destroy *in vivo* tumors.

The regulation of self-renewal of CSCs is maintained by DNA methylation and aberrant DNA methylation represents targets for cancer initiation. Li *et al.* administered DNA hypermethylation inhibitor, decitabine-loaded PEG-b-PCL NPs in combination with NPs loaded with doxorubicin [120]. In the mammospheres of breast cancer cells, the proportion of CSCs with high aldehyde dehydrogenase activity (ALDHhi) was reduced after combination therapy. In a breast cancer xenograft model, combined therapy resulted in increased DOX sensitivity and induction of apoptosis in both CSCs and non-CSCs. The encapsulation of differentiation agent of CSCs, all-trans-retinoic acid (ATRA) and doxorubicin (DOX) in the same PEG-b-PLA NPs demonstrated an efficient anticancer effect by delivering the drugs to both CSCs and non-CSCs, resulting the reduction self-renewal capacity by the differentiation of CSCs into non-CSCs and increase in the chemosensitivity [121]. Similar results were also seen *in vivo,* with suppressed tumor growth and reduced CSC percentage.

The combination or co-delivery of nucleic acid drugs with chemotherapeutic and targeted drugs can also be employed. Liu *et al.* co-delivered miR-200c that targets gene class III beta-tubulin and DOC in gelatinase-stimuli nanoparticles [122]. The co-delivery showed synergetic effects on the inhibition of both CSC and non-CSC populations, cell proliferation, migration, and invasion. The experiments in xenograft gastric cancer mice concluded high drug accumulation by this co-delivery system and promoted antitumor activity. In another study, a systemic nanodelivery platform was used for the delivery of wild-type (wt) p53, targeting O6-methylguanine-DNA methyltransferase that is upregulated in Temozolomide (TMZ)-resistant glioblastoma multiforme (GBM) [123]. The combination therapy increased TMZ sensitivity in both CSCs and non-CSCs, resulting in the induction of apoptosis and enhanced survival in the GBM mouse model.

Instead of conventional chemotherapeutics and nucleic acid-based drugs, recent studies showed anticancer monoclonal antibodies (mAbs) that induce tumor growth in clinical trials and have potency against CSCs. In addition, antibodies can guide therapeutic drugs to CSCs, resulting in enhanced specificity and efficiency. For example, Gu *et al.* used iron oxide NPs loaded with paclitaxel and ABC transporter monoclonal antibodies that target the ABCG2 transporter hat is overexpressedt in multiple myeloma CSCs [124]. The neutralization of the ABCG2 transporter improved chemosensitivity to PTX in CSCs both *in vitro* and *in vivo*. Panyam *et al.* conjugated anti-CD133 mAbs to paclitaxel-loaded NPs and targeted NPs efficiently delivered PAX to CSCs, resulting in decreases in the number of mammospheres, colonies and CSC proportions [125]. Wang *et al.* used another overexpressed CSC marker, CD44, and developed DOX-loaded CD44 mAb-directed liposomes. The results revealed that these liposomes selectively targeted hepatocellular CD44+ CSCs and induced apoptosis [126].

PERSPECTIVE AND DIRECTION

A promising strategy for treating cancer is to eliminate CSCs by focusing on the major signaling pathways that underlie CSC characteristics. Significant advancements have been made recently in the use of nanotechnology for the treatment of cancer and targeting CSCs. A number of nanocarriers with great therapeutic efficacy, bioavailability, and stability can now be produced in large amounts. Most nanocarriers have the ability to lower the dose and tolerability of cancer therapeutics. Specific targeting is also possible with certain carriers. Additionally, the majority of ingredients used in nanocarriers are biodegradable and thus environmentally beneficial. However, every new technology comes with its limitations or challenges. Due to the high surface area/volume of nanomaterials, problems like adhesion and friction are more significant than they are in larger systems [127]. These elements will influence how nanomaterials are used. Numerous nanocarriers have shown high toxicity and some others could trigger an immune response. In addition, the complexity of nanoformulations is a serious issue. Therefore, the development, improvement, and implementation of effective nanoparticle delivery systems require specific consideration.

Increasing Targeting Efficiency

Nanoparticles have proven great potential to eradicate CSCs, by accumulating the drug specifically in a tumor. However, several problems need to be considered in the design and development of NPs. For instance, in order to actively target tumor tissue, adding targeting molecules to a formulation increases complexity, which potentially makes it superior to passive targeting. However, it also raises production costs and poses a risk of up-scalability. The actual concentration of the drug targeted into the tumor or CSC population is also still under debate. Given their longer circulation period and enhanced permeability and retention, non-targeted liposomes have been demonstrated to have tumor-accumulating effects comparable to those of targeted liposomes [128]. This may be caused by the higher clearance rate of targeted NPs due to the presence of targeting moieties. Another problem with targeting is specificity; CSCs usually share common surface markers with normal stem cells and specific targeting without harming normal cells still is an issue [129]. More research is needed to better discriminate CSCs from bodies' own stem cells and thus reduce toxicity.

Increasing Cellular Internalization

Nanoparticle systems should enable longer circulation time, higher accumulation at tumor regions, and penetration into tumor cells, especially into CSCs. Although PEG-like modifications, usually used in formulations enhance blood circulation time, they decrease cellular internalization [130]. Alternative methods like nanoparticles responsive to the microenvironment, have been developed. These formulations are able to lose their PEG moieties when the NP reaches the tumor microenvironment, which possesses some common characteristics like low pH and altered MMP activity. One such successful example is a liposome formulation that loses its PEG coating in a low pH tumor microenvironment and penetrates cancer cells with the help of underlying specific peptide moieties [131]. Another approach for increasing cellular internalization is encapsulating efflux inhibitors together with a chemotherapeutic drug. Such formulations were shown to overcome drug resistance in several *in vitro* and *in vivo* studies [132 - 134].

Efficient Tumor Penetration

It is crucial that NPs get to every tumor cell for optimizing its potential. Hydrophilic modifications like PEGylation significantly enhance circulation time, minimise protein interactions and clearance by the immune system [135]. However, NPs are frequently restricted to areas just next to the blood arteries [136, 137]. Ironically, areas far from blood veins tend to have a high concentration CSCs [138, 139]. Thus, NPs' power to defeat resistance will only be recognized if only they could penetrate the areas with poor blood flow. The mobility of molecules in the tumor ECM has been enhanced by a number of combination treatments. One approach is targeting the leaky vasculature that causes high interstitial fluid pressure and thus low drug penetration into the tumor mass [140, 141]. A majority of the research suggests that inhibition of angiogenesis increases drug penetration into the tumor temporarily [142 - 144]. Inhibiting vascular endothelial growth factor (VEGF), phosphoinositol-3-kinase

(PI3K), and epidermal growth factor receptor (EGFR) are some of the approaches for anti-angiogenesis treatments [145].

Although vascular normalization is an efficient way to enhance intra-tumor drug accumulation of small molecules and macromolecules, these methods are usually not applicable for larger-sized drug-encapsulated nanoparticles. Modification of tumor ECM provides another way of enhancing tumor penetration. By altering cells directly influencing the behavior of ECM, or by employing enzymes to destroy particular ECM components, the tumor ECM can be altered. Covalent conjugation of ECM degrading enzymes on NPs reveals promising results for drug penetration [146]. An alternative way of modifying ECM is to inhibit the stromal cells' secretion of ECM components. Combination treatment with a stromal cell inhibiting agent and a drug-encapsulated formulation showed promising results in several studies [147, 148].

Genome Editing Delivery

Until recently, the two main genetic modification approaches for CSC therapy were siRNA and miRNA. In an orthotopic model of breast cancer, the use of doxorubicin-loaded silica nanoparticles plus siRNA, targeting an efflux transporter gene revealed a synergistic reduction of tumor development as opposed to the single agent-laden NPs [149]. However, RNAi-mediated gene silencing frequently fails to completely repress gene expression since it works by degrading mRNA or inhibiting its translation into but does not alter the genome. Several developments in genome editing technologies like zinc finger nucleases (ZFNs), transcription activator-like effector nucleases (TALENs), and the clustered regularly interspaced short palindromic repeat (CRISPR)-associated nuclease Cas9, now provide means of knocking down genes completely, inserting site specific correcting mutations or increasing expression of specified genes [5]. Even though genome editing technology is still in its infancy, it offers tantalizing possibilities for the treatment of many diseases, including cancer. It is also likely that the optimization of delivery formulations will further enhance the functionality of the CRISPR/Cas9 system.

Immunotherapy

Within the tumor microenvironment, immunomodulatory cells like T cells, macrophages, natural killer cells, dendritic cells have the ability to both suppress and stimulate CSCs [150]. Immunotherapy is a young field, and immunotherapeutic strategy is considered a promising method to enhance patient outcomes. Nanocarriers have recently been developed to enhance the effectiveness of cancer immunotherapy. Combination methods for immune checkpoint inhibitors with other medicines may result in enhanced immune responses, which could lead to better responses against cancer and recovery rates [151]. In order to activate the anticancer T-cell response, lipid-calcium-phosphate (LCP) nanoparticles have been employed effectively as a peptide vaccine delivery strategy [152]. Another method successfully integrates an LCP-based vaccine with targeted liposome nanoparticles for the suppression of TGF-b at the TME [153]. TGF- functions as an immune suppressor within the TME, therefore silencing it strengthens the body's overall defense against the tumor. In order to boost the immune response against malignancies, nanoparticles have also been linked to the surface of T cells [154]. Additionally, in order to modify TME, targeted nanoparticles were used to add inflammatory cytokines to CAFs [155]. Therefore, combining nanotechnology with immunotherapy may represent a particularly intriguing area that holds considerable promise for subsequent breakthroughs and is well worth careful examination.

CONCLUDING REMARKS

A tumor is a heterogeneous collection of several continuously communicating cell types that coordinate the growth and spread of the tumor. One of the main challenges to establishing good therapeutic efficacy in many cancers is the development of resistance to repeated treatments. Nanomedicine-based therapies provide a compelling platform to combat drug resistance. The dynamic CSC model currently in use suggests that the novel nanomedicine-based treatments being developed should not only concentrate on eliminating cancer cells but also take into account every element of the TME. Nanotechnology has made significant progress that will eradicate CSCs from tumors. However, multidisciplinary collaboration, taking advantage of cutting-edge biomedical technology, is necessary to overcome the aforementioned limitations and assure the clinical effectiveness of these formulations.

REFERENCES

[1]   D. Hanahan, and R.A. Weinberg, "Hallmarks of cancer: The next generation", *Cell,* vol. 144, no. 5, pp. 646-674, 2011.
       [http://dx.doi.org/10.1016/j.cell.2011.02.013] [PMID: 21376230]

[2]   X. Ye, and R.A. Weinberg, "Epithelial-mesenchymal plasticity: A central regulator of cancer progression", *Trends Cell Biol.,* vol. 25, no. 11, pp. 675-686, 2015.
       [http://dx.doi.org/10.1016/j.tcb.2015.07.012] [PMID: 26437589]

[3]   M.F. Clarke, J.E. Dick, P.B. Dirks, C.J. Eaves, C.H.M. Jamieson, D.L. Jones, J. Visvader, I.L. Weissman, and G.M. Wahl, "Cancer stem cells--perspectives on current status and future directions: AACR Workshop on cancer stem cells", *Cancer Res.,* vol. 66, no. 19, pp. 9339-9344, 2006.
       [http://dx.doi.org/10.1158/0008-5472.CAN-06-3126] [PMID: 16990346]

[4]   L.V. Nguyen, R. Vanner, P. Dirks, and C.J. Eaves, "Cancer stem cells: An evolving concept", *Nat. Rev. Cancer,* vol. 12, no. 2, pp. 133-143, 2012.
       [http://dx.doi.org/10.1038/nrc3184] [PMID: 22237392]

[5]   S. Shen, J.X. Xia, and J. Wang, "Nanomedicine-mediated cancer stem cell therapy", *Biomaterials,* vol. 74, pp. 1-18, 2016.
[http://dx.doi.org/10.1016/j.biomaterials.2015.09.037] [PMID: 26433488]

[6]   A. Kreso, and J.E. Dick, "Evolution of the cancer stem cell model", *Cell Stem Cell,* vol. 14, no. 3, pp. 275-291, 2014.
[http://dx.doi.org/10.1016/j.stem.2014.02.006] [PMID: 24607403]

[7]   Z.G. Chen, "Small-molecule delivery by nanoparticles for anticancer therapy", *Trends Mol. Med.,* vol. 16, no. 12, pp. 594-602, 2010.
[http://dx.doi.org/10.1016/j.molmed.2010.08.001] [PMID: 20846905]

[8]   M.E. Davis, Z. Chen, and D.M. Shin, "Nanoparticle therapeutics: An emerging treatment modality for cancer", *Nat. Rev. Drug Discov.,* vol. 7, no. 9, pp. 771-782, 2008.
[http://dx.doi.org/10.1038/nrd2614] [PMID: 18758474]

[9]   J.E. Dick, T. Lapidot, and F. Pflumio, "Transplantation of normal and leukemic human bone marrow into immune-deficient mice: development of animal models for human hematopoiesis", *Immunol. Rev.,* vol. 124, no. 1, pp. 25-43, 1991.
[http://dx.doi.org/10.1111/j.1600-065X.1991.tb00614.x] [PMID: 1804779]

[10]   M. Al-Hajj, M.S. Wicha, A. Benito-Hernandez, S.J. Morrison, and M.F. Clarke, "Prospective identification of tumorigenic breast cancer cells", *Proc. Natl. Acad. Sci.,* vol. 100, no. 7, pp. 3983-3988, 2003.
[http://dx.doi.org/10.1073/pnas.0530291100] [PMID: 12629218]

[11]   J.P. Medema, "Cancer stem cells: The challenges ahead", *Nat. Cell Biol.,* vol. 15, no. 4, pp. 338-344, 2013.
[http://dx.doi.org/10.1038/ncb2717] [PMID: 23548926]

[12]   Y. Bu, and D. Cao, "The origin of cancer stem cells", *Front. Biosci.,* vol. 4, no. 3, pp. 819-830, 2012.
[PMID: 22202093]

[13]   S.A. Mani, W. Guo, M.J. Liao, E.N. Eaton, A. Ayyanan, A.Y. Zhou, M. Brooks, F. Reinhard, C.C. Zhang, M. Shipitsin, L.L. Campbell, K. Polyak, C. Brisken, J. Yang, and R.A. Weinberg, "The epithelial-mesenchymal transition generates cells with properties of stem cells", *Cell,* vol. 133, no. 4, pp. 704-715, 2008.
[http://dx.doi.org/10.1016/j.cell.2008.03.027] [PMID: 18485877]

[14]   D. Bonnet, and J.E. Dick, "Human acute myeloid leukemia is organized as a hierarchy that originates from a primitive hematopoietic cell", *Nat. Med.,* vol. 3, no. 7, pp. 730-737, 1997.
[http://dx.doi.org/10.1038/nm0797-730] [PMID: 9212098]

[15]   P. Gener, P. Gonzalez Callejo, J. Seras-Franzoso, F. Andrade, D. Rafael, I. Abasolo, and S. Schwartz Jr, "The potential of nanomedicine to alter cancer stem cell dynamics: The impact of extracellular vesicles", *Nanomedicine,* vol. 15, no. 28, pp. 2785-2800, 2020.
[http://dx.doi.org/10.2217/nnm-2020-0099] [PMID: 33191837]

[16]   P. Gener, D. Rafael, J. Seras-Franzoso, A. Perez, L.A. Pindado, G. Casas, D. Arango, Y. Fernández, Z.V. Díaz-Riascos, I. Abasolo, and S. Schwartz Jr, "Pivotal role of AKT2 during dynamic phenotypic change of breast cancer stem cells", *Cancers,* vol. 11, no. 8, p. 1058, 2019.
[http://dx.doi.org/10.3390/cancers11081058] [PMID: 31357505]

[17]   J.E. Sarry, K. Murphy, R. Perry, P.V. Sanchez, A. Secreto, C. Keefer, C.R. Swider, A.C. Strzelecki, C. Cavelier, C. Récher, V. Mansat-De Mas, E. Delabesse, G. Danet-Desnoyers, and M. Carroll, "Human acute myelogenous leukemia stem cells are rare and heterogeneous when assayed in NOD/SCID/IL2Rγc-deficient mice", *J. Clin. Invest.,* vol. 121, no. 1, pp. 384-395, 2011.
[http://dx.doi.org/10.1172/JCI41495] [PMID: 21157036]

[18]   N. Goardon, E. Marchi, A. Atzberger, L. Quek, A. Schuh, S. Soneji, P. Woll, A. Mead, K.A. Alford, R. Rout, S. Chaudhury, A. Gilkes, S. Knapper, K. Beldjord, S. Begum, S. Rose, N. Geddes, M.

Griffiths, G. Standen, A. Sternberg, J. Cavenagh, H. Hunter, D. Bowen, S. Killick, L. Robinson, A. Price, E. Macintyre, P. Virgo, A. Burnett, C. Craddock, T. Enver, S.E.W. Jacobsen, C. Porcher, and P. Vyas, "Coexistence of LMPP-like and GMP-like leukemia stem cells in acute myeloid leukemia", *Cancer Cell,* vol. 19, no. 1, pp. 138-152, 2011.
[http://dx.doi.org/10.1016/j.ccr.2010.12.012] [PMID: 21251617]

[19]   T. Reya, S.J. Morrison, M.F. Clarke, and I.L. Weissman, "Stem cells, cancer, and cancer stem cells", *Nature,* vol. 414, no. 6859, pp. 105-111, 2001.
[http://dx.doi.org/10.1038/35102167] [PMID: 11689955]

[20]   P.B. Gupta, C.M. Fillmore, G. Jiang, S.D. Shapira, K. Tao, C. Kuperwasser, and E.S. Lander, "Stochastic state transitions give rise to phenotypic equilibrium in populations of cancer cells", *Cell,* vol. 146, no. 4, pp. 633-644, 2011.
[http://dx.doi.org/10.1016/j.cell.2011.07.026] [PMID: 21854987]

[21]   S. Sell, "Stem cell origin of cancer and differentiation therapy", *Crit. Rev. Oncol. Hematol.,* vol. 51, no. 1, pp. 1-28, 2004.
[http://dx.doi.org/10.1016/j.critrevonc.2004.04.007] [PMID: 15207251]

[22]   I. Pastushenko, and C. Blanpain, "EMT Transition states during tumor progression and metastasis", *Trends Cell Biol.,* vol. 29, no. 3, pp. 212-226, 2019.
[http://dx.doi.org/10.1016/j.tcb.2018.12.001] [PMID: 30594349]

[23]   D. Hanahan, and L.M. Coussens, "Accessories to the crime: Functions of cells recruited to the tumor microenvironment", *Cancer Cell,* vol. 21, no. 3, pp. 309-322, 2012.
[http://dx.doi.org/10.1016/j.ccr.2012.02.022] [PMID: 22439926]

[24]   N. Charles, T. Ozawa, M. Squatrito, A.M. Bleau, C.W. Brennan, D. Hambardzumyan, and E.C. Holland, "Perivascular nitric oxide activates notch signaling and promotes stem-like character in PDGF-induced glioma cells", *Cell Stem Cell,* vol. 6, no. 2, pp. 141-152, 2010.
[http://dx.doi.org/10.1016/j.stem.2010.01.001] [PMID: 20144787]

[25]   L. Vermeulen, F. De Sousa E Melo, M. van der Heijden, K. Cameron, J.H. de Jong, T. Borovski, J.B. Tuynman, M. Todaro, C. Merz, H. Rodermond, M.R. Sprick, K. Kemper, D.J. Richel, G. Stassi, and J.P. Medema, "Wnt activity defines colon cancer stem cells and is regulated by the microenvironment", *Nat. Cell Biol.,* vol. 12, no. 5, pp. 468-476, 2010.
[http://dx.doi.org/10.1038/ncb2048] [PMID: 20418870]

[26]   A.P. Thankamony, K. Saxena, R. Murali, M.K. Jolly, and R. Nair, "Cancer stem cell plasticity-a deadly deal", *Front. Mol. Biosci.,* vol. 7, p. 79, 2020.
[http://dx.doi.org/10.3389/fmolb.2020.00079] [PMID: 32426371]

[27]   S. Vinogradov, and X. Wei, "Cancer stem cells and drug resistance: The potential of nanomedicine", *Nanomedicine,* vol. 7, no. 4, pp. 597-615, 2012.
[http://dx.doi.org/10.2217/nnm.12.22] [PMID: 22471722]

[28]   B. Beck, and C. Blanpain, "Unravelling cancer stem cell potential", *Nat. Rev. Cancer,* vol. 13, no. 10, pp. 727-738, 2013.
[http://dx.doi.org/10.1038/nrc3597] [PMID: 24060864]

[29]   K. Chen, Y. Huang, and J. Chen, "Understanding and targeting cancer stem cells: Therapeutic implications and challenges", *Acta Pharmacol. Sin.,* vol. 34, no. 6, pp. 732-740, 2013.
[http://dx.doi.org/10.1038/aps.2013.27] [PMID: 23685952]

[30]   M. Dean, A. Rzhetsky, and R. Allikmets, "The human ATP-binding cassette (ABC) transporter superfamily", *Genome Res.,* vol. 11, no. 7, pp. 1156-1166, 2001.
[http://dx.doi.org/10.1101/gr.184901] [PMID: 11435397]

[31]   C.W. Scharenberg, M.A. Harkey, and B. Torok-Storb, "The ABCG2 transporter is an efficient Hoechst 33342 efflux pump and is preferentially expressed by immature human hematopoietic progenitors", *Blood,* vol. 99, no. 2, pp. 507-512, 2002.
[http://dx.doi.org/10.1182/blood.V99.2.507] [PMID: 11781231]

[32]   A.H. Schinkel, J.J.M. Smit, O. van Tellingen, J.H. Beijnen, E. Wagenaar, L. van Deemter, C.A.A.M. Mol, M.A. van der Valk, E.C. Robanus-Maandag, H.P.J. te Riele, A.J.M. Berns, and P. Borst, "Disruption of the mouse mdr1a P-glycoprotein gene leads to a deficiency in the blood-brain barrier and to increased sensitivity to drugs", *Cell,* vol. 77, no. 4, pp. 491-502, 1994.
[http://dx.doi.org/10.1016/0092-8674(94)90212-7] [PMID: 7910522]

[33]   L. Tang, S.M. Bergevoet, C. Gilissen, T. de Witte, J.H. Jansen, B.A. van der Reijden, and R.A.P. Raymakers, "Hematopoietic stem cells exhibit a specific ABC transporter gene expression profile clearly distinct from other stem cells", *BMC Pharmacol.,* vol. 10, no. 1, p. 12, 2010.
[http://dx.doi.org/10.1186/1471-2210-10-12] [PMID: 20836839]

[34]   K. Moitra, H. Lou, and M. Dean, "Multidrug efflux pumps and cancer stem cells: insights into multidrug resistance and therapeutic development", *Clin. Pharmacol. Ther.,* vol. 89, no. 4, pp. 491-502, 2011.
[http://dx.doi.org/10.1038/clpt.2011.14] [PMID: 21368752]

[35]   S.K. Rabindran, D.D. Ross, L.A. Doyle, W. Yang, and L.M. Greenberger, "Fumitremorgin C reverses multidrug resistance in cells transfected with the breast cancer resistance protein", *Cancer Res.,* vol. 60, no. 1, pp. 47-50, 2000.
[PMID: 10646850]

[36]   H. Woehlecke, H. Osada, A. Herrmann, and H. Lage, "Reversal of breast cancer resistance protein-mediated drug resistance by tryprostatin A", *Int. J. Cancer,* vol. 107, no. 5, pp. 721-728, 2003.
[http://dx.doi.org/10.1002/ijc.11444] [PMID: 14566821]

[37]   S. Nobili, I. Landini, B. Giglioni, and E. Mini, "Pharmacological strategies for overcoming multidrug resistance", *Curr. Drug Targets,* vol. 7, no. 7, pp. 861-879, 2006.
[http://dx.doi.org/10.2174/138945006777709593] [PMID: 16842217]

[38]   J.M. Angelastro, and M.W. Lamé, "Overexpression of CD133 promotes drug resistance in C6 glioma cells", *Mol. Cancer Res.,* vol. 8, no. 8, pp. 1105-1115, 2010.
[http://dx.doi.org/10.1158/1541-7786.MCR-09-0383] [PMID: 20663862]

[39]   N.Y. Frank, S.S. Pendse, P.H. Lapchak, A. Margaryan, D. Shlain, C. Doeing, M.H. Sayegh, and M.H. Frank, "Regulation of progenitor cell fusion by ABCB5 P-glycoprotein, a novel human ATP-binding cassette transporter", *J. Biol. Chem.,* vol. 278, no. 47, pp. 47156-47165, 2003.
[http://dx.doi.org/10.1074/jbc.M308700200] [PMID: 12960149]

[40]   N.Y. Frank, A. Margaryan, Y. Huang, T. Schatton, A.M. Waaga-Gasser, M. Gasser, M.H. Sayegh, W. Sadee, and M.H. Frank, "ABCB5-mediated doxorubicin transport and chemoresistance in human malignant melanoma", *Cancer Res.,* vol. 65, no. 10, pp. 4320-4333, 2005.
[http://dx.doi.org/10.1158/0008-5472.CAN-04-3327] [PMID: 15899824]

[41]   A.L. Harris, and D. Hochhauser, "Mechanisms of multidrug resistance in cancer treatment", *Acta Oncol.,* vol. 31, no. 2, pp. 205-213, 1992.
[http://dx.doi.org/10.3109/02841869209088904] [PMID: 1352455]

[42]   T.M. Grogan, C.M. Spier, S.E. Salmon, M. Matzner, J. Rybski, R.S. Weinstein, R.J. Scheper, and W.S. Dalton, "P-glycoprotein expression in human plasma cell myeloma: Correlation with prior chemotherapy", *Blood,* vol. 81, no. 2, pp. 490-495, 1993.
[http://dx.doi.org/10.1182/blood.V81.2.490.490] [PMID: 8093668]

[43]   A. Abolhoda, A.E. Wilson, H. Ross, P.V. Danenberg, M. Burt, and K.W. Scotto, "Rapid activation of MDR1 gene expression in human metastatic sarcoma after *in vivo* exposure to doxorubicin", *Clin. Cancer Res.,* vol. 5, no. 11, pp. 3352-3356, 1999.
[PMID: 10589744]

[44]   K.V. Chin, S.S. Chauhan, I. Pastan, and M.M. Gottesman, "Regulation of mdr RNA levels in response to cytotoxic drugs in rodent cells", *Cell Growth Differ.,* vol. 1, no. 8, pp. 361-365, 1990.
[PMID: 1703776]

[45]   M.M. Gottesman, T. Fojo, and S.E. Bates, "Multidrug resistance in cancer: Role of ATP–dependent transporters", *Nat. Rev. Cancer,* vol. 2, no. 1, pp. 48-58, 2002.
[http://dx.doi.org/10.1038/nrc706] [PMID: 11902585]

[46]   C. Hadjimichael, K. Chanoumidou, N. Papadopoulou, P. Arampatzi, J. Papamatheakis, and A. Kretsovali, "Common stemness regulators of embryonic and cancer stem cells", *World J. Stem Cells,* vol. 7, no. 9, pp. 1150-1184, 2015.
[http://dx.doi.org/10.4252/wjsc.v7.i9.1150] [PMID: 26516408]

[47]   L. Yang, P. Shi, G. Zhao, J. Xu, W. Peng, J. Zhang, G. Zhang, X. Wang, Z. Dong, F. Chen, and H. Cui, "Targeting cancer stem cell pathways for cancer therapy", *Signal Transduct. Target. Ther.,* vol. 5, no. 1, p. 8, 2020.
[http://dx.doi.org/10.1038/s41392-020-0110-5] [PMID: 32296030]

[48]   J.E. Visvader, and G.J. Lindeman, "Cancer stem cells: Current status and evolving complexities", *Cell Stem Cell,* vol. 10, no. 6, pp. 717-728, 2012.
[http://dx.doi.org/10.1016/j.stem.2012.05.007] [PMID: 22704512]

[49]   E.K. Ramos, A.D. Hoffmann, S.L. Gerson, and H. Liu, "New opportunities and challenges to defeat cancer stem cells", *Trends Cancer,* vol. 3, no. 11, pp. 780-796, 2017.
[http://dx.doi.org/10.1016/j.trecan.2017.08.007] [PMID: 29120754]

[50]   X. Zhang, and J. Hao, "Development of anticancer agents targeting the Wnt/β-catenin signaling", *Am. J. Cancer Res.,* vol. 5, no. 8, pp. 2344-2360, 2015.
[PMID: 26396911]

[51]   M. Karami fath, M. Ebrahimi, E. Nourbakhsh, A. Zia Hazara, A. Mirzaei, S. Shafieyari, A. Salehi, M. Hoseinzadeh, Z. Payandeh, and G. Barati, "PI3K/Akt/mTOR signaling pathway in cancer stem cells", *Pathol. Res. Pract.,* vol. 237, p. 154010, 2022.
[http://dx.doi.org/10.1016/j.prp.2022.154010] [PMID: 35843034]

[52]   D. Hanahan, and R.A. Weinberg, "The hallmarks of cancer", *Cell,* vol. 100, no. 1, pp. 57-70, 2000.
[http://dx.doi.org/10.1016/S0092-8674(00)81683-9] [PMID: 10647931]

[53]   M. Konopleva, S. Zhao, W. Hu, S. Jiang, V. Snell, D. Weidner, C.E. Jackson, X. Zhang, R. Champlin, E. Estey, J.C. Reed, and M. Andreeff, "The anti-apoptotic genes Bcl-X L and Bcl-2 are over-expressed and contribute to chemoresistance of non-proliferating leukaemic CD34 + cells", *Br. J. Haematol.,* vol. 118, no. 2, pp. 521-534, 2002.
[http://dx.doi.org/10.1046/j.1365-2141.2002.03637.x] [PMID: 12139741]

[54]   Z. Madjd, A.Z. Mehrjerdi, A.M. Sharifi, S. Molanaei, S.Z. Shahzadi, and M. Asadi-Lari, "CD44+ cancer cells express higher levels of the anti-apoptotic protein Bcl-2 in breast tumours", *Cancer Immun.,* vol. 9, p. 4, 2009.
[PMID: 19385591]

[55]   M.S. Soengas, R.M. Alarcón, H. Yoshida, A. J, Giaccia, R. Hakem, T.W. Mak, and S.W. Lowe, "Apaf-1 and caspase-9 in p53-dependent apoptosis and tumor inhibition", *Science,* vol. 284, no. 5411, pp. 156-159, 1999.
[http://dx.doi.org/10.1126/science.284.5411.156] [PMID: 10102818]

[56]   J. Domingo-Domenech, S.J. Vidal, V. Rodriguez-Bravo, M. Castillo-Martin, S.A. Quinn, R. Rodriguez-Barrueco, D.M. Bonal, E. Charytonowicz, N. Gladoun, J. de la Iglesia-Vicente, D.P. Petrylak, M.C. Benson, J.M. Silva, and C. Cordon-Cardo, "Suppression of acquired docetaxel resistance in prostate cancer through depletion of notch- and hedgehog-dependent tumor-initiating cells", *Cancer Cell,* vol. 22, no. 3, pp. 373-388, 2012.
[http://dx.doi.org/10.1016/j.ccr.2012.07.016] [PMID: 22975379]

[57]   M. Konopleva, D.A. Pollyea, J. Potluri, B. Chyla, L. Hogdal, T. Busman, E. McKeegan, A.H. Salem, M. Zhu, J.L. Ricker, W. Blum, C.D. DiNardo, T. Kadia, M. Dunbar, R. Kirby, N. Falotico, J. Leverson, R. Humerickhouse, M. Mabry, R. Stone, H. Kantarjian, and A. Letai, "Efficacy and biological correlates of response in a phase II study of venetoclax monotherapy in patients with acute

myelogenous leukemia", *Cancer Discov.,* vol. 6, no. 10, pp. 1106-1117, 2016.
[http://dx.doi.org/10.1158/2159-8290.CD-16-0313] [PMID: 27520294]

[58]   Q. Zhang, M.J. Lenardo, and D. Baltimore, "30 years of NF-κB: A blossoming of relevance to human pathobiology", *Cell,* vol. 168, no. 1-2, pp. 37-57, 2017.
[http://dx.doi.org/10.1016/j.cell.2016.12.012] [PMID: 28086098]

[59]   M.S. Hayden, and S. Ghosh, "Shared principles in NF-kappaB signaling", *Cell,* vol. 132, no. 3, pp. 344-362, 2008.
[http://dx.doi.org/10.1016/j.cell.2008.01.020] [PMID: 18267068]

[60]   C. Gonzalez-Torres, J. Gaytan-Cervantes, K. Vazquez-Santillan, E.A. Mandujano-Tinoco, G. Ceballos-Cancino, A. Garcia-Venzor, C. Zampedri, P. Sanchez-Maldonado, R. Mojica-Espinosa, L.E. Jimenez-Hernandez, and V. Maldonado, "NF-κB participates in the stem cell phenotype of ovarian cancer cells", *Arch. Med. Res.,* vol. 48, no. 4, pp. 343-351, 2017.
[http://dx.doi.org/10.1016/j.arcmed.2017.08.001] [PMID: 28886875]

[61]   J.P. Burnett, G. Lim, Y. Li, R.B. Shah, R. Lim, H.J. Paholak, S.P. McDermott, L. Sun, Y. Tsume, S. Bai, M.S. Wicha, D. Sun, and T. Zhang, "Sulforaphane enhances the anticancer activity of taxanes against triple negative breast cancer by killing cancer stem cells", *Cancer Lett.,* vol. 394, pp. 52-64, 2017.
[http://dx.doi.org/10.1016/j.canlet.2017.02.023] [PMID: 28254410]

[62]   J.U. Marquardt, L. Gomez-Quiroz, L.O. Arreguin Camacho, F. Pinna, Y.H. Lee, M. Kitade, M.P. Domínguez, D. Castven, K. Breuhahn, E.A. Conner, P.R. Galle, J.B. Andersen, V.M. Factor, and S.S. Thorgeirsson, "Curcumin effectively inhibits oncogenic NF-κB signaling and restrains stemness features in liver cancer", *J. Hepatol.,* vol. 63, no. 3, pp. 661-669, 2015.
[http://dx.doi.org/10.1016/j.jhep.2015.04.018] [PMID: 25937435]

[63]   G.L. Uy, M.P. Rettig, I.H. Motabi, K. McFarland, K.M. Trinkaus, L.M. Hladnik, S. Kulkarni, C.N. Abboud, A.F. Cashen, K.E. Stockerl-Goldstein, R. Vij, P. Westervelt, and J.F. DiPersio, "A phase 1/2 study of chemosensitization with the CXCR4 antagonist plerixafor in relapsed or refractory acute myeloid leukemia", *Blood,* vol. 119, no. 17, pp. 3917-3924, 2012.
[http://dx.doi.org/10.1182/blood-2011-10-383406] [PMID: 22308295]

[64]   J.D. Hainsworth, J.A. Reeves, J.R. Mace, E.J. Crane, O. Hamid, J.R. Stille, A. Flynt, S. Roberson, J. Polzer, and E.R. Arrowsmith, "A Randomized, open-label phase 2 study of the CXCR4 inhibitor LY2510924 in combination with sunitinib versus sunitinib alone in patients with metastatic renal cell carcinoma (RCC)", *Target. Oncol.,* vol. 11, no. 5, pp. 643-653, 2016.
[http://dx.doi.org/10.1007/s11523-016-0434-9] [PMID: 27154357]

[65]   Z. Li, S. Bao, Q. Wu, H. Wang, C. Eyler, S. Sathornsumetee, Q. Shi, Y. Cao, J. Lathia, R.E. McLendon, A.B. Hjelmeland, and J.N. Rich, "Hypoxia-inducible factors regulate tumorigenic capacity of glioma stem cells", *Cancer Cell,* vol. 15, no. 6, pp. 501-513, 2009.
[http://dx.doi.org/10.1016/j.ccr.2009.03.018] [PMID: 19477429]

[66]   M. Ghielmini, S.F.H. Schmitz, K. Bürki, G. Pichert, D.C. Betticher, R. Stupp, M. Wernli, A. Lohri, D. Schmitter, F. Bertoni, and T. Cerny, "The effect of rituximab on patients with follicular and mantle-cell lymphoma", *Ann. Oncol.,* vol. 11, no. 1, suppl. 1, pp. S123-S126, 2000.
[http://dx.doi.org/10.1093/annonc/11.suppl_1.S123] [PMID: 10707793]

[67]   D.R. Colnot, J.C. Roos, R. de Bree, A.J. Wilhelm, J.A. Kummer, G. Hanft, K.H. Heider, G. Stehle, G.B. Snow, and G.A.M.S. van Dongen, "Safety, biodistribution, pharmacokinetics, and immunogenicity of 99m Tc-labeled humanized monoclonal antibody BIWA 4 (bivatuzumab) in patients with squamous cell carcinoma of the head and neck", *Cancer Immunol. Immunother.,* vol. 52, no. 9, pp. 576-582, 2003.
[http://dx.doi.org/10.1007/s00262-003-0396-5] [PMID: 14627130]

[68]   B.Z. Tang, F.Y. Zhao, Y. Qu, and D.Z. Mu, "Hypoxia-inducible factor-1 alpha?: A promising target for tumor therapy", *Chin. J. Cancer,* vol. 28, no. 7, pp. 775-782, 2009.
[http://dx.doi.org/10.5732/cjc.008.10770] [PMID: 19624909]

[69]   XQ Ye, "Mitochondrial and energy metabolism-related properties as novel indicators of lung cancer stem cells", *Int J Cancer.,* vol. 129, no. 4, pp. 820-831, 2011.
[http://dx.doi.org/10.1002/ijc.25944]

[70]   W.A. Flavahan, Q. Wu, M. Hitomi, N. Rahim, Y. Kim, A.E. Sloan, R.J. Weil, I. Nakano, J.N. Sarkaria, B.W. Stringer, B.W. Day, M. Li, J.D. Lathia, J.N. Rich, and A.B. Hjelmeland, "Brain tumor initiating cells adapt to restricted nutrition through preferential glucose uptake", *Nat. Neurosci.,* vol. 16, no. 10, pp. 1373-1382, 2013.
[http://dx.doi.org/10.1038/nn.3510] [PMID: 23995067]

[71]   J.H. Kim, K.J. Lee, Y. Seo, J.H. Kwon, J.P. Yoon, J.Y. Kang, H.J. Lee, S.J. Park, S.P. Hong, J.H. Cheon, W.H. Kim, and T. Il Kim, "Effects of metformin on colorectal cancer stem cells depend on alterations in glutamine metabolism", *Sci. Rep.,* vol. 8, no. 1, p. 409, 2018.
[http://dx.doi.org/10.1038/s41598-017-18762-4] [PMID: 29323154]

[72]   O. Maehara, S. Ohnishi, A. Asano, G. Suda, M. Natsuizaka, K. Nakagawa, M. Kobayashi, N. Sakamoto, and H. Takeda, "Metformin regulates the expression of CD133 through the ampk-cebpβ pathway in hepatocellular carcinoma cell lines", *Neoplasia,* vol. 21, no. 6, pp. 545-556, 2019.
[http://dx.doi.org/10.1016/j.neo.2019.03.007] [PMID: 31042624]

[73]   T.I. Emeto, F.O. Alele, A.M. Smith, F.M. Smith, T. Dougan, and J. Golledge, "Use of nanoparticles as contrast agents for the functional and molecular imaging of abdominal aortic aneurysm", *Front. Cardiovasc. Med.,* vol. 4, p. 16, 2017.
[http://dx.doi.org/10.3389/fcvm.2017.00016] [PMID: 28386544]

[74]   N. Naseri, E. Ajorlou, F. Asghari, and Y. Pilehvar-Soltanahmadi, "An update on nanoparticle-based contrast agents in medical imaging", *Artif. Cells Nanomed. Biotechnol.,* vol. 46, no. 6, pp. 1111-1121, 2018.
[http://dx.doi.org/10.1080/21691401.2017.1379014] [PMID: 28933183]

[75]   M.J. Sands, and A. Levitin, "Basics of magnetic resonance imaging", *Semin. Vasc. Surg.,* vol. 17, no. 2, pp. 66-82, 2004.
[http://dx.doi.org/10.1053/j.semvascsurg.2004.03.011] [PMID: 15185173]

[76]   M. Silindir, A.Y. Özer, and S. Erdoğan, "The use and importance of liposomes in positron emission tomography", *Drug Deliv.,* vol. 19, no. 1, pp. 68-80, 2012.
[http://dx.doi.org/10.3109/10717544.2011.635721] [PMID: 22211758]

[77]   L.E. Cole, R.D. Ross, J.M.R. Tilley, T. Vargo-Gogola, and R.K. Roeder, "Gold nanoparticles as contrast agents in x-ray imaging and computed tomography", *Nanomedicine,* vol. 10, no. 2, pp. 321-341, 2015.
[http://dx.doi.org/10.2217/nnm.14.171] [PMID: 25600973]

[78]   O. Zitka, M. Ryvolova, J. Hubalek, T. Eckschlager, V. Adam, and R. Kizek, "From amino acids to proteins as targets for metal-based drugs", *Curr. Drug Metab.,* vol. 13, no. 3, pp. 306-320, 2012.
[http://dx.doi.org/10.2174/138920012799320437] [PMID: 22455554]

[79]   E.C. Dreaden, A.M. Alkilany, X. Huang, C.J. Murphy, and M.A. El-Sayed, "The golden age: Gold nanoparticles for biomedicine", *Chem. Soc. Rev.,* vol. 41, no. 7, pp. 2740-2779, 2012.
[http://dx.doi.org/10.1039/C1CS15237H] [PMID: 22109657]

[80]   A. Astefanei, O. Núñez, and M.T. Galceran, "Characterisation and determination of fullerenes: A critical review", *Anal. Chim. Acta,* vol. 882, pp. 1-21, 2015.
[http://dx.doi.org/10.1016/j.aca.2015.03.025] [PMID: 26043086]

[81]   K. Saeed, and I. Khan, "Preparation and properties of single-walled carbon nanotubes/poly(butylene terephthalate) nanocomposites", *Iran. Polym. J.,* vol. 23, no. 1, pp. 53-58, 2014.
[http://dx.doi.org/10.1007/s13726-013-0199-2]

[82]   K. Saeed, and I. Khan, "Preparation and characterization of single-walled carbon nanotube/nylon 6, 6 nanocomposites", *Instrum. Sci. Technol.,* vol. 44, no. 4, pp. 435-444, 2016.

[http://dx.doi.org/10.1080/10739149.2015.1127256]

[83]   J.M. Ngoy, N. Wagner, L. Riboldi, and O. Bolland, "CO2 capture technology using multi-walled carbon nanotubes with polyaspartamide surfactant", *Energy Procedia,* vol. 63, pp. 2230-2248, 2014.
[http://dx.doi.org/10.1016/j.egypro.2014.11.242]

[84]   L.F. Mabena, S. Sinha Ray, S.D. Mhlanga, and N.J. Coville, "Nitrogen-doped carbon nanotubes as a metal catalyst support", *Appl. Nanosci.,* vol. 1, no. 2, pp. 67-77, 2011.
[http://dx.doi.org/10.1007/s13204-011-0013-4]

[85]   S. Thomas, B.S.P. Harshita, P. Mishra, and S. Talegaonkar, "Ceramic nanoparticles: Fabrication methods and applications in drug delivery", *Curr. Pharm. Des.,* vol. 21, no. 42, pp. 6165-6188, 2015.
[http://dx.doi.org/10.2174/1381612821666151027153246] [PMID: 26503144]

[86]   T. Hisatomi, J. Kubota, and K. Domen, "Recent advances in semiconductors for photocatalytic and photoelectrochemical water splitting", *Chem. Soc. Rev.,* vol. 43, no. 22, pp. 7520-7535, 2014.
[http://dx.doi.org/10.1039/C3CS60378D] [PMID: 24413305]

[87]   M. Mansha, I. Khan, N. Ullah, and A. Qurashi, "Synthesis, characterization and visible-light-driven photoelectrochemical hydrogen evolution reaction of carbazole-containing conjugated polymers", *Int. J. Hydrogen Energy,* vol. 42, no. 16, pp. 10952-10961, 2017.
[http://dx.doi.org/10.1016/j.ijhydene.2017.02.053]

[88]   J.P. Rao, and K.E. Geckeler, "Polymer nanoparticles: Preparation techniques and size-control parameters", *Prog. Polym. Sci.,* vol. 36, no. 7, pp. 887-913, 2011.
[http://dx.doi.org/10.1016/j.progpolymsci.2011.01.001]

[89]   A. Puri, K. Loomis, B. Smith, J.H. Lee, A. Yavlovich, E. Heldman, and R. Blumenthal, "Lipid-based nanoparticles as pharmaceutical drug carriers: From concepts to clinic", *Crit. Rev. Ther. Drug Carrier Syst.,* vol. 26, no. 6, pp. 523-580, 2009.
[http://dx.doi.org/10.1615/CritRevTherDrugCarrierSyst.v26.i6.10] [PMID: 20402623]

[90]   M. Gujrati, A. Malamas, T. Shin, E. Jin, Y. Sun, and Z.R. Lu, "Multifunctional cationic lipid-based nanoparticles facilitate endosomal escape and reduction-triggered cytosolic siRNA release", *Mol. Pharm.,* vol. 11, no. 8, pp. 2734-2744, 2014.
[http://dx.doi.org/10.1021/mp400787s] [PMID: 25020033]

[91]   T.M. Sun, Y.C. Wang, F. Wang, J.Z. Du, C.Q. Mao, C.Y. Sun, R.Z. Tang, Y. Liu, J. Zhu, Y.H. Zhu, X.Z. Yang, and J. Wang, "Cancer stem cell therapy using doxorubicin conjugated to gold nanoparticles via hydrazone bonds", *Biomaterials,* vol. 35, no. 2, pp. 836-845, 2014.
[http://dx.doi.org/10.1016/j.biomaterials.2013.10.011] [PMID: 24144908]

[92]   K. Wang, L. Liu, T. Zhang, Y.L. Zhu, F. Qiu, X.G. Wu, X.L. Wang, F.Q. Hu, and J. Huang, "Oxaliplatin-incorporated micelles eliminate both cancer stem-like and bulk cell populations in colorectal cancer", *Int. J. Nanomedicine,* vol. 6, pp. 3207-3218, 2011.
[PMID: 22238509]

[93]   P.B. Gupta, T.T. Onder, G. Jiang, K. Tao, C. Kuperwasser, R.A. Weinberg, and E.S. Lander, "Identification of selective inhibitors of cancer stem cells by high-throughput screening", *Cell,* vol. 138, no. 4, pp. 645-659, 2009.
[http://dx.doi.org/10.1016/j.cell.2009.06.034] [PMID: 19682730]

[94]   X. Wang, X.C. Low, W. Hou, L.N. Abdullah, T.B. Toh, M. Mohd Abdul Rashid, D. Ho, and E.K.H. Chow, "Epirubicin-adsorbed nanodiamonds kill chemoresistant hepatic cancer stem cells", *ACS Nano,* vol. 8, no. 12, pp. 12151-12166, 2014.
[http://dx.doi.org/10.1021/nn503491e] [PMID: 25437772]

[95]   P. Zhao, S. Dong, J. Bhattacharyya, and M. Chen, "iTEP nanoparticle-delivered salinomycin displays an enhanced toxicity to cancer stem cells in orthotopic breast tumors", *Mol. Pharm.,* vol. 11, no. 8, pp. 2703-2712, 2014.
[http://dx.doi.org/10.1021/mp5002312] [PMID: 24960465]

[96]   K. Wang, T. Zhang, L. Liu, X. Wang, P. Wu, Z. Chen, C. Ni, J. Zhang, F. Hu, and J. Huang, "Novel micelle formulation of curcumin for enhancing antitumor activity and inhibiting colorectal cancer stem cells", *Int. J. Nanomedicine,* vol. 7, pp. 4487-4497, 2012.
       [PMID: 22927762]

[97]   X. Wei, T.H. Senanayake, G. Warren, and S.V. Vinogradov, "Hyaluronic acid-based nanogel-drug conjugates with enhanced anticancer activity designed for the targeting of CD44-positive and drug-resistant tumors", *Bioconjug. Chem.,* vol. 24, no. 4, pp. 658-668, 2013.
       [http://dx.doi.org/10.1021/bc300632w] [PMID: 23547842]

[98]   S. Krishnamurthy, V.W.L. Ng, S. Gao, M.H. Tan, and Y.Y. Yang, "Phenformin-loaded polymeric micelles for targeting both cancer cells and cancer stem cells *in vitro* and *in vivo*", *Biomaterials,* vol. 35, no. 33, pp. 9177-9186, 2014.
       [http://dx.doi.org/10.1016/j.biomaterials.2014.07.018] [PMID: 25106770]

[99]   Y. Zhou, J. Yang, and J. Kopeček, "Selective inhibitory effect of HPMA copolymer-cyclopamine conjugate on prostate cancer stem cells", *Biomaterials,* vol. 33, no. 6, pp. 1863-1872, 2012.
       [http://dx.doi.org/10.1016/j.biomaterials.2011.11.029] [PMID: 22138033]

[100] S. Shen, X.J. Du, J. Liu, R. Sun, Y.H. Zhu, and J. Wang, "Delivery of bortezomib with nanoparticles for basal-like triple-negative breast cancer therapy", *J. Control. Release,* vol. 208, pp. 14-24, 2015.
       [http://dx.doi.org/10.1016/j.jconrel.2014.12.043] [PMID: 25575864]

[101] A. Khan, E. Ahmed, N. Elareer, K. Junejo, M. Steinhoff, and S. Uddin, "Role of miRNA-regulated cancer stem cells in the pathogenesis of human malignancies", *Cells,* vol. 8, no. 8, p. 840, 2019.
       [http://dx.doi.org/10.3390/cells8080840] [PMID: 31530793]

[102] M. Ishida, and F.M. Selaru, "miRNA-based therapeutic strategies", *Curr. Anesthesiol. Rep.,* vol. 1, no. 1, pp. 63-70, 2013.
       [PMID: 23524956]

[103] S. Hager, F.J. Fittler, E. Wagner, and M. Bros, "Nucleic acid-based approaches for tumor therapy", *Cells,* vol. 9, no. 9, p. 2061, 2020.
       [http://dx.doi.org/10.3390/cells9092061] [PMID: 32917034]

[104] Y.P. Yang, Y. Chien, G.Y. Chiou, J.Y. Cherng, M.L. Wang, W.L. Lo, Y.L. Chang, P.I. Huang, Y.W. Chen, Y.H. Shih, M.T. Chen, and S.H. Chiou, "Inhibition of cancer stem cell-like properties and reduced chemoradioresistance of glioblastoma using microRNA145 with cationic polyurethane-short branch PEI", *Biomaterials,* vol. 33, no. 5, pp. 1462-1476, 2012.
       [http://dx.doi.org/10.1016/j.biomaterials.2011.10.071] [PMID: 22098779]

[105] G.Y. Chiou, J.Y. Cherng, H.S. Hsu, M.L. Wang, C.M. Tsai, K.H. Lu, Y. Chien, S.C. Hung, Y.W. Chen, C.I. Wong, L.M. Tseng, P.I. Huang, C.C. Yu, W.H. Hsu, and S.H. Chiou, "Cationic polyurethanes-short branch PEI-mediated delivery of Mir145 inhibited epithelial–mesenchymal transdifferentiation and cancer stem-like properties and in lung adenocarcinoma", *J. Control. Release,* vol. 159, no. 2, pp. 240-250, 2012.
       [http://dx.doi.org/10.1016/j.jconrel.2012.01.014] [PMID: 22285547]

[106] F.B. Cui, Q. Liu, R.T. Li, J. Shen, P.Y. Wu, L.X. Yu, W.J. Hu, F.L. Wu, C.P. Jiang, G.F. Yue, X.P. Qian, X.Q. Jiang, and B.R. Liu, "Enhancement of radiotherapy efficacy by miR-200c-loaded gelatinase-stimuli PEG-Pep-PCL nanoparticles in gastric cancer cells", *Int. J. Nanomedicine,* vol. 9, pp. 2345-2358, 2014.
       [PMID: 24872697]

[107] L. Piao, M. Zhang, J. Datta, X. Xie, T. Su, H. Li, T.N. Teknos, and Q. Pan, "Lipid-based nanoparticle delivery of Pre-miR-107 inhibits the tumorigenicity of head and neck squamous cell carcinoma", *Mol. Ther.,* vol. 20, no. 6, pp. 1261-1269, 2012.
       [http://dx.doi.org/10.1038/mt.2012.67] [PMID: 22491216]

[108] S. Shi, L. Han, T. Gong, Z. Zhang, and X. Sun, Systemic delivery of microRNA-34a for cancer stem cell therapy.*Angew. Chem. In. Ed.* vol. 52. wiley., 2013, pp. 3901-3905.

[109] W. Alshaer, H. Zureigat, A. Al Karaki, A. Al-Kadash, L. Gharaibeh, M.M. Hatmal, A.A.A. Aljabali, and A. Awidi, "siRNA: Mechanism of action, challenges, and therapeutic approaches", *Eur. J. Pharmacol.,* vol. 905, p. 174178, 2021.
[http://dx.doi.org/10.1016/j.ejphar.2021.174178] [PMID: 34044011]

[110] T. Andey, S. Marepally, A. Patel, T. Jackson, S. Sarkar, M. O'Connell, R.C. Reddy, S. Chellappan, P. Singh, and M. Singh, "Cationic lipid guided short-hairpin RNA interference of annexin A2 attenuates tumor growth and metastasis in a mouse lung cancer stem cell model", *J. Control. Release,* vol. 184, pp. 67-78, 2014.
[http://dx.doi.org/10.1016/j.jconrel.2014.03.049] [PMID: 24727000]

[111] W.L. Lo, Y. Chien, G.Y. Chiou, L.M. Tseng, H.S. Hsu, Y.L. Chang, K.H. Lu, C.S. Chien, M.L. Wang, Y.W. Chen, P.I. Huang, F.W. Hu, C.C. Yu, P.Y. Chu, and S.H. Chiou, "Nuclear localization signal-enhanced RNA interference of EZH2 and Oct4 in the eradication of head and neck squamous Cell carcinoma-derived cancer stem cells", *Biomaterials,* vol. 33, no. 14, pp. 3693-3709, 2012.
[http://dx.doi.org/10.1016/j.biomaterials.2012.01.016] [PMID: 22361100]

[112] H. Gul-Uludağ, J. Valencia-Serna, C. Kucharski, L.A. Marquez-Curtis, X. Jiang, L. Larratt, A. Janowska-Wieczorek, and H. Uludağ, "Polymeric nanoparticle-mediated silencing of CD44 receptor in CD34+ acute myeloid leukemia cells", *Leuk. Res.,* vol. 38, no. 11, pp. 1299-1308, 2014.
[http://dx.doi.org/10.1016/j.leukres.2014.08.008] [PMID: 25262448]

[113] C.F. Xu, Y. Liu, S. Shen, Y.H. Zhu, and J. Wang, "Targeting glucose uptake with siRNA-based nanomedicine for cancer therapy", *Biomaterials,* vol. 51, pp. 1-11, 2015.
[http://dx.doi.org/10.1016/j.biomaterials.2015.01.068] [PMID: 25770992]

[114] C. Liu, G. Zhao, J. Liu, N. Ma, P. Chivukula, L. Perelman, K. Okada, Z. Chen, D. Gough, and L. Yu, "Novel biodegradable lipid nano complex for siRNA delivery significantly improving the chemosensitivity of human colon cancer stem cells to paclitaxel", *J. Control. Release,* vol. 140, no. 3, pp. 277-283, 2009.
[http://dx.doi.org/10.1016/j.jconrel.2009.08.013] [PMID: 19699770]

[115] Y. Zhang, H. Zhang, X. Wang, J. Wang, X. Zhang, and Q. Zhang, "The eradication of breast cancer and cancer stem cells using octreotide modified paclitaxel active targeting micelles and salinomycin passive targeting micelles", *Biomaterials,* vol. 33, no. 2, pp. 679-691, 2012.
[http://dx.doi.org/10.1016/j.biomaterials.2011.09.072] [PMID: 22019123]

[116] Y. Zhou, J. Yang, J.S. Rhim, and J. Kopeček, "HPMA copolymer-based combination therapy toxic to both prostate cancer stem/progenitor cells and differentiated cells induces durable anti-tumor effects", *J. Control. Release,* vol. 172, no. 3, pp. 946-953, 2013.
[http://dx.doi.org/10.1016/j.jconrel.2013.09.005] [PMID: 24041709]

[117] X.Y. Ke, V.W. Lin Ng, S.J. Gao, Y.W. Tong, J.L. Hedrick, and Y.Y. Yang, "Co-delivery of thioridazine and doxorubicin using polymeric micelles for targeting both cancer cells and cancer stem cells", *Biomaterials,* vol. 35, no. 3, pp. 1096-1108, 2014.
[http://dx.doi.org/10.1016/j.biomaterials.2013.10.049] [PMID: 24183698]

[118] D. Wang, J. Huang, X. Wang, Y. Yu, H. Zhang, Y. Chen, J. Liu, Z. Sun, H. Zou, D. Sun, G. Zhou, G. Zhang, Y. Lu, and Y. Zhong, "The eradication of breast cancer cells and stem cells by 8-hydroxyquinoline-loaded hyaluronan modified mesoporous silica nanoparticle-supported lipid bilayers containing docetaxel", *Biomaterials,* vol. 34, no. 31, pp. 7662-7673, 2013.
[http://dx.doi.org/10.1016/j.biomaterials.2013.06.042] [PMID: 23859657]

[119] Y. Liu, W.L. Lu, J. Guo, J. Du, T. Li, J.W. Wu, G.L. Wang, J.C. Wang, X. Zhang, and Q. Zhang, "A potential target associated with both cancer and cancer stem cells: A combination therapy for eradication of breast cancer using vinorelbine stealthy liposomes plus parthenolide stealthy liposomes", *J. Control. Release,* vol. 129, no. 1, pp. 18-25, 2008.
[http://dx.doi.org/10.1016/j.jconrel.2008.03.022] [PMID: 18466993]

[120] S.Y. Li, R. Sun, H.X. Wang, S. Shen, Y. Liu, X.J. Du, Y.H. Zhu, and W. Jun, "Combination therapy with epigenetic-targeted and chemotherapeutic drugs delivered by nanoparticles to enhance the chemotherapy response and overcome resistance by breast cancer stem cells", *J. Control. Release,* vol. 205, pp. 7-14, 2015.
[http://dx.doi.org/10.1016/j.jconrel.2014.11.011] [PMID: 25445694]

[121] R. Sun, Y. Liu, S.Y. Li, S. Shen, X.J. Du, C.F. Xu, Z.T. Cao, Y. Bao, Y.H. Zhu, Y.P. Li, X.Z. Yang, and J. Wang, "Co-delivery of all-trans-retinoic acid and doxorubicin for cancer therapy with synergistic inhibition of cancer stem cells", *Biomaterials,* vol. 37, pp. 405-414, 2015.
[http://dx.doi.org/10.1016/j.biomaterials.2014.10.018] [PMID: 25453968]

[122] Q. Liu, R.T. Li, H.Q. Qian, J. Wei, L. Xie, J. Shen, M. Yang, X.P. Qian, L.X. Yu, X.Q. Jiang, and B.R. Liu, "Targeted delivery of miR-200c/DOC to inhibit cancer stem cells and cancer cells by the gelatinases-stimuli nanoparticles", *Biomaterials,* vol. 34, no. 29, pp. 7191-7203, 2013.
[http://dx.doi.org/10.1016/j.biomaterials.2013.06.004] [PMID: 23806972]

[123] S.S. Kim, A. Rait, E. Kim, K.F. Pirollo, M. Nishida, N. Farkas, J.A. Dagata, and E.H. Chang, "A nanoparticle carrying the p53 gene targets tumors including cancer stem cells, sensitizes glioblastoma to chemotherapy and improves survival", *ACS Nano,* vol. 8, no. 6, pp. 5494-5514, 2014.
[http://dx.doi.org/10.1021/nn5014484] [PMID: 24811110]

[124] C. Yang, F. Xiong, J. Wang, J. Dou, J. Chen, D. Chen, Y. Zhang, S. Luo, and N. Gu, "Anti-ABCG2 monoclonal antibody in combination with paclitaxel nanoparticles against cancer stem-like cell activity in multiple myeloma", *Nanomedicine,* vol. 9, no. 1, pp. 45-60, 2014.
[http://dx.doi.org/10.2217/nnm.12.216] [PMID: 23534833]

[125] S.K. Swaminathan, E. Roger, U. Toti, L. Niu, J.R. Ohlfest, and J. Panyam, "CD133-targeted paclitaxel delivery inhibits local tumor recurrence in a mouse model of breast cancer", *J. Control. Release,* vol. 171, no. 3, pp. 280-287, 2013.
[http://dx.doi.org/10.1016/j.jconrel.2013.07.014] [PMID: 23871962]

[126] L. Wang, W. Su, Z. Liu, M. Zhou, S. Chen, Y. Chen, D. Lu, Y. Liu, Y. Fan, Y. Zheng, Z. Han, D. Kong, J.C. Wu, R. Xiang, and Z. Li, "CD44 antibody-targeted liposomal nanoparticles for molecular imaging and therapy of hepatocellular carcinoma", *Biomaterials,* vol. 33, no. 20, pp. 5107-5114, 2012.
[http://dx.doi.org/10.1016/j.biomaterials.2012.03.067] [PMID: 22494888]

[127] N. Tabassum, V. Verma, M. Kumar, A. Kumar, and B. Singh, "Nanomedicine in cancer stem cell therapy: From fringe to forefront", *Cell Tissue Res.,* vol. 374, no. 3, pp. 427-438, 2018.
[http://dx.doi.org/10.1007/s00441-018-2928-5] [PMID: 30302547]

[128] K.M. McNeeley, A. Annapragada, and R.V. Bellamkonda, "Decreased circulation time offsets increased efficacy of PEGylated nanocarriers targeting folate receptors of glioma", *Nanotechnology,* vol. 18, no. 38, p. 385101, 2007.
[http://dx.doi.org/10.1088/0957-4484/18/38/385101]

[129] P. Xia, "Surface markers of cancer stem cells in solid tumors", *Curr. Stem Cell Res. Ther.,* vol. 9, no. 2, pp. 102-111, 2014.
[http://dx.doi.org/10.2174/1574888X09666131217003709] [PMID: 24359139]

[130] S. Mishra, P. Webster, and M.E. Davis, "PEGylation significantly affects cellular uptake and intracellular trafficking of non-viral gene delivery particles", *Eur. J. Cell Biol.,* vol. 83, no. 3, pp. 97-111, 2004.
[http://dx.doi.org/10.1078/0171-9335-00363] [PMID: 15202568]

[131] A.A. Kale, and V.P. Torchilin, "Environment-responsive multifunctional liposomes", *Methods Mol. Biol.,* vol. 605, pp. 213-242, 2010.
[http://dx.doi.org/10.1007/978-1-60327-360-2_15] [PMID: 20072884]

[132] X.R. Song, Y. Zheng, Y. Zheng, G. He, L. Yang, Y.F. Luo, Z.Y. He, S.Z. Li, J.M. Li, S. Yu, X. Luo, S.X. Hou, and Y.Q. Wei, "Development of PLGA nanoparticles simultaneously loaded with vincristine and verapamil for treatment of hepatocellular carcinoma", *J. Pharm. Sci.,* vol. 99, no. 12,

pp. 4874-4879, 2010.
[http://dx.doi.org/10.1002/jps.22200] [PMID: 20821385]

[133] Y. Patil, T. Sadhukha, L. Ma, and J. Panyam, "Nanoparticle-mediated simultaneous and targeted delivery of paclitaxel and tariquidar overcomes tumor drug resistance", *J. Control. Release,* vol. 136, no. 1, pp. 21-29, 2009.
[http://dx.doi.org/10.1016/j.jconrel.2009.01.021] [PMID: 19331851]

[134] H.L. Wong, R. Bendayan, A.M. Rauth, and X.Y. Wu, "Simultaneous delivery of doxorubicin and GG918 (Elacridar) by new Polymer-Lipid hybrid nanoparticles (PLN) for enhanced treatment of multidrug-resistant breast cancer", *J. Control. Release,* vol. 116, no. 3, pp. 275-284, 2006.
[http://dx.doi.org/10.1016/j.jconrel.2006.09.007] [PMID: 17097178]

[135] K. Knop, R. Hoogenboom, D. Fischer, and U.S. Schubert, "Poly(ethylene glycol) in drug delivery: Pros and cons as well as potential alternatives", *Angew. Chem. Int. Ed.,* vol. 49, no. 36, pp. 6288-6308, 2010.
[http://dx.doi.org/10.1002/anie.200902672] [PMID: 20648499]

[136] R.K. Jain, "The next frontier of molecular medicine: Delivery of therapeutics", *Nat. Med.,* vol. 4, no. 6, pp. 655-657, 1998.
[http://dx.doi.org/10.1038/nm0698-655] [PMID: 9623964]

[137] A.J. Primeau, "The distribution of the anticancer drug doxorubicin in relation to blood vessels in solid tumors", *Clin. Cancer. Res.,* vol. 11, no. 24, pp. 8782-8788, 2005.
[http://dx.doi.org/10.1158/1078-0432.CCR-05-1664]

[138] L. Milane, Z. Duan, and M. Amiji, "Role of hypoxia and glycolysis in the development of multi-drug resistance in human tumor cells and the establishment of an orthotopic multi-drug resistant tumor model in nude mice using hypoxic pre-conditioning", *Cancer Cell Int.,* vol. 11, no. 1, p. 3, 2011.
[http://dx.doi.org/10.1186/1475-2867-11-3] [PMID: 21320311]

[139] S. Liu, Y. Cong, D. Wang, Y. Sun, L. Deng, Y. Liu, R. Martin-Trevino, L. Shang, S.P. McDermott, M.D. Landis, S. Hong, A. Adams, R. D'Angelo, C. Ginestier, E. Charafe-Jauffret, S.G. Clouthier, D. Birnbaum, S.T. Wong, M. Zhan, J.C. Chang, and M.S. Wicha, "Breast cancer stem cells transition between epithelial and mesenchymal states reflective of their normal counterparts", *Stem Cell Reports,* vol. 2, no. 1, pp. 78-91, 2014.
[http://dx.doi.org/10.1016/j.stemcr.2013.11.009] [PMID: 24511467]

[140] B.D. Curti, W.J. Urba, W.G. Alvord, J.E. Janik, J.W. Smith II, K. Madara, and D.L. Longo, "Interstitial pressure of subcutaneous nodules in melanoma and lymphoma patients: Changes during treatment", *Cancer Res.,* vol. 53, no. 10, pp. 2204-2207, 1993.
[PMID: 8485703]

[141] Y. Boucher, and R.K. Jain, "Microvascular pressure is the principal driving force for interstitial hypertension in solid tumors: Implications for vascular collapse", *Cancer Res.,* vol. 52, no. 18, pp. 5110-5114, 1992.
[PMID: 1516068]

[142] P.V. Dickson, J.B. Hamner, T.L. Sims, C.H. Fraga, C.Y.C. Ng, S. Rajasekeran, N.L. Hagedorn, M.B. McCarville, C.F. Stewart, and A.M. Davidoff, "Bevacizumab-induced transient remodeling of the vasculature in neuroblastoma xenografts results in improved delivery and efficacy of systemically administered chemotherapy", *Clin. Cancer Res.,* vol. 13, no. 13, pp. 3942-3950, 2007.
[http://dx.doi.org/10.1158/1078-0432.CCR-07-0278] [PMID: 17606728]

[143] F. Yuan, Y. Chen, M. Dellian, N. Safabakhsh, N. Ferrara, and R.K. Jain, "Time-dependent vascular regression and permeability changes in established human tumor xenografts induced by an anti-vascular endothelial growth factor/vascular permeability factor antibody", *Proc. Natl. Acad. Sci.,* vol. 93, no. 25, pp. 14765-14770, 1996.
[http://dx.doi.org/10.1073/pnas.93.25.14765] [PMID: 8962129]

[144] D.W. Miles, A. Chan, L.Y. Dirix, J. Cortés, X. Pivot, P. Tomczak, T. Delozier, J.H. Sohn, L. Provencher, F. Puglisi, N. Harbeck, G.G. Steger, A. Schneeweiss, A.M. Wardley, A. Chlistalla, and G. Romieu, "Phase III study of bevacizumab plus docetaxel compared with placebo plus docetaxel for the first-line treatment of human epidermal growth factor receptor 2-negative metastatic breast cancer", *J. Clin. Oncol.,* vol. 28, no. 20, pp. 3239-3247, 2010.
[http://dx.doi.org/10.1200/JCO.2008.21.6457] [PMID: 20498403]

[145] A.R. Kirtane, S.M. Kalscheuer, and J. Panyam, "Exploiting nanotechnology to overcome tumor drug resistance: Challenges and opportunities", *Adv. Drug Deliv. Rev.,* vol. 65, no. 13-14, pp. 1731-1747, 2013.
[http://dx.doi.org/10.1016/j.addr.2013.09.001] [PMID: 24036273]

[146] S.J. Kuhn, S.K. Finch, D.E. Hallahan, and T.D. Giorgio, "Proteolytic surface functionalization enhances *in vitro* magnetic nanoparticle mobility through extracellular matrix", *Nano Lett.,* vol. 6, no. 2, pp. 306-312, 2006.
[http://dx.doi.org/10.1021/nl052241g] [PMID: 16464055]

[147] A. Jimeno, G.J. Weiss, W.H. Miller Jr, S. Gettinger, B.J.C. Eigl, A.L.S. Chang, J. Dunbar, S. Devens, K. Faia, G. Skliris, J. Kutok, K.D. Lewis, R. Tibes, W.H. Sharfman, R.W. Ross, and C.M. Rudin, "Phase I study of the Hedgehog pathway inhibitor IPI-926 in adult patients with solid tumors", *Clin. Cancer Res.,* vol. 19, no. 10, pp. 2766-2774, 2013.
[http://dx.doi.org/10.1158/1078-0432.CCR-12-3654] [PMID: 23575478]

[148] B. Diop-Frimpong, V.P. Chauhan, S. Krane, Y. Boucher, and R.K. Jain, "Losartan inhibits collagen I synthesis and improves the distribution and efficacy of nanotherapeutics in tumors", *Proc. Natl. Acad. Sci.,* vol. 108, no. 7, pp. 2909-2914, 2011.
[http://dx.doi.org/10.1073/pnas.1018892108] [PMID: 21282607]

[149] H. Meng, W.X. Mai, H. Zhang, M. Xue, T. Xia, S. Lin, X. Wang, Y. Zhao, Z. Ji, J.I. Zink, and A.E. Nel, "Codelivery of an optimal drug/siRNA combination using mesoporous silica nanoparticles to overcome drug resistance in breast cancer *in vitro* and *in vivo*", *ACS Nano,* vol. 7, no. 2, pp. 994-1005, 2013.
[http://dx.doi.org/10.1021/nn3044066] [PMID: 23289892]

[150] H. Korkaya, S. Liu, and M.S. Wicha, "Breast cancer stem cells, cytokine networks, and the tumor microenvironment", *J. Clin. Invest.,* vol. 121, no. 10, pp. 3804-3809, 2011.
[http://dx.doi.org/10.1172/JCI57099] [PMID: 21965337]

[151] C.D. Phung, H.T. Nguyen, T.H. Tran, H.G. Choi, C.S. Yong, and J.O. Kim, "Rational combination immunotherapeutic approaches for effective cancer treatment", *J. Control. Release,* vol. 294, pp. 114-130, 2019.
[http://dx.doi.org/10.1016/j.jconrel.2018.12.020] [PMID: 30553850]

[152] Z. Xu, S. Ramishetti, Y.C. Tseng, S. Guo, Y. Wang, and L. Huang, "Multifunctional nanoparticles co-delivering Trp2 peptide and CpG adjuvant induce potent cytotoxic T-lymphocyte response against melanoma and its lung metastasis", *J. Control. Release,* vol. 172, no. 1, pp. 259-265, 2013.
[http://dx.doi.org/10.1016/j.jconrel.2013.08.021] [PMID: 24004885]

[153] Z. Xu, Y. Wang, L. Zhang, and L. Huang, "Nanoparticle-delivered transforming growth factor-β siRNA enhances vaccination against advanced melanoma by modifying tumor microenvironment", *ACS Nano,* vol. 8, no. 4, pp. 3636-3645, 2014.
[http://dx.doi.org/10.1021/nn500216y] [PMID: 24580381]

[154] M.T. Stephan, J.J. Moon, S.H. Um, A. Bershteyn, and D.J. Irvine, "Therapeutic cell engineering with surface-conjugated synthetic nanoparticles", *Nat. Med.,* vol. 16, no. 9, pp. 1035-1041, 2010.
[http://dx.doi.org/10.1038/nm.2198] [PMID: 20711198]

[155] Y. Lu, Y. Wang, L. Miao, M. Haynes, G. Xiang, and L. Huang, "Exploiting in situ antigen generation and immune modulation to enhance chemotherapy response in advanced melanoma: A combination nanomedicine approach", *Cancer Lett.,* vol. 379, no. 1, pp. 32-38, 2016.
[http://dx.doi.org/10.1016/j.canlet.2016.05.025] [PMID: 27235608]

CHAPTER 9

Novel Nanotechnological Therapy Approaches to Glioblastoma

Bakiye Goker Bagca[1] and **Cigir Biray Avci**[2,*]

[1] *Department of Medical Biology, Aydin Adnan Menderes University, Aydin, Turkey*

[2] *Department of Medical Biology, Ege University, Izmir, Turkey*

Abstract: Glioblastoma is one of the most aggressive and deadly types of cancer. The blood-brain barrier is the biggest obstacle to overcome in glioblastoma treatment. Nanomedicine, which describes the use of nanostructures in medicine, has significant potential for glioblastoma. Nanomedicine provides advantages in crossing the blood-brain barrier, increasing the amount and effectiveness of drugs reaching the cancer site, monitoring diagnosis and treatment through imaging agents, and increasing the effectiveness of treatments in combination applications. This chapter reviews current nanotechnology research in glioblastoma over the past few years.

Keywords: Blood-brain barrier, Cancer, Chemotherapy, Cubosome, Dendrimers, Glioblastoma, *In vitro*, *In vivo*, Lipid nanoparticle, Liposomes, Micelles, Nanodiamonds, Nanomedicine, Nanoparticles, Nanostructures, Nanotubes, PAMAM, Polymersomes, Quantum dots, Radiotherapy.

INTRODUCTION

The grade IV cancer of the brain, glioblastoma, is the most aggressive and lethal subtype of the primary brain tumor. It is responsible for approximately half of all malignant central nervous system cancers. Its mean incidence varies between 3.19 and 4.17 per person per 100,000 a year. With a median survival of 15 months and a 5-year survival of approximately 5%, glioblastoma has a poor prognosis [1].

The fifth edition of the World Health Organization Classification of Tumors of the Central Nervous System (CNS5) defines glioblastoma under The "Gliomas, glioneuronal tumors, and neuronal tumors, Adult-type diffuse gliomas" group. According to the classification, glioblastoma is genetically defined as IDH-wildtype and occurs *de novo* at approximately 60 years of age [2].

* **Corresponding author Cigir Biray Avci:** Department of Medical Biology, Ege University, Izmir, Turkey; Tel: +90 (232) 390 22 62; E-mail: cigir.biray@ege.edu.tr

Habibe Yılmaz (Ed.)

Various risk factors have been identified, including tobacco, smoking, nitrosamines, inflammation, ionizing radiation, electromagnetic radiation, obesity, metal ions, nutritional factors, chemical exposure, and genetic factors [1].

Glioblastoma treatment includes surgical resection, radiotherapy, and chemotherapy. An alkylating agent, temozolomide, is the major chemotherapeutic in glioblastoma treatment. This lipophilic molecule crosses the blood-brain barrier (BBB) and causes irreversible mutations that trigger cancer cell apoptosis. In addition to temozolomide, FDA-approved glioblastoma chemotherapy contains lomustine, carmustine, and bevacizumab agents [3].

However, the BBB is the primary hurdle to chemotherapy. It limits chemotherapeutic options to only BBB-crossing agents and forces dose increase. Increased chemotherapy doses cause systemic toxicity and drug resistance in cancer cells over time. Therefore, enhancing the quantity and effectiveness of the drug reaching the cancer site by overcoming BBB constitutes one of the main focus points in glioblastoma research [4].

Nanotechnology utilization has enormous power to break down the therapy limitations originating from the BBB.

CURRENT NANOMEDICINE PLATFORMS RESEARCHED IN GLIOBLASTOMA

Nanotechnology utilizes the surface, electrical, magnetic, and optical properties of nanosized materials (<100 nm) to produce designable, controllable, and smarter nanodevices. Nanomedicine defines the development of new-generation diagnostic, imaging, and treatment approaches using nanotechnology tools [5]. The widespread nanoparticles in nanomedicine originate from inert and low-toxicity molecules such as gold, silica, iron, and zinc. Biocompatible molecules such as chitosan are also among the potential nanomaterials in glioblastoma. Nanoparticle shapes are variable, such as cubes, spheres, plates, stars, and tubes. They are powerful drugs delivered through encapsulation or conjugation mechanisms. Encapsulation or conjugation allows drugs to pass the BBB or to increase the amount and effectiveness of agents reaching glioblastoma tissue. Likewise, imaging agent transport enables online monitoring of diagnosis and treatment. It also empowers the genetic regulation of glioblastoma by nucleic acid delivery. The ability of nanoparticles to directly target glioblastoma cells or tumor tissue through surface modifications minimizes toxicity in healthy cells [6 - 12]. In addition to nanoparticles, a wide variety of nanostructures are evaluated in glioblastoma according to usage purposes Fig. (1).

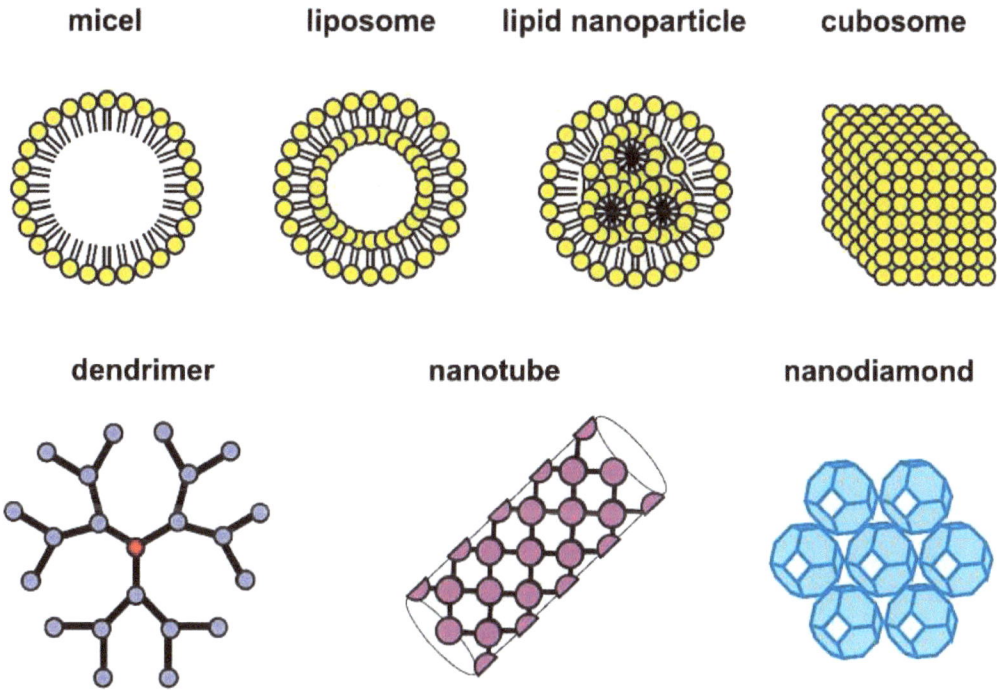

Fig. (1). Examples of nanostructures commonly used in nanomedicine.

Micelles

Micelles are single-layered vesicles composed mainly of amphiphilic molecules, especially proteins or lipids. These molecules form compartments by clustering in the aqueous medium such that their hydrophilic surfaces remain outside and their hydrophobic surfaces remain in the lumen. The vesicular structure allows the safe delivery of high-dose chemotherapeutic agents that may cause systemic toxicity in routine practice, efficient delivery of hydrophobic drugs or nucleic acids, targeting glioblastoma cells directly through surface modifications, crossing the BBB, and simultaneous regulation of multiple signaling pathways by dual drug loading. Micelles with conventional applications can increase the effectiveness of chemoradiotherapy and immunotherapy in glioblastoma (Table **1**). They can also increase photothermal, photodynamic, and sonodynamic therapy effectiveness [13, 29].

Table 1. Current studies evaluating the potential of micelles in glioblastoma.

Nanoparticle	Design	Models	Drug	Effect	Refs.
p(MAG-co-HEMA)-b-PBAE micelles	*in vitro*	U87MG and HUVEC cell lines	Doxorubicin	Selective cytotoxic effect on glioblastoma cells	[13]
pFTMC$_{16}$-b-pDMAEMA$_{131}$ micelles	*in vitro*	U87MG cell line	Nucleic acid	Effective gene delivery	[14]
CH-K$_5$(s-s)R$_8$-An micelles	*in vitro* and *in vivo*	U251 cell line and mice	Radiosensitizer and doxorubicin	Cytotoxic effect on glioblastoma cells and increased sensitivity of the tumor to chemotherapy and radiotherapy	[15]
IDP-2Yx2A micelles	*in vivo*	mice	SN38	Increased drug concentration without harming the organism	[16]
PEG-PBPA micelles	*in vitro*	U87MG cell line	Curcumin and sorafenib tosylate	Enhanced anticancer effect	[17]
siRNA-SS-PNIPAM micelles	*in vivo*	mice	siRNA (STAT3) and temozolomide	Both drug delivery and RNA interference	[18]
LDLR-Specific-PEG PTMC-PCL-mefenamate micelles	*in vitro* and *in vivo*	U87MG cell line and mice	Sorafenib	Enhanced BBB penetration by cell-specific targeting	[19]
Nanomicelle	*in vitro*	U87MG cell line	Curcumin and erlotinib	Enhanced anticancer effect	[20]
SiO$_2$-coated polymeric micelles	*in vitro*	Patient derived cells	Pitavastatin	Enhanced anticancer effect by cell-specific targeting	[21]
Rabies virus-derived polypeptide -modified nanomicelles	*in vitro* and *in vivo*	C6 cell line and mice	Doxorubicin	Enhanced anticancer effect by tissue-specific targeting	[22]
CD133 aptamer coated-PS-b-PEO and PLGA micelles	*in vitro*	Glioblastoma stem cells	Temozolomide and RG7388	Enhanced anticancer effect by cancer stem cell-specific targeting	[23]
Miktoarm Star Polymers	*in vitro*	U251N cell line	Curcumin	Enhanced ROS-scavenging ability	[24]

(Table 1) cont.....

Nanoparticle	Design	Models	Drug	Effect	Refs.
PEG- poly(amino acid)	*in vitro* and *in vivo*	U251, U373, U87, CT2A, GL261 cell lines, and mice	Desacetylvinblastine hydrazide	Extended survival by pH- triggered drug release	[25]
TfR-Targeted PEG-PLA micelles	*in vitro* and *in vivo*	U87MG cell line and mice	Paclitaxel	Cytotoxic effect and extended survival by brain cells transferrin receptor targeting	[26]
Dodecylamine -- -phosphotyrosine conjugated micelles	*in vitro*	T98G cell line	-	Cytotoxic effect by increasing cancer cell membrane tension	[27]
ROS-Responsive-mPEG-TK micelles	*in vitro*	C6, U251 cell lines	Melphalan	Enhanced anticancer effect by ROS-responsive drug release	[28]
Folate modified mPEG and PLA micelles	*in vitro* and *in vivo*	GL261 cell line and mice	Curcumin	Enhanced anticancer effect	[29]

p(MAG-co-HEMA)-b-PBAE,poly(2-deoxy-2-methacrylamido-d-glucose-co-2-hydroxyethyl methacrylate)---poly(β-amino ester); pFTMC$_{16}$-b-pDMAEMA$_{131}$, poly(fluorenetrimethylenecarbonate)-b-poly(2-(dimethyl-amino)ethyl methacrylate); CH, cholesterol; ch-K$_5$(s-s)R$_8$-An, cholesterol- Angiopep-2; IDP, Intrinsically Disordered Protein; PEG, poly(ethylene glycol); PBPA, poly(l-boronophenylalanine); SS, disulfide; PNIPAM, poly(N-isopropylacrylamide); LDLR, low density lipoprotein receptor; PTMC-PCL, poly(ε-caprolactone-co-dithiolane trimethylene carbonate); SiO$_2$, silica; PS-b-PEO, poly(styrene-b-ethylene oxide); PLGA, poly(lactic-co-glycolic) acid; ROS, Reactive Oxygen Species; TfR, Transferrin Receptor; PLA, Polylactic acid; mPEG, Methoxy PEG; TK, thioketal.

Lipids in Nanoscale

Lipids are popular tools in nanomedicine with their ability to form vesicles in aqueous environments due to their amphiphilic characteristics. They can form bilayer liposomes in aqueous media. In addition, they form various nanostructures, which include nucleic acid-carrying or peptide-carrying lipid nanoparticles (lipoplexes), cationic lipid nanoparticles, solid lipid nanoparticles, nanostructured lipid particles, cubosomes, and ethosomes. Lipid-based nanostructures have been in the field of nanomedicine for many years, including cancer, due to their advantages such as drug encapsulation efficacy, targeting cells or tissue by allowing surface modifications, stimuli-responsive controlled release (pH, temperature, ultrasound, light, magnetic field, laser irritation). They can increase the anti-cancer activity of ultrasound by applying it in combination with sonosensitive formulations and microbubbles. Lipid-based nanostructures are also accomplished tools in overcoming BBB, which is the main challenge in glioblastoma treatment [30–59] (Table **2**).

Table 2. Current studies evaluating the potential of lipid-based nanoparticles in glioblastoma.

Nanoparticle	Design	Models	Drug	Effect	Refs.
RGDK-lipopeptide liposome	*in vivo*	mice	WP1066 and STAT3siRNA	Enhanced anticancer effect by targeting integrin receptor and inhibiting JAK/STAT pathway	[32]
PEGylated liposome	*in vitro and in vivo*	C6 cell line, rats	Doxorubicin and Carboplatin	Enhanced anticancer effect	[33]
PEGylated liposome	*in vitro*	GL261 cell line	-	Enhanced cellular uptake by pH-responsive release	[34]
CH-SM liposome	*in vitro and in vivo*	Gli36ΔEGFR2 and U87MG cell lines, healthy mice	Givinostat	Enhanced anticancer effect, and improved cellular uptake	[35]
ApoE-functionalized-ARTPC liposome	*in vitro and in vivo*	Drug resistant U251 cell line, mice	Temozolomide	Enhanced anticancer effect and reduced systemic toxicity by targeting LDLRs receptors	[36]
Glucose-CH and Biotin-CH liposome	*in vitro and in vivo*	C6 and bEnd.3 cell lines, mice	-	Dual cell specific targeting both glucose and biotin molecules	[37]
Rabies virus glycoprotein-lactoferrin-liposome	*in vitro*	U87MG cell line and hBCSCs	AZD5582 and SM164	Tissue-specific targeting and dual effect mechanism	[38]
Magnetic temperature-sensitive liposome	*in vitro*	U87MG and U251N cell lines	Temozolomide	Controlled drug release	[39]
PC-CH liposome	*in vitro and ex vivo*	U87, NIH/3T3, and A2780 cell lines, Rabbit nasal mucosa	Lomustine and n-Propyl Gallate	Dual effect mechanism	[40]
DPPC-CH liposome	*in vitro and in vivo*	C6 and bEnd.3 cell lines, rats	Temozolomide	Enhanced BBB-crossing by ultrasound	[41]

(Table 2) cont.....

Nanoparticle	Design	Models	Drug	Effect	Refs.
PR_b-functionalized PEGylated liposome	*in vitro*	patient-derived glioblastoma stem cells	miR-603 and polyethylenimine	Cell-specific targeting by PR_b, a fibronectin-mimetic peptide	[42]
ApoE- peptide modificated liposome	*in vitro* and *in vivo*	Patient-derived GSCs, hCMEC/D3 cell line, mice	Doxorubicin	Enhanced BBB-crossing by LDLR receptor targeting and combinational treatment with radiation	[43]
DSPC-DDAB-CH-DSPE-PEG liposome	*in vitro* and *in vivo*	U87MG cell line, mice	IR-780	Enhanced anticancer activity by delivering a photosensitizer agent	[44]
DOPC/CH/DODA-GLY-PEG$_{2000}$ liposome	*in vitro*	EA.hy926 and U87MG cell lines	Fisetin and Cisplatin	Enhanced anticancer activity by combinational drug delivery	[45]
PC-PS liposome	*in vitro*	U87, GL261, and bEnd.3 cell lines	Chalcone 1	Overcome low water-solubility of drug	[46]
Tf- Paclitaxel nanoparticles	*in vitro* and *in vivo*	U87MG cell line, mice	Paclitaxel and miltefosine	Enhanced anticancer activity by combinational drug delivery, tissue specific targeting and BBB-crossing ability	[47]
Tumor penetrating terpolymer-lipid nanoparticle	*in vitro* and *in vivo*	U87MG cell line, mice	Doxorubicin	Enhanced drug delivery by tissue specific targeting	[48]
Cyclic RGD peptide-lipid nanoparticle	*in vitro* and *in vivo*	U87MG cell line, rats	Paclitaxel and naringenin	Enhanced drug delivery by tissue specific targeting and enhanced chemoprotective effect	[49]
sgPLK1-lipid nanoparticle	*in vitro* and *in vivo*	GBM005 cells and mice	sgPLK1	Enhanced anticancer activity and survival	[50]
DODAP lipid nanoparticle	*in vitro*	U251, LN229 and 42MGBA cell lines	siSAT1	Cytotoxic effect by SAT1 knockdown	[51]

(Table 2) cont.....

Nanoparticle	Design	Models	Drug	Effect	Refs.
Cationic lipid nanoparticle with ionizable amine headgroups	*in vivo*	mice	siRNA against CD47 and PD-L1	Enhanced immunotherapy efficacy	[52]
Cell-penetrating peptide –cationic peptide cationic lipoplex	*in vitro, in vivo, and ex vivo*	U87MG, bEnd.3 and Calu-3 cell lines, mice and rats	c-Myc siRNA	Enhanced glial internalization and targeted gene silencing	[53]
Arginine-tocopherol bioconjugated lipoplex	*in vitro*	U87MG cell line	TRAIL plasmid	Cytotoxic effect	[54]
Bevacizumab-modified nanostructured lipid	*in vitro* and *in vivo*	U87MG and A172 cell lines and rats	Docetaxel	Enhanced anticancer effect	[55]
Tf decorated nanostructured lipid	*in vitro* and *in vivo*	U87MG cell line and mice	Rapamycin	Enhanced anticancer effect by tissue specific targeting	[56]
Nanostructured lipid	*in vitro*	U87MG cell line	SN38	Enhanced cytotoxic effect	[57]
Monoolein cubosomes	*in vitro*	A172 and T98G cell lines	miR-7-5p and doxorubicin	Enhanced cytotoxic effect and impaired MDR	[58]
Glyceryl monooleate cubosome	*in vitro*	A172 and LN229 cell lines	AT101	Cytotoxic effect	[59]

PEG, poly(ethylene glycol); ARTPC, artesunate-phosphatidylcholine; LDLR, density lipoprotein receptor; hBCSC, human brain cancer stem cells; CH, cholesterol; SM, sphingomyelin; ApoE, Apolipoprotein E; PC, Phosphatidylcholine; DPPC, 1,2-Dipalmitoyl-sn-glycero-3-phosphocholine; GSCs, glioblastoma stem cells; DSPC, 1,2-Distearoyl-sn-glycero-3-phosphocholine; DDAB, didodecyldimethylammonium bromide; DSPE, 1,2-dipalmitoyl-sn-glycero-3-phosphoethanolamine sodium salt; DOPC, Dioleoylphosphatidylcholine; DODA-GLY, 2-dioctadecylcarbamoyl-methoxyacetylamino) acetic acid-(ω-methoxy); PS, phosphatidyl-serine; Tf, Transferrin; DODAP, 1,2-dioleoyl-3-dimethylammonium propane; sg, single guide RNA.

Polymersomes

Another group of artificial nanovesicles formed by utilizing the amphiphilic properties of molecules is polymersomes. They are produced using both hydrophilic and hydrophobic polymers such as poly (ethylene glycol) (PEG), poly (ethyl ethylene) (PEE), poly (butadiene) (PBD), poly (dimethylsiloxane) (PDMS), poly (2-methyl-2-oxazoline) (PMOXA), poly (lactic acid) (PLA), poly (ϵ-caprolactone) (PCL), poly (trimethylene carbonate) (PTMC), and poly (N-vinylpyrrolidone) (PVP). Although they can carry drugs, imaging agents, or genetic materials, and target-delivery by surface modifications similar to lipid-based nanostructures, they also have the advantage of being more stable and impermeable than lipids [60 - 65] (Table **3**).

Table 3. Current studies evaluating the potential of polymersomes in glioblastoma.

Nanoparticle	Design	Models	Drug	Effect	Refs.
An-conjugated pH-sensitive Au - PCL-PEOX polymersomes	*in vitro*	U87MG and G422 cell lines	Doxorubicin	Enhanced immunogenic cell death by targeting LRP1	[61, 62]
ApoE directed PEG-P(TMC-DTC) polymersomes	*in vitro* and *in vivo*	U87MG cell line and mice	Doxorubicin	Enhanced anticancer effect by tissue specific targeting	[63]
An-conjugated PEG-b-poly(L-tyrosine)-b-poly(L-aspartic acid)	*in vitro* and *in vivo*	U87MG cell line and mice	Volasertib	Enhanced anticancer effect by tissue specific targeting	[64]
PB-PEO polymersomes	*in vitro*	U87MG and hCMEC/D3 cell lines	Binimetinib	Cytotoxic effect and BBB-crossing	[65]

An, Angiopep-2; Au, gold; PCL, polycaprolactone; PEOX, poly(2-ethyl-2-oxazoline); ApoE, Apolipoprotein E; PEG, poly(ethylene glycol); P(TMC-DTC), poly(trimethylene carbonate- co-dithiolane trimethylene carbonate); PB-PEO, poly(butadiene-b-ethylene oxide).

Dendrimers

Dendrimers are macromolecules containing symmetrically branched chains around a core structure. Branching capacities around the nucleus are called generation, and this capacity enhances the utilization potential of the nanostructure. Although dendrimers can originate from different molecules such as polyesters, polyamines, phosphorus, and poly (L-lysine), polyamidoamines (PAMAM) dendrimers are the most preferred in nanomedicine because of their high generation capacity. Dendrimers are among the most preferred nanostructures in nanomedicine with terminal modifications and conjugation or encapsulation of drugs, imaging agents, genetic materials, or other nanostructures [66, 75] (Table **4**).

Table 4. Current studies evaluating the potential of dendrimers in glioblastoma.

Nanoparticle	Design	Models	Drug	Effect	Refs.
PAMAM dendrimer	*in vitro* and *in vivo*	GL261 cell line and mice	siRNA	Improved siRNA efficacy	[67]

(Table 4) cont.....

Nanoparticle	Design	Models	Drug	Effect	Refs.
GO- 6-armed dendrimer	*in vitro*	U87MG cell line	CPI444 and vatalanib	Enhanced anticancer activity by dual targeting	[68]
N¹-alkyl-tryptophan functionalized dendrimer	*in vitro*	U87MG, LN229, T98G cell lines	-	Cytotoxic and ros scavenging effect	[69]
Polycationic phosphorus dendrimer	*in vitro*	U87MG, JHH520, NCH644, and BTSC233 cell lines	Lyn-siRNA	Cell specific siRNA targeting	[70]
CXCR4-targeted PAMAM dendrimer	*in vitro*	U87MG cell line	-	Enhanced cellular uptake	[71]
PAMAM dendrimer	*in vitro*	U87MG cell line	Etoposide and protoporphyrin IX	Enhanced cellular uptake	[72]
PAMAM dendrimer	*in vitro*	SNB19 cell line	5-fluorouracil	Increased drug stability	[73]
PAMAM dendrimer	*in vitro*	GL261, F98, and U87 cell lines	Curcumin	Enhanced cytotoxic effect	[74]
B19 aptamer- conjugated PAMAM dendrimer	*in vitro*	U87MG derived stem cells	Paclitaxel and temozolomide	Enhanced cytotoxic and apoptotoic effects	[75]

PAMAM, polyamidoamines; GO, Graphene oxide; PEG, poly(ethylene glycol).

Quantum Dots

Quantum dots, one of the most current topics in nanotechnology, are semiconductor particles that carry electrons that rise to the upper energy level when excited by UV light. As the excited electron returns to its previous energy level, it radiates at different wavelengths depending on the material. This excitation and emission feature supply potential in nanomedicine as it allows high-resolution monitoring of the reactions in target cells. Quantum dots have the potential to guide surgery, monitor drugs, and increase the effectiveness of photodynamic therapy, in glioblastoma. Carbon, cadmium, selenium, titanium, neodymium, phosphorus, and graphene quantum dots are among the structures investigated in glioblastoma. However, the fact that they usually originate from toxic molecules limits their use. Therefore, studies have generally focused on evaluating their toxicity and biocompatibility on healthy cells [76 - 85] (Table **5**).

Nanodiamonds

Nanodiamonds, are nanosized carbon structures, that have potential in the field of nanomedicine with their advantages such as non-toxicity, biocompatible, stable and inert structures, inexpensive and abundant availability, and effortless surface modifications due to the nature of carbon. Since they can cross the BBB, they may also play a crucial role as a drug carrier in glioblastoma [86, 89] (Table **6**).

Table 5. Current studies evaluating the potential of quantum dots in glioblastoma.

Nanoparticle	Design	Models	Drug	Effect	Refs.
CMC biofunctionalized Ag-In-S quantum dot	*in vitro*	U87MG cell line	KLA peptide-conjugated	Enhanced apoptotic effect by proapoptotic KLA peptide	[83]
Graphene dots	*in vitro*	U87MG and E1518cell lines, C57BL/6 mice embryo primary cells	Doxorubicin loaded	Enhanced chemotherapy on glioblastoma cells	[80]
Tf conjugated-carbon nitride dots	*in vitro*	SJGBM2 cell line	Gemcitabine loaded	Enhanced drug delivery and tissue specific targeting	[84]
Carbon dots	*in vitro*	Glioblastoma stem cells, U87 MG, SJGBM2 cell lines	Chalcone conjugated	Enhanced cytotoxic effect	[85]

CMC, Carboxymethylcellulose; Tr, transferrin.

Table 6. Current studies evaluating the potential of nanodiamonds in glioblastoma.

Nanoparticle	Design	Models	Drug	Effect	Refs.
Carboxylic acid functionalized nanodiamond	*in vitro*	U87MG cell line	Cabazitaxel	Enhanced cellular uptake and drug release	[87]
Lectin conjugated fluorescent nanodiamond	*in vitro*	U87MG, PC12, and BV2 cell lines	-	Enhanced cellular uptake	[88]
Polyglycerol-conjugated nanodiamond	*in vitro* and *in vivo*	U87MG, U251, and GL261 cell lines, mice	Doxorubicin	Enhanced anticancer effect by suppressing STAT3/IL6 signaling	[89]

Nanotubes

Nanotubes are nano-sized tubes. Although they are generally carbon-based, they can also be produced from various polymers and nucleic acids. Similar to other nanostructures, they are promising drugs, imaging agents, or nucleic acid carriers to glioblastoma [90, 92] (Table **7**).

Current Clinical Studies and Approved Therapies

Nanostructures with different properties have been in medicine for many years. Today, approximately 30 nanomedicine have been approved by the FDA for varied conditions or diseases. About ten of these have been approved for cancer therapy. Liposomal formulations of chemotherapeutics predominate among approved nanomedicines [93]. Although there are phase I or phase II clinical studies in glioblastoma (Table **8**), there is no approved nanomedicine yet (www. clinicaltrials.gov; cited in November 2022).

Table 7. Current studies evaluating the potential of nanodiamonds in glioblastoma.

Nanoparticle	Design	Models	Drug	Effect	Refs.
PEG nanotubes	*in vitro*	C6, U87, and U251 cell lines	Doxorubicin	Enhanced cytotoxic activity	[91]
ssDNA nanotube	*in vitro* and *in vivo*	GL261 and C8D1A cell lines, mice	Doxorubicin	Enhanced survival and decreased systemic toxicity	[92]

PEG, poly(ethylene glycol); ss, single strand

Table 8. Clinical studies of nanomedicines in glioblastoma.

Study ID	Study Title	Status	Phase
NCT03603379	Doxorubicin-loaded Anti-EGFR-immunoliposomes (C225-IL--dox) in High-grade Gliomas	Completed	Phase 1
NCT00944801	Pegylated Liposomal Doxorubicine and Prolonged Temozolomide in Addition to Radiotherapy in Newly Diagnosed Glioblastoma	Completed	Phase 1 Phase 2
NCT04590664	Verteporfin (liposomal) for the Treatment of Recurrent High Grade EGFR-Mutated Glioblastoma	Recruiting	Phase 1 Phase 2
NCT04573140	A Study of RNA-lipid Particle (RNA-LP) Vaccines for Newly Diagnosed Pediatric High-Grade Gliomas (pHGG) and Adult Glioblastoma (GBM) (PNOC020)	Recruiting	Phase 1
NCT00734682	A Phase I Trial of Nanoliposomal CPT-11 (NL CPT-11) in Patients With Recurrent High-Grade Gliomas	Completed	Phase 1
NCT05460507	Safety & Efficacy/Tolerability of Rhenium-186 NanoLiposomes (186RNL) for Patients Who Received a Prior 186RNL Treatment	Not yet recruiting	Phase 1
NCT01906385	Maximum Tolerated Dose, Safety, and Efficacy of Rhenium Nanoliposomes in Recurrent Glioma (ReSPECT)	Recruiting	Phase 1 Phase 2
NCT04881032	AGuIX Nanoparticles With Radiotherapy Plus Concomitant Temozolomide in the Treatment of Newly Diagnosed Glioblastoma (NANO-GBM)	Recruiting	Phase 1 Phase 2
NCT03020017	NU-0129 in Treating Patients With Recurrent Glioblastoma or Gliosarcoma Undergoing Surgery (gold nanoparticle)	Completed	Early Phase 1

(Table 8) cont.....

Study ID	Study Title	Status	Phase
NCT03463265	ABI-009 (Nab-rapamycin) in Recurrent High Grade Glioma and Newly Diagnosed Glioblastoma (nanoparticle albumin-bound rapamycin)	Active, not recruiting	Phase 2
NCT02766699	A Study to Evaluate the Safety, Tolerability and Immunogenicity of EGFR(V)-EDV-Dox in Subjects With Recurrent Glioblastoma Multiforme (GBM) (CerebralEDV) (nanocell)	Unknown	Phase 1

CONCLUDING REMARKS

Nanomedicine approaches have promise in glioblastoma, among the most aggressive and deadly cancer types. As evaluated above through current studies, nanostructures offer advantages especially in overcoming the BBB. Nanomedicine can ensure that chemotherapeutics are delivered directly to the cancer site without losing their effectiveness by encapsulation or conjugation, and it can achieve cancer-specific targeting through surface modifications. Thus, while the efficacy of cancer treatment increases, systemic toxicity decreases. Choosing the right materials for the needs is critical when designing a nanomedicine tool. In this process, the biocompatibility of the materials, drug-carrying capacity, flexibility, surface charge, hydrophobic or hydrophilic character, and ability to cross the BBB are critical. Although there is no approved nanomedicine application for glioblastoma yet, it is clear that liposomal formulations that are approved for use in other cancers also have potential in glioblastoma. In particular, it can be suggested that CRISPR/Cas9-based nanomedicine approaches will be groundbreaking in glioblastoma treatment through unique molecular regulation mechanisms [94, 95]. Lastly, it seems possible that soon, nanorobots and nanoreactors produced by mimicking defense cells may play a role in both targeting glioblastoma cells and triggering specific signaling mechanisms in these cells [96, 97]. Thus, in addition to diagnosis, the glioblastoma treatment will also be improved.

REFERENCES

[1]   S. Grochans, A.M. Cybulska, D. Simińska, J. Korbecki, K. Kojder, D. Chlubek, and I. Baranowska-Bosiacka, "Epidemiology of glioblastoma multiforme–literature review", *Cancers,* vol. 14, no. 10, p. 2412, 2022.
[http://dx.doi.org/10.3390/cancers14102412] [PMID: 35626018]

[2]   D.N. Louis, A. Perry, P. Wesseling, D.J. Brat, I.A. Cree, D. Figarella-Branger, C. Hawkins, H.K. Ng, S.M. Pfister, G. Reifenberger, R. Soffietti, A. von Deimling, and D.W. Ellison, "The 2021 WHO classification of tumors of the central nervous system: A summary", *Neuro-oncol.,* vol. 23, no. 8, pp. 1231-1251, 2021.
[http://dx.doi.org/10.1093/neuonc/noab106] [PMID: 34185076]

[3]   J.P. Fisher, and D.C. Adamson, "Current FDA-approved therapies for high-grade malignant Gliomas", *Biomedicines,* vol. 9, no. 3, p. 324, 2021.
[http://dx.doi.org/10.3390/biomedicines9030324] [PMID: 33810154]

[4]   D.H. Upton, C. Ung, S.M. George, M. Tsoli, M. Kavallaris, and D.S. Ziegler, "Challenges and opportunities to penetrate the blood-brain barrier for brain cancer therapy", *Theranostics,* vol. 12, no. 10, pp. 4734-4752, 2022.
[http://dx.doi.org/10.7150/thno.69682] [PMID: 35832071]

[5]   S. Sim, and N. Wong, "Nanotechnology and its use in imaging and drug delivery (Review)", *Biomed. Rep.,* vol. 14, no. 5, p. 42, 2021.
[http://dx.doi.org/10.3892/br.2021.1418] [PMID: 33728048]

[6]   B. Gabold, "Transferrin-modified chitosan nanoparticles for targeted nose-to-brain delivery of proteins", *Drug Deliv. Transl. Res.,* vol. 13, no. 3, pp. 822-838, 2022.
[http://dx.doi.org/10.1007/s13346-022-01245-z] [PMID: 36207657]

[7]   O. Gal, *Antibody delivery into the brain by radiosensitizer nanoparticles for targeted glioblastoma therapy.,* vol. 3, no. 4, pp. 177-188, 2022.*J. nanotheranostics,* vol. 3, no. 4, pp. 177-188, 2022.
[http://dx.doi.org/10.3390/jnt3040012]

[8]   N.C. Allen, R. Chauhan, P.J. Bates, and M.G. O'Toole, "Optimization of tumor targeting gold nanoparticles for glioblastoma applications", *Nanomaterials,* vol. 12, no. 21, p. 3869, 2022.
[http://dx.doi.org/10.3390/nano12213869] [PMID: 36364644]

[9]   Y. Cao, L. Jin, S. Zhang, Z. Lv, N. Yin, H. Zhang, T. Zhang, Y. Wang, Y. Chen, X. Liu, and G. Zhao, "Blood-brain barrier permeable and multi-stimuli responsive nanoplatform for orthotopic glioma inhibition by synergistic enhanced chemo-/chemodynamic/photothermal/starvation therapy", *Eur. J. Pharm. Sci.,* vol. 180, p. 106319, 2023.
[http://dx.doi.org/10.1016/j.ejps.2022.106319] [PMID: 36328086]

[10]   B.A. da Silva, M. Nazarkovsky, H.I. Padilla-Chavarría, E.A.C. Mendivelso, H.L. Mello, C.S.C. Nogueira, R.S. Carvalho, M. Cremona, V. Zaitsev, Y. Xing, R.C. Bisaggio, L.A. Alves, and J. Kai, "Novel scintillating nanoparticles for potential application in photodynamic cancer therapy", *Pharmaceutics,* vol. 14, no. 11, p. 2258, 2022.
[http://dx.doi.org/10.3390/pharmaceutics14112258] [PMID: 36365077]

[11]   T. Hou, S. Sankar Sana, H. Li, X. Wang, Q. Wang, V.K.N. Boya, R. Vadde, R. Kumar, D.V. Kumbhakar, Z. Zhang, and N. Mamidi, "Development of plant protein derived Tri angular shaped nano zinc oxide particles with inherent antibacterial and neurotoxicity properties", *Pharmaceutics,* vol. 14, no. 10, p. 2155, 2022.
[http://dx.doi.org/10.3390/pharmaceutics14102155] [PMID: 36297590]

[12]   A.A.P. Mansur, S.M. Carvalho, L.C.A. Oliveira, E.M. Souza-Fagundes, Z.I.P. Lobato, M.F. Leite, and H.S. Mansur, "Bioengineered carboxymethylcellulose–peptide hybrid nanozyme cascade for targeted intracellular biocatalytic–magnetothermal therapy of brain cancer cells", *Pharmaceutics,* vol. 14, no. 10, p. 2223, 2022.
[http://dx.doi.org/10.3390/pharmaceutics14102223] [PMID: 36297660]

[13]   E.L. Sahkulubey Kahveci, M.U. Kahveci, A. Celebi, T. Avsar, and S. Derman, "Glycopolymer and poly(β-amino ester)-based amphiphilic block copolymer as a drug carrier", *Biomacromolecules,* vol. 23, no. 11, pp. 4896-4908, 2022.
[http://dx.doi.org/10.1021/acs.biomac.2c01076] [PMID: 36317475]

[14]   S.T.G. Street, J. Chrenek, R.L. Harniman, K. Letwin, J.M. Mantell, U. Borucu, S.M. Willerth, and I. Manners, "Length-controlled nanofiber micelleplexes as efficient nucleic acid delivery vehicles", *J. Am. Chem. Soc.,* vol. 144, no. 43, pp. 19799-19812, 2022.
[http://dx.doi.org/10.1021/jacs.2c06695] [PMID: 36260789]

[15]   S. Zhang, X. Jiao, M. Heger, S. Gao, M. He, N. Xu, J. Zhang, M. Zhang, Y. Yu, B. Ding, and X. Ding, "A tumor microenvironment-responsive micelle co-delivered radiosensitizer dbait and doxorubicin for the collaborative chemo-radiotherapy of glioblastoma", *Drug Deliv.,* vol. 29, no. 1, pp. 2658-2670, 2022.
[http://dx.doi.org/10.1080/10717544.2022.2108937] [PMID: 35975300]

[16] J.M. Gleason, S.H. Klass, P. Huang, T. Ozawa, R.A. Santos, M.M. Fogarty, D.R. Raleigh, M.S. Berger, and M.B. Francis, "Intrinsically disordered protein micelles as vehicles for convection-enhanced drug delivery to glioblastoma multiforme", *ACS Appl. Bio Mater.,* vol. 5, no. 8, pp. 3695-3702, 2022.
[http://dx.doi.org/10.1021/acsabm.2c00215] [PMID: 35857070]

[17] Q. Zhang, Y. Liu, Y. Fei, J. Xie, X. Zhao, Z. Zhong, and C. Deng, "Phenylboronic acid-functionalized copolypeptides: Facile synthesis and responsive dual anticancer drug release", *Biomacromolecules,* vol. 23, no. 7, pp. 2989-2998, 2022.
[http://dx.doi.org/10.1021/acs.biomac.2c00482] [PMID: 35758844]

[18] T. Jiang, Y. Qiao, W. Ruan, D. Zhang, Q. Yang, G. Wang, Q. Chen, F. Zhu, J. Yin, Y. Zou, R. Qian, M. Zheng, and B. Shi, "Cation–Free siRNA micelles as effective drug delivery platform and potent RNAi nanomedicines for glioblastoma therapy (Adv. Mater. 45/2021)", *Adv. Mater.,* vol. 33, no. 45, p. 2170357, 2021.
[http://dx.doi.org/10.1002/adma.202170357]

[19] J. Wei, Y. Xia, F. Meng, D. Ni, X. Qiu, and Z. Zhong, "Small, Smart, and LDLR-Specific micelles augment sorafenib therapy of glioblastoma", *Biomacromolecules,* vol. 22, no. 11, pp. 4814-4822, 2021.
[http://dx.doi.org/10.1021/acs.biomac.1c01103] [PMID: 34677048]

[20] A. Bagherian, B. Roudi, N. Masoudian, and H. Mirzaei, "Anti-glioblastoma effects of nanomicelle-curcumin plus erlotinib", *Food Funct.,* vol. 12, no. 21, pp. 10926-10937, 2021.
[http://dx.doi.org/10.1039/D1FO01611C] [PMID: 34647945]

[21] P.S. Chauhan, M. Kumarasamy, A.M. Carcaboso, A. Sosnik, and D. Danino, "Multifunctional silica-coated mixed polymeric micelles for integrin-targeted therapy of pediatric patient-derived glioblastoma", *Mater. Sci. Eng. C,* vol. 128, p. 112261, 2021.
[http://dx.doi.org/10.1016/j.msec.2021.112261] [PMID: 34474820]

[22] J. Xu, X. Yang, J. Ji, Y. Gao, N. Qiu, Y. Xi, A. Liu, and G. Zhai, "RVG-functionalized reduction sensitive micelles for the effective accumulation of doxorubicin in brain", *J. Nanobiotechnology,* vol. 19, no. 1, p. 251, 2021.
[http://dx.doi.org/10.1186/s12951-021-00997-z] [PMID: 34419071]

[23] S.B. Smiley, Y. Yun, P. Ayyagari, H.E. Shannon, K.E. Pollok, M.W. Vannier, S.K. Das, and M.C. Veronesi, "Development of CD133 targeting multi-drug polymer micellar nanoparticles for glioblastoma - *In Vitro* evaluation in glioblastoma stem cells", *Pharm. Res.,* vol. 38, no. 6, pp. 1067-1079, 2021.
[http://dx.doi.org/10.1007/s11095-021-03050-8] [PMID: 34100216]

[24] V. Lotocki, H. Yazdani, Q. Zhang, E.R. Gran, A. Nyrko, D. Maysinger, and A. Kakkar, "Miktoarm star polymers with environment–selective ros/gsh responsive locations: From modular synthesis to tuned drug release through micellar partial corona shedding and/or core disassembly", *Macromol. Biosci.,* vol. 21, no. 2, p. 2000305, 2021.
[http://dx.doi.org/10.1002/mabi.202000305] [PMID: 33620748]

[25] S. Quader, X. Liu, K. Toh, Y.L. Su, A.R. Maity, A. Tao, W.K.D. Paraiso, Y. Mochida, H. Kinoh, H. Cabral, and K. Kataoka, "Supramolecularly enabled pH- triggered drug action at tumor microenvironment potentiates nanomedicine efficacy against glioblastoma", *Biomaterials,* vol. 267, p. 120463, 2021.
[http://dx.doi.org/10.1016/j.biomaterials.2020.120463] [PMID: 33130321]

[26] P. Sun, Y. Xiao, Q. Di, W. Ma, X. Ma, Q. Wang, and W. Chen, "Transferrin receptor-targeted peg-pla polymeric micelles for chemotherapy against glioblastoma multiforme", *Int. J. Nanomedicine,* vol. 15, pp. 6673-6687, 2020.
[http://dx.doi.org/10.2147/IJN.S257459] [PMID: 32982226]

[27] J. Wang, W. Tan, G. Li, D. Wu, H. He, J. Xu, M. Yi, Y. Zhang, S.A. Aghvami, S. Fraden, and B. Xu,

"Enzymatic insertion of lipids increases membrane tension for inhibiting drug resistant cancer cells", *Chemistry,* vol. 26, no. 66, pp. 15116-15120, 2020.
[http://dx.doi.org/10.1002/chem.202002974] [PMID: 32579262]

[28]   N. Oddone, F. Boury, E. Garcion, A.M. Grabrucker, M.C. Martinez, F. Da Ros, A. Janaszewska, F. Forni, M.A. Vandelli, G. Tosi, B. Ruozi, and J.T. Duskey, "Synthesis, characterization, and *in vitro* studies of an reactive oxygen species (ROS)-Responsive methoxy polyethylene glycol-thioketa--melphalan prodrug for glioblastoma treatment", *Front. Pharmacol.,* vol. 11, p. 574, 2020.
[http://dx.doi.org/10.3389/fphar.2020.00574] [PMID: 32425795]

[29]   Y. He, C. Wu, J. Duan, J. Miao, H. Ren, and J. Liu, "Anti-glioma effect with targeting therapy using folate modified nano-micelles delivery curcumin", *J. Biomed. Nanotechnol.,* vol. 16, no. 1, pp. 1-13, 2020.
[http://dx.doi.org/10.1166/jbn.2020.2878] [PMID: 31996281]

[30]   R. Tenchov, R. Bird, A.E. Curtze, and Q. Zhou, "Lipid nanoparticles—from liposomes to mRNA vaccine delivery, a landscape of research diversity and advancement", *ACS Nano,* vol. 15, no. 11, pp. 16982-17015, 2021.
[http://dx.doi.org/10.1021/acsnano.1c04996] [PMID: 34181394]

[31]   H. Moon, K. Hwang, K.M. Nam, Y.S. Kim, M.J. Ko, H.R. Kim, H.J. Lee, M.J. Kim, T.H. Kim, K.S. Kang, N.G. Kim, S.W. Choi, and C.Y. Kim, "Enhanced delivery to brain using sonosensitive liposome and microbubble with focused ultrasound", *Biomat. Adv.,* vol. 141, p. 213102, 2022.
[http://dx.doi.org/10.1016/j.bioadv.2022.213102] [PMID: 36103796]

[32]   V. Vangala, N.V. Nimmu, S. Khalid, M. Kuncha, R. Sistla, R. Banerjee, and A. Chaudhuri, "Combating glioblastoma by codelivering the small-molecule inhibitor of STAT3 and STAT3siRNA with α5β1 integrin receptor-selective liposomes", *Mol. Pharm.,* vol. 17, no. 6, pp. 1859-1874, 2020.
[http://dx.doi.org/10.1021/acs.molpharmaceut.9b01271] [PMID: 32343904]

[33]   M. Ghaferi, A. Raza, M. Koohi, W. Zahra, A. Akbarzadeh, H. Ebrahimi Shahmabadi, and S.E. Alavi, "Impact of PEGylated liposomal doxorubicin and carboplatin combination on glioblastoma", *Pharmaceutics,* vol. 14, no. 10, p. 2183, 2022.
[http://dx.doi.org/10.3390/pharmaceutics14102183] [PMID: 36297618]

[34]   E.A.L. Rustad, S. von Hofsten, R. Kumar, E.A. Lænsman, G. Berge, and N. Škalko-Basnet, "The pH-Responsive Liposomes—The effect of pegylation on release kinetics and cellular uptake in glioblastoma cells", *Pharmaceutics,* vol. 14, no. 6, p. 1125, 2022.
[http://dx.doi.org/10.3390/pharmaceutics14061125] [PMID: 35745698]

[35]   L. Taiarol, C. Bigogno, S. Sesana, M. Kravicz, F. Viale, E. Pozzi, L. Monza, V.A. Carozzi, C. Meregalli, S. Valtorta, R.M. Moresco, M. Koch, F. Barbugian, L. Russo, G. Dondio, C. Steinkühler, and F. Re, "Givinostat-Liposomes: Anti-tumor effect on 2D and 3D glioblastoma models and pharmacokinetics", *Cancers,* vol. 14, no. 12, p. 2978, 2022.
[http://dx.doi.org/10.3390/cancers14122978] [PMID: 35740641]

[36]   M. Ismail, W. Yang, Y. Li, T. Chai, D. Zhang, Q. Du, P. Muhammad, S. Hanif, M. Zheng, and B. Shi, "Targeted liposomes for combined delivery of artesunate and temozolomide to resistant glioblastoma", *Biomaterials,* vol. 287, p. 121608, 2022.
[http://dx.doi.org/10.1016/j.biomaterials.2022.121608] [PMID: 35690021]

[37]   S. Wang, Z. Yang, C. Yang, J. Chen, L. Zhou, Y. Wu, and R. Lu, "Investigation of functionalised nanoplatforms using branched-ligands with different chain lengths for glioblastoma targeting", *J. Drug Target.,* vol. 30, no. 9, pp. 992-1005, 2022.
[http://dx.doi.org/10.1080/1061186X.2022.2077948] [PMID: 35549968]

[38]   Y. C. Kuo, Y. J. Lee, and R. Rajesh, "Enhanced activity of AZD5582 and SM-164 in rabies virus glycoprotein-lactoferrin-liposomes to downregulate inhibitors of apoptosis proteins in glioblastoma", *Biomater. Adv,* vol. 133, p. 112615, 2022.
[http://dx.doi.org/10.1016/j.msec.2021.112615]

[39]   J. Yao, X. Feng, X. Dai, G. Peng, Z. Guo, Z. Liu, M. Wang, W. Guo, P. Zhang, and Y. Li, "TMZ magnetic temperature-sensitive liposomes-mediated magnetothermal chemotherapy induces pyroptosis in glioblastoma", *Nanomedicine,* vol. 43, p. 102554, 2022.
[http://dx.doi.org/10.1016/j.nano.2022.102554] [PMID: 35358733]

[40]   G. Katona, F. Sabir, B. Sipos, M. Naveed, Z. Schelz, I. Zupkó, and I. Csóka, "Development of lomustine and n-Propyl gallate co-encapsulated liposomes for targeting glioblastoma multiforme via intranasal administration", *Pharmaceutics,* vol. 14, no. 3, p. 631, 2022.
[http://dx.doi.org/10.3390/pharmaceutics14030631] [PMID: 35336006]

[41]   Z. Song, X. Huang, J. Wang, F. Cai, P. Zhao, and F. Yan, "Targeted delivery of liposomal temozolomide enhanced anti-glioblastoma efficacy through ultrasound-mediated blood–brain barrier opening", *Pharmaceutics,* vol. 13, no. 8, p. 1270, 2021.
[http://dx.doi.org/10.3390/pharmaceutics13081270] [PMID: 34452234]

[42]   A.M. Shabana, B. Xu, Z. Schneiderman, J. Ma, C.C. Chen, and E. Kokkoli, "Targeted liposomes encapsulating mir-603 complexes enhance radiation sensitivity of patient-derived glioblastoma stem-like cells", *Pharmaceutics,* vol. 13, no. 8, p. 1115, 2021.
[http://dx.doi.org/10.3390/pharmaceutics13081115] [PMID: 34452076]

[43]   M. Pizzocri, "Radiation and adjuvant drug-loaded liposomes target glioblastoma stem cells and trigger in-situ immune response", *Neuro-Oncology Adv,* vol. 3, no. 1, p. vdab076, 2021.
[http://dx.doi.org/10.1093/noajnl/vdab076]

[44]   Y.J. Lu, A.T. S, C.C. Chuang, and J.P. Chen, "Liposomal ir-780 as a highly stable nanotheranostic agent for improved photothermal/photodynamic therapy of brain tumors by convection-enhanced delivery", *Cancers,* vol. 13, no. 15, p. 3690, 2021.
[http://dx.doi.org/10.3390/cancers13153690] [PMID: 34359590]

[45]   M. Renault-Mahieux, V. Vieillard, J. Seguin, P. Espeau, D.T. Le, R. Lai-Kuen, N. Mignet, M. Paul, and K. Andrieux, "Co-encapsulation of fisetin and cisplatin into liposomes for glioma therapy: From formulation to cell evaluation", *Pharmaceutics,* vol. 13, no. 7, p. 970, 2021.
[http://dx.doi.org/10.3390/pharmaceutics13070970] [PMID: 34206986]

[46]   D. Mendanha, J. Vieira de Castro, J. Moreira, B.M. Costa, H. Cidade, M. Pinto, H. Ferreira, and N.M. Neves, "A new chalcone derivative with promising antiproliferative and anti-invasion activities in glioblastoma cells", *Molecules,* vol. 26, no. 11, p. 3383, 2021.
[http://dx.doi.org/10.3390/molecules26113383] [PMID: 34205043]

[47]   P. Sandbhor, "Targeted nano-delivery of chemotherapy *via* intranasal route suppresses *in vivo* glioblastoma growth and prolongs survival in the intracranial mouse model", *Drug Deliv. Transl. Res.,* vol. 13, no. 2, pp. 608-626, 2022.
[http://dx.doi.org/10.1007/s13346-022-01220-8] [PMID: 36245060]

[48]   T. Ahmed, F.C.F. Liu, C. He, A.Z. Abbasi, P. Cai, A.M. Rauth, J.T. Henderson, and X.Y. Wu, "Optimizing the design of blood–brain barrier-penetrating polymer-lipid-hybrid nanoparticles for delivering anticancer drugs to glioblastoma", *Pharm. Res.,* vol. 38, no. 11, pp. 1897-1914, 2021.
[http://dx.doi.org/10.1007/s11095-021-03122-9] [PMID: 34655006]

[49]   L. Wang, X. Wang, L. Shen, M. Alrobaian, S.K. Panda, H.A. Almasmoum, M.M. Ghaith, R.A. Almaimani, I.A.A. Ibrahim, T. Singh, A.A. Baothman, H. Choudhry, and S. Beg, "Paclitaxel and naringenin-loaded solid lipid nanoparticles surface modified with cyclic peptides with improved tumor targeting ability in glioblastoma multiforme", *Biomed. Pharmacother.,* vol. 138, p. 111461, 2021.
[http://dx.doi.org/10.1016/j.biopha.2021.111461] [PMID: 33706131]

[50]   D. Rosenblum, A. Gutkin, R. Kedmi, S. Ramishetti, N. Veiga, A.M. Jacobi, M.S. Schubert, D. Friedmann-Morvinski, Z.R. Cohen, M.A. Behlke, J. Lieberman, and D. Peer, "CRISPR-Cas9 genome editing using targeted lipid nanoparticles for cancer therapy", *Sci. Adv.,* vol. 6, no. 47, p. eabc9450, 2020.
[http://dx.doi.org/10.1126/sciadv.abc9450] [PMID: 33208369]

[51]   V. Yathindranath, N. Safa, B.V. Sajesh, K. Schwinghamer, M.I. Vanan, R. Bux, D.S. Sitar, M. Pitz, T.J. Siahaan, and D.W. Miller, "Spermidine/Spermine N1-Acetyltransferase 1 (SAT1)—A potential gene target for selective sensitization of glioblastoma cells using an ionizable lipid nanoparticle to deliver siRNA", *Cancers,* vol. 14, no. 21, p. 5179, 2022.
[http://dx.doi.org/10.3390/cancers14215179] [PMID: 36358597]

[52]   S. Liu, J. Liu, H. Li, K. Mao, H. Wang, X. Meng, J. Wang, C. Wu, H. Chen, X. Wang, X. Cong, Y. Hou, Y. Wang, M. Wang, Y.G. Yang, and T. Sun, "An optimized ionizable cationic lipid for brain tumor-targeted siRNA delivery and glioblastoma immunotherapy", *Biomaterials,* vol. 287, p. 121645, 2022.
[http://dx.doi.org/10.1016/j.biomaterials.2022.121645] [PMID: 35779480]

[53]   Y. Hu, K. Jiang, D. Wang, S. Yao, L. Lu, H. Wang, J. Song, J. Zhou, X. Fan, Y. Wang, W. Lu, J. Wang, and G. Wei, "Core-shell lipoplexes inducing active macropinocytosis promote intranasal delivery of c-Myc siRNA for treatment of glioblastoma", *Acta Biomater.,* vol. 138, pp. 478-490, 2022.
[http://dx.doi.org/10.1016/j.actbio.2021.10.042] [PMID: 34757231]

[54]   V. Ravula, Y.L. Lo, Y.T. Wu, C.W. Chang, S.V. Patri, and L.F. Wang, "Arginine-tocopherol bioconjugated lipid vesicles for selective pTRAIL delivery and subsequent apoptosis induction in glioblastoma cells", *Mater. Sci. Eng. C,* vol. 126, p. 112189, 2021.
[http://dx.doi.org/10.1016/j.msec.2021.112189] [PMID: 34082988]

[55]   L.D. Di Filippo, J. Lobato Duarte, J. Hofstätter Azambuja, R. Isler Mancuso, M. Tavares Luiz, V. Hugo Sousa Araújo, I. Delbone Figueiredo, L. Barretto-de-Souza, R. Miguel Sábio, E. Sasso-Cerri, A. Martins Baviera, C.C. Crestani, S. Teresinha Ollala Saad, and M. Chorilli, "Glioblastoma multiforme targeted delivery of docetaxel using bevacizumab-modified nanostructured lipid carriers impair *in vitro* cell growth and *in vivo* tumor progression", *Int. J. Pharm.,* vol. 618, p. 121682, 2022.
[http://dx.doi.org/10.1016/j.ijpharm.2022.121682] [PMID: 35307470]

[56]   F. Khonsari, M. Heydari, R. Dinarvand, M. Sharifzadeh, and F. Atyabi, "Correction: Brain targeted delivery of rapamycin using transferrin decorated nanostructured lipid carriers", *Bioimpacts,* vol. 12, no. 1, pp. 21-32, 2022.
[http://dx.doi.org/10.34172/bi.2022.27678] [PMID: 35087713]

[57]   A.S. Shirazi, R. Varshochian, M. Rezaei, Y.H. Ardakani, and R. Dinarvand, "SN38 loaded nanostructured lipid carriers (NLCs); preparation and *in vitro* evaluations against glioblastoma", *J. Mater. Sci. Mater. Med.,* vol. 32, no. 7, p. 78, 2021.
[http://dx.doi.org/10.1007/s10856-021-06538-2] [PMID: 34191134]

[58]   E. Gajda, M. Godlewska, Z. Mariak, E. Nazaruk, and D. Gawel, "Combinatory treatment with mir-7-5p and drug-loaded cubosomes effectively impairs cancer cells", *Int. J. Mol. Sci.,* vol. 21, no. 14, p. 5039, 2020.
[http://dx.doi.org/10.3390/ijms21145039] [PMID: 32708846]

[59]   D.K. Flak, V. Adamski, G. Nowaczyk, K. Szutkowski, M. Synowitz, S. Jurga, and J. Held-Feindt, "At101-loaded cubosomes as an alternative for improved glioblastoma therapy", *Int. J. Nanomedicine,* vol. 15, pp. 7415-7431, 2020.
[http://dx.doi.org/10.2147/IJN.S265061] [PMID: 33116479]

[60]   E. Hernández Becerra, J. Quinchia, C. Castro, and J. Orozco, "Light-triggered polymersome-based anticancer therapeutics delivery", *Nanomaterials,* vol. 12, no. 5, p. 836, 2022.
[http://dx.doi.org/10.3390/nano12050836] [PMID: 35269324]

[61]   C. He, H. Ding, J. Chen, Y. Ding, R. Yang, C. Hu, Y. An, D. Liu, P. Liu, Q. Tang, and Z. Zhang, "Immunogenic cell death induced by chemoradiotherapy of novel pH-sensitive cargo-loaded polymersomes in glioblastoma", *Int. J. Nanomedicine,* vol. 16, pp. 7123-7135, 2021.
[http://dx.doi.org/10.2147/IJN.S333197] [PMID: 34712045]

[62]   C. He, Z. Zhang, Y. Ding, K. Xue, X. Wang, R. Yang, Y. An, D. Liu, C. Hu, and Q. Tang, "LRP1-mediated pH-sensitive polymersomes facilitate combination therapy of glioblastoma *in vitro* and *in*

vivo", *J. Nanobiotechnology,* vol. 19, no. 1, p. 29, 2021.
[http://dx.doi.org/10.1186/s12951-020-00751-x] [PMID: 33482822]

[63]   J. Ouyang, Y. Jiang, C. Deng, Z. Zhong, and Q. Lan, "Doxorubicin delivered via apoe-directed reduction-sensitive polymersomes potently inhibit orthotopic human glioblastoma xenografts in nude mice", *Int. J. Nanomedicine,* vol. 16, pp. 4105-4115, 2021.
[http://dx.doi.org/10.2147/IJN.S314895] [PMID: 34163162]

[64]   Q. Fan, Y. Liu, G. Cui, Z. Zhong, and C. Deng, "Brain delivery of Plk1 inhibitor *via* chimaeric polypeptide polymersomes for safe and superb treatment of orthotopic glioblastoma", *J. Control. Release,* vol. 329, pp. 1139-1149, 2021.
[http://dx.doi.org/10.1016/j.jconrel.2020.10.043] [PMID: 33131697]

[65]   F. Bikhezar, R.M. de Kruijff, A.J.G.M. van der Meer, G. Torrelo Villa, S.M.A. van der Pol, G. Becerril Aragon, A. Gasol Garcia, R.S. Narayan, H.E. de Vries, B.J. Slotman, A.G. Denkova, and P. Sminia, "Preclinical evaluation of binimetinib (MEK162) delivered via polymeric nanocarriers in combination with radiation and temozolomide in glioma", *J. Neurooncol.,* vol. 146, no. 2, pp. 239-246, 2020.
[http://dx.doi.org/10.1007/s11060-019-03365-y] [PMID: 31875307]

[66]   Z. Bober, D. Bartusik-Aebisher, and D. Aebisher, "Application of dendrimers in anticancer diagnostics and therapy", *Molecules,* vol. 27, no. 10, p. 3237, 2022.
[http://dx.doi.org/10.3390/molecules27103237] [PMID: 35630713]

[67]   W. Liyanage, T. Wu, S. Kannan, and R.M. Kannan, "Dendrimer–siRNA conjugates for targeted intracellular delivery in glioblastoma animal models", *ACS Appl. Mater. Interfaces,* vol. 14, no. 41, pp. 46290-46303, 2022.
[http://dx.doi.org/10.1021/acsami.2c13129] [PMID: 36214413]

[68]   V.S. Mishra, S. Patil, P.C. Reddy, and B. Lochab, "Combinatorial delivery of CPI444 and vatalanib loaded on PEGylated graphene oxide as an effective nanoformulation to target glioblastoma multiforme: *In vitro* evaluation", *Front. Oncol.,* vol. 12, p. 953098, 2022.
[http://dx.doi.org/10.3389/fonc.2022.953098] [PMID: 36052261]

[69]   M. Sowińska, M. Szeliga, M. Morawiak, B. Zabłocka, and Z. Urbanczyk-Lipkowska, "Design, synthesis and activity of new N1-Alkyl tryptophan functionalized dendrimeric peptides against glioblastoma", *Biomolecules,* vol. 12, no. 8, p. 1116, 2022.
[http://dx.doi.org/10.3390/biom12081116] [PMID: 36009010]

[70]   N. Knauer, V. Arkhipova, G. Li, M. Hewera, E. Pashkina, P.H. Nguyen, M. Meschaninova, V. Kozlov, W. Zhang, R. Croner, A.M. Caminade, J.P. Majoral, E. Apartsin, and U. Kahlert, "*In Vitro* validation of the therapeutic potential of dendrimer-based nanoformulations against tumor stem cells", *Int. J. Mol. Sci.,* vol. 23, no. 10, p. 5691, 2022.
[http://dx.doi.org/10.3390/ijms23105691] [PMID: 35628503]

[71]   W.G. Lesniak, B.B. Azad, S. Chatterjee, A. Lisok, and M.G. Pomper, "An evaluation of CXCR4 targeting with PAMAM dendrimer conjugates for oncologic applications", *Pharmaceutics,* vol. 14, no. 3, p. 655, 2022.
[http://dx.doi.org/10.3390/pharmaceutics14030655] [PMID: 35336029]

[72]   M.H.C. Lin, L.C. Chang, C.Y. Chung, W.C. Huang, M.H. Lee, K.T. Chen, P.S. Lai, and J.T. Yang, "Photochemical internalization of etoposide using dendrimer nanospheres loaded with etoposide and protoporphyrin ix on a glioblastoma cell line", *Pharmaceutics,* vol. 13, no. 11, p. 1877, 2021.
[http://dx.doi.org/10.3390/pharmaceutics13111877] [PMID: 34834292]

[73]   M. Szota, K. Reczyńska-Kolman, E. Pamuła, O. Michel, J. Kulbacka, and B. Jachimska, "Poly(Amidoamine) dendrimers as nanocarriers for 5-fluorouracil: Effectiveness of complex formation and cytotoxicity studies", *Int. J. Mol. Sci.,* vol. 22, no. 20, p. 11167, 2021.
[http://dx.doi.org/10.3390/ijms222011167] [PMID: 34681827]

[74]   J. Gallien, B. Srinageshwar, K. Gallo, G. Holtgrefe, S. Koneru, P.S. Otero, C.A. Bueno, J. Mosher, A.

Roh, D.S. Kohtz, D. Swanson, A. Sharma, G. Dunbar, and J. Rossignol, "Curcumin loaded dendrimers specifically reduce viability of glioblastoma cell lines", *Molecules,* vol. 26, no. 19, p. 6050, 2021.
[http://dx.doi.org/10.3390/molecules26196050] [PMID: 34641594]

[75]   A.B. Behrooz, R. Vazifehmand, A.A. Tajudin, M.J. Masarudin, Z. Sekawi, M. Masomian, and A. Syahir, "Tailoring drug co-delivery nanosystem for mitigating U-87 stem cells drug resistance", *Drug Deliv. Transl. Res.,* vol. 12, no. 5, pp. 1253-1269, 2022.
[http://dx.doi.org/10.1007/s13346-021-01017-1] [PMID: 34405338]

[76]   E. Fuster, H. Candela, J. Estévez, E. Vilanova, and M.A. Sogorb, "A transcriptomic analysis of T98G human glioblastoma cells after exposure to cadmium-selenium quantum dots mainly reveals alterations in neuroinflammation processes and hypothalamus regulation", *Int. J. Mol. Sci.,* vol. 23, no. 4, p. 2267, 2022.
[http://dx.doi.org/10.3390/ijms23042267] [PMID: 35216387]

[77]   M.A. Al-Duais, Z.M. Mohammedsaleh, H.S. Al-Shehri, Y.S. Al-Awthan, S.A. Bani-Atta, A.A. Keshk, S.K. Mustafa, A.D. Althaqafy, J.N. Al-Tweher, H.A. Al-Aoh, and C. Panneerselvam, "Bovine serum albumin functionalized blue emitting Ti$_3$C$_2$ MXene quantum dots as a sensitive fluorescence probe for Fe^{3+} ion detection and its toxicity analysis", *Luminescence,* vol. 37, no. 4, pp. 633-641, 2022.
[http://dx.doi.org/10.1002/bio.4204] [PMID: 35102681]

[78]   Y. Zhao, Y. Xie, Y. Liu, X. Tang, and S. Cui, "Comprehensive exploration of long-wave emission carbon dots for brain tumor visualization", *J. Mater. Chem. B Mater. Biol. Med.,* vol. 10, no. 18, pp. 3512-3523, 2022.
[http://dx.doi.org/10.1039/D2TB00322H] [PMID: 35416232]

[79]   Z. Li, C. Zhao, Q. Fu, J. Ye, L. Su, X. Ge, L. Chen, J. Song, and H. Yang, "Neodymium (3+)–Coordinated black phosphorus quantum dots with retrievable NIR/X–Ray optoelectronic switching effect for anti–glioblastoma", *Small,* vol. 18, no. 5, p. 2105160, 2022.
[http://dx.doi.org/10.1002/smll.202105160] [PMID: 34821027]

[80]   G. Perini, V. Palmieri, G. Ciasca, M. D'Ascenzo, A. Primiano, J. Gervasoni, F. De Maio, M. De Spirito, and M. Papi, "Enhanced chemotherapy for glioblastoma multiforme mediated by functionalized graphene quantum dots", *Materials,* vol. 13, no. 18, p. 4139, 2020.
[http://dx.doi.org/10.3390/ma13184139] [PMID: 32957607]

[81]   I. Arduino, N. Depalo, F. Re, R. Dal Magro, A. Panniello, E. Fanizza, A. Lopalco, V. Laquintana, A. Cutrignelli, A.A. Lopedota, M. Franco, and N. Denora, "PEGylated solid lipid nanoparticles for brain delivery of lipophilic kiteplatin Pt(IV) prodrugs: An *in vitro* study", *Int. J. Pharm.,* vol. 583, p. 119351, 2020.
[http://dx.doi.org/10.1016/j.ijpharm.2020.119351] [PMID: 32339634]

[82]   Q.L. Wu, H.L. Xu, C. Xiong, Q.H. Lan, M.L. Fang, J.H. Cai, H. Li, S.T. Zhu, J.H. Xu, F.Y. Tao, C.T. Lu, Y.Z. Zhao, and B. Chen, "c(RGDyk)-modified nanoparticles encapsulating quantum dots as a stable fluorescence probe for imaging-guided surgical resection of glioma under the auxiliary UTMD", *Artif. Cells Nanomed. Biotechnol.,* vol. 48, no. 1, pp. 143-158, 2020.
[http://dx.doi.org/10.1080/21691401.2019.1699821] [PMID: 32207347]

[83]   A.A.P. Mansur, M.R.B. Paiva, O.A.L. Cotta, L.M. Silva, I.C. Carvalho, N.S.V. Capanema, S.M. Carvalho, É.A. Costa, N.R. Martin, R. Ecco, B.S. Santos, S.L. Fialho, Z.I.P. Lobato, and H.S. Mansur, "Carboxymethylcellulose biofunctionalized ternary quantum dots for subcellular-targeted brain cancer nanotheranostics", *Int. J. Biol. Macromol.,* vol. 210, pp. 530-544, 2022.
[http://dx.doi.org/10.1016/j.ijbiomac.2022.04.207] [PMID: 35513094]

[84]   P.Y. Liyanage, Y. Zhou, A.O. Al-Youbi, A.S. Bashammakh, M.S. El-Shahawi, S. Vanni, R.M. Graham, and R.M. Leblanc, "Pediatric glioblastoma target-specific efficient delivery of gemcitabine across the blood–brain barrier via carbon nitride dots", *Nanoscale,* vol. 12, no. 14, pp. 7927-7938, 2020.
[http://dx.doi.org/10.1039/D0NR01647K] [PMID: 32232249]

[85] E.A. Veliz, A. Kaplina, S.D. Hettiarachchi, A.L. Yoham, C. Matta, S. Safar, M. Sankaran, E.L. Abadi, E.K. Cilingir, F.A. Vallejo, W.M. Walters, S. Vanni, R.M. Leblanc, and R.M. Graham, "Chalcones as anti-glioblastoma stem cell agent alone or as nanoparticle formulation using carbon dots as nanocarrier", *Pharmaceutics,* vol. 14, no. 7, p. 1465, 2022.
[http://dx.doi.org/10.3390/pharmaceutics14071465] [PMID: 35890360]

[86] J.X. Qin, X-G. Yang, C-F. Lv, Y-Z. Li, K-K. Liu, J-H. Zang, X. Yang, L. Dong, and C-X. Shan, "Nanodiamonds: Synthesis, properties, and applications in nanomedicine", *Mater. Des.,* vol. 210, p. 110091, 2021.
[http://dx.doi.org/10.1016/j.matdes.2021.110091]

[87] S. Patil, V.S. Mishra, N. Yadav, P.C. Reddy, and B. Lochab, "Dendrimer-functionalized nanodiamonds as safe and efficient drug carriers for cancer therapy: Nucleus penetrating nanoparticles", *ACS Appl. Bio Mater.,* vol. 5, no. 7, pp. 3438-3451, 2022.
[http://dx.doi.org/10.1021/acsabm.2c00373] [PMID: 35754387]

[88] M. Ghanimi Fard, Z. Khabir, P. Reineck, N.M. Cordina, H. Abe, T. Ohshima, S. Dalal, B.C. Gibson, N.H. Packer, and L.M. Parker, "Targeting cell surface glycans with lectin-coated fluorescent nanodiamonds", *Nanoscale Adv.,* vol. 4, no. 6, pp. 1551-1564, 2022.
[http://dx.doi.org/10.1039/D2NA00036A] [PMID: 36134370]

[89] Z. Chen, S.J. Yuan, K. Li, Q. Zhang, T.F. Li, H.C. An, H.Z. Xu, Y. Yue, M. Han, Y.H. Xu, N. Komatsu, L. Zhao, and X. Chen, "Doxorubicin-polyglycerol-nanodiamond conjugates disrupt STAT3/IL-6-mediated reciprocal activation loop between glioblastoma cells and astrocytes", *J. Control. Release,* vol. 320, pp. 469-483, 2020.
[http://dx.doi.org/10.1016/j.jconrel.2020.01.044] [PMID: 31987922]

[90] M.V. Kharlamova, M. Paukov, and M.G. Burdanova, "Nanotube functionalization: Investigation, methods and demonstrated applications", *Materials,* vol. 15, no. 15, p. 5386, 2022.
[http://dx.doi.org/10.3390/ma15155386] [PMID: 35955321]

[91] M. Alghamdi, F. Chierchini, D. Eigel, C. Taplan, T. Miles, D. Pette, P.B. Welzel, C. Werner, W. Wang, C. Neto, M. Gumbleton, and B. Newland, "Poly(ethylene glycol) based nanotubes for tuneable drug delivery to glioblastoma multiforme", *Nanoscale Adv.,* vol. 2, no. 10, pp. 4498-4509, 2020.
[http://dx.doi.org/10.1039/D0NA00471E] [PMID: 36132909]

[92] M.A. Harris, H. Kuang, Z. Schneiderman, M.L. Shiao, A.T. Crane, M.R. Chrostek, A.F. Tăbăran, T. Pengo, K. Liaw, B. Xu, L. Lin, C.C. Chen, M.G. O'Sullivan, R.M. Kannan, W.C. Low, and E. Kokkoli, "ssDNA nanotubes for selective targeting of glioblastoma and delivery of doxorubicin for enhanced survival", *Sci. Adv.,* vol. 7, no. 49, p. eabl5872, 2021.
[http://dx.doi.org/10.1126/sciadv.abl5872] [PMID: 34851666]

[93] A.C. Anselmo, and S. Mitragotri, "Nanoparticles in the clinic: An update", *Bioeng. Transl. Med.,* vol. 4, no. 3, p. e10143, 2019.
[http://dx.doi.org/10.1002/btm2.10143] [PMID: 31572799]

[94] W. Ruan, M. Jiao, S. Xu, M. Ismail, X. Xie, Y. An, H. Guo, R. Qian, B. Shi, and M. Zheng, "Brain-targeted CRISPR/Cas9 nanomedicine for effective glioblastoma therapy", *J. Control. Release,* vol. 351, pp. 739-751, 2022.
[http://dx.doi.org/10.1016/j.jconrel.2022.09.046] [PMID: 36174804]

[95] Y. Zou, X. Sun, Q. Yang, M. Zheng, O. Shimoni, W. Ruan, Y. Wang, D. Zhang, J. Yin, X. Huang, W. Tao, J.B. Park, X.J. Liang, K.W. Leong, and B. Shi, "Blood-brain barrier–penetrating single CRISPR-Cas9 nanocapsules for effective and safe glioblastoma gene therapy", *Sci. Adv.,* vol. 8, no. 16, p. eabm8011, 2022.
[http://dx.doi.org/10.1126/sciadv.abm8011] [PMID: 35442747]

[96] V. Sunil, A. Mozhi, W. Zhan, J.H. Teoh, P.B. Ghode, N.V. Thakor, and C.H. Wang, "*In-situ* vaccination using dual responsive organelle targeted nanoreactors", *Biomaterials,* vol. 290, p. 121843, 2022.

[http://dx.doi.org/10.1016/j.biomaterials.2022.121843] [PMID: 36228516]

[97]   G. Deng, X. Peng, Z. Sun, W. Zheng, J. Yu, L. Du, H. Chen, P. Gong, P. Zhang, L. Cai, and B.Z. Tang, "Natural-killer-cell-inspired nanorobots with aggregation-induced emission characteristics for near-infrared-II fluorescence-guided glioma theranostics", *ACS Nano,* vol. 14, no. 9, pp. 11452-11462, 2020.
[http://dx.doi.org/10.1021/acsnano.0c03824] [PMID: 32820907]

CHAPTER 10

Biocompatibility of Nanomedicines and Relation with Protein Corona

Yakup Kolcuoglu[1,*], **Fulya Oz Tuncay**[1] and **Ummuhan Cakmak**[1]

[1] *Karadeniz Technical University, Faculty of Science, Department of Chemistry, Trabzon, Türkiye*

Abstract: When NPs are included in a Biological environment, they associate with a large number of circulating proteins. As a result, they interact dynamically with each other. This structure, which is defined as PC, affects the physical parameters of NPs and causes positive or negative effects on them. PC composition is affected by many properties of NPs, such as size, shape, and surface charge. Therefore, various surface modifications on NPs directly affect PC formation and nature. Although many studies have been carried out to understand the formation and composition of the resulting PC structure, this area still maintains its popularity as a research topic. This review aims to briefly give an idea about the effect of proteins in metabolism on NPs designed as carrier molecules, the determination of these protein structures and the final fate of NPs after PC formation.

Keywords: Atomic force microscopy, Biocompatibility, Differential centrifugal sedimentation, Dynamic light scattering, Fluorescence correlation spectroscopy, Hard corona, Isothermal titration calorimetry, Nanomaterial shape, Nanoparticle, Nanoparticle application, Protein adsorption, Protein corona, Protein corona characterization, Protein corona dispersion, Protein corona stability, Protein corona formation, Small-angle X-ray scattering, Soft corona, Transmission electron microscopy, UV-visible spectroscopy.

INTRODUCTION

Nanomaterials are defined as substances up to several hundred nanometers in size. With these features, nanoparticles (NPs) have a similar size range with the structures in the cell, and therefore they have many applications in nanomedicine. Because of their physical and structural properties, well-designed nanomaterials have the power to significantly improve the treatment and diagnosis of diseases [1, 2]. In fact, humanity began to produce nano-structured materials since the time of the Lycurgus Cup during the Roman Empire and was impressed by their unique

* **Corresponding author Yakup Kolcuoglu:** Karadeniz Technical University, Faculty of Science, Department of Chemistry, Trabzon, Turkey; Tel: +90 462 377 2495; E-mail: yakupkolcuoglu@ktu.edu.tr

Habibe Yılmaz (Ed.)

nature [3]. NP have attracted increasing attention in the last 20 years, due to the development of new synthesis, characterization and analysis methods and the increase in investments in this field [4 - 6]. Due to the increasing use of nanoparticles, it has become inevitable for scientists to investigate how nanoparticles are involved in human health, their effects on the environment, and how they are involved in metabolism. As a result of the researches, it is reported that besides the positive aspects of nanoparticles, they can cause toxic effects on living things as a result of excessive misuse [7, 8]. At the point reached today, the knowledge we have about the factors affecting the safety of nanoparticles is not enough for the development of reliable and effective nanomedicine. Due to the ability of nanoparticles to penetrate different cells and tissues after entering the body, the need for further investigation of the risks of nanoparticles on health arises [9]. Biocompatibility is defined as the ability of a used material to produce a host response under certain conditions. Metabolism reacts to NPs, as it does to all foreign substances, in order to clean the NPs it sees as foreign matter. The response of the living organism is determined by the level of interaction of the NP with various biological substances in the environment. Biocompatibility is achieved when a substance involved in metabolism enters the circulation without causing carcinogenic, immunogenic, thrombogenic or toxic responses. However, toxicity or bio-incompatibility occurs if there are undesirable responses in biological processes.

The current applications of nanomaterials in medicine as drug carriers are increasing day by day and, at the same time, gaining importance [10]. The main purpose of nanodrug carriers in oncological applications is to prevent serious side effects of toxic compounds used in treatments. As nano-sized biomaterials have high free energy when incorporated into the system, due to large surface areas, they tend to interact dynamically with the surrounding molecules due to these energies [11]. When nanomaterials are included in the biological environment, proteins are the main molecules that bind tightly with the surface of these NPs, and as a result of this interaction, the proteins form a layer surrounding the surface. Although the formation of this layer (about 30 sec) is thought to be tight, the process is reversible. In this case, there is a dynamic protein exchange with the microenvironment. This protein structure is called protein corona (PC). The PC layer consists of two different structures characterized by slow change (hard corona) and fast protein exchange (soft corona). PC formation has numerous biological effects, such as cell interaction control of NPs, induction of their cytotoxicity, optimal targeting, and possible modulation of drug pharmacokinetics [12 - 14]. Remarkably, only a fraction (some of them are defined as opsonins and some others are dysopsonins) of the approximately 3,400 proteins of human plasma could be detected to interact directly with nanoparticles. Although small in number, these proteins produce impressive results in terms of cellular uptake [15 -

18]. For example, the adsorption of some proteins on the surface of nanoparticles may lead to the deterioration of the protein structure depending on the surface charge of this material [19]. For similar reasons, the physiological folding process of nanomolecules by proteins is the main part of the complex that determines their cellular uptake [20].

Protein adsorption has been extensively investigated in biomedical NPs to determine the modification and cellular uptake of nano-sized drug carriers [21]. The binding of proteins to the surface of NPs is an unpredictable complex process that alters the toxicological properties and efficacy of nanomaterials [22]. In the protein adsorption of nanomaterials, the type, geometry and conformation of this material are significantly effective [23]. Numerous studies have been carried out to modify the molecular surfaces of such molecules so that they do not interact with the protein. Examples of these studies are PEGylation [24, 25], colloidal silica nanoparticle production by adding PEG and Pluronic-F127 with different molecular weights [26] and silver nanoparticle production [27]. However, PEGylation remains an important standard for modifications of nanocarriers designed for drug delivery [28].

FACTORS AFFECTING PC FORMATION

After being metabolized, NPs interact with physiological biomolecules as a result of combining with blood and other biological fluids. As a result, the PC layer comes into existence. PC formation is a dynamic process described as the "Vroman effect" involving different forces (such as Van Der Waals forces, $\pi-\pi$ stacking bond, H-bonds, electrostatic and hydrophobic interactions) between nanomaterials and proteins [29, 30]. Initially, proteins with high abundance but low affinity in the medium of NP inclusion are rapidly adsorbed on the surface of these molecules; they are then replaced by low amount and high affinity proteins [31]. As a result of all these formations, two different layers are formed as hard corona (HC) and soft corona (SC). HC-type binding, which is responsible for the behavior of NPs in metabolism, affects the membrane interaction and biodistribution of these molecules. Proteins in HC bind directly and with high affinity to the surface of the molecule. As a result, they form a stable layer [32, 33]. The proteins forming the SC layer are a replaceable layer depending on the environmental conditions and they are indirectly located on the surface of the NPs as they interact with the proteins in the HC layer [34]. The process of formation of PC and, consequently, the cellular behavior of NPs is highly dependent on factors such as size, morphology, surface properties, the type and composition of the biological fluid (cytoplasm, blood, extracellular matrix) in which the nanomaterials are contained, pH, and temperature [35].

Size

The shape of NPs is highly influential on PC formation. Besides, the thickness of the PC layer, the proteins to be bound, and the conformational change of these proteins also depend on the NP structure. It has been observed that proteins bind more to nanorod-shaped NPs than to nanosphere-shaped structures [36]. This is because rod-type NPs provide more packaging space for proteins. Beyond that, studies have shown that the material constituting the core of the molecules can have an effect on the type and concentration of proteins that provide PC formation [34]. Some molecules added to NPs, such as PEG or citrate-stabilized, have been observed to interact with less protein compared to NPs in the pure form [37, 38]. In addition, as a result of different studies, it has been reported that PC formation becomes more complex by reducing the NP diameter from 70 nm to 40 nm [39]. In light of this information, it is understood that the NP size not only contributes to the interaction with the protein but also to the selectivity in the composition of the formed layers and the control of their thickness.

Surface

The physicochemical property of an NP can change the probability that this molecule interacts with proteins. One of the reasons for this is that the surface charge of NPs can change or be neutralized depending on the increase or decrease in the binding affinity of physiological proteins. As a possible case, the presence or absence of a specific charge in the medium may contribute to the formation of specific interactions with proteins by supporting electrostatic and ionic forces or hydrophobic and π–π interactions. Accordingly, the formation and content of PC are highly dependent on the type of weak interactions that occur between the proteins and the NP surface. With the formation of these specific bonds as a result of this process, the stability of adsorbed proteins can vary considerably. Studies on the subject have shown that proteins can change their structures when interacting with the charged surface of NPs, but on the contrary, proteins retain their structure as a result of binding the surface of NPs with neutral ligands [40 - 42].

For the purpose of stabilizing the NPs colloidally, the surface structure of these particles can be arranged to contain the desired groups. One of the modification types followed to prevent aggregation is the attachment of citrate groups to the surface. It has been shown in different studies that this method gives positive results [43 - 46].

Another preferred method for surface modification of NPs is to coat them with polymers as a protective layer that can act against uncontrolled protein binding. Chitosan coated NPs can be given as an example of this method. The positively charged and hydrophilic surface enhanced the absorption of these molecules from the gut and the homogeneous distribution of NPs [47 - 49]. On the other hand, it has been shown that the use of cysteine and zwitter ion polymers increases the targeting efficiency in cancer cells, reduces the surface adsorption of proteins and promotes cellular uptake [50 - 52]. Also, regarding the polymeric coating, polyethylene glycol (PEG) is mostly used because of its biocompatibility and ability to prevent non-specific protein adsorption [53].

Biological Environment

Another important factor in PC formation is the biological environment in which the NP is located and its composition. The components of the formed PC vary according to the biological environment in which it is incubated [54]. The concentration of proteins in the medium is highly effective in this binding. In addition, one of the important factors is the composition, which can vary individually according to gender and health status. Beyond this, studies have shown that NPs with different structures can have different PC formations as a result of incubation in plasma and serum [55, 56]. Ultimately, the contributing factors in biocorona formation vary in the case of *in Vitro* or *in Vivo* incubation, but become more complex in the case of the *in Vivo* biological environment [57 - 59].

Environment: Temperature, pH and Exposure Time

Although temperature and pH are in a certain range in metabolism, pH and temperature can vary within this range. Accordingly, scientists were curious about the effect of these changes on protein corona formation and composition, and researches showed that PC formation was affected by the aforementioned environmental conditions [60]. Two possible factors are thought to play a role in PC formation caused by temperature changes. These are expressed as the control of protein adsorption on the NP surface and the modification of proteins in the composition of biological fluids, including NPs [61].

In a study, it was stated that the PC thickness of polymer-coated iron-platinum NPs changes with temperature. It was observed that while a monolayer was formed by protein adsorption at low temperatures, the thickness of the protein shell decreased at high temperatures [62]. In a study on the nature of the bonding between BSA and Cu NPs at different temperatures, it was reported that there was an increase in protein adsorption on the surface of the NPs when the temperature ranged from 14°C to 42°C [63].

Another important factor that changes the interaction of proteins with NPs is the pH of the environment. The reason for this is that changes in pH can cause changes in the conformation of proteins. Biological fluids have specific pH values. For example, blood may have a neutral pH, while its intracellular matrix may have a pH of 6.8. In studies conducted with different spectroscopic methods, it has been reported that there are differences in the binding affinities of proteins as a result of conformational changes in protein structure with pH change [64, 65].

The exposure time is highly effective on the proteins adsorbed to the NP surfaces. For example, it has been reported that complex PCs consisting of about 300 different proteins can form on the NP 30 seconds after application to human plasma, and a significant increase in the number of proteins is observed when the exposure time is increased to over 30 minutes [66]. In another study, changes in PC composition were investigated as a function of exposure time, and significant changes in PC composition were detected during 48 hours of residence time [67]. Interestingly, in a time-dependent study on PEGylated liposomal NPs, it was reported that there was no change in the amount of protein involved in PC formation. However, it has been observed that there are fluctuations in the protein structure due to the dynamic structure of PC [68].

Biocompatibility

The uses of NPs for medical diagnostics and therapeutics are increasing, but their potential risks to the environment and human health raise concerns about their use. In order to design safe and efficient nanostructures to eliminate such risks, a comprehensive investigation of NP toxicity is a very important issue.

PC composition, thickness, and protein binding pattern can give NPs a different identity beyond their production purpose, improving their biocompatibility or increasing their toxic properties. It was observed that cellular internalization decreased and biocompatibility increased when the graphene oxide nanosheet was coated with blood proteins [62]. In the case of zeolite NPs coated with fibrinogen, proinflammatory effects were observed and cytotoxicity increased [69]. In a study with gold nanospheres, it was reported that the permeability and retention effect in tumor tissue increased as a result of apolipoprotein E coating, and in a different study using polystyrene nanoparticles, an increase in bioavailability with hydrophobin coating [70, 71]. If biocompatibility is observed with the restructuring resulting from the interaction of NPs with PC, it is possible to achieve the production purposes of NPs, such as drug delivery. However, if cytotoxicity is observed, clearance and systemic side effects may increase. At this point, it is clear that the interaction of NPs with plasma proteins plays a crucial role in controlling the biological fate of these molecules *in vivo*. The main factors in this process are the change in the permeability of the vessels to NPs, the activation of immune cells such as monocytes and lymphocytes against NPs, and ultimately the destruction by macrophages. However, NPs can control such defense mechanisms by binding to surface physiological molecules that can inhibit them. For example, it has been reported that the immune response and cytokine expression depend on the hydrophobicity of NPs [72]. Both native and engineered NPs may have the capacity to directly bind immune receptors. For this reason, some NPs have also been used as immunosuppressants, as they can directly kill immune cells or reduce their response [73, 74]. Conversely, in some NPs, it can activate cytokine production by inducing inflammation, leading to immunity as well as different side effects [75, 76]. Consequently, the interaction between NPs and the immune system may be desirable in both cases. For example, while immune activation is desirable in vaccines and anticancer therapy, depending on the purpose of NP administration, immunosuppression is required to treat inflammation and autoimmune diseases [77].

Interactions of NPs with disopsonines (*e.g.*, albumin and apolipoproteins) can prolong the circulation time of these molecules in the bloodstream by inhibiting cell membrane adhesion and cell entry. Contrary to this situation, the presence of opsonins (*e.g.*, complement, immunoglobulins, scavenger receptor) in the biocorona can induce internalization mechanisms and clearance of NPs. In studies conducted for this purpose, the surfaces of NPs are masked with disopsonines in order to increase the blood circulation time and prevent their accumulation in the tissue [78 - 80].

COMMON CHARACTERIZATION TECHNIQUES

Proteins that can interact with NPs affect the biological functionality and interfacial interactions of these molecules, causing them to acquire a new physicochemical feature in their behavior in biological fluids. In addition, by triggering conformational changes in the bound proteins, they may cause the structures on the NP surface to lose their function. In order to further understand the relationship between the corona content and the properties of NPs, detailed characterizations are needed, such as evaluation of the level of interaction, dispersion stability, and estimation of the protein/nanoparticle ratio in corona complexes. Many studies in the literature seem to focus on the characterization of NP-PC complexes. Here, techniques suitable for characterizing protein corona formation and their functional aspects are reviewed.

There are various analytical methods that are frequently used and continue to be developed for the physiochemical identification of NPs. Dynamic light scattering (DLS), differential centrifugal sedimentation (DCS), fluorescence correlation spectroscopy (FCS), transmission electron microscopy (TEM) or atomic force microscopy (AFM) and small-angle X-ray scattering (SAXS) are generally used to determine the parameters like size, surface charge and shape of NP-PC complexes.

DLS is a widely used technique, it can precisely detect the NP distribution and is used to study the hydrodynamic size distribution of NPs before and after PC adsorption. With this method, PC adsorption can be determined from the scattering intensity caused by Brownian motion [56]. Although DLS and FCS analysis principles are extremely relevant, DLS measures fluctuations in scattered light, whereas FCS measures fluctuations in fluorescence emitted from NPs [81]. While DLS is more suitable for use on small NPs (<100 nm), especially to provide precise results about the hydrodynamic diameter of NP-PC complexes [82, 83], particle sizes measured by DLS may be affected by particle agglomeration [84]. For this reason, TEM or AFM visuals of NPs and NP-PC complexes are often preferred to detect whether NPs are aggregated prior to DLS assessments [85, 86]. Besides all this, TEM and AFM are the most comprehensive, clear and powerful techniques for characterizing the thickness of the protein layer and the sizes of particles interacting with NPs. The advantage of TEM or AFM over DLS is that these techniques provide precise data on the sizes-shapes of NPs and the PC layer [87]. Besides, sample preparation for AFM is simpler as TEM requires a counterstaining that can change the morphology of NP-PC complexes and cause a thermal shift in the sample [88]. DCS is also an effective method for determining NP size distribution by measuring the sedimentation time of NPs through a density gradient subjected to a centrifugal force [89]. DCS has the advantage of ultra-high resolution capability—detecting and measuring particle size ranging from about 2 nm to 80 μm, but causes significant degradation of the NP-PC complex and often underestimates protein shell thicknesses [90].

Another common approach, SAXS, can be used to determine the scattering density of NPs and, specifically, particles ranging in size from 5 to 25 nm [91]. However, SAXS is a low-resolution structural technique, providing information on the shapes, evolution of interactions, and morphologies of NPs and NP-PC complexes [92]. Compared to DLS, SAXS provides cases of particles with larger size ranges and information on the structure and thermodynamics of PC dissociation under physiological conditions [84].

Because proteins have different binding energies, they bind to NPs with different affinities during corona formation. The elucidation of the binding patterns between proteins and NPs plays an important role in evaluating the applications of NPs. The characterization techniques commonly used today are given below.

UV-visible (UV-vis) spectroscopy gives the ratio intensity of light coming from and refracted from the sample at a given wavelength. Proteins absorb UV-vis spectroscopy light at a wavelength of 280 nm due to the aromatic amino acids in their structure. The binding of NPs to proteins alters the absorption spectra of proteins as well as NPs, which is due to the optical properties and electron transfer abilities of proteins [93, 94]. While UV-vis spectroscopy is one of the common techniques providing easy sample preparation, a fast, simple applicable approach for measuring protein corona (PC) and a qualitative description of NP-protein interactions [95, 96], It is also very sensitive to environmental factors. Therefore, characterizing the binding affinities of proteins to NPs alone is not sufficient [97].

Fluorescence correlation spectroscopy (FCS) measures the fluorescence of a compound when excited at a specific wavelength. Due to the interaction of NPs with proteins, the sizes of NPs generally increase accordingly, resulting in increased diffusion times. With FCS, changes in diffusion time can be detected when using fluorescently labeled NPs or proteins, and consequently, information about protein binding affinity can be generated. FCS can be easily used in the measurements of the binding kinetics and thermodynamic properties of proteins in relation to NPs [82, 98]. However, the insensitivity of this technique to protein conformation causes a lack of use of the technique [99].

Isothermal titration calorimetry (ITC) is used for the determination of thermodynamics parameters in a sample solution like binding stoichiometry, binding affinity, enthalpy (ΔHb), free energy (ΔGb), association constant (Ka), and entropy (ΔSb) of NP–protein interaction [56, 100, 101]. In this technique, the qualitative determination of the amount of protein on the surface of NPs is obtained by titration of the protein in an NP solution, and the heat generated simultaneously is recorded. Thermodynamic component measurement is performed with the help of isothermal functions created from the obtained data [56]. Without separation and isolation, ITC would not be sufficient to characterize the capacity of proteins to bind to NPs. Also, although suitable for a single protein, this method can be problematic for corona structures with complex protein content. Besides, this method has some limitations as it requires molar concentrations of NPs as well as the protein to be analyzed [102].

Nuclear magnetic resonance (NMR) spectroscopy is another technique that provides information about the structures, reaction states, conformational changes, thermodynamics and dynamics of molecules [84, 100]. It is a method that eliminates the need to separate the NP-PC complex and information about the binding sites and hydrodynamic diameters can be obtained easily [84]. The main problem experienced in the application of NMR is the uncertain effect on the protein adsorption of the NP surface.

Determining the content of a PC is a very important element to understand the properties of NP-PC complexes. Some of the common methods used in the detailed characterization of NP-PC complexes are gel electrophoresis, mass spectrometry (MS), and size exclusion chromatography (SEC).

Gel electrophoresis (one-dimensional gel electrophoresis (1-DE) or sodium dodecyl sulfate polyacrylamide gel electrophoresis (SDS-PAGE) and two-dimensional gel electrophoresis, 2-DE) is one of the most important methods used in the determination of proteins in a sample and estimation of molecular weights [103]. In these methods, depending on the electrophoretic mobility of the proteins, separation is provided depending on the size or charge of the proteins. 1-DE provides separation of the protein mixture in the electric field depending on their molecular weight. This method is an inexpensive, fast and, at the same time, effective approach to separating and identifying the composition of PC. Molecules with the same molecular weight are difficult to identify with this method because they have similar migration rates. Because of these weaknesses, this method is preferred more for comparison purposes. Another important disadvantage is that it has low sensitivity. 1-DE has the ability to detect between 1 and 50 ng for a single protein band. On the other hand, since protein mixtures are very rich, some proteins migrate together and as a result, separation is not achieved at the desired level [104]. Because 2-DE separates proteins based on their isoelectric points and molecular weights, it takes place in two stages, isoelectric focusing and SDS-PAGE. Following the procedure, the 2-DE protein map is analyzed by staining the proteins [105]. Because the process is time consuming, it is not useful for monitoring the kinetics of dynamic PC adsorption [106]. Since gel electrophoresis can provide qualitative and semi-quantitative information, mass spectrometry analysis is usually performed following this procedure to determine the identity of the separated proteins.

MS is used for the qualitative and quantitative characterization of protein corona (up to 100 kDa) formation on any nanomaterial surface and for the identification of a single protein in the sample [107]. This method is important because it is possible to analyze the protein composition over time [108]. This method is a stand-alone method that mainly depends on the analysis of fragmented protein ion and is most suitable for biomarker analysis. In particular, the use of the fragmented protein ion provides information about the components of the hard PC. It can also be combined with other methods such as chromatography and ICP-MS to achieve better quantitative and qualitative results [109].

SEC can be used to determine NP-PC interaction and lifetime. It provides separation of molecules and aggregates depending on the molecular size [110]. Following the incubation period, it is used to separate and characterize the NP-PC structure from free proteins. Large molecules/structures that cannot enter the pores decompose in the first place [103, 111]. To determine the residence time of proteins on the NP surface or to determine the affinity of a protein to NP and the rate of change of adsorbed proteins and to examine the formation of NP-PC [112], it is widely used to assess the strength of protein interaction in NPs, measure PC thickness [111, 113], determine NP size distribution, and purify them as a function of their size [114].

In addition to the above techniques, circular dichroism (CD) spectroscopy, Fourier transform infrared spectroscopy (FTIR), and differential scanning calorimetry (DSC) can also be used to detect conformational changes in NP-PC complexes. CD is used to describe proteins according to their binding activity, folding and secondary structure and is based on the principle of circularly polarized light, which distinguishes between the absorption of left-handed and right-handed polarized light by proteins [56, 100]. Secondary structures such as an α-helix, β-sheets or loops of the protein possess their own characteristic CD spectrum in the UV region (250–350 nm) [115]. CD can provide information on protein structure changes that interact with NPs, although it requires a relatively high concentration of the sample used and is ineffective in complex protein mixtures [116].

Surface properties of the NPs can also be determined using FTIR in aqueous solution. It is a rational method to detect vibrational shifts and shape changes corresponding to amide bonds, and this method is also used to monitor the binding of cysteine-containing proteins from the corona to NPs over time [81]. When FTIR is used in conjunction with Raman Spectroscopy, the detailed secondary structure of the protein can be obtained by utilizing vibration and rotational parameters [117]. Besides all the techniques mentioned above, the DSC method can be used to detect conformational changes in PCs [87]. It is by far the most common used and reliable method of thermal analysis since it provides to measure the heat change associated with the thermal denaturation of a molecule by phase transition as the temperature increases and it therefore maintain enthalpy change information after integration [118]. Thus, information about protein stability can be obtained after the NP adsorption process has taken place, and the protein structure after PC formation can be defined.

CONCLUDING REMARKS

Today, although NPs have many applications, there are limitations in their applications in the medical field. One of the most important reasons for this is the formation of PC on NP. Despite manipulations in the surface structure of NPs in current studies, the complex nature of PC continues to influence the behavior of these molecules in the biological environment. The tight interaction between NPs and PC causes differences in their uptake and drug release from the targeted ones. Since the nano-biome mechanism has not been fully understood in the studies carried out to date, there is a need for more extensive research in this area. identification of the proteins involved in the PC structure becomes important. If this mechanism is understood, it will be possible to synthesize carrier NPs with different surface structures and to direct them to the target more easily.

REFERENCES

[1]   C.D. Walkey, J.B. Olsen, H. Guo, A. Emili, and W.C.W. Chan, "Nanoparticle size and surface chemistry determine serum protein adsorption and macrophage uptake", *J. Am. Chem. Soc.,* vol. 134, no. 4, pp. 2139-2147, 2012.
        [http://dx.doi.org/10.1021/ja2084338] [PMID: 22191645]

[2]   B. Fadeel, N. Feliu, C. Vogt, A.M. Abdelmonem, and W.J. Parak, "Bridge over troubled waters: Understanding the synthetic and biological identities of engineered nanomaterials", *Wiley Interdiscip. Rev. Nanomed. Nanobiotechnol.,* vol. 5, no. 2, pp. 111-129, 2013.
        [http://dx.doi.org/10.1002/wnan.1206] [PMID: 23335558]

[3]   I. Freestone, N. Meeks, M. Sax, and C. Higgitt, "The lycurgus cup: A Roman nanotechnology", *Gold Bull.,* vol. 40, no. 4, pp. 270-277, 2007.
        [http://dx.doi.org/10.1007/BF03215599]

[4]   W.J. Stark, P.R. Stoessel, W. Wohlleben, and A. Hafner, "Industrial applications of nanoparticles", *Chem. Soc. Rev.,* vol. 44, no. 16, pp. 5793-5805, 2015.
        [http://dx.doi.org/10.1039/C4CS00362D] [PMID: 25669838]

[5]   TG Laure Brice, "Nanoparticles toxicity and biocompatibility tests", *Nanoparticle.,* vol. 2, no. 1, pp. 1-7, 2020.

[6]   M. Hoseinnejad, S.M. Jafari, and I. Katouzian, "Inorganic and metal nanoparticles and their antimicrobial activity in food packaging applications", *Crit. Rev. Microbiol.,* vol. 44, no. 2, pp. 161-181, 2018.
        [http://dx.doi.org/10.1080/1040841X.2017.1332001] [PMID: 28578640]

[7]   X. Hu, D. Li, Y. Gao, L. Mu, and Q. Zhou, "Knowledge gaps between nanotoxicological research and nanomaterial safety", *Environ. Int.,* vol. 94, pp. 8-23, 2016.
        [http://dx.doi.org/10.1016/j.envint.2016.05.001] [PMID: 27203780]

[8]   A.V. Samrot, C. Justin, S. Padmanaban, and U. Burman, "A study on the effect of chemically synthesized magnetite nanoparticles on earthworm: Eudrilus eugeniae", *Appl. Nanosci.,* vol. 7, no. 1-2, pp. 17-23, 2017.
        [http://dx.doi.org/10.1007/s13204-016-0542-y]

[9]   Y. Yoshioka, K. Higashisaka, and Y. Tsutsumi, "Biocompatibility of nanomaterials", In: *Nanomaterials in Pharmacology.,* Z-R. Lu, S. Sakuma, Eds., Springer New York: NY, 2016, pp. 185-199.
        [http://dx.doi.org/10.1007/978-1-4939-3121-7_9]

[10]   E. Hutter, and D. Maysinger, "Gold nanoparticles and quantum dots for bioimaging", *Microsc. Res. Tech.,* vol. 74, no. 7, pp. 592-604, 2011.
        [http://dx.doi.org/10.1002/jemt.20928] [PMID: 20830812]

[11]   M.P. Monopoli, D. Walczyk, A. Campbell, G. Elia, I. Lynch, F. Baldelli Bombelli, and K.A. Dawson, "Physical-chemical aspects of protein corona: Relevance to *in vitro* and *in vivo* biological impacts of nanoparticles", *J. Am. Chem. Soc.,* vol. 133, no. 8, pp. 2525-2534, 2011.
        [http://dx.doi.org/10.1021/ja107583h] [PMID: 21288025]

[12]   S. Wilhelm, A.J. Tavares, Q. Dai, S. Ohta, J. Audet, H.F. Dvorak, and W.C.W. Chan, "Analysis of nanoparticle delivery to tumours", *Nat. Rev. Mater.,* vol. 1, no. 5, p. 16014, 2016.
[http://dx.doi.org/10.1038/natrevmats.2016.14]

[13]   A. Cox, P. Andreozzi, R. Dal Magro, F. Fiordaliso, A. Corbelli, L. Talamini, C. Chinello, F. Raimondo, F. Magni, M. Tringali, S. Krol, P. Jacob Silva, F. Stellacci, M. Masserini, and F. Re, "Evolution of nanoparticle protein corona across the blood–brain barrier", *ACS Nano,* vol. 12, no. 7, pp. 7292-7300, 2018.
[http://dx.doi.org/10.1021/acsnano.8b03500] [PMID: 29953205]

[14]   K. Choi, J.E. Riviere, and N.A. Monteiro-Riviere, "Protein corona modulation of hepatocyte uptake and molecular mechanisms of gold nanoparticle toxicity", *Nanotoxicology,* vol. 11, no. 1, pp. 64-75, 2017.
[http://dx.doi.org/10.1080/17435390.2016.1264638] [PMID: 27885867]

[15]   R. Gref, M. Lück, P. Quellec, M. Marchand, E. Dellacherie, S. Harnisch, T. Blunk, and R.H. Müller, "Stealth corona-core nanoparticles surface modified by polyethylene glycol (PEG): Influences of the corona (PEG chain length and surface density) and of the core composition on phagocytic uptake and plasma protein adsorption", *Colloids Surf. B Biointerfaces,* vol. 18, no. 3-4, pp. 301-313, 2000.
[http://dx.doi.org/10.1016/S0927-7765(99)00156-3] [PMID: 10915952]

[16]   V.C.F. Mosqueira, P. Legrand, A. Gulik, O. Bourdon, R. Gref, D. Labarre, and G. Barratt, "Relationship between complement activation, cellular uptake and surface physicochemical aspects of novel PEG-modified nanocapsules", *Biomaterials,* vol. 22, no. 22, pp. 2967-2979, 2001.
[http://dx.doi.org/10.1016/S0142-9612(01)00043-6] [PMID: 11575471]

[17]   T. Cedervall, I. Lynch, S. Lindman, T. Berggård, E. Thulin, H. Nilsson, K.A. Dawson, and S. Linse, "Understanding the nanoparticle–protein corona using methods to quantify exchange rates and affinities of proteins for nanoparticles", *Proc. Natl. Acad. Sci.,* vol. 104, no. 7, pp. 2050-2055, 2007.
[http://dx.doi.org/10.1073/pnas.0608582104] [PMID: 17267609]

[18]   M. Lundqvist, J. Stigler, T. Cedervall, T. Berggård, M.B. Flanagan, I. Lynch, G. Elia, and K. Dawson, "The evolution of the protein corona around nanoparticles: A test study", *ACS Nano,* vol. 5, no. 9, pp. 7503-7509, 2011.
[http://dx.doi.org/10.1021/nn202458g] [PMID: 21861491]

[19]   V. Silin, H. Weetall, and D.J. Vanderah, "SPR Studies of the nonspecific adsorption kinetics of human IgG and BSA on gold surfaces modified by self-assembled monolayers (SAMs)", *J. Colloid Interface Sci.,* vol. 185, no. 1, pp. 94-103, 1997.
[http://dx.doi.org/10.1006/jcis.1996.4586] [PMID: 9056309]

[20]   S. Tenzer, D. Docter, J. Kuharev, A. Musyanovych, V. Fetz, R. Hecht, F. Schlenk, D. Fischer, K. Kiouptsi, C. Reinhardt, K. Landfester, H. Schild, M. Maskos, S.K. Knauer, and R.H. Stauber, "Rapid formation of plasma protein corona critically affects nanoparticle pathophysiology", *Nat. Nanotechnol.,* vol. 8, no. 10, pp. 772-781, 2013.
[http://dx.doi.org/10.1038/nnano.2013.181] [PMID: 24056901]

[21]   M.P. Monopoli, C. Åberg, A. Salvati, and K.A. Dawson, "Biomolecular coronas provide the biological identity of nanosized materials", *Nat. Nanotechnol.,* vol. 7, no. 12, pp. 779-786, 2012.
[http://dx.doi.org/10.1038/nnano.2012.207] [PMID: 23212421]

[22]   M.A. Dobrovolskaia, B.W. Neun, S. Man, X. Ye, M. Hansen, A.K. Patri, R.M. Crist, and S.E. McNeil, "Protein corona composition does not accurately predict hematocompatibility of colloidal gold nanoparticles", *Nanomedicine,* vol. 10, no. 7, pp. 1453-1463, 2014.
[http://dx.doi.org/10.1016/j.nano.2014.01.009] [PMID: 24512761]

[23]   K. Hamad-Schifferli, "Exploiting the novel properties of protein coronas: Emerging applications in nanomedicine", *Nanomedicine,* vol. 10, no. 10, pp. 1663-1674, 2015.
[http://dx.doi.org/10.2217/nnm.15.6] [PMID: 26008198]

[24]　K. Natte, J.F. Friedrich, S. Wohlrab, J. Lutzki, R. von Klitzing, W. Österle, and G. Orts-Gil, "Impact of polymer shell on the formation and time evolution of nanoparticle–protein corona", *Colloids Surf. B Biointerfaces,* vol. 104, pp. 213-220, 2013.
[http://dx.doi.org/10.1016/j.colsurfb.2012.11.019] [PMID: 23318220]

[25]　C. Sacchetti, K. Motamedchaboki, A. Magrini, G. Palmieri, M. Mattei, S. Bernardini, N. Rosato, N. Bottini, and M. Bottini, "Surface polyethylene glycol conformation influences the protein corona of polyethylene glycol-modified single-walled carbon nanotubes: Potential implications on biological performance", *ACS Nano,* vol. 7, no. 3, pp. 1974-1989, 2013.
[http://dx.doi.org/10.1021/nn400409h] [PMID: 23413928]

[26]　R. Petry, V.M. Saboia, L.S. Franqui, C.A. Holanda, T.R.R. Garcia, M.A. de Farias, A.G. de Souza Filho, O.P. Ferreira, D.S.T. Martinez, and A.J. Paula, "On the formation of protein corona on colloidal nanoparticles stabilized by depletant polymers", *Mater. Sci. Eng. C,* vol. 105, p. 110080, 2019.
[http://dx.doi.org/10.1016/j.msec.2019.110080] [PMID: 31546390]

[27]　C.C.S. Batista, L.J.C. Albuquerque, A. Jäger, P. Stepánek, and F.C. Giacomelli, "Probing protein adsorption onto polymer-stabilized silver nanocolloids towards a better understanding on the evolution and consequences of biomolecular coronas", *Mater. Sci. Eng. C,* vol. 111, p. 110850, 2020.
[http://dx.doi.org/10.1016/j.msec.2020.110850] [PMID: 32279743]

[28]　X. Wang, C. Yang, C. Wang, L. Guo, T. Zhang, Z. Zhang, H. Yan, and K. Liu, "Polymeric micelles with α-glutamyl-terminated PEG shells show low non-specific protein adsorption and a prolonged *in vivo* circulation time", *Mater. Sci. Eng. C,* vol. 59, pp. 766-772, 2016.
[http://dx.doi.org/10.1016/j.msec.2015.10.084] [PMID: 26652431]

[29]　D.F. Moyano, K. Saha, G. Prakash, B. Yan, H. Kong, M. Yazdani, and V.M. Rotello, "Fabrication of corona-free nanoparticles with tunable hydrophobicity", *ACS Nano,* vol. 8, no. 7, pp. 6748-6755, 2014.
[http://dx.doi.org/10.1021/nn5006478] [PMID: 24971670]

[30]　ST Yang, Y Liu, YW Wang, and A Cao, "Biosafety and bioapplication of nanomaterials by designing protein-nanoparticle interactions", *Small,* vol. 9, no. 9-10, pp. 1635-1653, 2013.
[http://dx.doi.org/10.1002/smll.201201492]

[31]　E. Fasoli, "Protein corona: Dr. Jekyll and Mr. Hyde of nanomedicine", *Biotechnol. Appl. Biochem.,* vol. 68, no. 6, pp. 1139-1152, 2021.
[PMID: 33007792]

[32]　D. Maiolo, P. Bergese, E. Mahon, K.A. Dawson, and M.P. Monopoli, "Surfactant titration of nanoparticle-protein corona", *Anal. Chem.,* vol. 86, no. 24, pp. 12055-12063, 2014.
[http://dx.doi.org/10.1021/ac5027176] [PMID: 25350777]

[33]　V.A. Senapati, K. Kansara, R. Shanker, A. Dhawan, and A. Kumar, "Monitoring characteristics and genotoxic effects of engineered nanoparticle–protein corona", *Mutagenesis,* vol. 32, no. 5, pp. 479-490, 2017.
[http://dx.doi.org/10.1093/mutage/gex028] [PMID: 29048576]

[34]　W. Lai, Q. Wang, L. Li, Z. Hu, J. Chen, and Q. Fang, "Interaction of gold and silver nanoparticles with human plasma: Analysis of protein corona reveals specific binding patterns", *Colloids Surf. B Biointerfaces,* vol. 152, pp. 317-325, 2017.
[http://dx.doi.org/10.1016/j.colsurfb.2017.01.037] [PMID: 28131092]

[35]　D. Chen, S. Ganesh, W. Wang, and M. Amiji, "Plasma protein adsorption and biological identity of systemically administered nanoparticles", *Nanomedicine,* vol. 12, no. 17, pp. 2113-2135, 2017.
[http://dx.doi.org/10.2217/nnm-2017-0178] [PMID: 28805542]

[36]　J.E. Gagner, M.D. Lopez, J.S. Dordick, and R.W. Siegel, "Effect of gold nanoparticle morphology on adsorbed protein structure and function", *Biomaterials,* vol. 32, no. 29, pp. 7241-7252, 2011.
[http://dx.doi.org/10.1016/j.biomaterials.2011.05.091] [PMID: 21705074]

[37]   W. Xiao, J. Xiong, S. Zhang, Y. Xiong, H. Zhang, and H. Gao, "Influence of ligands property and particle size of gold nanoparticles on the protein adsorption and corresponding targeting ability", *Int. J. Pharm.,* vol. 538, no. 1-2, pp. 105-111, 2018.
[http://dx.doi.org/10.1016/j.ijpharm.2018.01.011] [PMID: 29341915]

[38]   J. Piella, N.G. Bastús, and V. Puntes, "Size-dependent protein–nanoparticle interactions in citrate-stabilized gold nanoparticles: The emergence of the protein corona", *Bioconjug. Chem.,* vol. 28, no. 1, pp. 88-97, 2017.
[http://dx.doi.org/10.1021/acs.bioconjchem.6b00575] [PMID: 27997136]

[39]   R. García-Álvarez, M. Hadjidemetriou, A. Sánchez-Iglesias, L.M. Liz-Marzán, and K. Kostarelos, "*In vivo* formation of protein corona on gold nanoparticles. The effect of their size and shape", *Nanoscale,* vol. 10, no. 3, pp. 1256-1264, 2018.
[http://dx.doi.org/10.1039/C7NR08322J] [PMID: 29292433]

[40]   M.E. Aubin-Tam, and K. Hamad-Schifferli, "Gold nanoparticle-cytochrome C complexes: The effect of nanoparticle ligand charge on protein structure", *Langmuir,* vol. 21, no. 26, pp. 12080-12084, 2005.
[http://dx.doi.org/10.1021/la052102e] [PMID: 16342975]

[41]   Y. Kim, S.M. Ko, and J.M. Nam, "Protein-nanoparticle interaction-induced changes in protein structure and aggregation", *Chem. Asian J.,* vol. 11, no. 13, pp. 1869-1877, 2016.
[http://dx.doi.org/10.1002/asia.201600236] [PMID: 27062521]

[42]   A. Bekdemir, S. Liao, and F. Stellacci, "On the effect of ligand shell heterogeneity on nanoparticle/protein binding thermodynamics", *Colloids Surf. B Biointerfaces,* vol. 174, pp. 367-373, 2019.
[http://dx.doi.org/10.1016/j.colsurfb.2018.11.027] [PMID: 30472623]

[43]   D. Kumar, I. Mutreja, K. Chitcholtan, and P. Sykes, "Cytotoxicity and cellular uptake of different sized gold nanoparticles in ovarian cancer cells", *Nanotechnology,* vol. 28, no. 47, p. 475101, 2017.
[http://dx.doi.org/10.1088/1361-6528/aa935e] [PMID: 29027909]

[44]   E. Lavagna, J. Barnoud, G. Rossi, and L. Monticelli, "Size-dependent aggregation of hydrophobic nanoparticles in lipid membranes", *Nanoscale,* vol. 12, no. 17, pp. 9452-9461, 2020.
[http://dx.doi.org/10.1039/D0NR00868K] [PMID: 32328605]

[45]   J.W. Park, and J.S. Shumaker-Parry, "Structural study of citrate layers on gold nanoparticles: role of intermolecular interactions in stabilizing nanoparticles", *J. Am. Chem. Soc.,* vol. 136, no. 5, pp. 1907-1921, 2014.
[http://dx.doi.org/10.1021/ja4097384] [PMID: 24422457]

[46]   R Stein, and B. Friedrich, *Synthesis and characterization of citrate-stabilized gold-coated superparamagnetic iron oxide nanoparticles for biomedical applications.,* vol. 25, no. 19, 2020.
[http://dx.doi.org/10.3390/molecules25194425]

[47]   M. Hood, M. Mari, and R. Muñoz-Espí, "Synthetic strategies in the preparation of polymer/inorganic hybrid nanoparticles", *Materials,* vol. 7, no. 5, pp. 4057-4087, 2014.
[http://dx.doi.org/10.3390/ma7054057] [PMID: 28788665]

[48]   Y. Herdiana, N. Wathoni, S. Shamsuddin, and M. Muchtaridi, "Drug release study of the chitosan-based nanoparticles", *Heliyon,* vol. 8, no. 1, p. e08674, 2022.
[http://dx.doi.org/10.1016/j.heliyon.2021.e08674] [PMID: 35028457]

[49]   J. Sharifi-Rad, C. Quispe, M. Butnariu, L.S. Rotariu, O. Sytar, S. Sestito, S. Rapposelli, M. Akram, M. Iqbal, A. Krishna, N.V.A. Kumar, S.S. Braga, S.M. Cardoso, K. Jafernik, H. Ekiert, N. Cruz-Martins, A. Szopa, M. Villagran, L. Mardones, M. Martorell, A.O. Docea, and D. Calina, "Chitosan nanoparticles as a promising tool in nanomedicine with particular emphasis on oncological treatment", *Cancer Cell Int.,* vol. 21, no. 1, p. 318, 2021.
[http://dx.doi.org/10.1186/s12935-021-02025-4] [PMID: 34167552]

[50]   M. Debayle, E. Balloul, F. Dembele, X. Xu, M. Hanafi, F. Ribot, C. Monzel, M. Coppey, A. Fragola, M. Dahan, T. Pons, and N. Lequeux, "Zwitterionic polymer ligands: An ideal surface coating to totally suppress protein-nanoparticle corona formation?", *Biomaterials,* vol. 219, p. 119357, 2019.
[http://dx.doi.org/10.1016/j.biomaterials.2019.119357] [PMID: 31351245]

[51]   B.M. King, and J. Fiegel, "Zwitterionic polymer coatings enhance gold nanoparticle stability and uptake in various biological environments", *AAPS J.,* vol. 24, no. 1, p. 18, 2022.
[http://dx.doi.org/10.1208/s12248-021-00652-3] [PMID: 34984558]

[52]   R. Safavi-Sohi, S. Maghari, M. Raoufi, S.A. Jalali, M.J. Hajipour, A. Ghassempour, and M. Mahmoudi, "Bypassing protein corona issue on active targeting: zwitterionic coatings dictate specific interactions of targeting moieties and cell receptors", *ACS Appl. Mater. Interfaces,* vol. 8, no. 35, pp. 22808-22818, 2016.
[http://dx.doi.org/10.1021/acsami.6b05099] [PMID: 27526263]

[53]   P. Mohapatra, D. Singh, and S.K. Sahoo, "PEGylated nanoparticles as a versatile drug delivery system", In: *Nanoeng. Biomater.*, 2022, pp. 309-341.
[http://dx.doi.org/10.1002/9783527832095.ch10]

[54]   S.M. Ahsan, C.M. Rao, and M.F. Ahmad, "Nanoparticle-protein interaction: The significance and role of protein corona", *Adv. Exp. Med. Biol.,* vol. 1048, pp. 175-198, 2018.
[http://dx.doi.org/10.1007/978-3-319-72041-8_11] [PMID: 29453539]

[55]   JH Shannahan, KS Fritz, AJ Raghavendra, R Podila, I Persaud, and JM Brown, "From the cover: Disease-induced disparities in formation of the nanoparticle-biocorona and the toxicological consequences", *Toxicol. Sci.,* vol. 152, no. 2, pp. 406-416, 2016.

[56]   J. Shah, and S. Singh, *CHAPTER 1 nanoparticle–protein corona complex: Composition, kinetics, physico–chemical characterization, and impact on biomedical applications. nanoparticle–protein corona: Biophysics to biology.* The Royal Society of Chemistry, 2019, pp. 1-30.

[57]   X. Bai, J. Wang, Q. Mu, and G. Su, "*In vivo* protein corona formation: Characterizations, effects on engineered nanoparticles' biobehaviors, and applications", *Front. Bioeng. Biotechnol.,* vol. 9, p. 646708, 2021.
[http://dx.doi.org/10.3389/fbioe.2021.646708] [PMID: 33869157]

[58]   M García Vence, MDP Chantada-Vázquez, S Vázquez-Estévez, J Manuel Cameselle-Teijeiro, SB Bravo, and C Núñez, "Potential clinical applications of the personalized, disease-specific protein corona on nanoparticles", *Clin. Chim. Acta.,* vol. 501, pp. 102-111, 2020.
[http://dx.doi.org/10.1016/j.cca.2019.10.027]

[59]   M.J. Hajipour, J. Raheb, O. Akhavan, S. Arjmand, O. Mashinchian, M. Rahman, M. Abdolahad, V. Serpooshan, S. Laurent, and M. Mahmoudi, "Personalized disease-specific protein corona influences the therapeutic impact of graphene oxide", *Nanoscale,* vol. 7, no. 19, pp. 8978-8994, 2015.
[http://dx.doi.org/10.1039/C5NR00520E] [PMID: 25920546]

[60]   R. Rampado, S. Crotti, P. Caliceti, S. Pucciarelli, and M. Agostini, "Recent advances in understanding the protein corona of nanoparticles and in the formulation of "stealthy" nanomaterials", *Front. Bioeng. Biotechnol.,* vol. 8, p. 166, 2020.
[http://dx.doi.org/10.3389/fbioe.2020.00166] [PMID: 32309278]

[61]   A. Lesniak, A. Campbell, M.P. Monopoli, I. Lynch, A. Salvati, and K.A. Dawson, "Serum heat inactivation affects protein corona composition and nanoparticle uptake", *Biomaterials,* vol. 31, no. 36, pp. 9511-9518, 2010.
[http://dx.doi.org/10.1016/j.biomaterials.2010.09.049] [PMID: 21059466]

[62]   M. Mahmoudi, A.M. Abdelmonem, S. Behzadi, J.H. Clement, S. Dutz, M.R. Ejtehadi, R. Hartmann, K. Kantner, U. Linne, P. Maffre, S. Metzler, M.K. Moghadam, C. Pfeiffer, M. Rezaei, P. Ruiz-Lozano, V. Serpooshan, M.A. Shokrgozar, G.U. Nienhaus, and W.J. Parak, "Temperature: The ignored factor at the NanoBio interface", *ACS Nano,* vol. 7, no. 8, pp. 6555-6562, 2013.
[http://dx.doi.org/10.1021/nn305337c] [PMID: 23808533]

[63]   A. Bhogale, N. Patel, J. Mariam, P.M. Dongre, A. Miotello, and D.C. Kothari, "Comprehensive studies on the interaction of copper nanoparticles with bovine serum albumin using various spectroscopies", *Colloids Surf. B Biointerfaces,* vol. 113, pp. 276-284, 2014.
[http://dx.doi.org/10.1016/j.colsurfb.2013.09.021] [PMID: 24121071]

[64]   I Yadav, S Kumar, and VK Aswal, "Structure and interaction in the ph-dependent phase behavior of nanoparticle-protein systems", *Langmuir,* vol. 33, no. 5, pp. 1227-1238, 2017.
[http://dx.doi.org/10.1021/acs.langmuir.6b04127]

[65]   M. Raoufi, M.J. Hajipour, S.M. Kamali Shahri, I. Schoen, U. Linn, and M. Mahmoudi, "Probing fibronectin conformation on a protein corona layer around nanoparticles", *Nanoscale,* vol. 10, no. 3, pp. 1228-1233, 2018.
[http://dx.doi.org/10.1039/C7NR06970G] [PMID: 29292453]

[66]   S. Tenzer, D. Docter, S. Rosfa, A. Wlodarski, J. Kuharev, A. Rekik, S.K. Knauer, C. Bantz, T. Nawroth, C. Bier, J. Sirirattanapan, W. Mann, L. Treuel, R. Zellner, M. Maskos, H. Schild, and R.H. Stauber, "Nanoparticle size is a critical physicochemical determinant of the human blood plasma corona: A comprehensive quantitative proteomic analysis", *ACS Nano,* vol. 5, no. 9, pp. 7155-7167, 2011.
[http://dx.doi.org/10.1021/nn201950e] [PMID: 21866933]

[67]   N.P. Mortensen, G.B. Hurst, W. Wang, C.M. Foster, P.D. Nallathamby, and S.T. Retterer, "Dynamic development of the protein corona on silica nanoparticles: Composition and role in toxicity", *Nanoscale,* vol. 5, no. 14, pp. 6372-6380, 2013.
[http://dx.doi.org/10.1039/c3nr33280b] [PMID: 23736871]

[68]   M. Hadjidemetriou, Z. Al-Ahmady, M. Mazza, R.F. Collins, K. Dawson, and K. Kostarelos, "*In Vivo* biomolecule corona around blood-circulating, clinically used and antibody-targeted lipid bilayer nanoscale vesicles", *ACS Nano,* vol. 9, no. 8, pp. 8142-8156, 2015.
[http://dx.doi.org/10.1021/acsnano.5b03300] [PMID: 26135229]

[69]   M. Falahati, F. Attar, M. Sharifi, T. Haertlé, J.F. Berret, R.H. Khan, and A.A. Saboury, "A health concern regarding the protein corona, aggregation and disaggregation", *Biochim. Biophys. Acta, Gen. Subj.,* vol. 1863, no. 5, pp. 971-991, 2019.
[http://dx.doi.org/10.1016/j.bbagen.2019.02.012] [PMID: 30802594]

[70]   K. Park, "To PEGylate or not to PEGylate, that is not the question", *J. Control. Release.,* vol. 142, no. 2, pp. 147-148, 2010.
[http://dx.doi.org/10.1016/j.jconrel.2010.01.025]

[71]   B.S. Zolnik, Á. González-Fernández, N. Sadrieh, and M.A. Dobrovolskaia, "Nanoparticles and the immune system", *Endocrinology,* vol. 151, no. 2, pp. 458-465, 2010.
[http://dx.doi.org/10.1210/en.2009-1082] [PMID: 20016026]

[72]   D.F. Moyano, M. Goldsmith, D.J. Solfiell, D. Landesman-Milo, O.R. Miranda, D. Peer, and V.M. Rotello, "Nanoparticle hydrophobicity dictates immune response", *J. Am. Chem. Soc.,* vol. 134, no. 9, pp. 3965-3967, 2012.
[http://dx.doi.org/10.1021/ja2108905] [PMID: 22339432]

[73]   M.A. Dobrovolskaia, "Dendrimers effects on the immune system: Insights into toxicity and therapeutic utility", *Curr. Pharm. Des.,* vol. 23, no. 21, pp. 3134-3141, 2017.
[PMID: 28294045]

[74]   T.A. Ngobili, and M.A. Daniele, "Nanoparticles and direct immunosuppression", *Exp. Biol. Med.,* vol. 241, no. 10, pp. 1064-1073, 2016.
[http://dx.doi.org/10.1177/1535370216650053] [PMID: 27229901]

[75]   C.T. Ng, L.Q. Yong, M.P. Hande, C.N. Ong, L. Yu, B.H. Bay, and G.H. Baeg, "Zinc oxide nanoparticles exhibit cytotoxicity and genotoxicity through oxidative stress responses in human lung fibroblasts and Drosophila melanogaster", *Int. J. Nanomed.,* vol. 12, pp. 1621-1637, 2017.
[http://dx.doi.org/10.2147/IJN.S124403] [PMID: 28280330]

[76]   R.A. Yokel, S. Hussain, S. Garantziotis, P. Demokritou, V. Castranova, and F.R. Cassee, "The yin: An adverse health perspective of nanoceria: uptake, distribution, accumulation, and mechanisms of its toxicity", *Environ. Sci. Nano,* vol. 1, no. 5, pp. 406-428, 2014.
[http://dx.doi.org/10.1039/C4EN00039K] [PMID: 25243070]

[77]   L. García-Fernández, J. Garcia-Pardo, O. Tort, I. Prior, M. Brust, E. Casals, J. Lorenzo, and V.F. Puntes, "Conserved effects and altered trafficking of Cetuximab antibodies conjugated to gold nanoparticles with precise control of their number and orientation", *Nanoscale,* vol. 9, no. 18, pp. 6111-6121, 2017.
[http://dx.doi.org/10.1039/C7NR00947J] [PMID: 28447703]

[78]   P Grenier, IMO Viana, EM Lima, and N Bertrand, "Anti-polyethylene glycol antibodies alter the protein corona deposited on nanoparticles and the physiological pathways regulating their fate *in vivo*", *J. Control. Release.,* vol. 287, pp. 121-131, 2018.
[http://dx.doi.org/10.1016/j.jconrel.2018.08.022]

[79]   G.M. Mortimer, N.J. Butcher, A.W. Musumeci, Z.J. Deng, D.J. Martin, and R.F. Minchin, "Cryptic epitopes of albumin determine mononuclear phagocyte system clearance of nanomaterials", *ACS Nano,* vol. 8, no. 4, pp. 3357-3366, 2014.
[http://dx.doi.org/10.1021/nn405830g] [PMID: 24617595]

[80]   R. Cai, and C. Chen, "The crown and the scepter: Roles of the protein corona in nanomedicine", *Adv. Mater.,* vol. 31, no. 45, p. 1805740, 2019.
[http://dx.doi.org/10.1002/adma.201805740] [PMID: 30589115]

[81]   C. Carrillo-Carrion, M. Carril, and W.J. Parak, "Techniques for the experimental investigation of the protein corona", *Curr. Opin. Biotechnol.,* vol. 46, pp. 106-113, 2017.
[http://dx.doi.org/10.1016/j.copbio.2017.02.009] [PMID: 28301820]

[82]   S. Dominguez-Medina, S. Chen, J. Blankenburg, P. Swanglap, C.F. Landes, and S. Link, "Measuring the hydrodynamic size of nanoparticles using fluctuation correlation spectroscopy", *Annu. Rev. Phys. Chem.,* vol. 67, no. 1, pp. 489-514, 2016.
[http://dx.doi.org/10.1146/annurev-physchem-040214-121510] [PMID: 27215820]

[83]   T Liedl, S Keller, FC Simmel, JO Rädler, and WJ Parak, "Fluorescent nanocrystals as colloidal probes in complex fluids measured by fluorescence correlation spectroscopy", *Small,* vol. 1, no. 10, pp. 997-1003, 2005.
[http://dx.doi.org/10.1002/smll.200500108]

[84]   Y Li, and JS Lee, "Insights into characterization methods and biomedical applications of nanoparticle-protein corona", *Materials,* vol. 13, no. 14, p. 3093, 2020.
[http://dx.doi.org/10.3390/ma13143093]

[85]   M. Kokkinopoulou, J. Simon, K. Landfester, V. Mailänder, and I. Lieberwirth, "Visualization of the protein corona: Towards a biomolecular understanding of nanoparticle-cell-interactions", *Nanoscale,* vol. 9, no. 25, pp. 8858-8870, 2017.
[http://dx.doi.org/10.1039/C7NR02977B] [PMID: 28632260]

[86]   B.D. Johnston, W.G. Kreyling, C. Pfeiffer, M. Schäffler, H. Sarioglu, S. Ristig, S. Hirn, N. Haberl, S. Thalhammer, S.M. Hauck, M. Semmler-Behnke, M. Epple, J. Hühn, P. Del Pino, and W.J. Parak, "Colloidal stability and surface chemistry are key factors for the composition of the protein corona of inorganic gold nanoparticles", *Adv. Funct. Mater.,* vol. 27, no. 42, p. 1701956, 2017.
[http://dx.doi.org/10.1002/adfm.201701956]

[87]   V. Vergaro, I. Pisano, R. Grisorio, F. Baldassarre, R. Mallamaci, and A. Santoro, "CaCO(3) as an environmentally friendly renewable material for drug delivery systems: Uptake of HSA-CaCO(3) nanocrystals conjugates in cancer cell lines", *Materials,* vol. 12, no. 9, p. 1481, 2019.

[88]   K. Natte, J.F. Friedrich, S. Wohlrab, J. Lutzki, R. von Klitzing, W. Österle, and G. Orts-Gil, "Impact of polymer shell on the formation and time evolution of nanoparticle–protein corona", *Colloids Surf. B Biointerfaces,* vol. 104, pp. 213-220, 2013.

[http://dx.doi.org/10.1016/j.colsurfb.2012.11.019] [PMID: 23318220]

[89]   P.M. Kelly, C. Åberg, E. Polo, A. O'Connell, J. Cookman, J. Fallon, Ž. Krpetić, and K.A. Dawson, "Mapping protein binding sites on the biomolecular corona of nanoparticles", *Nat. Nanotechnol.*, vol. 10, no. 5, pp. 472-479, 2015.
[http://dx.doi.org/10.1038/nnano.2015.47] [PMID: 25822932]

[90]   B. Kharazian, N.L. Hadipour, and M.R. Ejtehadi, "Understanding the nanoparticle–protein corona complexes using computational and experimental methods", *Int. J. Biochem. Cell Biol.*, vol. 75, pp. 162-174, 2016.
[http://dx.doi.org/10.1016/j.biocel.2016.02.008] [PMID: 26873405]

[91]   J. Mukherjee, and M.N. Gupta, "Protein aggregates: Forms, functions and applications", *Int. J. Biol. Macromol.*, vol. 97, pp. 778-789, 2017.
[http://dx.doi.org/10.1016/j.ijbiomac.2016.11.014] [PMID: 27825997]

[92]   S. Kumar, I. Yadav, and V.K. Aswal, "Structure and interaction of nanoparticle-protein complexes", *Langmuir*, vol. 34, no. 20, pp. 5679-5695, 2018.

[93]   N. Jain, A. Bhargava, M. Rathi, R.V. Dilip, and J. Panwar, "Removal of protein capping enhances the antibacterial efficiency of biosynthesized silver nanoparticles", *PLoS One*, vol. 10, no. 7, p. e0134337, 2015.
[http://dx.doi.org/10.1371/journal.pone.0134337] [PMID: 26226385]

[94]   N. Dasgupta, S. Ranjan, D. Patra, P. Srivastava, A. Kumar, and C. Ramalingam, "Bovine serum albumin interacts with silver nanoparticles with a side-on or end on conformation", *Chem. Biol. Interact.*, vol. 253, pp. 100-111, 2016.
[http://dx.doi.org/10.1016/j.cbi.2016.05.018] [PMID: 27180205]

[95]   K Furumoto, K Ogawara, S Nagayama, Y Takakura, M Hashida, and K Higaki, "Important role of serum proteins associated on the surface of particles in their hepatic disposition", *J. Control. Release.*, vol. 83, no. 1, pp. 89-96, 2002.
[http://dx.doi.org/10.1016/S0168-3659(02)00196-7]

[96]   T. Kopac, "Protein corona, understanding the nanoparticle–protein interactions and future perspectives: A critical review", *Int. J. Biol. Macromol.*, vol. 169, pp. 290-301, 2021.
[http://dx.doi.org/10.1016/j.ijbiomac.2020.12.108] [PMID: 33340622]

[97]   A Tomak, S Cesmeli, BD Hanoglu, and D Winkler, "Nanoparticle-protein corona complex: Understanding multiple interactions between environmental factors, corona formation, and biological activity", *Nanotoxicology*, vol. 15, no. 10, pp. 1331-1357, 2021.
[http://dx.doi.org/10.1080/17435390.2022.2025467]

[98]   O. Vilanova, J.J. Mittag, P.M. Kelly, S. Milani, K.A. Dawson, and J.O. Rädler, "Understanding the kinetics of protein-nanoparticle corona formation", *ACS. Nano.*, vol. 10, no. 12, pp. 10842-10850, 2016.

[99]   C.C. Fleischer, and C.K. Payne, "Nanoparticle-cell interactions: Molecular structure of the protein corona and cellular outcomes", *Acc. Chem. Res.*, vol. 47, no. 8, pp. 2651-2659, 2014.
[http://dx.doi.org/10.1021/ar500190q] [PMID: 25014679]

[100] R.K. Mishra, A. Ahmad, A. Vyawahare, P. Alam, T.H. Khan, and R. Khan, "Biological effects of formation of protein corona onto nanoparticles", *Int. J. Biol. Macromol.*, vol. 175, pp. 1-18, 2021.
[http://dx.doi.org/10.1016/j.ijbiomac.2021.01.152] [PMID: 33508360]

[101] X. Zhang, J. Zhang, F. Zhang, and S. Yu, "Probing the binding affinity of plasma proteins adsorbed on Au nanoparticles", *Nanoscale*, vol. 9, no. 14, pp. 4787-4792, 2017.
[http://dx.doi.org/10.1039/C7NR01523B] [PMID: 28345718]

[102] D. Prozeller, S. Morsbach, and K. Landfester, "Isothermal titration calorimetry as a complementary method for investigating nanoparticle–protein interactions", *Nanoscale*, vol. 11, no. 41, pp. 19265-19273, 2019.

[http://dx.doi.org/10.1039/C9NR05790K] [PMID: 31549702]

[103] A.L. Capriotti, G. Caracciolo, C. Cavaliere, V. Colapicchioni, S. Piovesana, D. Pozzi, and A. Laganà, "Analytical methods for characterizing the nanoparticle–protein corona", *Chromatographia*, vol. 77, no. 11-12, pp. 755-769, 2014.
[http://dx.doi.org/10.1007/s10337-014-2677-x]

[104] P. Aggarwal, J.B. Hall, C.B. McLeland, M.A. Dobrovolskaia, and S.E. McNeil, "Nanoparticle interaction with plasma proteins as it relates to particle biodistribution, biocompatibility and therapeutic efficacy", *Adv. Drug Deliv. Rev.*, vol. 61, no. 6, pp. 428-437, 2009.
[http://dx.doi.org/10.1016/j.addr.2009.03.009] [PMID: 19376175]

[105] T.M. Göppert, and R.H. Müller, "Plasma protein adsorption of Tween 80- and poloxamer 188-stabilized solid lipid nanoparticles", *J. Drug Target.*, vol. 11, no. 4, pp. 225-231, 2003.
[http://dx.doi.org/10.1080/1061186031000161 5956] [PMID: 14578109]

[106] H.R. Kim, K. Andrieux, S. Gil, M. Taverna, H. Chacun, D. Desmaële, F. Taran, D. Georgin, and P. Couvreur, "Translocation of poly(ethylene glycol-co-hexadecyl)cyanoacrylate nanoparticles into rat brain endothelial cells: Role of apolipoproteins in receptor-mediated endocytosis", *Biomacromolecules*, vol. 8, no. 3, pp. 793-799, 2007.
[http://dx.doi.org/10.1021/bm060711a] [PMID: 17309294]

[107] M. Magro, M. Zaccarin, G. Miotto, L. Da Dalt, D. Baratella, P. Fariselli, G. Gabai, and F. Vianello, "Analysis of hard protein corona composition on selective iron oxide nanoparticles by MALDI-TOF mass spectrometry: Identification and amplification of a hidden mastitis biomarker in milk proteome", *Anal. Bioanal. Chem.*, vol. 410, no. 12, pp. 2949-2959, 2018.
[http://dx.doi.org/10.1007/s00216-018-0976-z] [PMID: 29532191]

[108] P. Pino, B. Pelaz, Q. Zhang, P. Maffre, G.U. Nienhaus, and W.J. Parak, "Protein corona formation around nanoparticles: From the past to the future", *Mater. Horiz.*, vol. 1, no. 3, pp. 301-313, 2014.
[http://dx.doi.org/10.1039/C3MH00106G]

[109] D. Pozzi, G. Caracciolo, L. Digiacomo, V. Colapicchioni, S. Palchetti, A.L. Capriotti, C. Cavaliere, R. Zenezini Chiozzi, A. Puglisi, and A. Laganà, "The biomolecular corona of nanoparticles in circulating biological media", *Nanoscale*, vol. 7, no. 33, pp. 13958-13966, 2015.
[http://dx.doi.org/10.1039/C5NR03701H] [PMID: 26222625]

[110] F. Pederzoli, G. Tosi, M.A. Vandelli, D. Belletti, F. Forni, and B. Ruozi, "Protein corona and nanoparticles: How can we investigate on?", *Wiley Interdiscip. Rev. Nanomed. Nanobiotechnol.*, vol. 9, no. 6, 2017.
[http://dx.doi.org/10.1002/wnan.1467] [PMID: 28296346]

[111] C.D. Walkey, and W.C.W. Chan, "Understanding and controlling the interaction of nanomaterials with proteins in a physiological environment", *Chem. Soc. Rev.*, vol. 41, no. 7, pp. 2780-2799, 2012.
[http://dx.doi.org/10.1039/C1CS15233E] [PMID: 22086677]

[112] M. Mahmoudi, I. Lynch, M.R. Ejtehadi, M.P. Monopoli, F.B. Bombelli, and S. Laurent, "Protein-nanoparticle interactions: Opportunities and challenges", *Chem. Rev.*, vol. 111, no. 9, pp. 5610-5637, 2011.
[http://dx.doi.org/10.1021/cr100440g] [PMID: 21688848]

[113] J. Klein, "Probing the interactions of proteins and nanoparticles", *Proc. Natl. Acad. Sci.*, vol. 104, no. 7, pp. 2029-2030, 2007.
[http://dx.doi.org/10.1073/pnas.0611610104] [PMID: 17284585]

[114] R. Lévy, N.T.K. Thanh, R.C. Doty, I. Hussain, R.J. Nichols, D.J. Schiffrin, M. Brust, and D.G. Fernig, "Rational and combinatorial design of peptide capping ligands for gold nanoparticles", *J. Am. Chem. Soc.*, vol. 126, no. 32, pp. 10076-10084, 2004.
[http://dx.doi.org/10.1021/ja0487269] [PMID: 15303884]

[115] M.F. Zarabi, A. Farhangi, S.K. Mazdeh, Z. Ansarian, D. Zare, M.R. Mehrabi, and A. Akbarzadeh, "Synthesis of gold nanoparticles coated with aspartic acid and their conjugation with FVIII Protein and

FVIII Antibody", *Indian J. Clin. Biochem.,* vol. 29, no. 2, pp. 154-160, 2014.
[http://dx.doi.org/10.1007/s12291-013-0323-2] [PMID: 24757296]

[116] M Rahman, S Laurent, N Tawil, LH Yahia, and M Mahmoudi, "Protein-nanoparticle interactions", In: *Bio-Nano. Inter.* Springer-Verlag: Berlin Heidelberg, 2013, pp. 1-84.

[117] D.M. Ridgley, E.C. Claunch, and J.R. Barone, "Characterization of large amyloid fibers and tapes with Fourier transform infrared (FT-IR) and Raman spectroscopy", *Appl. Spectrosc.,* vol. 67, no. 12, pp. 1417-1426, 2013.
[http://dx.doi.org/10.1366/13-07059] [PMID: 24359656]

[118] S Wang, X Sha, S Yu, and Y. Zhao, "Nanocalorimeters for biomolecular analysis and cell metabolism monitoring", *Biomicrofluidics,* vol. 14, no. 1, p. 011503, 2020.
[http://dx.doi.org/10.1063/1.5134870]

CHAPTER 11

Role of Nanoparticular/Nanovesicular Systems as Biosensors

Özlem Çoban[1,*] and **Emine Taşhan[2]**

[1] *Department of Pharmaceutical Technology, Faculty of Pharmacy, Karadeniz Technical University, Trabzon, Türkiye*

[2] *Zoleant LLC, Zoeuticals, Regulatory Affairs and Quality, New York, USA*

Abstract: Biosensors are analytical apparatus utilized for the qualitative and quantitative detection of various biological or non-biological analytes. Early diagnosis of diseases (cancer, infectious disease), monitoring environmental pollution, and ensuring food safety are very important in terms of individual and public health. Therefore, it is also crucial to detect these markers sensitively and accurately, with cheap and simple methods, especially despite limited resources. Nanoparticles, thanks to their nano size, provide wide areas of biosensing and amplify signals. In most of the works, it was observed that the limit of detection (LOD) value decreased and the selectivity improved in biosensors prepared using nanosystems compared to conventional sensors. In this respect, the results give us hope for the use of nanosystems in biosensors. In this section, the subject of biosensors is briefly mentioned and mainly studies on the use of nanoparticular/nanovesicular systems in the field of biosensors are included.

Keywords: Analyte, Bioassay, Biosensor, Bioreceptor, Biosensing, Conjugate, Detection, Genosensor, Gold nanoparticle, Immunosensor, Limit of detection, Liposome, Nanobiosensor, Nanomaterials, Nanosystems, Nanotechnology, Polymeric nanoparticle, Quantum dots, Signal amplification, Transducer.

INTRODUCTION

Biosensors basically consist of two components as bioreceptor and transducer [1]. More recently, it has been stated that biosensors consist of three components: The sensor, which is a membrane with varied biological structures, the transducer and the electronic system that magnifies and saves the signal for data presentation (Fig. **1**) [2]. A bioreceptor is a biomolecule that recognizes and can selectively bind to target analytes. The transducer, on the other hand, converts the binding

---

* **Corresponding author Özlem Çoban:** Department of Pharmaceutical Technology, Faculty of Pharmacy, Karadeniz Technical University, Trabzon, Türkiye; Tel: +905057642367; E-mail: o.coban88@gmail.com

Habibe Yılmaz (Ed.)

event into electrical or optical signals, usually *via* electrochemical or fluorescent techniques [1].

It is very important to follow biological or biochemical processes for medical and biological applications [3]. In addition, the detection of microorganisms, dangerous chemicals and other hazardous wastes in water and soil, drugs and toxic substances in food, trace gases in mining regions, and the decrease in the ozone layer become more of an issue for environmental health [4]. There is great attention in the work of sensitive, selective and economical biosensors because of their contribution to the realization of high-precision diagnostics and personalized medicine. Various biosensors have been extensively investigated since the development of the first generation biosensors, in which glucose oxidase is immobilized on an amperometric oxygen electrode for glucose sensing, which was developed by Clark and Lyons in 1962 [3, 5].

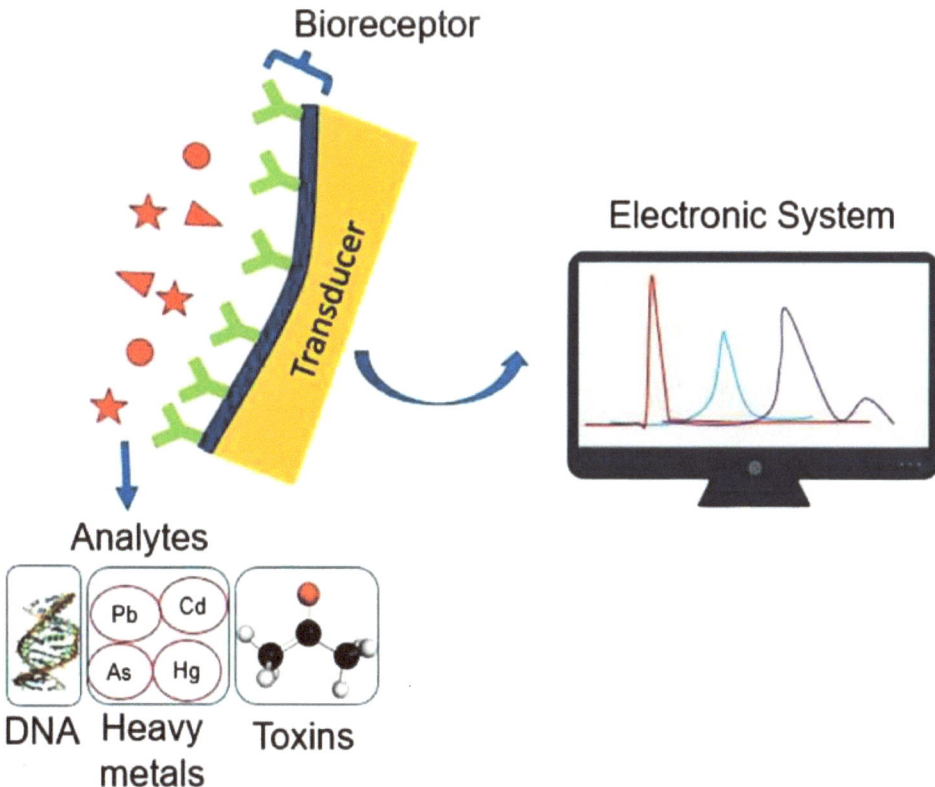

Fig. (1). Schematic image of the biosensor and its components.

Biosensors have many advantages over other biological sensing methods, such as high test speeds and flexibility. For example, they provide information that can be useful in patient care planning to healthcare providers thanks to their ability to perform fast and real-time analysis. They can do multi-target analysis. The system is suitable for automation and thus testing costs can be reduced. They make contributions to the betimes detection of cancer and other diseases due to their fast, selective and high sensitivity properties and thus to the improvement of the prognosis. It may be useful for communities where healthcare delivery is inadequate since its ease of use and portability [6].

Classifications of biosensors are based on the physicochemical transduction mode used for the detection and analysis of signals, or type of biorecognition material [4, 5].

According to the transducer mode, biosensors are divided into 4 groups:

• Electrochemical biosensor

• Optical biosensor

• Thermal biosensor

• Piezoelectric biosensor [5]

Electrochemical biosensors: They measure fluctuations in current, potential, conductivity or impedance in the test sample induced by the interaction among the biological material and the analyte. In this case, we can classify electrochemical biosensors into amperometric, potentiometric, conductometric or impedimetric biosensors [4]. Amperometric biosensors meter the current arised during the oxidation or reduction of the material, which is electrically active, whereas potentiometric biosensors quantify the potential of the biosensor electrode relative to the reference one. On the other hand, conductometric biosensors gauge the change in conductivity resulting from the biochemical reaction [5], and impedimetric biosensors meter variances in charge conductivity and impedance on the sensor surface resulting from specific binding to the target [7]. There are many studies on the detection of glucose, cholesterol, amino acids, urea, alcohol, neurotransmitters, carbon dioxide, pesticides, heavy metals, chemicals, possible markers and endocrine disrupting hormones, polychlorinated biphenyls and milk toxins in biological or environmental media using ampoterimetric, potentiometric, conductometric or impedimetric biosensors [4].

Optical biosensors: As a result of the interaction between the biocatalyst and the sample, the optical properties of the sample may change, resulting in an alteration in the intensity of the absorbed or emitted light [4, 5]. Optical biosensors rest on several methods, namely absorption, fluorescence, luminescence, and surface plasmon resonance (SPR) [5], and accordingly, they are classified as bioluminescence/chemiluminescence, fluorescence, colorimetric (alteration in UV/visible absorption) or SPR based biosensors [4]. Different classifications can be observed in different sources. For example, in addition to the ones mentioned above, optical waveguides, optical resonators, photonic crystal, refraktif index, Raman scattering [8], localized surface plasmon resonance, interferometric, ellipsometric, reflectometric interference spectroscopy [9], reflection, refraction, flow cytometry [10] types are also mentioned. Bioluminescent biosensors are based on the fundamental of measuring light emission by living bacteria in reaction to any biological, physical or chemical difference in the analyte, and with this mechanism, heavy metal pollution, food and environmental conditions can be monitored. Fluorescently labeled biomolecules are used in fluorescence biosensors, and fluorescence is emitted by these biomolecules as a consequence of interaction with an analyte. In this way, biochemical oxygen demand, iron or water presence in plants or microbial habitats, and cell populations can be measured. Colorimetric biosensors, on the other hand, measure the alteration in color or optical intensity of the analysis sample as a result of a chemical reaction [4]. These biosensors are being extensively investigated to detect food and waterborne pathogens, as visible color change can be easily detected by these biosensors [4, 5]. For SPR-based biosensors, the plasmonic characteristics of noble metals are exploited [5]. These biosensors quantitatively measure binding occurrences in concurrent without labeling interacting molecules. It was first shown in 1983 that changes in the refractive index occur meanwhile molecules interact with the sensing surface [9, 11]. Today, this technology is widely used to investigate biomolecular interactivities in pharmaceutical engineering, food analysis, antigen-antibody specification and basic science research [11]. Apart from these, the development of various optical biosensors has gained momentum with the emergence of fiber optic technology [5], because this technology provides high detection speed, low cost and ease of use [12]. Biosensors using fiber optic technology are fiber optic-derived devices that use an optical field to measure the analyte, and an optical fiber consists of a core and covering formed of silica or plastic that allows incoming light to be transmitted through the fiber axis with the lowest loss [8, 12, 13]. In these sensors, the refractive index of the core must be moderately higher than the covering to direct the incoming light [8]. Fiber optic biosensors can be parted into two classes, intrinsic and extrinsic. In intrinsic biosensors, the interaction with the analyte occurs within a component of the optical fiber, while in extrinsic biosensors, the optical fiber combines with light, usually from the zone where the light beam is influenced by the measured magnitude [13].

Thermal biosensors: The mechanism of these biosensors is based on measuring the thermal energy absorbed or released during biochemical recognition [5, 14]. A change in enthalpy occurs in most biochemical reactions, and changes in temperature can be measured by sensitive thermistors [5]. The advantages it provides are long-term stability (no contact between the transducer and sample so less damage to thermistors), low cost (thermistors are inexpensive products) and accuracy (changes in optical or ionic sample properties do not affect the measurement). On the other hand, there are drawbacks: Complex structure (due to the complex nature of the thermostating system), poor responsiveness, and poor reputation for non-specific heating effects [14].

Piezoelectric biosensors: Piezoelectric or gravimetric biosensors, which detect the difference in the resonance frequency of the piezoelectric material due to the adsorption/desorption of molecules from the surface of the material, gauge the mass change depending on the alteration in the resonance frequency [4, 5]. In these sensors, piezoelectric crystals are used to correlate the mass change with the alteration in the oscillation frequency of the piezo crystal [5]. It has uses in the detection of immunity and nucleic acid and in cellular studies [4].

Biosensors outside these categories are baroxymeter and Field-effect transistor (FET) biosensors. The baroxymeter detects the change in microbial respiration by measuring the pressure, thus providing a rapid measurement of water toxicity (respiratory inhibition indicates toxicity) [2, 4]. FET-based biosensors, on the other part, measure the change in the conductivity of the FET as a result of a biological reaction and have been used in clinical research and for recording the intravenous blood pH value [4].

According to the unit of biorecognition, biosensors are divided into 5 groups:

• Enzyme-based biosensors

• Nucleic acid-based biosensors

• Aptamer-based biosensors

• Antibody-based biosensors

• Whole cell-based biosensors [5]

Enzyme-based biosensors: They consist of three parts: a biorecognition element (enzyme), a transducer and a digital signal processor. Their effectiveness depends on the selection of suitable enzymes as bioreceptor molecules, techniques to highly immobilize enzymes to the identified carriers, sensitive transducers, and the appropriate integration of these components to develop different types of biosensors [15]. Techniques used for immobilization include physical adsorption, covalent binding, physical entrapment, and cross-linking [5]. In addition to this, in an enzyme-based biosensor, the biocatalyst must be combined with conductive electrodes in order to transmit the enzyme catalytic conversion data electronically [15].

Nucleic acid-based biosensors: These biosensors are described as analytical devices designed by integrating an oligonucleotide with a base sequence or a complex nucleic acid structure (for example, DNA obtained from calf thymus) integrated into or closely related to the signal transducer [16]. It can be used to ascertain DNA/RNA segments or biological or chemical species. They are also called genosensors, as most of them rest on highly specific hybridization of complementary strands of DNA or RNA molecules. In these biosensors, nucleic acids act as receptors for analytes. As with enzyme-based biosensors, optical, electrochemical [15, 16], mass, magnetic and micromechanical systems [16] can be used as transducers. DNA-based biosensors can be used to detect environmental pollutants such as pesticides and toxic metal ions, since nucleic acid bases (adenine and thymine) form complexes with metal ions such as Hg^{2+} due to their high stability constants [5].

Aptamer-based biosensors: Aptamers, which are single-stranded DNA or RNA molecules that can be obtained through an *in vitro* technique known as the systematic evolution of ligands by exponential enrichment (SELEX), were first used as identification materials in biosensors in 2004. Biosensors in which aptamers are used in this way are also called aptasensors [5, 17]. However, due to its structure, it is also considered as a sub-title of nucleic acid-based biosensors in some sources [10, 18]. They have many advantages: High affinity, simple and high capacity to form Watson-Crick base pairs (thus inorganic ions (K^+, Hg^{2+}), organic molecules (cocaine, adenosine triphosphate-ATP), large biomolecules (proteins) and whole organisms (cells, bacteria) can be detected), compatibility with a variety of detection strategies, virtually any detection requirements and reading techniques, amenable to modification, high selectivity and chemical stability [5, 17].

Antibody-based biosensors: In this group of biosensors, also called immunosensors, antibodies are used as biosensors to take advantage of the high affinity among antibodies and antigens for recognition [5]. The first commercial immunoassay developed based on antigen-antibody interactions was the detection of human chorionic gonadotropin (hCG), also known as a pregnancy test. While in the early versions of the test, only hCG was detected (pregnant or not); it is possible to determine hCG levels and, thus the time elapsed since pregnancy by a colorimetric mechanism in current models [10].

Whole cell-based biosensors: Whole cells are used as an option for enzymes, especially when enzymes become inactive during isolation, purification or direct immobilization. It offers several advantages over enzyme-based biosensors. For example, whole-cell-based biosensors are also highly stable, as the natural structures of enzymes in whole cells can be preserved for a long time. Since it contains more than one enzyme, simultaneous determination of various analytes can be made. It can be genetically modified to improve the specificity of the analyte [5]. It also offers high sensitivity and selectivity, fast reagent-free detection, and economical testing at routine inspection or care points. As a result of these advantages, they have a high potential for use in biomedical diagnosis and investigation of environmental/food samples. For example, the detection of *Staphylococcus aureus*, a pathogen responsible for foodborne illness, was achieved with a developed whole-cell-based biosensor [10].

Biosensors are utilized for the qualitative and quantitative analysis of several analytes in environmental, clinical, agricultural, food and defense applications. Although many biosensors are highly effective in the analysis of synthetic samples and pure laboratory samples, their practical application is unfortunately not common due to some limitations. Some of these restrictions can be listed as follows [19]:

• Insufficient operational and long-term stability of the bio-receptor and transducer.

• Poor repeatability between biosensors.

• Low selectivity in complex sample environments.

• The requirement to be used outside of laboratory conditions in practical applications and to monitor real samples *in situ* [19].

• Conventional biosensors have low sensitivity, less selectivity, low concentration range and high energy requirement [20].

• Unreliable dynamic measurement range in electrochemical biosensors due to enzyme saturation kinetics.

• Possible interference of another substance in solutions.

• Especially in glucose measurement, oxygen necessity and concentration fluctuation in solution.

• The effect of pH values or ionic forces on enzymatic reactions.

• Biochemical and insufficient electron transfer processes that limit efficiency and speed [21].

• The interference of incident beam in optical biosensors.

• Having a large volume/mass.

• Needing a high amount of enzyme [20].

These problems can be overcome by developing immobilization techniques, using nanotechnology, and getting miniaturization and multi-sensor array determinations [19]. In addition, smart materials can be used to improve biorecognition and transduction processes. In this context, designing biosensors with high sensitivity and selectivity, rapid response and low LOD value is a multidisciplinary field and requires different scientific and technological knowledge such as electrochemistry, surface chemistry, biochemistry, solid state physics [3]. Regarding our topic, in this part of the book, only the use of nanotechnology in biosensors is mentioned.

USE OF NANOPARTICULAR/NANOVESICULAR SYSTEMS AS BIOSENSORS

Advances in nanotechnology have led to the emergence of nanoparticles with unique physical and chemical properties, thus leading researchers to new studies on the application of nanoparticles in biosensors and bioanalysis with their excellent results in chemical and biological sensing. In this regard, the use of nanoparticles with different contents and sizes has become widespread in recent years, especially in the biomolecular recognition process for transduction by electronic, optical and microgravimetric methods [22]. Because the dimensions of nanosystems are between 1 and 100 nanometers, most of the atoms are at or near the surface due to their nanosize [23, 24]. In addition, it has physical/chemical features, namely large specific surface area, magnetism, electrical, optical and catalytic properties [24]. This situation allows nanosystems to display different physicochemical properties according to the bulk type. Thus they improve the detection mechanism of biosensor technology thanks to their large surface areas [23].

Liposomes and other lipid structures, polymeric nanoparticles (chitosan, silicon), nanofibers and nanocrystals are used in addition to colloidal gold nanoparticles and semiconductor quantum dots in the field of biosensors [22, 23, 25, 26]. In some biosensors, hybrid systems with carbon nanotubes and polymers, sol-gel matrix and layer-by-layer structures are also used [27]. In this way, nanoparticle-biomolecule assemblies created using nanoparticles in the biosensor field have greatly improved the signal generation and ultra-sensitive optical/electrical biosensors with polymerase chain reaction (PCR)-like sensitivity have been designed. Moreover, multiple amplification protocols have been developed, containing together various nanomaterial-based amplification units and processes, to ensure the high sensitivity of modern bioassays [22].

Biosensors using nanosystems are also called nanobiosensors, however nanobiosensors are not special sensors that can detect nanoscale events and formations [23]. These sensors can be split into two groups in respect of their functions: Nanoparticle-based transducers for bioanalytical techniques, and sensors using biomolecule-nanoparticle conjugates as markers for biosensing and bioassays [22]. Nevertheless, in this chapter of the book, the focus is on only the applications of nanoparticles/nanovesicular prepared using metals, lipids or polymers in the field of biosensors and studies with these sensors without any functional classification.

Gold Nanoparticles and Their Applications as Biosensors

Gold nanoparticles (AuNPs) are one of the most durable metal nanoparticles. They can be prepared through different methods, however, the most used way is citrate reduction. In the meantime, the preparation method might affect the size, shape, stability, solubility and function of AuNPs. They are widely used as biosensors owing to their inertness, inimitable optical properties, large surface area, and being suitable for various surface modifications (polymer, nucleic acid, protein, *etc.*) [28]. Their optical properties have made them extensively used in several optical biosensors get involving colorimetry, scanometry, dry reactive strip, surface plasmon resonance (SPR), surface enhanced Raman scattering (SERS), chemiluminescence and fluorescence [29 - 31]. In addition, they have been used in electrochemical and piezoelectric biosensors to improve analytical performance due to their biocompatibility, conductivity, catalytic particulars, elevated surface-to-volume ratio and high density [24, 32].

AuNPs facilitate the recognition process in these biosensors with different mechanisms. For example, a shift becomes in the SPR of AuNPs when AuNPs are aggregated, which contributes to the variation in the distance between the discrete AgNPs and produces color changes from pink to purple to pale blue in solution depending on their size [29, 33, 34]. By means of AuNP's distance-dependent optical property, AuNP-based colorimetric biosensors can detect biomacro-molecules, metal ions and small molecules even with the naked eye [29, 35, 36]. AuNPs' have been used in SPR-based optical biosensors in order to amplify the signal and increase the sensitivity, and the first study on this subject was carried out by Natan's group in 1998, getting results with 25 times higher sensitivity than the conventional SPR method [37]. If Raman molecules are conjugated to the surface of AuNPs, the aggregation of these nanoparticles changes the interparticle spacing and induces surface-enhanced Raman scattering, which is used for highly sensitive detection. In addition, ultra-small AuNPs and Au nanoclusters have fluorescence properties that enable fluorescence reading [36].

Due to the poor electrical conductivity of biomolecules, the inability to fully detect the electron exchange that occurs on the electrode depending on the analyte concentration as a result of the electrochemical redox reaction reduces the sensitivity of the analysis [24]. In contrast, AuNPs make good the sensitivity and specificity of electrochemical biosensors by changing the sensing surface to increase conductivity, improving the bioimmobilization and inducing electrochemical reactions. When they are fixed on the electrode surface, they both increase the electron transfer rate and widen the detection area, thus improving the sensitivity of the test [24, 38, 39]. In addition, due to the redox reaction between Au^0 and Au^{3+}, AuNPs are also used as electrochemical indicators [24]. In electrochemical immunosensors, on the other hand, the affinity reaction is used to increase the electrode transduction and immobilization efficiency of immunoreactives [27].

One of the points to be considered in the use of AuNPs in the field of biosensors is to ensure their colloidal stability and to keep the changes in aggregation under control [33]. The colloidal stability of AuNPs depends on many factors, such as solvent, stabilizer, surfactant, and pH [34, 40]. On the other hand, various biomolecules such as nucleic acids, enzymes, receptors, lectins, antibodies and superantigens can be conjugated to the surface of AuNPs by binding and/or electrostatic interactions to provide their colloidal stability [33, 41, 42]. In order to achieve controlled aggregation, it can benefit from cross-linkers or strong interaction between ligand and target analyte [33].

AuNPs have been used alone or in combination with structures such as chitosan and graphene to improve analysis sensitivity and selectivity in biosensors. Although there are many studies conducted in this context, only a few of the studies in which AuNPs are used alone in biosensors have been mentioned in this section. Other studies with AuNPs are summarized in Table **1**.

Table 1. Studies on the use of nanosystems in biosensors.

Nanosystem	Analyte	Analysis Method Used	Results	Developed Biosensor System	References
Gold Nanoparticles (AuNPs)					
Polyclonal Abs bound AuNPs	*Escherichia coli* (*E. coli*)	Electrochemical (Voltammetric)	LOD: 10^1 cfu/mL Range: 10^1-10^6 cfu/mL	AuNP-labeled Biosensor	[68]
Blank AuNPs	Hydrogen peroxide (H_2O_2) and glucose	Optical (Resonance Light Scattering-RLS)	LOD: 6.8×10^{-7} M Range: 1×10^{-6}-1.1×10^{-4} M	Optical Biosensor (Resonance Light Scattering-RLS)	[69]
Blank AuNPs or AuNP functionalized with 11-mercaptoundecanoicacid	Biotin or Bovine serum albumin	Localized Surface Plasmon Resonance (LSPR)	LOD: 8 pM or 11 pM	Optical Fiber Biosensor	[70]
AuNP/Graphene composites	Uric acid	Electrochemical	LOD: 2×10^{-7} M Range: 2×10^{-6}-6.2×10^{-5} M	Uric Acid Amperometric Sensor	[71]
DNAzyme-functionalized AuNPs	Lead	UV-Vis Spectroscopy	Range: 0.1-4 μM	Colorimetric Lead Biosensor	[72]
DNA-modified AuNPs	Thrombin	Fluorescence	LOD: 0.14-0.46 nM	Aptamer Biosensor	[73]
Horseradish peroxidase-functionalized AuNPs and AuNPs/carbon nanotube hybrid combination	Human IgG (HIgG)	Electrochemical	LOD: 40 pg/mL Range: 0.125-80 ng/mL	Amperometric Immunosensor	[74]
Sol-gel silica network with dehydrogenase enzyme and AuNPs	Nicotinamide adenine dinucleotide (NADH), lactate and ethanol	Electrochemical	LOD: 5 nM (NADH)	Amperometric Dehydrogenase Biosensor	[75]
Sol-gel silica network with horseradish peroxidas bound AuNPs	H_2O_2	Amperometric	LOD: 2 μM Range: 5 μM-10 mM	Third-generation Horseradish Peroxidase Biosensor	[76]
Sol-gel silica network with glucose oxidase bound AuNPs	Glucose	Cyclic Voltammetry (CV) and Electrochemical Impedance Spectroscopy (EIS)	LOD: 23 μM Range: Up to 6 mM	Glucose Biosensor	[77]
Aptazyme-functionalized AuNPs	Adenosine	UV-Vis Spectroscopy	Range: 100 μM-1 mM	Colorimetric Biosensor	[78]
Cystamine-AuNPs-single stranded DNA layer	Aflatoxin M1	CV and EIS	Range: 1-14 ng/mL	Impedimetric Aflatoxin M1 Biosensor	[79]
Aptamer-functionalized AuNPs	Thrombin	Optical Intensity and Photo Images	LOD: 2.5 nM Range: 5-100 nM	Dry-reagent Strip Biosensor	[80]
Thiol-coated AuNPs	*E. coli* O157	NanoDrop-UV–Vis Spectroscopy	LOD: 2.5 ng/μL Range: 2.5-10 ng/μL	Colorimetric DNA Biosensors	[81]
IgG-capped AuNPs	SARS-CoV-2 spike protein	Dynamic Light Scattering (DLS)	LOD: 13 ng/mL	Immunosensor	[82]
Thiol-linked oligonucleotide bound AuNPs	Foot and mouth disease virus (FMDV)	Real-time Polymerase Chain Reaction (RT-PCR)	LOD: 1 copy number Range: -3.544	AuNPs-FMDV Biosensor	[83]
Enzyme electro-crosslinked tannic acid capped AuNPs	H_2O_2 and glucose	CV	For H_2O_2 LOD: 10 mM Range: 10-250 mM For glucose LOD: 0.3 mM Range: Up to 10 mM	Nanohybrid Enzymatic Biosensor	[84]

(Table 1) cont.....

Nanosystem	Analyte	Analysis Method Used	Results	Developed Biosensor System	References
Gold Nanoparticles (AuNPs)					
Cystamine-AuNPs-cholesterol oxidase layer	Cholesterol and H_2O_2	CV	LOD: $5x10^{-9}$ M Range: $7.5x10^{-8}$-$1x10^{-6}$ and $1x10^{-6}$-$5x10^{-5}$ M	Electrochemical Cholesterol Biosensor	[85]
Cystamine-AuNPs-Fullerene-*Trametes versicolor* Laccase layer	Gallic acid	Chronoamperometric	LOD: 0.006 mM Range: 0.03-0.30 mM	Nanostructured Enzymatic Biosensor	[86]
Horseradish peroxidase-AuNPs-DNA conjugates	DNA	Lateral Flow Strip Biosensor	LOD: 0.01 pM Range: 0.5 pM-0.1 nM and 0.1-50 nM	Ultrasensitive Nucleic Acid Biosensor	[87]
Streptavidin/BSA bound AuNPs	Carcinoembryonic antigen	Surface Plasmon Resonance (SPR)	LOD: 0.1 ng/mL Range: 0.4-25 ng/mL	SPR Biosensor	[88]
Blank AuNPs	Copper (II)	Colorimetric	LOD: 250 nM Range: 0.5-10 μM	Clickable Biosensor	[89]
Blank AuNPs	Sucrose or goat anti-rabbit IgG	LSPR	For goat anti-rabbit IgG LOD: 11.1 ng/mL	Optical Fiber LSPR Biosensor	[90]
Polypyrrole coated AuNPs	MicroRNA-21	CV and Differential Pulse Voltammetry (DPV)	LOD: 78 aM Range: 100 aM-1 nM	Ultrasensitive Electrochemical Biosensor	[91]
1-Dodecanethiol bound AuNPs	Concanavalin A	SPR	LOD: 0.1 nM	Label-free Optical Biosensor	[92]
Organic-inorganic hybrid silsesquioxane/AuNPs nanoconjugates	Anti-*Trypanosoma cruzi* antibodies	CV	-	Electrochemical Immunosensor	[93]
2,3-diaminophenazine-AuNP and Toluidine blue-AuNP	Cancer antigen 15-3 (CA15-3) and microRNA-21 (miRNA-21)	DPV	For CA 15-3 LOD: 0.14 U/mL For miRNA-21 LOD: 1.2 fM	Label-free Dual Electrochemical Biosensor	94
Antibody-conjugated AuNP	Thyroglobulin	LSPR	LOD: 6.6 fg/mL	Fiber-optic Localized Surface Plasmon Resonance Immunosensor	[95]
Blank AuNPs	Genetically modified soybean	EIS	LOD: 1.792 ng/mL Range: 1.792-$1.922x10^1$ ng/mL	Genosensor	[96]
Liposomes					
Acetylcholinesterase loaded liposome	Pesticides (dichlorvos and paraoxon)	Fluorescence	For dichlorvos LOD: $2x10^{-10}$ M For paraoxon LOD: $6.7x10^{-10}$ M	Liposome-based Nanobiosensor	[97]
Dopamine loaded liposome	Oligonucleotides	DPV	LOD: 5.68 fM Range: 0.01 pM-10 nM	Electrochemical DNA Biosensor	[98]
Acetylcholinesterase loaded liposome	Acetylthiocholine chloride	Fluorescence	LOD: 1 mM Range: 1-13.3 mM	Liposome-based Nanobiosensor	[99]

Nanosystem	Analyte	Analysis Method Used	Results	Developed Biosensor System	References
Gold Nanoparticles (AuNPs)					
Horseradish Peroxidase/Monosialoganglioside functionalized liposome	Cholera toxin	Chemiluminescence	LOD: 0.8 pg/mL Range: 1 pg/mL-1 ng/mL	Ultrasensitive Chemiluminescence Biosensor	[100]
Glucose oxidase loaded liposome	H_2O_2 and glucose	Amperometric	For glucose LOD: 8.6±1.1 µM For H_2O_2 LOD: 5.7±1.4 µM	Amperometric Glucose Biosensor	[101]
Tyrosinase loaded liposome	Phenolic compounds	CV	LOD: 0.091 nM Range: 0.25 nM-25 µM	Amperometric Phenolic Compounds Biosensor	[102]
Sulforhodamine B loaded liposome	Dengue virus	Electrochemiluminescence	LOD: 10 pfu/mL	Serotype-specific RNA Biosensor	[103]
Polydiacetyelene-Phosphatidylinositol-4,5-bisphosphate liposome	Neomycin	Naked-eye observation and Fluorescence	LOD: 1 ppm	Paper-based Colorimetric Biosensor	[104]
Sulforhodamine B loaded and streptavidin coupled liposome	RNA and DNA sequences [*E. coli, Cryptosporidium parvum (C. parvum), Bacillus anthracis (B. anthracis)*]	Reflectometric	For *E. coli* LOD: 50 nM For *C. parvum* and *B. anthracis* LOD: 10 nM	Generic Sandwich-type Biosensor	[105]
Potassium ferri/Ferrohexacyanide loaded liposome	Nucleic acid (RNA)	Amperometric	LOD: 0.01 µM	Electrochemical Microfluidic Biosensor	[106]
Blank liposome	Protein-Membrane Interaction (target protein: lysozyme and carbonic anhydrase from bovine)	Micro-cantilever	-	Micromechanical Cantilever-based Liposome Biosensor	[107]
AuNP bound polydiacetylene liposome	Thrombin	UV-Vis Spectroscopy	-	Label-free Realtime Colorimetric Biosensor	[108]
Sulforhodamine B loaded liposome	*E. coli*	Reflectometric	LOD: 40 cfu/mL	RNA Biosensor	[109]
Calcein loaded liposome	Listeriolysin O	Fluorescence	Range: 0-6 ng/µL	Artificial-Cell-based Biosensors	[110]
Targeted lipopolyplexes bearing epidermal growth factor receptor (EGFR)-binding peptides	Intratumoral heterogeneity of EGFR activity	Fluorescence Lifetime Microscopy (FLIM)	-	FLIM-based Biosensing	[111]
Polymeric Nanoparticles					
Poly (γ-glutamic acid) modified with 3-aminothiophene copolymer	Lysozyme	DPV	Range: 1×10^{-10}-1×10^{-5} mg/mL	Molecularly Imprinted Biosensor	[112]
Methoxy-poly(ethylene glycol)-block-poly lactic acid (mPEG-b-PLA) NP	Avian influenza virus (AIV) (Highly pathogenic-HPAIV or Low pathogenic-LPAIV)	Fluorescence Resonance Energy Transfer (FRET)	For LPAIV Sensitivity: 96.4% Specificity: 100% For HPAIV Sensitivity: 100% Specificity: 100%	Cell-mimetic Biosensors	[113]
CdTe quantum dots coated silica cross-linked molecularly imprinted polymer (MIP)	Lysozyme	Fluorescence	LOD: 3.2 µg/mL Range: 10-120 µg/mL	Molecularly Imprinted Polymers-based Novel Optical Biosensor	[114]

(Table 1) cont.....

Nanosystem	Analyte	Analysis Method Used	Results	Developed Biosensor System	References
Gold Nanoparticles (AuNPs)					
Magnetic MIP NP	N-acyl-homoserine-lactones	DPV	LOD: 8×10^{-10} mol/L Range: 2.5×10^{-9} mol/L-1×10^{-7} mol/L	Electrochemical Sensor	[115]
Electroactive molecularly imprinted polymer nanoparticles (nanoMIP)	Insulin	DPV	LOD: 26 fM (buffer) and 81 fM (human plasma) Range: 50-2000 pM	NanoMIP-based Sensor	[116]
Polyvinyl chloride (PVC) membrane bound polyphenol oxidase	Polyphenols	Amperometric	LOD: 7.5×10^{-7} M Range: 4-8.4 µM (tea leaves), 1.9-3 µM (alcoholic beverages), 0.7-7.6 µM (waste water)	Amperometric Polyphenol Biosensor	[117]
PVC membrane bound xanthine oxidase	Xanthine	Amperometric	LOD: 2.5×10^{-8} M Range: 0.025-0.4×10^{-6} M	Amperometric Xanthine Biosensor	[118]
Chromogenic polymeric NP	Influenza virus A subtype H1N1 (IV/A/H1N1) and H3N2 (IV/A/H3N2)	Colorimetric	For IV/A/H1N1 LOD: 27.56 fg/mL Range: 10^0 fg/mL-10^6 fg/mL For IV/A/H3N2 LOD: 28.38 fg/mL	pH-sensitive Polymeric Nanoparticle-laden Nanocarriers (PNLN)-based Biosensor	[119]
Chitosan-glucose oxidase-AuNP biocomposite	Glucose	Amperometric	LOD: 2.7 µM Range: 5.0 µM-2.4 mM	Glucose Biosensor	[120]
DNA-functionalized dye-loaded polymeric NP	Nucleic acids	FRET	LOD: 0.25 pM	FRET Biosensor	[121]
Polyaniline NP	H_2O_2	Amperometric	-	Amperometric Enzyme Biosensor	[122]
MIP contained polymer nanofibers	Fluorescent amino acid derivative dansyl-L-phenylalanine	Fluorescence microscopy	-	Fluorescence-based Biosensor	[123]
Ultrabright silicon NP	Cholesterol	Fluorescence	LOD: 0.018 µM Range: 0.025-10 µM	Inner Filter Effect (IFE)-based Fluorescent Biosensor	[124]
L-asparaginase entrapped chitosan alginate NP	Asparagine	Electrochemical	LOD: 1.69×10^{-10} M Range: 10^{-10}-10^{-1} M	Enzyme-based Electrochemical Biosensor	[125]
Triethylene glycol dimethacrylate (TEGDMA) polymer containing AuNPs	Thrombin	LSPR	LOD: 0.01 ng/mL	Plasmon-active Polymer-Nanoparticle Composites LSPR Biosensor	[126]
Chitosan micro-membranes containing AuNPs	-	High-Resolution Localized Surface Plasmon Resonance (HR-LSPR)	LOD: 1.2×10^{-5} refractive index unit (RIU)	Nanoplasmonic Biosensor	[127]

(Table 1) cont.....

Nanosystem	Analyte	Analysis Method Used	Results	Developed Biosensor System	References
Gold Nanoparticles (AuNPs)					
MIP NP	α-casein	SPR	LOD: 127±97.6 ng/mL	Label-free Surface Plasmon Resonance	[128]
Iron oxide NP/Poly(vinyl alcohol) nanocomposite film containing glucose oxidase	Glucose	CV	Range: 30-400 mg/dL	Electrochemical Glucose Biosensor	[129]
Chitosan-gold nanoparticle film containing glucose oxidase	Glucose	Amperometric	LOD: 13 μM Range: $5x10^{-5}$ - $1.3x10^{-3}$ M	Amperometric Glucose Biosensor	[130]
Graphene nanoribbon/chitosan nanocomposite	Sarcosine	CV	LOD: 0.001 μM Range: 0.001-100 μM	Amperometric Sarcosine Biosensor	[131]
Chitosan NP containing glucose oxidase	Glucose	Amperometric	LOD: 1.1 μM Range: 0.001-1 mM	Amperometric Glucose Biosensor	[132]
Zinc oxide (ZnO)-polyvinyl alcohol hybrid film containing urease	Urea	CV and EIS	LOD: 3 mg/dL Range: 5-125 mg/dL	Impedimetric Urea Biosensor	[133]
Poly(vinyl alcohol) capped silver NP	H_2O_2	LSPR	LOD: 10^{-6} M	Colorimetric Hydrogen Peroxide Sensor	[134]
Mesoporous silica-chitosan-AuNP containing curcumin	MUC-1 positive tumor cells	Fluorescence Imaging	-	On/Off Optical Biosensor	[135]
Polyaniline-silicon wafer	Acetylcholine	Ion-Sensitive Field-Effect Transistors (ISFET)	LOD: 1 μM Range: Down to 10 μM	Polyaniline ISFET Acetylcholine Biosensors	[136]
ZnO/chitosan-graft-poly(vinylalcohol) core-shell nanocomposite	Glucose	Potentiometric	LOD: 0.2 μM Range: 2 μM-1.2 mM	Tunable Glucose Biosensor	[137]
ZnO NP-chitosan composite film containing cholesterol oxidase	Cholesterol	Photometric and Amperometric	LOD: 5 mg/dL Range: 5-300 mg/dL	Cholesterol Biosensor	[138]
Silicon NP	β-Glucuronidase activity	Fluorescence	LOD: 0.02 U/L	Fluorescent Nanosensors	[139]
Platinum-porphyrin-poly-lactic acid-alginate hybrid system containing glucose oxidase (GPP-AM) or Platinum-porphyri--poly-lactic acid-alginate hybrid system (PP-AM)	Glucose and Oxygen	Fluorescence	GPP-AM-For glucose LOD: 1.5 mM Range: 0-10 mM PP-AM-For oxygen Range: 0-6 mM and 0-14 mM	Implantable Glucose Biosensors	[140]

AuNPs: Gold nanoparticles, *E.coli: Escherichia coli*, LOD: Limit of detection, cfu: Colony-forming unit, mL: Mililiter, H_2O_2: Hydrogen peroxide, RLS: Resonance light scattering, M: Molarity, LSPR: Localized Surface Plasmon Resonance, pM: Picomolar, μM: Micromolar, nM: Nanomolar, HIgG: Human IgG, pg:Picogram, ng: Nanogram, NADH: Nicotinamide adenine dinucleotide, mM: Milimolar, CV: Cyclic voltammetry, EIS: Electrochemical impedance spectroscopy, mM: Milimolar, μL: Microliter, DLS: Dynamic light scattering, FMDV: Foot and mouth disease virus, RT-PCR: Real-time polymerase chain reaction, SPR: Surface plasmon resonance, DPV: Differential pulse voltammetry, aM: Attomolar, CA15-3: Cancer antigen 15-3, miRNA-21: microRNA-21, U: Unit, fM: Femtomolar, fg: Femtogram, pfu: Plaque forming units, ppm: Parts per million, *C. parvum: Cryptosporidium parvum, B. anthracis: Bacillus anthracis*, EGFR: Epidermal growth factor receptor, FLIM: Fluorescence lifetime microscopy, mPEG-b-PLA: Methoxy-poly(ethylene glycol)-bloc--poly lactic acid, NP: Nanoparticle, AIV: Avian influenza virus, HPAIV: Highly pathogenic avian influenza virus, LPAIV: Low pathogenic avian influenza virus, FRET: Fluorescence resonance energy transfer, MIP: Molecularly imprinted polymer, nanoMIP: Electroactive molecularly imprinted polymer nanoparticles, PVC: Polyvinyl chloride, IV/A/H1N1: Influenza virus A subtype H1N1, IV/A/H3N2: Influenza virus A subtype

H3N2, PNLN: pH-sensitive polymeric nanoparticle-laden nanocarriers, IFE: Inner filter effect, HR-LSPR: High-resolution localized surface plasmon resonance, RIU: Refractive index unit, ZnO: Zinc oxide, ISFET: Ion-sensitive field-effect transistors, GPP-AM: Platinum-porphyrin-poly-lactic acid-alginate hybrid system containing glucose oxidase, PP-AM: Platinum-porphyrin-poly-lactic acid-alginate hybrid system

In their study, Tabibi *et al.* developed an electrochemical biosensor that can measure arsenite sensitively and selectively using gold nanoparticles. The purpose of using AuNPs in this study was to provide an appropriate nano-environment for enzyme activity and thus improve the analytical performance of the biosensor. First of all, GC-AuNP-ArOx electrode was prepared by electrostatically depositing arsenite oxidase (ArOx) and AuNPs on glass carbon (GC). For this, the GC electrode was polished using alumina (0.3 μm), then washed with pure water and sonicated for 5 minutes. The electrode was dried with a nitrogen stream. AuNPs were collected on the electrode by applying a potential of 1000-0 mV to a 0.5 M sulfuric acid (H_2SO_4) solution containing 0.15 mM sodium tetrachloroaurate [Na(AuCl$_4$)] for 1 minute. Finally, ArOx enzyme was deposited galvanostatically (10 μA) on the GC-AuNP electrode. In addition, an enzyme-free electrode containing AuNP was prepared. Various studies have been carried out on the designed biosensor. These studies were to examine the effects of sample pH on the activity of the biosensor (pH varies between 4-10), to determine the most appropriate enzyme concentration used in the biosensor, to evaluate the stability of the biosensor (by examining the effects of storage temperature and time on the biosensor response for seven days), to determine undesirable effects that may come from interfering agents (heavy metals such as copper, cadmium, zinc, iron and dissolved oxygen) on the response of arsenite in the biosensor, to examine the reproducibility of the biosensor (arsenite concentration in the waters of various regions was compared with the conventional spectroscopic method). The working mechanism of the designed biosensor was that when the ArOx enzyme was alone, it cannot directly undergo heterogeneous electron transfer with the electrode, whereas the efficiency of the redox reaction in the presence of AuNP increased and it did not need external electron transfer mediators while doing this. The results of the previous studies also supported this situation through no redox activity observed in systems that did not contain enzymes or that contained enzymes but without co-deposition agents. The limit of detection (LOD) value of the designed biosensor for arsenite was calculated as 5 nM, which is approximately 100 times lower than the recommended DNA-based sensors for arsenic amount determination. As a result of the studies mentioned above and performed on the sensor, it was determined that the pH value of the sample should be 7 and the enzyme concentration should be 7 μM in order to get an optimum response against arsenite. The developed biosensor was stable for three days at 4°C. In the presence of various heavy metals, the selectivity for arsenite was quite high (10 nM arsenite versus 1 μM heavy metal presence could be detected).

Finally, the designed biosensor exhibited higher reproducibility than the conventional method [43].

Li and co-workers have designed an optical fiber SPR biosensor using AuNPs to detect DNA hybridization. The sandwich model was used to increase the sensitivity of the analysis. For this, first of all, probe DNA (pDNA) was covalently attached to the optical fiber detection area. Biotinylated target DNA (tDNA) was attached to the surface of AuNPs. Finally, the tDNA on the AuNP surface specifically hybridized with the pDNA on the optical fiber to create a sandwich structure (Fig. **2**). As a result of the studies, it has been shown that the sandwich-type fiber SPR sensor developed was linear for DNA in the concentration range of 1 pM-10 nM. The LOD and sensitivity were 1 pM and 4.04 µM, respectively. It was also more sensitive than the other non-sandwich method (1.44 fold) and existing fiber optic DNA sensors, so it can be said that the sensor can detect DNA with low concentration or small fragments. In addition, the biosensor had good specificity and stability [44].

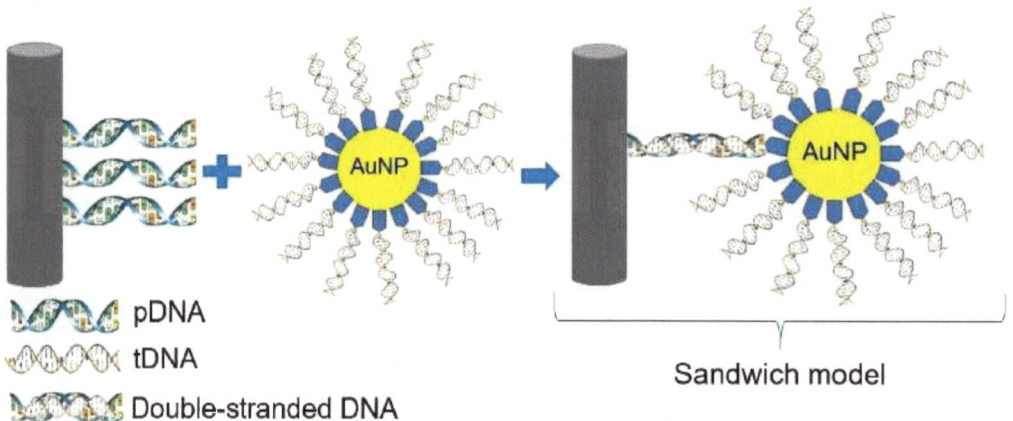

pDNA
tDNA
Double-stranded DNA

Sandwich model

Fig. (2). Schematic image of the sandwich structure model with AuNP and optic fiber.

Finally, Hasan *et al.* have developed a colorimetric biosensor that can be seen with the naked eye, using citrate-modified AuNPs for the ovarian cancer marker (platelet derived growth factor-PDGF). In this context, AuNPs were prepared by core growth method at first: Chloroauric acid ($HAuCl_4$) and tri-sodium citrate solutions, which are the same concentration (0.0005 M) were mixed, and 0.1 M sodium borohydride ($NaBH_4$) solution was added into this solution. To prepare the growth solution, cetyltrimethylammonium bromide (CTAB) powder was added to the $HAuCl_4$ solution and heated to 50-60°C until the solution was orange. In the last step, the ascorbic acid solution was added to the growth solution, and then the seed solution was added dropwise to this mixture. The sensor was prepared by adding different concentrations of PDGF to the aptamer-containing solution and putting AuNPs obtained on it after 10 minutes of incubation. After a reaction time of 5 minutes, sodium chloride (NaCl) solution was added to the mixture so that the AuNPs precipitated, and the color change could be seen with the naked-eye. The absorbance of this prepared mixture was measured in a UV-Vis spectrophotometer. In addition, in this study, the formation of aggregates due to strong van der Waals attraction between AuNPs was prevented by coating AuNPs using citrate. Analysis mechanism of the developed biosensor; when aptamers bind to the surface of AuNPs, they remain dispersed despite the addition of NaCl (therefore, no color change), whereas when PDGF is added to the medium, the aptamer separates from the AuNP surface and forms a tight PDGF-aptamer structure, and when NaCl is added, AuNPs change color because they were aggregated. The colorimetric biosensor developed has detected PDGF sensitively and specifically with a concentration range of 0.01–10 µg/mL and a LOD of 0.01 µg/mL. Moreover, concentration differences in PDGF could be noticed with the naked eye. Therefore, the study offers a simple system that does not need expensive equipment and diagnoses ovarian cancer with good sensitivity and specificity [45].

Liposomes and Their Applications as Biosensors

Composed of two Greek words, "Lipos" and "Soma", liposomes are artificial vesicles whose basic building blocks are phospholipids [46]. They form a vesicular structure which has a lipid double layer and an inner aqueous phase by self-assembly of phospholipids in the aqueous medium [47]. Liposomes were first found out by Dr. Alec D Bangham in 1961 during the institute's new electron microscope testing by adding a negative dye to dry phospholipids by Bangham and R.W. Horne at the Babraham Institute-Cambridge [46, 48].

Liposomes are similar in structure and shape to the cell membrane. A liposome may be formed in a monolayer or multilayer structure and various sizes [46]. A wide variety of substances, hydrophilic or hydrophobic, small or large molecular structures, can be loaded into liposomes. Because of their big inner volume, substances can be entrapped in the aqueous interior or within the lipid tails of the bilayer. It can be used in the diagnosis and therapy of cancer and other diseases [25, 46, 47, 49, 50]. Today, there are commercialized liposomal products for various uses [25, 51].

The relatively low cost of lipids and the successful experience from commercializing liposomal drugs have offered significant advantages for promoting the application and commercialization steps of liposome-based biosensors. Liposomes can be used to transduce and amplify signals because they can encapsulate a variety of signal markers, including dye, enzyme, salt, chelate, DNA, electrochemical, and chemiluminescent labels [25]. Since they have alike components and structures to cell membranes, liposomes can also function as cell membrane models. In this respect, liposomes can procure a simple platform to vulgarise the system in the investigation of cell-related interactions and physiological events [25, 49]. Thanks to these advantages, there are many studies on the application of liposomes in the field of biosensors. These studies were briefly mentioned in the rest of the page. Other studies with liposomes are summarized in Table **1**.

In a study by Lin and co-workers, disease markers as if thrombin and C-reactive protein (CRP) could be measured with high sensitivity with a portative glucose meter (PGM) by an immunosorbent assay method developed using enzyme-encapsulated liposomes. In this study, liposomes non-covalently encapsulated large amounts of amyloglucosidase or invertase enzymes for signal transduction and amplification were prepared by the thin lipid film hydration method. According to this method, cholesterol, 1,2-distearoyl-sn-glyce ro-3-phosphoe-thanolamine-poly(ethylene glycol) (DSPE-PEG) and dipalmitoylphosphatidyl-choline (DPPC) dissolved in chloroform, and incubated with phosphate buffer (1 mL, 0.01 M, pH 7.4) including 60 mg/mL invertase or amyloglucosidase at 35°C for two hours following to at 45°C for 20 minutes. Finally passed through 0.4 μm membrane filters. Amyloglucosidase, which is not loaded on liposomes and is in free form, was removed by dialysis. Afterward, to determine the properties of prepared liposomes, the average size, liposome concentration and encapsulation efficiency were measured. To prepare the biosensor, the capture probe was linked onto magnetic beads (MB) by streptavidin-biotin coupling and the detection probe was conjugated to the liposome surface by DNA-liposome incubation. For detection of thrombin, 40 μL of MB (6 mg/mL) were incubated with various concentrations (0 to 250 nM) of thrombin for one hour at 25°C and then washed twice with PBS buffer (pH 7.4) to remove free thrombin. MBs prepared in this way were incubated with 40 μL of amyloglucosidase-encapsulated liposome for one hour at 25°C and then washed six times with PBS buffer (pH 7.4) to form a sandwich structure (Fig. 3). This construct was incubated with 30 μL of amylose solution (3%, pH 4.5 0.1 M PBS buffer containing 1% Triton X-100) for one hour at 40°C and 1 μL of the sample taken from this medium was analyzed with PGM. The analysis mechanism in this biosensor was based on the fact that the enzyme released from liposomes degraded by the presence of Triton X-100 converts amylose to glucose, and thus, the released glucose can be measured with a glucose meter. As a result, the entire analysis process for thrombin took three hours. Thrombin detection has been performed with high sensitivity (LOD: 1.7 nM) compared to conventional detection methods (such as electrochemical and fluorescent). The method developed demonstrated high selectivity against thrombin in the presence of other proteins (such as lysozyme and human serum albumin) that can be found in the blood. Since results comparable to the conventional ELISA method were obtained, this study has shown that the developed biosensor will provide great convenience in areas with limited resources and can be used for effective signal amplification of liposomes that can encapsulate a wide variety of enzyme molecules [52].

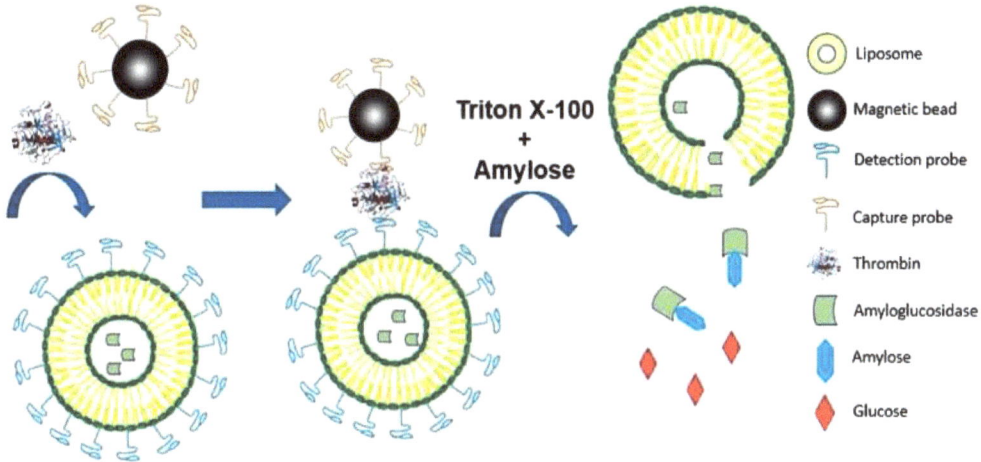

Fig. (3). Schematic image of the sandwich structure created with liposomes and magnetic beads.

Phytase-encapsulated-liposomes immobilized on polypyrrole (PPy) films were prepared for the detection of phytic acid by Rodrigues *et al.* [53]. The thin film method was used in this study as in the previous study. Accordingly, the dipalmitoyl phosphatidylglycerol (DPPG) was dissolved in the methanol/chloroform mixture and the solution of phytase in methanol was added. After the organic solvent was evaporated, the obtained dry film was hydrated with distilled water and then passed through a 50 nm polycarbonate membrane [53, 54]. PPy films were prepared by cyclic voltammetry in which 10 full cycles were applied on platinum (Pt) electrodes in an aqueous solution, including 0.07 mol/L pyrrole and 0.1 mol/L NaCl at 50 mVs^{-1} and 0.0 V-0.8 V. In this way, an electroactive polymer film was formed on the electrode. The PPy films were then incubated in a liposome solution with 0.2 mg/mL phytase at 40°C for 2 hours to immobilize phytase-loaded liposome on the electrode and obtain liposome/PPy films. In order to compare the efficiency of the developed system, phytase-loaded PPy films were also prepared using the same incubation method. Phytic acid detection was fulfilled by chronoamperometry at a constant operational potential of 0.0 V. The basis of the method was to enhance the sensitivity of amperometric analysis of phytic acid in the presence of phytase enzyme, and this has been further enhanced with enzyme-loaded liposomal systems. As a result, when the films were exposed to increased phytic acid, for phytase-encapsulated liposome-loaded films, compared to the reference film, the LOD value was lower, and the measurement range was wider. In this respect, it can be said that liposomal biosensors provide better detection of phytic acid in amperometric and voltammetric analyses with higher sensitivity and wider concentration. It can also be expected that the system will be more stable than in solution due to the encapsulation process [53].

Finally, unlabeled visible colorimetric biosensors using polydiacetylene (PDA) liposomes for multiple detections of pathogens were prepared by Zhou *et al*. In this study, different PDA, phospholipid and cholesterol ratios and different phospholipid types (phospholipids used: 1,2-dimyristoyl-sn-glyce ro-3-phospho-choline (DMPC, 14:0), 1,2-dipalmitoyl-sn-glycerro-3- phosphocholine (DPPC, 16:0) and 1,2-distearoyl-sn-glycero-3-phosphocholine (DSPC, 18:0). Liposomes were produced by the thin lipid film technique. Accordingly, PDA, cholesterol and phospholipid were dissolved in chloroform, and the solvent was evaporated. The obtained dry film was hydrated by adding saline buffer. The liposomal colloidal dispersion was extruded by passing through a polycarbonate membrane with a pore diameter of 200 nm and polymerized by exposure to UV (λ=254 nm) in the final stage. In addition, PDA liposomes containing dioctadecyl glycerylether-β-glucosides (DGG), which act as receptors, were prepared using the same method (only this time, PDA matrix lipid containing DGG, not PDA, was used) to compare biosensor sensitivity. In order to determine the colorimetric response, liposome dispersion was added to the medium containing bacteria at a certain concentration (1:2 v/v medium:liposome), and the colorimetric response was measured at certain time intervals. In order to determine the colorimetric dose-response relationship, liposome dispersion was added to a series of the medium containing bacteria at different concentrations and kept for a certain period of time, and then UV-Vis spectroscopy measurements were made at OD_{640} nm and OD_{550} nm. Experiments were carried out at 37°C. As a result, three different phospholipids with different Tm temperatures and, therefore, liposome rigidity were used. As a general rule, increasing the rigidity of the liposome may prevent the degradation of the lipid bilayer and the formation of pores in the liposome by toxins, which will adversely affect the sensitivity of the analysis method. However, in this study, the fluidity/permeability of the lipid increased, adding the cholesterol, and liposome degradation was facilitated. The sensitivity of the assay method has further improved, as cholesterol also acts as a recognition site for some toxins, such as cholesterol-dependent cytolysis. When PDA is added to liposomes, a color change (colorless to blue color change) occurs at 650 nm. However, for this change to be high, PDA must be well integrated into the liposome structure. In this respect, liposomes with DSPC gave more successful results and further studies were continued with this liposome. For the method with high sensitivity, high color change (from blue to red) should occur as a result of the incubation of liposomes and bacteria. High cholesterol-containing PDA-DSPC liposomes and DGG-PDA-DSPC liposomes were compared in terms of color change during bacteria-liposome incubation, and the color change was observed in PDA-DSPC liposomes within four hours, while in liposomes with DGG-PDA-DSPC within 30 minutes. Therefore, in DGG liposomes, the sensitivity of the biosensor was superior to high-cholesterol liposomes. However, in PDA-DSPC liposomes, a detection limit range of $1\text{-}10^5$ CFU was obtained for 6 different bacterial strains (*P. aeruginosa*, *S. aureus*, *E. coli*, *K. pneumoniae*, *S. pneumoniae*, *E.faecalis*) except *E.faecalis* which is much lower than the infection doses of these strains [55].

Polymeric Nanoparticles and Their Applications as Biosensors

Polymer-based nanoparticles (P-NPs) is a general term for nanoparticles prepared with any type of polymer. It was first developed by Paul Ehrlich and extensively studied between 1960 and 1970. Its usage areas have grown rapidly, and today it has started to act an important role in wide areas such as photonics, electronics, sensors, medicine, pollution control and environmental technology [56].

Polymers have become one of the major materials used in the development of biosensors, thanks to their ease of synthesis and processing, flexibility and molecular printing ability. In addition, biocompatible and natural polymers are preferred for the design of injectable and implantable biosensors. Because it contains homing peptides that increase the stability of the nanoparticle *in vivo* and its retention time enough to allow real-time monitoring [57].

Conductive polymers, a class of polymers used in biosensors, have changing single and double carbon-carbon bonds throughout the polymeric chain and highly unpaired (π) conjugated polymeric chains [56, 58]. Therefore, it displays reversible chemical, electrochemical and physical features depending on the doping/de-doping process [58]. In addition, compared to bulk forms, nanostructured forms offer many advantages such as large surface area, increased mechanical properties for strain housing, shortened paths for charge/mass/ion transport, high degree of conductivity and increased specificity [56, 59]. These properties allow these polymers to be used as transducer materials in various sensors [58]. Electrically conductive polymers show less toxic/non-toxic properties than their metallic alternatives, thus reducing the risk of immune reaction when they come into contact with living tissues [57].

In addition to conductive polymers, conjugated polymers (CP) [poly(fluorene), poly(p-phenylenevinylene), poly(p-phenyleneethynylene), poly(thiophene)] are also used in the biological field due to their incomparable optical-electronic properties. CP nanoparticles (CP-NPs) offer bennies such as high brightness, outstanding photostability, low cytotoxicity, high quantum yield and multiple surface modification. Also, thanks to their large π-conjugated backbone and delocalized electronic structure, CPs act as a kind of conductive material. Apart from optoelectronic equipments such as light-emitting diodes, photovoltaic cells, and field-effect transistors, they also have uses in *in vitro/in vivo* imaging since they recognize the target molecule when conjugated with specific recognition elements [60].

Apart from these polymers, chitosan, polyurethane and silicone are also used as main materials in biosensors. Enough immobilization of biological detection elements is very important to manufacture effective and reliable biosensors. In this respect, chitosan appears as a perfect immobilization matrix on the biosensor surface. Chitosan, a natural polysaccharide, has many advantages such as high permeability and mechanical strength, biological compatibility and non-toxicity, availability and inexpensive. Thanks to the amino and hydroxyl groups in its structure, it can easily form crosslinks with various nanomaterials [61]. It also has the ability to form a hard hydrogel thin film, which will be used for electrodeposition on electrode surfaces in the production of micro and nano biosensors [62]. Owing to these properties of chitosan, chitosan-based biosensors have been developed that show high sensitivity, selectivity and stability, and can detect various targets such as protein, glucose, bacteria, DNA and some small biomolecules [63]. Similarly, polyurethanes show synthetic versatility, glorious mechanical properties, good biocompatibility and biodegradability, and in this respect, they can be used in biosensors. For example, hemoglobin-loaded multi-walled carbon nanotubes attached to bare polyurethane nanoparticles by non-covalent bonds have been used in biosensors [64].

Polymers used in biosensor design are polymers with different properties, such as polypyrrole, polyaniline, boronate polyethyleneglycol acrylate, chitosan, polyurethane, silicone, gelatin, cellulose, agarose [57, 62, 65, 66]. However, under this title, only a few studies in which nanoparticles prepared with simple polymers such as chitosan, gelatin and silicon are used in biosensor designs were summarized. The studies in which the polymer and its various composites are used in biosensor design are presented in Table **1**.

Ahmed and co-workers have developed an amperometric biosensor for dopamine detection using novel porous silicon nanoparticles. In this study, a simple stain etching method was used to obtain single-crystalline mesoporous silicon (PSi) nanoparticles. Accordingly, 2 g of silicon powder (approximately 40 μm in diameter) was dispersed in a mixture of 20 mL of 48% hydrofluoric acid:80 mL distilled water, and 10 mL of 70% nitric acid was added dropwise under continuous stirring at room temperature. PSi NPs were obtained with the formation of nitrogen oxide vapor. A decantation method was used to collect the formed Psi NPs and after rinsing with pure water, the PSi NPs were dried at 65°C for 8 hours. FESEM and TEM analyzes were performed on the obtained PSi NP and it was determined that the pores in the PSi NPs were randomly distributed and their diameters were less than 25 nm. In addition, X-ray diffraction (XRD), Raman spectroscopy, Fourier transform infrared spectroscopy (FTIR) and X-ray photoelectron spectroscopy (XPS) analyses were performed. In the design of the biosensor, firstly, the GC electrode was polished with diamond (1 μm) and alumina (0.05 μm), and then sonicated in ethanol and distilled water. For electrode modification, 4 mg of PSi NP was put into a mixture of 0.05 mL 5% nafion: 0.45 mL isopropyl alcohol and sonicated for 20 minutes. 3 μL of this suspension was taken and placed on the GC electrode surface, dried at room temperature for 5 minutes and then at 60°C for 20 minutes to obtain a homogeneous PSi layer on the GC electrode surface. Analyzes were performed under ambient conditions. In conclusion, it has been proven that the designed PSi NP-modified GC electrode biosensor detects dopamine in a wide concentration range (0.5-333.3 μM) in PBS buffer. In addition, for dopamine, the LOD value and the sensitivity limit of this biosensor were measured as 3.2 nM and 0.2715 $μAμM^{-1}cm^{-2}$, respectively. It displayed good selectivity towards dopamine in the presence of interference substances. In addition, the developed biosensor exhibited very good reproducibility and long-term stability properties [65].

A fiber optic biosensor was designed by Singh *et al.* using gelatin alginate nanoparticles loaded with L-asparaginase (L-ASP) to detect asparagine. L-ASP entrapped gelatin alginate nanoparticles were prepared by the ionic gelation method. Gelatin (a cationic polymer), alginate (an anionic polymer) and calcium chloride (a crosslinker) were used to achieve L-ASP immobilization. To prepare blank NPs, 0.2% calcium chloride (2 mL) solution was added to sodium alginate (0.3%, 10 mL) solution within 40 minutes under continuous stirring. Afterwards, the gelatin solution (0.05%, 1 mL) was dropped into the resulting calcium alginate microgel and stirred for another 60 minutes. After standing overnight, it was centrifuged at 20 000 g for 40 minutes. L-ASP loaded NPs were prepared similarly, only L-ASP was added to the calcium chloride solution in the first stage. The size of the NPs obtained was 396 nm and the loading efficiency of L-ASP was 79.7%, and the NPs were found to be stable for 50 days. Obtained L-ASP loaded NPs were fixed on nylon membrane discs (0.2 μm pore size) simultaneously with Rhodamine 6G (2%), and this system was affixed to the fiber optic probe. The basis of the method was that the developed biosensor detects the change in fluorescence intensity of Rhodamine 6G due to ammonia release in the presence of asparagine. As a result of the studies on the biosensor, it has been proven that the designed biosensor is very sensitive to asparagine at different concentrations (10^{-10}-10^{-1} M) with a detection limit of 0.05×10^{-10} M within 2 minutes response time. Moreover, the biosensor showed good repeatability for asparagine concentration of 10^{-10}-10^{-1} M as a result of three consecutive experiments. The biosensor was evaluated for reproducibility for 10^{-1}, 10^{-5} and 10^{-10} M asparagine concentrations and the relative deviation was less than 3%. Finally, the selectivity of the sensor was investigated and high selectivity was obtained for asparagine in the presence of other components. As a result, the developed biosensor could potentially be applied for the detection and quantification of L-asparagine in the diagnosis and treatment of leukemia [66].

Finally, an innovative amperometric biosensor based on tyrosinase/chitosan nanoparticles was designed by Gigli *et al.* for the sensitive and selective detection of catecholamine in human urine samples (Fig. **4**). Chitosan NPs (Ch-NPs) were prepared according to the ionic gelation method. Hereof, chitosan dispersion (9 mL) was added to TPP solution (3 mL) at the same concentration (1 mg/mL) and stirred for 2 hours at 25°C with mild magnetic stirring up to an opalescent solution was formed. Then it was centrifuged at 6000 rpm for 15 minutes and the formed pellets were washed with distilled water and dried. Two different methods (layer-by-layer and nanoprecipitation) were used to prepare the modified electrode. According to the layer-by-layer method, Ch-NP and tyrosinase (Try) solutions were added on the electrode and dried, respectively. In the nanoprecipitation technique, Tyr and Ch-NP were precipitated in the same solution at the same time, and this mixture was added to the electrode and dried. Optimization of the electrochemical sensor was evaluated with regard to the loading method of the Try and Ch-NPs onto the electrode, enzyme concentration and crosslinking agents' [1-ethyl-3-(3-dimethylaminopropyl)-carbodiimide-EDC and N-hydroxysuccinimide-NHS] effect on enzyme immobilization. As a result of the studies, it was determined that the Tyr/EDC-NHS/Ch-NPs nanosystem exhibited good conductivity and biocompatibility. Biosensor also showed high biocatalytic activity for the electrochemical reduction of dopamine, epinephrine and norepinephrine to o-quinone derivatives by oxidation at the redesigned electrode. The LOD value and and sensitivity of the developed biosensor for dopamine were measured as 0.17 μM and 0.583 $\mu A \mu M^{-1} cm^{-2}$, respectively. In addition, the response time is quite fast (3 seconds). The biosensor has shown sufficient feasibility to analyze total catecholamines in physiological samples, as the performance of the biosensor has been repeatedly evaluated in human urine samples, with satisfying results [67].

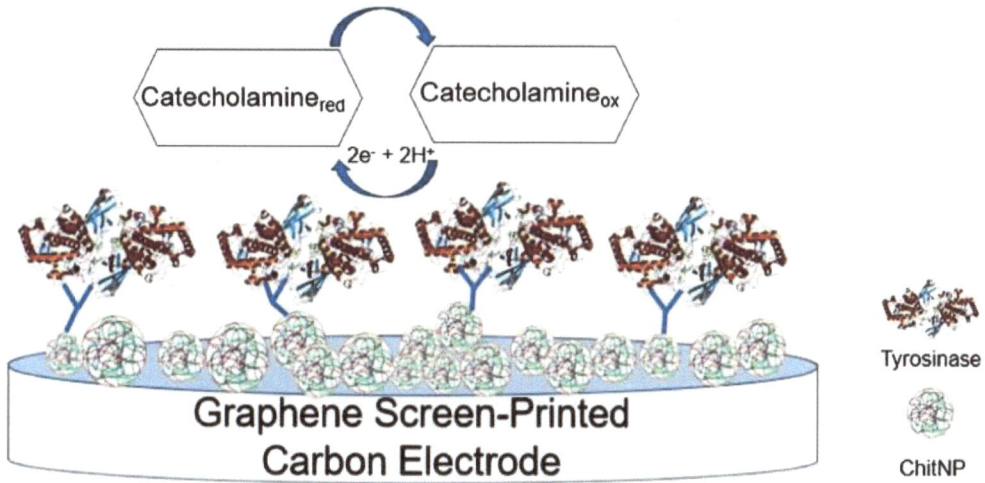

Fig. (4). Schematic image of graphene electrode designed using tyrosinase and chitosan nanoparticle.

CONCLUDING REMARKS

Biosensors are analytical devices that can qualitatively and quantitatively detect various target components, and consist of bioreceptor and transducer parts. Analytes detected by biosensors can be biological [DNA of bacteria or viruses, proteins produced from the immune system of infected/contaminated alive organisms (antibodies, antigens)] or small molecules (glucose or contaminants) [141].

Biosensors are used in a wide variety of fields, such as health and food, air, water and soil analysis. Because effective control of infectious diseases is crucial for public health [28]. In addition, other main problems faced in today's world are environmental pollution and food safety. For example, substances commonly found in foods (foodborne pathogens, heavy metals, mycotoxins, pesticides, herbicides, veterinary drugs, allergens and illegal additives) pose a major threat to human health. Therefore, there is a haste need to develop electro-analytical techniques that can detect and monitor pathogens, pollutants and allergens quickly, cost-effectively, portable and *in situ* and sensitive, specific and selective to detect infectious diseases and environmental pollutants and ensure food safety. However, traditional methods have different disadvantages, such as calibration, sample preparation, blind determination, qualified operator, time-consuming procedure, high cost and non-universal approach. In this respect, biosensors are promising as they can detect infectious diseases or environmental pollutants quickly, cheaply and sensitively, especially in regions with limited resources [28, 36, 59]. For example, in biosensors designed using AuNPs, based on color changes that are visible to the naked eye, antigens and nucleic acids can be detected simply, without the need for any device or power source, so that infectious diseases can be detected worldwide, but especially in some countries and remote areas with limited resources [35]. Electrochemical biosensors, in which conductive polymer nanoparticles are used, allow continuous monitoring of the environment [59].

Biosensors are evaluated in terms of 8 different properties. These; (1) sensitivity is the sensor's response to a unit alteration in the analyte amount, (2) selectivity is the sensor's capability to respond only to the target analyte, *i.e.*, it is desirable not to react to other interfering chemicals, (3) the measuring range is the concentration range where the sensor's sensitivity is good-sometimes this is called dynamic range or linearity, (4) response time is the time it takes for the sensor to show 63% of its final response to variation in analyte amount, (5) repeatability is the accuracy at which the sensor output can be achieved, (6) limit of detection is the analyte to which there is a measurable response is the lowest concentration, (7) lifetime is the time during which the sensor can work without a notable change in performance properties, 8) stability signalizes the alteration in baseline or sensitivity over a period of time [1]. For this reason, it is desired that biosensors have high sensitivity and selectivity, wide measuring range, fast response (short analysis time), high reproducibility and stability.

However, one of the difficulties encountered in biosensor design is that the biological recognition event (transduction) cannot capture signals efficiently. Today, nanomaterials are used to increase signal generation efficiency. Nanomaterials increase the biosensing area and provide signal amplification thanks to their nano size [141]. Nanoparticles, on the other hand, provide an ideal platform for biosensors in that they are easily dispersed on the electrode surface during the modification process in an aqueous medium, provide a large active area due to their maximum surface-to-volume ratio, and can entrap bioreactive materials at a high rate [58]. In this respect, nanotechnology can improve analysis sensitivity and selectivity through analyte-specific surface modifications. Gold nanoparticles and semiconductor quantum dots are the leading nanomaterials used in sensor technology due to their optical properties. However, studies have also focused on liposomes, polymeric nanoparticles, carbon nanotubes, nanodiamonds and graphene. In fact, various studies are underway to develop this biosensing platform into wearable/implantable biosensors and incorporate it into smartphones or similar portable electronic devices [33].

In addition to these advantages that nanosystems provide to biosensor technology, there are also some problems. For example, the inability to sufficiently immobilize the bio-specific material on the nanomaterial negatively affects the analysis feature of the biosensor [141]. With AuNP-based biosensors, method stability, sensitivity, and selectivity have been increased, and reusability and detection of complex analytes have been achieved. However, there are still challenges to overcome before colorimetric AuNPs can provide powerful diagnostic tests that can be transported into the laboratory or allow testing near the patient. These; (1) stability issues encountered in analysis samples with high ionic strength: Samples such as serum and urine cause non-specific AuNP aggregation and consequent color change. This is particularly observed in bare AuNPs, (2) lack of integration into simple platforms, which can eliminate the need for sample preprocessing and avoid the possibility of ligand exchange and non-specific interaction in complex matrices [33], (3) observing low biomolecule-AuNP conjugates stability in complex matrices [35], (4) the possibility of ligand exchange, non-specific interactions with biomacromolecules and the difficulties posed by *in vivo* applications [142], (5) the resulting nanoparticle concentration cannot be known during multiple washes to remove unbound analytes during preparation, (6) Low shelf life of AuNPs [40]. There are also some challenges in the *in vivo* application of polymers as biosensors. For example, when exposed to different temperature and humidity conditions, their mechanical properties may change, gas diffusion through the polymeric sheets causes the chemical structure of the electrode to change, and this may affect the sensitivity of the biosensor. The long stay of the nanosensors in the blood may cause surface modification, which will result in the generation of non-specific background signals [57].

Efforts are still being made to eliminate these negatives and to develop biosensors that can respond sensitively and quickly, allowing multiple analyses and on-site testing. As a result of these studies, it is thought that nanosystems will be used in most of the biosensors on the market in the near future.

REFERENCES

[1]   Y.H. Lee, and R. Mutharasan, "Biosensors", In: *Sensor technology handbook.,* J.S. Wilson, Ed., Elsevier, 2005, pp. 161-180.
       [http://dx.doi.org/10.1016/B978-075067729-5/50046-X]

[2]   S. Singh, V. Kumar, D.S. Dhanjal, S. Datta, R. Prasad, and J. Singh, "Biological biosensors for monitoring and diagnosis", In: *Microbial biotechnology: Basic research and applications.,* J. Singh, A. Vyas, S. Wang, R. Prasad, Eds., 1st. Springer Singapore, 2020, pp. 317-335.
       [http://dx.doi.org/10.1007/978-981-15-2817-0_14]

[3]   A. Kawamura, and T. Miyata, "Biosensors", In: *Biomaterials nanoarchitectonics.,* M. Ebara, Ed., William Andrew, 2016, pp. 157-176.
       [http://dx.doi.org/10.1016/B978-0-323-37127-8.00010-8]

[4]   P. Parkhey, and S.V. Mohan, "Biosensing applications of microbial fuel cell: Approach toward miniaturization", In: *Microbial electrochemical technology: sustainable platform for fuels, chemicals and remediation, biomass, biofuels and biochemicals.,* S.V. Mohan, S. Varjani, A. Pandey, Eds., Elsevier, 2018, pp. 977-997.

[5]   S.N. Sawant, Development of biosensors from biopolymer composites.K.K. Sadasivuni, D. Ponnamma, J. Kim, J.J. Cabibihan, and MA. AlMaadeed, *Biopolymer composites in electronics*

Elsevier, 2017, pp. 353-383.
[http://dx.doi.org/10.1016/B978-0-12-809261-3.00013-9]

[6]   Prickril B, Rasooly A, Eds. Biosensors and biodetection, Methods and protocols volume 2: Electrochemical, bioelectronic, piezoelectric, cellular and molecular biosensors. 2nd ed. New York: Humana Press 2017.

[7]   M. Kim, R. Iezzi Jr, B.S. Shim, and D.C. Martin, "Impedimetric biosensors for detecting vascular endothelial growth factor (VEGF) based on poly (3, 4-ethylene dioxythiophene) (PEDOT)/gold nanoparticle (Au NP) composites", *Front Chem.,* vol. 7, p. 234, 2019.
[http://dx.doi.org/10.3389/fchem.2019.00234] [PMID: 31058131]

[8]   C. Chen, and J. Wang, "Optical biosensors: An exhaustive and comprehensive review", *Analyst,* vol. 145, no. 5, pp. 1605-1628, 2020.
[http://dx.doi.org/10.1039/C9AN01998G] [PMID: 31970360]

[9]   P. Damborský, J. Švitel, and J. Katrlík, "Optical biosensors", *Essays Biochem.,* vol. 60, no. 1, pp. 91-100, 2016.
[http://dx.doi.org/10.1042/EBC20150010] [PMID: 27365039]

[10]   L. Carvajal Barbosa, D. Insuasty Cepeda, A.F. León Torres, M.M. Arias Cortes, Z.J. Rivera Monroy, and J.E. Garcia Castaneda, "Nucleic acid-based biosensors: Analytical devices for prevention, diagnosis and treatment of diseases", *Vitae,* vol. 28, no. 3, p. 347259, 2021.
[http://dx.doi.org/10.17533/udea.vitae.v28n3a347259]

[11]   P.P. Vachali, B. Li, A. Bartschi, and P.S. Bernstein, "Surface plasmon resonance (SPR)-based biosensor technology for the quantitative characterization of protein–carotenoid interactions", *Arch. Biochem. Biophys.,* vol. 572, pp. 66-72, 2015.
[http://dx.doi.org/10.1016/j.abb.2014.12.005] [PMID: 25513962]

[12]   T. Pasinszki, and M. Krebsz, "Advances in celiac disease testing", *Adv. Clin. Chem.,* vol. 91, pp. 1-29, 2019.
[http://dx.doi.org/10.1016/bs.acc.2019.03.001] [PMID: 31331486]

[13]   B.D. Malhotra, M.A. Ali, Ed., *Plasmonic nanostructures: Fiber-optic biosensors, nanomaterials for biosensors: Fundamentals and applications.* 1st. Elsevier, 2017, pp. 161-181.

[14]   S. Vasuki, V. Varsha, and R. Mithra, "Thermal biosensors and their applications", *Am. Int. J. Res. Sci. Tech. Eng. Math.,* pp. 262-264, 2019.

[15]   Y.C. Zhu, L.P. Mei, Y.F. Ruan, N. Zhang, W.W. Zhao, J.J. Xu, and H.Y. Chen, Enzyme-based biosensors and their applications.*Advances in enzyme technology,* R.S. Singh, RR. Singhania, A. Pandey, C. Larroche, Eds., Elsevier, 2019, pp. 201-234.
[http://dx.doi.org/10.1016/B978-0-444-64114-4.00008-X]

[16]   I. Palchetti, and M. Mascini, "Nucleic acid biosensors for environmental pollution monitoring", *Analyst,* vol. 133, no. 7, pp. 846-854, 2008.
[http://dx.doi.org/10.1039/b802920m] [PMID: 18575633]

[17]   Y. Du, and S. Dong, "Nucleic acid biosensors: Recent advances and perspectives", *Anal. Chem.,* vol. 89, no. 1, pp. 189-215, 2017.
[http://dx.doi.org/10.1021/acs.analchem.6b04190] [PMID: 28105831]

[18]   U. Bora, A. Sett, and D. Singh, "Nucleic acid based biosensors for clinical applications", *Biosensors. J.,* vol. 2, no. 1, pp. 1-8, 2013.
[http://dx.doi.org/10.4172/2090-4967.1000104]

[19]   B.D. Malhotra, C.M. Pandey, Ed., *Biosensors: Fundamentals and applications.* Smithers Rapra Publishing: United Kingdom, 2017.

[20]   L. Chhiba, B. Zaher, M. Sidqui, and A. Marzak, "Glucose sensing for diabetes monitoring: From invasive to wearable device", *The proceedings of the third international conference on smart city applications,* 2020 pp. 350-364.

[21]   E.O. Polat, M.M. Cetin, A.F. Tabak, E. Bilget Güven, B.Ö. Uysal, T. Arsan, A. Kabbani, H. Hamed, and S.B. Gül, "Transducer technologies for biosensors and their wearable applications", *Biosensors,* vol. 12, no. 6, p. 385, 2022.
[http://dx.doi.org/10.3390/bios12060385] [PMID: 35735533]

[22]   G. Liu, J. Wang, Y. Lin, and J. Wang, Nanoparticle-based biosensors and bioassays.*Electrochemical sensors, biosensors and their biomedical applications.*, X. Zhang, H. Ju, J. Wang, Eds., Academic Press, 2008, pp. 441-457.
[http://dx.doi.org/10.1016/B978-012373738-0.50016-7]

[23]   P. Malik, V. Katyal, V. Malik, A. Asatkar, G. Inwati, and T.K. Mukherjee, "Nanobiosensors: Concepts and variations", *Int. Sch. Res. Notices,* p. 327435, 2013.

[24]   P. Jiang, Y. Wang, L. Zhao, C. Ji, D. Chen, and L. Nie, "Applications of gold nanoparticles in non-optical biosensors", *Nanomaterials,* vol. 8, no. 12, p. 977, 2018.
[http://dx.doi.org/10.3390/nano8120977] [PMID: 30486293]

[25]   Q. Liu, and B.J. Boyd, "Liposomes in biosensors", *Analyst,* vol. 138, no. 2, pp. 391-409, 2013.
[http://dx.doi.org/10.1039/C2AN36140J] [PMID: 23072757]

[26]   X. Zhang, Q. Guo, and D. Cui, "Recent advances in nanotechnology applied to biosensors", *Sensors,* vol. 9, no. 2, pp. 1033-1053, 2009.
[http://dx.doi.org/10.3390/s90201033] [PMID: 22399954]

[27]   J.M. Pingarrón, P. Yáñez-Sedeño, and A. González-Cortés, "Gold nanoparticle-based electrochemical biosensors", *Electrochim. Acta,* vol. 53, no. 19, pp. 5848-5866, 2008.
[http://dx.doi.org/10.1016/j.electacta.2008.03.005]

[28]   X. Yu, Y. Jiao, and Q. Chai, "Applications of gold nanoparticles in biosensors", *Nano Life,* vol. 6, no. 2, p. 1642001, 2016.
[http://dx.doi.org/10.1142/S1793984416420010]

[29]   L. Nie, F. Liu, P. Ma, and X. Xiao, "Applications of gold nanoparticles in optical biosensors", *J. Biomed. Nanotechnol.,* vol. 10, no. 10, pp. 2700-2721, 2014.
[http://dx.doi.org/10.1166/jbn.2014.1987] [PMID: 25992415]

[30]   J.H. Lee, H.Y. Cho, H. Choi, J.Y. Lee, and J.W. Choi, "Application of gold nanoparticle to plasmonic biosensors", *Int. J. Mol. Sci.,* vol. 19, no. 7, p. 2021, 2018.
[http://dx.doi.org/10.3390/ijms19072021] [PMID: 29997363]

[31]   J. Zhao, B. Bo, Y.M. Yin, and G.X. Li, "Gold nanoparticles-based biosensors for biomedical application", *Nano Life,* vol. 2, no. 4, p. 1230008, 2012.
[http://dx.doi.org/10.1142/S1793984412300087]

[32]   Y. Li, H.J. Schluesener, and S. Xu, "Gold nanoparticle-based biosensors", *Gold Bull.,* vol. 43, no. 1, pp. 29-41, 2010.
[http://dx.doi.org/10.1007/BF03214964]

[33]   H. Aldewachi, T. Chalati, M.N. Woodroofe, N. Bricklebank, B. Sharrack, and P. Gardiner, "Gold nanoparticle-based colorimetric biosensors", *Nanoscale,* vol. 10, no. 1, pp. 18-33, 2018.
[http://dx.doi.org/10.1039/C7NR06367A] [PMID: 29211091]

[34]   S. Zeng, K.T. Yong, I. Roy, X.Q. Dinh, X. Yu, and F. Luan, "A review on functionalized gold nanoparticles for biosensing applications", *Plasmonics,* vol. 6, no. 3, pp. 491-506, 2011.
[http://dx.doi.org/10.1007/s11468-011-9228-1]

[35]   M. Lin, H. Pei, F. Yang, C. Fan, and X. Zuo, "Applications of gold nanoparticles in the detection and identification of infectious diseases and biothreats", *Adv. Mater.,* vol. 25, no. 25, pp. 3490-3496, 2013.
[http://dx.doi.org/10.1002/adma.201301333] [PMID: 23977699]

[36]   Z. Hua, T. Yu, D. Liu, and Y. Xianyu, "Recent advances in gold nanoparticles-based biosensors for food safety detection", *Biosens. Bioelectron.,* vol. 179, p. 113076, 2021.

[http://dx.doi.org/10.1016/j.bios.2021.113076] [PMID: 33601132]

[37]   E.E. Bedford, J. Spadavecchia, C.M. Pradier, and F.X. Gu, "Surface plasmon resonance biosensors incorporating gold nanoparticles", *Macromol. Biosci.,* vol. 12, no. 6, pp. 724-739, 2012.
[http://dx.doi.org/10.1002/mabi.201100435] [PMID: 22416018]

[38]   P. Yáñez-Sedeño, and J.M. Pingarrón, "Gold nanoparticle-based electrochemical biosensors", *Anal. Bioanal. Chem.,* vol. 382, no. 4, pp. 884-886, 2005.
[http://dx.doi.org/10.1007/s00216-005-3221-5] [PMID: 15864491]

[39]   T.T. Bezuneh, T.H. Fereja, S.A. Kitte, H. Li, and Y. Jin, "Gold nanoparticle-based signal amplified electrochemiluminescence for biosensing applications", *Talanta,* vol. 248, p. 123611, 2022.
[http://dx.doi.org/10.1016/j.talanta.2022.123611] [PMID: 35660995]

[40]   J. Satija, R. Bharadwaj, V. Sai, and S. Mukherji, "Emerging use of nanostructure films containing capped gold nanoparticles in biosensors", *Nanotechnol. Sci. Appl.,* vol. 3, pp. 171-188, 2010.
[PMID: 24198481]

[41]   L. Zhang, Y. Mazouzi, M. Salmain, B. Liedberg, and S. Boujday, "Antibody-gold nanoparticle bioconjugates for biosensors: Synthesis, characterization and selected applications", *Biosens. Bioelectron.,* vol. 165, p. 112370, 2020.
[http://dx.doi.org/10.1016/j.bios.2020.112370] [PMID: 32729502]

[42]   X-M. Ma, M. Sun, Y. Lin, Y-J. Liu, F. Luo, L-H. Guo, B. Qiu, Z-Y. Lin, and G-N. Chen, "Progress of visual biosensor based on gold nanoparticles", *Chin. J. Anal. Chem.,* vol. 46, no. 1, pp. 1-10, 2018.
[http://dx.doi.org/10.1016/S1872-2040(17)61061-2]

[43]   Z. Tabibi, J. Massah, and K. Asefpour Vakilian, "A biosensor for the sensitive and specific measurement of arsenite using gold nanoparticles", *Measurement,* vol. 187, p. 110281, 2022.
[http://dx.doi.org/10.1016/j.measurement.2021.110281]

[44]   L. Li, Y. Zhang, W. Zheng, X. Li, and Y. Zhao, "Optical fiber SPR biosensor based on gold nanoparticle amplification for DNA hybridization detection", *Talanta,* vol. 247, p. 123599, 2022.
[http://dx.doi.org/10.1016/j.talanta.2022.123599] [PMID: 35653863]

[45]   MR Hasan, P Sharma, R Pilloton, M Khanuja, and J Narang, "Colorimetric biosensor for the naked-eye detection of ovarian cancer biomarker PDGF using citrate modified gold nanoparticles", *Biosensors and Bioelectronics: X,* vol. 11, p. 100142, 2022.
[http://dx.doi.org/10.1016/j.biosx.2022.100142]

[46]   V.K. Sharma, and M.K. Agrawal, "A historical perspective of liposomes: A bio nanomaterial", *Mater. Today Proc.,* vol. 45, pp. 2963-2966, 2021.
[http://dx.doi.org/10.1016/j.matpr.2020.11.952]

[47]   M.K. Lee, "Liposomes for enhanced bioavailability of water-insoluble drugs: *In vivo* evidence and recent approaches", *Pharmaceutics,* vol. 12, no. 3, p. 264, 2020.
[http://dx.doi.org/10.3390/pharmaceutics12030264] [PMID: 32183185]

[48]   A.D. Bangham, and R.W. Horne, "Negative staining of phospholipids and their structural modification by surface-active agents as observed in the electron microscope", *J. Mol. Biol.,* vol. 8, no. 5, pp. 660-IN10, 1964.
[http://dx.doi.org/10.1016/S0022-2836(64)80115-7] [PMID: 14187392]

[49]   F. Mazur, M. Bally, B. Städler, and R. Chandrawati, "Liposomes and lipid bilayers in biosensors", *Adv. Colloid. Interface. Sci.,* vol. 249, pp. 88-99, 2017.
[http://dx.doi.org/10.1016/j.cis.2017.05.020] [PMID: 28602208]

[50]   N. Hamano, and Y. Negishi, "Liposome-based biosensors and diagnosis imaging agents", *Sens. Mater.,* vol. 34, no. 3, pp. 961-970, 2022.
[http://dx.doi.org/10.18494/SAM3611]

[51]   S. Kalepu, K.T. Sunilkumar, and S. Betha, "Liposomal drug delivery system-a comprehensive review", *Int. J. Drug. Dev. Res.,* vol. 5, no. 4, pp. 62-75, 2013.

[52]   B. Lin, D. Liu, J. Yan, Z. Qiao, Y. Zhong, J. Yan, Z. Zhu, T. Ji, and C.J. Yang, "Enzyme-encapsulated liposome-linked immunosorbent assay enabling sensitive personal glucose meter readout for portable detection of disease biomarkers", *ACS Appl. Mater. Interfaces,* vol. 8, no. 11, pp. 6890-6897, 2016.
[http://dx.doi.org/10.1021/acsami.6b00777] [PMID: 26918445]

[53]   V.C. Rodrigues, M.L. Moraes, J.C. Soares, A.L. Souza, A.C. Soares, O.N. Oliveira Jr, and D. Gonçalves, "Liposome-based biosensors using phytase immobilized on polypyrrole films for phytic acid determination", *Bull. Chem. Soc. Jpn.,* vol. 92, no. 4, pp. 847-851, 2019.
[http://dx.doi.org/10.1246/bcsj.20180369]

[54]   M.L. Moraes, P.J. Gomes, P.A. Ribeiro, P. Vieira, A.A. Freitas, R. Köhler, O.N. Oliveira Jr, and M. Raposo, "Polymeric scaffolds for enhanced stability of melanin incorporated in liposomes", *J. Colloid Interface Sci.,* vol. 350, no. 1, pp. 268-274, 2010.
[http://dx.doi.org/10.1016/j.jcis.2010.06.043] [PMID: 20633887]

[55]   J. Zhou, M. Duan, D. Huang, H. Shao, Y. Zhou, and Y. Fan, "Label-free visible colorimetric biosensor for detection of multiple pathogenic bacteria based on engineered polydiacetylene liposomes", *J. Colloid Interface Sci.,* vol. 606, no. Pt 2, pp. 1684-1694, 2022.
[http://dx.doi.org/10.1016/j.jcis.2021.07.155] [PMID: 34500167]

[56]   S. Mallakpour, and V. Behranvand, "Polymeric nanoparticles: Recent development in synthesis and application", *Express Polym. Lett.,* vol. 10, no. 11, pp. 895-913, 2016.
[http://dx.doi.org/10.3144/expresspolymlett.2016.84]

[57]   S. Gupta, A. Sharma, and R.S. Verma, "Polymers in biosensor devices for cardiovascular applications", *Curr. Opin. Biomed. Eng.,* vol. 13, pp. 69-75, 2020.
[http://dx.doi.org/10.1016/j.cobme.2019.10.002]

[58]   L. Xia, Z. Wei, and M. Wan, "Conducting polymer nanostructures and their application in biosensors", *J. Colloid Interface Sci.,* vol. 341, no. 1, pp. 1-11, 2010.
[http://dx.doi.org/10.1016/j.jcis.2009.09.029] [PMID: 19837415]

[59]   H. Kumar, N. Kumari, and R. Sharma, "Nanocomposites (conducting polymer and nanoparticles) based electrochemical biosensor for the detection of environment pollutant: Its issues and challenges", *Environ. Impact Assess. Rev.,* vol. 85, p. 106438, 2020.
[http://dx.doi.org/10.1016/j.eiar.2020.106438]

[60]   L. Feng, C. Zhu, H. Yuan, L. Liu, F. Lv, and S. Wang, "Conjugated polymer nanoparticles: Preparation, properties, functionalization and biological applications", *Chem. Soc. Rev.,* vol. 42, no. 16, pp. 6620-6633, 2013.
[http://dx.doi.org/10.1039/c3cs60036j] [PMID: 23744297]

[61]   Y. Jiang, and J. Wu, "Recent development in chitosan nanocomposites for surface–based biosensor applications", *Electrophoresis,* vol. 40, no. 16-17, pp. 2084-2097, 2019.
[http://dx.doi.org/10.1002/elps.201900066] [PMID: 31081120]

[62]   N. Bedi, D.K. Srivastava, A. Srivastava, S. Mahapatra, D.S. Dkhar, P. Chandra, and A. Srivastava, "Marine biological macromolecules as matrix material for biosensor fabrication", *Biotechnol. Bioeng.,* vol. 119, no. 8, pp. 2046-2063, 2022.
[http://dx.doi.org/10.1002/bit.28122] [PMID: 35470439]

[63]   Y. Gao, and Y. Wu, "Recent advances of chitosan-based nanoparticles for biomedical and biotechnological applications", *Int. J. Biol. Macromol.,* vol. 203, pp. 379-388, 2022.
[http://dx.doi.org/10.1016/j.ijbiomac.2022.01.162] [PMID: 35104473]

[64]   Morral-Ruíz G, Melgar-Lesmes P, Solans C, García-Celma MJ. Polyurethane nanoparticles, a new tool for biomedical applications?. In: Cooper SL, Guan J, Eds. *Advances in polyurethane biomaterials.* Woodhead Publishing 2016; pp: 195-216.
[http://dx.doi.org/10.1016/B978-0-08-100614-6.00007-X]

[65]   J. Ahmed, M. Faisal, F.A. Harraz, M. Jalalah, and S.A. Alsareii, "Development of an amperometric biosensor for dopamine using novel mesoporous silicon nanoparticles fabricated via a facile stain etching approach", *Physica E,* vol. 135, p. 114952, 2022.
[http://dx.doi.org/10.1016/j.physe.2021.114952]

[66]   A. Singh, N. Verma, and K. Kumar, "Fabrication and characterization of L-asparaginase entrapped polymeric nanoparticles for asparagine biosensor construction", *Mater. Today Proc.,* vol. 67, pp. 591-597, 2022.
[http://dx.doi.org/10.1016/j.matpr.2022.05.227]

[67]   V. Gigli, C. Tortolini, E. Capecchi, A. Angeloni, A. Lenzi, and R. Antiochia, "Novel amperometric biosensor based on tyrosinase/chitosan nanoparticles for sensitive and interference-free detection of total catecholamine", *Biosensors,* vol. 12, no. 7, p. 519, 2022.
[http://dx.doi.org/10.3390/bios12070519] [PMID: 35884322]

[68]   Y. Wang, and E.C. Alocilja, "Gold nanoparticle-labeled biosensor for rapid and sensitive detection of bacterial pathogens", *J. Biol. Eng.,* vol. 9, no. 1, p. 16, 2015.
[http://dx.doi.org/10.1186/s13036-015-0014-z] [PMID: 26435738]

[69]   L. Shang, H. Chen, L. Deng, and S. Dong, "Enhanced resonance light scattering based on biocatalytic growth of gold nanoparticles for biosensors design", *Biosens. Bioelectron.,* vol. 23, no. 7, pp. 1180-1184, 2008.
[http://dx.doi.org/10.1016/j.bios.2007.10.024] [PMID: 18068347]

[70]   S. Lepinay, A. Staff, A. Ianoul, and J. Albert, "Improved detection limits of protein optical fiber biosensors coated with gold nanoparticles", *Biosens. Bioelectron.,* vol. 52, pp. 337-344, 2014.
[http://dx.doi.org/10.1016/j.bios.2013.08.058] [PMID: 24080213]

[71]   W. Hong, H. Bai, Y. Xu, Z. Yao, Z. Gu, and G. Shi, "Preparation of gold nanoparticle/graphene composites with controlled weight contents and their application in biosensors", *J. Phys. Chem. C,* vol. 114, no. 4, pp. 1822-1826, 2010.
[http://dx.doi.org/10.1021/jp9101724]

[72]   J. Liu, and Y. Lu, "A colorimetric lead biosensor using DNAzyme-directed assembly of gold nanoparticles", *J. Am. Chem. Soc.,* vol. 125, no. 22, pp. 6642-6643, 2003.
[http://dx.doi.org/10.1021/ja034775u] [PMID: 12769568]

[73]   W. Wang, C. Chen, M. Qian, and X.S. Zhao, "Aptamer biosensor for protein detection using gold nanoparticles", *Anal. Biochem.,* vol. 373, no. 2, pp. 213-219, 2008.
[http://dx.doi.org/10.1016/j.ab.2007.11.013] [PMID: 18054771]

[74]   R. Cui, H. Huang, Z. Yin, D. Gao, and J.J. Zhu, "Horseradish peroxidase-functionalized gold nanoparticle label for amplified immunoanalysis based on gold nanoparticles/carbon nanotubes hybrids modified biosensor", *Biosens. Bioelectron.,* vol. 23, no. 11, pp. 1666-1673, 2008.
[http://dx.doi.org/10.1016/j.bios.2008.01.034] [PMID: 18359217]

[75]   B.K. Jena, and C.R. Raj, "Electrochemical biosensor based on integrated assembly of dehydrogenase enzymes and gold nanoparticles", *Anal. Chem.,* vol. 78, no. 18, pp. 6332-6339, 2006.
[http://dx.doi.org/10.1021/ac052143f] [PMID: 16970306]

[76]   J. Jia, B. Wang, A. Wu, G. Cheng, Z. Li, and S. Dong, "A method to construct a third-generation horseradish peroxidase biosensor: Self-assembling gold nanoparticles to three-dimensional sol-gel network", *Anal. Chem.,* vol. 74, no. 9, pp. 2217-2223, 2002.
[http://dx.doi.org/10.1021/ac011116w] [PMID: 12033329]

[77]   S. Zhang, N. Wang, Y. Niu, and C. Sun, "Immobilization of glucose oxidase on gold nanoparticles modified Au electrode for the construction of biosensor", *Sens. Actuators B Chem.,* vol. 109, no. 2, pp. 367-374, 2005.
[http://dx.doi.org/10.1016/j.snb.2005.01.003]

[78]   J. Liu, and Y. Lu, "Adenosine-dependent assembly of aptazyme-functionalized gold nanoparticles and its application as a colorimetric biosensor", *Anal. Chem.,* vol. 76, no. 6, pp. 1627-1632, 2004.
[http://dx.doi.org/10.1021/ac0351769] [PMID: 15018560]

[79]   E. Dinçkaya, Ö. Kınık, M.K. Sezgintürk, Ç. Altuğ, and A. Akkoca, "Development of an impedimetric aflatoxin M1 biosensor based on a DNA probe and gold nanoparticles", *Biosens. Bioelectron.,* vol. 26, no. 9, pp. 3806-3811, 2011.
[http://dx.doi.org/10.1016/j.bios.2011.02.038] [PMID: 21420290]

[80]   H. Xu, X. Mao, Q. Zeng, S. Wang, A.N. Kawde, and G. Liu, "Aptamer-functionalized gold nanoparticles as probes in a dry-reagent strip biosensor for protein analysis", *Anal. Chem.,* vol. 81, no. 2, pp. 669-675, 2009.
[http://dx.doi.org/10.1021/ac8020592] [PMID: 19072289]

[81]   E. Dester, K. Kao, and E.C. Alocilja, "Detection of unamplified E. coli o157 DNA extracted from large food samples using a gold nanoparticle colorimetric biosensor", *Biosensors,* vol. 12, no. 5, p. 274, 2022.
[http://dx.doi.org/10.3390/bios12050274] [PMID: 35624575]

[82]   C.B.P. Ligiero, T.S. Fernandes, D.L. D'Amato, F.V. Gaspar, P.S. Duarte, M.A. Strauch, J.G. Fonseca, L.G.R. Meirelles, P. Bento da Silva, R.B. Azevedo, G. Aparecida de Souza Martins, B.S. Archanjo, C.D. Buarque, G. Machado, A.M. Percebom, and C.M. Ronconi, "Influence of particle size on the SARS-CoV-2 spike protein detection using IgG-capped gold nanoparticles and dynamic light scattering", *Mater. Today Chem.,* vol. 25, p. 100924, 2022.
[http://dx.doi.org/10.1016/j.mtchem.2022.100924] [PMID: 35475288]

[83]   M.E. Hamdy, M. Del Carlo, H.A. Hussein, T.A. Salah, A.H. El-Deeb, M.M. Emara, G. Pezzoni, and D. Compagnone, "Development of gold nanoparticles biosensor for ultrasensitive diagnosis of foot and mouth disease virus", *J. Nanobiotechnology,* vol. 16, no. 1, p. 48, 2018.
[http://dx.doi.org/10.1186/s12951-018-0374-x] [PMID: 29751767]

[84]   R. Savin, N.O. Benzaamia, C. Njel, S. Pronkin, C. Blanck, M. Schmutz, and F. Boulmedais, "Nanohybrid biosensor based on mussel-inspired electro-cross-linking of tannic acid capped gold nanoparticles and enzymes", *Mat. Adv.,* vol. 3, no. 4, pp. 2222-2233, 2022.
[http://dx.doi.org/10.1039/D1MA01193F]

[85]   N. Zhou, J. Wang, T. Chen, Z. Yu, and G. Li, "Enlargement of gold nanoparticles on the surface of a self-assembled monolayer modified electrode: A mode in biosensor design", *Anal. Chem.,* vol. 78, no. 14, pp. 5227-5230, 2006.
[http://dx.doi.org/10.1021/ac0605492] [PMID: 16841954]

[86]   C. Lanzellotto, G. Favero, M.L. Antonelli, C. Tortolini, S. Cannistraro, E. Coppari, and F. Mazzei, "Nanostructured enzymatic biosensor based on fullerene and gold nanoparticles: Preparation, characterization and analytical applications", *Biosens. Bioelectron.,* vol. 55, pp. 430-437, 2014.
[http://dx.doi.org/10.1016/j.bios.2013.12.028] [PMID: 24441023]

[87]   Y. He, S. Zhang, X. Zhang, M. Baloda, A.S. Gurung, H. Xu, X. Zhang, and G. Liu, "Ultrasensitive nucleic acid biosensor based on enzyme–gold nanoparticle dual label and lateral flow strip biosensor", *Biosens. Bioelectron.,* vol. 26, no. 5, pp. 2018-2024, 2011.
[http://dx.doi.org/10.1016/j.bios.2010.08.079] [PMID: 20875950]

[88]   T. Špringer, and J. Homola, "Biofunctionalized gold nanoparticles for SPR-biosensor-based detection of CEA in blood plasma", *Anal. Bioanal. Chem.,* vol. 404, no. 10, pp. 2869-2875, 2012.
[http://dx.doi.org/10.1007/s00216-012-6308-9] [PMID: 22895740]

[89]   Q. Shen, W. Li, S. Tang, Y. Hu, Z. Nie, Y. Huang, and S. Yao, "A simple "clickable" biosensor for colorimetric detection of copper(II) ions based on unmodified gold nanoparticles", *Biosens. Bioelectron.,* vol. 41, pp. 663-668, 2013.
[http://dx.doi.org/10.1016/j.bios.2012.09.032] [PMID: 23089325]

[90]   Y. Shao, S. Xu, X. Zheng, Y. Wang, and W. Xu, "Optical fiber LSPR biosensor prepared by gold nanoparticle assembly on polyelectrolyte multilayer", *Sensors,* vol. 10, no. 4, pp. 3585-3596, 2010. [http://dx.doi.org/10.3390/s100403585] [PMID: 22319313]

[91]   L. Tian, K. Qian, J. Qi, Q. Liu, C. Yao, W. Song, and Y. Wang, "Gold nanoparticles superlattices assembly for electrochemical biosensor detection of microRNA-21", *Biosens. Bioelectron.,* vol. 99, pp. 564-570, 2018. [http://dx.doi.org/10.1016/j.bios.2017.08.035] [PMID: 28826000]

[92]   C. Guo, P. Boullanger, L. Jiang, and T. Liu, "Highly sensitive gold nanoparticles biosensor chips modified with a self-assembled bilayer for detection of Con A", *Biosens. Bioelectron.,* vol. 22, no. 8, pp. 1830-1834, 2007. [http://dx.doi.org/10.1016/j.bios.2006.09.006] [PMID: 17045470]

[93]   D. Lima, A. Ribicki, L. Gonçalves, A.C.M. Hacke, L.C. Lopes, R.P. Pereira, K. Wohnrath, S.T. Fujiwara, and C.A. Pessôa, "Nanoconjugates based on a novel organic-inorganic hybrid silsesquioxane and gold nanoparticles as hemocompatible nanomaterials for promising biosensing applications", *Colloids Surf. B Biointerfaces,* vol. 213, p. 112355, 2022. [http://dx.doi.org/10.1016/j.colsurfb.2022.112355] [PMID: 35158220]

[94]   C. Pothipor, S. Bamrungsap, J. Jakmunee, and K. Ounnunkad, "A gold nanoparticle-dye/poly(3-aminobenzylamine)/two dimensional MoSe2/graphene oxide electrode towards label-free electrochemical biosensor for simultaneous dual-mode detection of cancer antigen 15-3 and microRNA-21", *Colloids Surf. B Biointerfaces,* vol. 210, p. 112260, 2022. [http://dx.doi.org/10.1016/j.colsurfb.2021.112260] [PMID: 34894598]

[95]   H.M. Kim, H.J. Kim, J.H. Park, and S.K. Lee, "High-performance biosensor using a sandwich assay via antibody-conjugated gold nanoparticles and fiber-optic localized surface plasmon resonance", *Anal. Chim. Acta,* vol. 1213, p. 339960, 2022. [http://dx.doi.org/10.1016/j.aca.2022.339960] [PMID: 35641064]

[96]   C.C. Chou, Y.T. Lin, I. Kuznetsova, and G.J. Wang, "Genetically modified soybean detection using a biosensor electrode with a self-assembled monolayer of gold nanoparticles", *Biosensors,* vol. 12, no. 4, p. 207, 2022. [http://dx.doi.org/10.3390/bios12040207] [PMID: 35448267]

[97]   V. Vamvakaki, and N.A. Chaniotakis, "Pesticide detection with a liposome-based nano-biosensor", *Biosens. Bioelectron.,* vol. 22, no. 12, pp. 2848-2853, 2007. [http://dx.doi.org/10.1016/j.bios.2006.11.024] [PMID: 17223333]

[98]   T. Mahmoudi-Badiki, E. Alipour, H. Hamishehkar, and S.M. Golabi, "Dopamine-loaded liposome and its application in electrochemical DNA biosensor", *J. Biomater. Appl.,* vol. 31, no. 2, pp. 273-282, 2016. [http://dx.doi.org/10.1177/0885328216650378] [PMID: 27194602]

[99]   V. Vamvakaki, D. Fournier, and N.A. Chaniotakis, "Fluorescence detection of enzymatic activity within a liposome based nano-biosensor", *Biosens. Bioelectron.,* vol. 21, no. 2, pp. 384-388, 2005. [http://dx.doi.org/10.1016/j.bios.2004.10.028] [PMID: 16023967]

[100] H. Chen, Y. Zheng, J.H. Jiang, H.L. Wu, G.L. Shen, and R.Q. Yu, "An ultrasensitive chemiluminescence biosensor for cholera toxin based on ganglioside-functionalized supported lipid membrane and liposome", *Biosens. Bioelectron.,* vol. 24, no. 4, pp. 684-689, 2008. [http://dx.doi.org/10.1016/j.bios.2008.06.031] [PMID: 18672355]

[101] J.S. Graça, R.F. de Oliveira, M.L. de Moraes, and M. Ferreira, "Amperometric glucose biosensor based on layer-by-layer films of microperoxidase-11 and liposome-encapsulated glucose oxidase", *Bioelectrochemistry,* vol. 96, pp. 37-42, 2014. [http://dx.doi.org/10.1016/j.bioelechem.2014.01.001] [PMID: 24491835]

[102] H. Guan, X. Liu, and W. Wang, "Encapsulation of tyrosinase within liposome bioreactors for developing an amperometric phenolic compounds biosensor", *J. Solid State Electrochem.,* vol. 17, no.

11, pp. 2887-2893, 2013.
[http://dx.doi.org/10.1007/s10008-013-2181-5]

[103] A.J. Baeumner, N.A. Schlesinger, N.S. Slutzki, J. Romano, E.M. Lee, and R.A. Montagna, "Biosensor for dengue virus detection: sensitive, rapid, and serotype specific", *Anal. Chem.,* vol. 74, no. 6, pp. 1442-1448, 2002.
[http://dx.doi.org/10.1021/ac015675e] [PMID: 11922316]

[104] D.H. Kang, K. Kim, Y. Son, P.S. Chang, J. Kim, and H.S. Jung, "Design of a simple paper-based colorimetric biosensor using polydiacetylene liposomes for neomycin detection", *Analyst,* vol. 143, no. 19, pp. 4623-4629, 2018.
[http://dx.doi.org/10.1039/C8AN01097H] [PMID: 30207329]

[105] A.J. Baeumner, C. Jones, C.Y. Wong, and A. Price, "A generic sandwich-type biosensor with nanomolar detection limits", *Anal. Bioanal. Chem.,* vol. 378, no. 6, pp. 1587-1593, 2004.
[http://dx.doi.org/10.1007/s00216-003-2466-0] [PMID: 15214421]

[106] S. Kwakye, V.N. Goral, and A.J. Baeumner, "Electrochemical microfluidic biosensor for nucleic acid detection with integrated minipotentiostat", *Biosens. Bioelectron.,* vol. 21, no. 12, pp. 2217-2223, 2006.
[http://dx.doi.org/10.1016/j.bios.2005.11.017] [PMID: 16386889]

[107] Z. Zhang, M. Sohgawa, K. Yamashita, and M. Noda, "A Micromechanical cantilever–based liposome biosensor for characterization of protein–membrane interaction", *Electroanalysis,* vol. 28, no. 3, pp. 620-625, 2016.
[http://dx.doi.org/10.1002/elan.201500412]

[108] J. Kim, B.S. Moon, E. Hwang, S. Shaban, W. Lee, D.G. Pyun, D.H. Lee, and D.H. Kim, "Solid-state colorimetric polydiacetylene liposome biosensor sensitized by gold nanoparticles", *Analyst,* vol. 146, no. 5, pp. 1682-1688, 2021.
[http://dx.doi.org/10.1039/D0AN02375B] [PMID: 33449063]

[109] A.J. Baeumner, R.N. Cohen, V. Miksic, and J. Min, "RNA biosensor for the rapid detection of viable Escherichia coli in drinking water", *Biosens. Bioelectron.,* vol. 18, no. 4, pp. 405-413, 2003.
[http://dx.doi.org/10.1016/S0956-5663(02)00162-8] [PMID: 12604258]

[110] J. Zhao, S.S. Jedlicka, J.D. Lannu, A.K. Bhunia, and J.L. Rickus, "Liposome-doped nanocomposites as artificial-cell-based biosensors: Detection of listeriolysin O", *Biotechnol. Prog.,* vol. 22, no. 1, pp. 32-37, 2006.
[http://dx.doi.org/10.1021/bp050154o] [PMID: 16454489]

[111] G. Weitsman, N.J. Mitchell, R. Evans, A. Cheung, T.L. Kalber, R. Bofinger, G.O. Fruhwirth, M. Keppler, Z.V.F. Wright, P.R. Barber, P. Gordon, T. de Koning, W. Wulaningsih, K. Sander, B. Vojnovic, S. Ameer-Beg, M. Lythgoe, J.N. Arnold, E. Årstad, F. Festy, H.C. Hailes, A.B. Tabor, and T. Ng, "Detecting intratumoral heterogeneity of EGFR activity by liposome-based *in vivo* transfection of a fluorescent biosensor", *Oncogene,* vol. 36, no. 25, pp. 3618-3628, 2017.
[http://dx.doi.org/10.1038/onc.2016.522] [PMID: 28166195]

[112] X. Yang, H. Liu, Y. Ji, S. Xu, C. Xia, R. Zhang, C. Zhang, and Z. Miao, "A molecularly imprinted biosensor based on water-compatible and electroactive polymeric nanoparticles for lysozyme detection", *Talanta,* vol. 236, p. 122891, 2022.
[http://dx.doi.org/10.1016/j.talanta.2021.122891] [PMID: 34635270]

[113] G. Park, J.W. Lim, C. Park, M. Yeom, S. Lee, K.S. Lyoo, D. Song, and S. Haam, "Cell-mimetic biosensors to detect avian influenza virus via viral fusion", *Biosens. Bioelectron.,* vol. 212, p. 114407, 2022.
[http://dx.doi.org/10.1016/j.bios.2022.114407] [PMID: 35623252]

[114] L Wang, H Wang, X Tang, and L Zhao, "Molecularly imprinted polymers-based novel optical biosensor for the detection of cancer marker lysozyme", *Sens. Actuator. A.Phys.,* vol. 334, p. 113324, 2022.

[http://dx.doi.org/10.1016/j.sna.2021.113324]

[115] H. Jiang, D. Jiang, J. Shao, and X. Sun, "Magnetic molecularly imprinted polymer nanoparticles based electrochemical sensor for the measurement of Gram-negative bacterial quorum signaling molecules (N-acyl-homoserine-lactones)", *Biosens. Bioelectron.,* vol. 75, pp. 411-419, 2016.
[http://dx.doi.org/10.1016/j.bios.2015.07.045] [PMID: 26344904]

[116] A. Garcia Cruz, I. Haq, T. Cowen, S. Di Masi, S. Trivedi, K. Alanazi, E. Piletska, A. Mujahid, and S.A. Piletsky, "Design and fabrication of a smart sensor using in silico epitope mapping and electro-responsive imprinted polymer nanoparticles for determination of insulin levels in human plasma", *Biosens. Bioelectron.,* vol. 169, p. 112536, 2020.
[http://dx.doi.org/10.1016/j.bios.2020.112536] [PMID: 32980804]

[117] S. Chawla, J. Narang, and C.S. Pundir, "An amperometric polyphenol biosensor based on polyvinyl chloride membrane", *Anal. Methods.,* vol. 2, no. 8, pp. 1106-1111, 2010.
[http://dx.doi.org/10.1039/c0ay00165a]

[118] C.S. Pundir, R. Devi, J. Narang, S. Singh, J. Nehra, and S. Chaudhry, "Fabrication of an amperometric xanthine biosensor based on polyvinylchloride membrane", *J. Food. Biochem.,* vol. 36, no. 1, pp. 21-27, 2012.
[http://dx.doi.org/10.1111/j.1745-4514.2010.00499.x]

[119] I.M. Khoris, A.B. Ganganboina, and E.Y. Park, "Self-assembled chromogenic polymeric nanoparticle-laden nanocarrier as a signal carrier for derivative binary responsive virus detection", *ACS Appl. Mater. Interfaces,* vol. 13, no. 31, pp. 36868-36879, 2021.
[http://dx.doi.org/10.1021/acsami.1c08813] [PMID: 34328304]

[120] X.L. Luo, J.J. Xu, Y. Du, and H.Y. Chen, "A glucose biosensor based on chitosan–glucose oxidase–gold nanoparticles biocomposite formed by one-step electrodeposition", *Anal. Biochem.,* vol. 334, no. 2, pp. 284-289, 2004.
[http://dx.doi.org/10.1016/j.ab.2004.07.005] [PMID: 15494135]

[121] N. Melnychuk, and A.S. Klymchenko, "DNA-functionalized dye-loaded polymeric nanoparticles: ultrabright FRET platform for amplified detection of nucleic acids", *J. Am. Chem. Soc.,* vol. 140, no. 34, pp. 10856-10865, 2018.
[http://dx.doi.org/10.1021/jacs.8b05840] [PMID: 30067022]

[122] A. Morrin, O. Ngamna, A.J. Killard, S.E. Moulton, M.R. Smyth, and G.G. Wallace, "An amperometric enzyme biosensor fabricated from polyaniline nanoparticles", *Int. J. Dev. Fund. Prac. Asp. Electroanal.,* vol. 17, no. 5–6, pp. 423-430, 2005.

[123] S. Piperno, B. Tse Sum Bui, K. Haupt, and L.A. Gheber, "Immobilization of molecularly imprinted polymer nanoparticles in electrospun poly(vinyl alcohol) nanofibers", *Langmuir,* vol. 27, no. 5, pp. 1547-1550, 2011.
[http://dx.doi.org/10.1021/la1041234] [PMID: 21222445]

[124] X. Ye, Y. Jiang, X. Mu, Y. Sun, P. Ma, P. Ren, and D. Song, "Ultrabright silicon nanoparticle fluorescence probe for sensitive detection of cholesterol in human serum", *Anal. Bioanal. Chem.,* vol. 414, no. 13, pp. 3827-3836, 2022.
[http://dx.doi.org/10.1007/s00216-022-04024-4] [PMID: 35347354]

[125] A. Singh, N. Verma, and K. Kumar, "Fabrication and construction of highly sensitive polymeric nanoparticle-based electrochemical biosensor for asparagine detection", *Curr. Pharmacol. Rep.,* vol. 8, no. 1, pp. 62-71, 2022.
[http://dx.doi.org/10.1007/s40495-021-00271-8]

[126] A. Mishra, A.R. Ferhan, C.M.B. Ho, J.H. Lee, D-H. Kim, Y-J. Kim, and Y-J. Yoon, "Fabrication of plasmon-active polymer-nanoparticle composites for biosensing applications", *Int. J. Prec. Eng. Manufac. Green Tech.,* vol. 8, no. 3, pp. 945-954, 2021.
[http://dx.doi.org/10.1007/s40684-020-00257-9]

[127] D.I. Meira, M. Proença, R. Rebelo, A.I. Barbosa, M.S. Rodrigues, J. Borges, F. Vaz, R.L. Reis, and V.M. Correlo, "Chitosan micro-membranes with integrated gold nanoparticles as an LSPR-based sensing platform", *Biosensors,* vol. 12, no. 11, p. 951, 2022.
[http://dx.doi.org/10.3390/bios12110951] [PMID: 36354460]

[128] J. Ashley, Y. Shukor, and R. D'Aurelio, "Synthesis of MIP nanoparticles for α-casein detection using SPR as a milk allergen sensor", *ACS Sens.,* vol. 3, no. 2, pp. 418-424, 2018.
[http://dx.doi.org/10.1021/acssensors.7b00850] [PMID: 29333852]

[129] 129. Sanaeifar N, Rabiee M, Abdolrahim M, Monfared AH. A novel glucose biosensor based on immobilization of glucose oxidase in iron oxide nanoparticles/poly (vinyl alcohol) nanocomposite film. *Proceedings of the 23rd Iranian Conference on Biomedical Engineering and 1st International Iranian Conference on Biomedical Engineering (ICBME);* 2016 Nov 23-25; Tehran, Iran. 2016; pp. 252-6.
[http://dx.doi.org/10.1109/ICBME.2016.7890966]

[130] Y. Du, X.L. Luo, J.J. Xu, and H.Y. Chen, "A simple method to fabricate a chitosan-gold nanoparticles film and its application in glucose biosensor", *Bioelectrochemistry,* vol. 70, no. 2, pp. 342-347, 2007.
[http://dx.doi.org/10.1016/j.bioelechem.2006.05.002] [PMID: 16793348]

[131] R. Deswal, V. Narwal, P. Kumar, V. Verma, A.S. Dang, and C.S. Pundir, "An improved amperometric sarcosine biosensor based on graphene nanoribbon/chitosan nanocomposite for detection of prostate cancer", *Sens. Int.,* vol. 3, p. 100174, 2022.
[http://dx.doi.org/10.1016/j.sintl.2022.100174]

[132] J.R. Anusha, C.J. Raj, B.B. Cho, A.T. Fleming, K.H. Yu, and B.C. Kim, "Amperometric glucose biosensor based on glucose oxidase immobilized over chitosan nanoparticles from gladius of Uroteuthis duvauceli", *Sens. Actuators B Chem.,* vol. 215, pp. 536-543, 2015.
[http://dx.doi.org/10.1016/j.snb.2015.03.110]

[133] R. Rahmanian, and S.A. Mozaffari, "Electrochemical fabrication of ZnO-polyvinyl alcohol nanostructured hybrid film for application to urea biosensor", *Sens. Actuators B Chem.,* vol. 207, pp. 772-781, 2015.
[http://dx.doi.org/10.1016/j.snb.2014.10.129]

[134] E. Filippo, A. Serra, and D. Manno, "Poly(vinyl alcohol) capped silver nanoparticles as localized surface plasmon resonance-based hydrogen peroxide sensor", *Sens. Actuators B Chem.,* vol. 138, no. 2, pp. 625-630, 2009.
[http://dx.doi.org/10.1016/j.snb.2009.02.056]

[135] Y. Esmaeili, M. Khavani, A. Bigham, A. Sanati, E. Bidram, L. Shariati, A. Zarrabi, N.A. Jolfaie, and M. Rafienia, "Mesoporous silica@chitosan@gold nanoparticles as on/off optical biosensor and pH-sensitive theranostic platform against cancer", *Int. J. Biol. Macromol.,* vol. 202, pp. 241-255, 2022.
[http://dx.doi.org/10.1016/j.ijbiomac.2022.01.063] [PMID: 35041881]

[136] Y Liu, AG Erdman, and T Cui, "Acetylcholine biosensors based on layer-by-layer self-assembled polymer/nanoparticle ion-sensitive field-effect transistors", *Sens. Actuator. A. Phys.,* vol. 136, no. 2, pp. 540-545, 2007.

[137] S.K. Shukla, S.R. Deshpande, S.K. Shukla, and A. Tiwari, "Fabrication of a tunable glucose biosensor based on zinc oxide/chitosan-graft-poly(vinyl alcohol) core-shell nanocomposite", *Talanta,* vol. 99, pp. 283-287, 2012.
[http://dx.doi.org/10.1016/j.talanta.2012.05.052] [PMID: 22967553]

[138] R. Khan, A. Kaushik, P.R. Solanki, A.A. Ansari, M.K. Pandey, and B.D. Malhotra, "Zinc oxide nanoparticles-chitosan composite film for cholesterol biosensor", *Anal. Chim. Acta.,* vol. 616, no. 2, pp. 207-213, 2008.
[http://dx.doi.org/10.1016/j.aca.2008.04.010] [PMID: 18482605]

[139] S. Nsanzamahoro, W.F. Wang, and Y. Zhang, "Designing biosensing platform for β-glucuronidase determination based on the formation of fluorescent silicon nanoparticles", *SSRN,* p. 4089579, 2022.

[http://dx.doi.org/10.2139/ssrn.4089579]

[140] G. Pandey, R. Chaudhari, B. Joshi, S. Choudhary, J. Kaur, and A. Joshi, "Fluorescent biocompatible platinum-porphyrin-doped polymeric hybrid particles for oxygen and glucose biosensing", *Sci. Rep.,* vol. 9, no. 1, p. 5029, 2019.
[http://dx.doi.org/10.1038/s41598-019-41326-7] [PMID: 30903010]

[141] M. Holzinger, A. Le Goff, and S. Cosnier, "Nanomaterials for biosensing applications: A review", *Front Chem.,* vol. 2, p. 63, 2014.
[http://dx.doi.org/10.3389/fchem.2014.00063] [PMID: 25221775]

[142] E. Hutter, and D. Maysinger, "Gold-nanoparticle-based biosensors for detection of enzyme activity", *Trends Pharmacol. Sci.,* vol. 34, no. 9, pp. 497-507, 2013.
[http://dx.doi.org/10.1016/j.tips.2013.07.002] [PMID: 23911158]

CHAPTER 12

Role of Nano and Biopharmaceutics in Precision Medicine

Habibe Yılmaz[1,*] and **Ayça Erek**[1]

[1] *Department of Pharmaceutical Biotechnology, Faculty of Pharmacy, Trakya University, Edirne, Türkiye*

Abstract: As our knowledge of developing technology and human biology increases, the need for changes in our perspectives on diseases and treatment modalities has emerged. The individual variation of diseases at the molecular level has long led to the abandonment of the one-fits-to-all approach. These changes at the molecular level are illuminated using -omics technologies and are among the most powerful tools in precision medicine. The discovery of new drug targets and biomarkers results in the structural elucidation of targets. Thus, it has been possible to develop new drug molecules as well as to select the appropriate drug for the target, the appropriate dose, and, when necessary, the appropriate drug combination. Awareness of the changes in diseases at the molecular level has also updated clinical research designs to make precision medicine applicable. In this section, information and examples of developments in precision medicine, diagnosis and treatment in precision medicine, as well as -omics technologies and other technologies are presented.

Keywords: Basket design, Biobank, Biopharmaceutics, Biosensors, Diagnostics, Epigenomics, Genomics, Metabolomics, Nanomedicine, Omic technologies, Pharmaceutics, Precision medicine, Precision medicine tools, Personalized medicine, Proteomics, Transcriptomics, Targeted therapy, Theranostic, Therapeutics, Umbrella design.

INTRODUCTION

Precision medicine, which is formerly known as personalized medicine, has attracted many researchers in recent years. This chapter provides an overview of precisions medicine's history, tools, diagnosis, therapeutics, recent updates, and the future. Traditional medicine focused on diseases and epidemics, and it was thought that the treatment methods applied would be suitable for all patients. However, some patients can tolerate a certain drug without any side effects while the same dose can be toxic to other patients. With the development of medical

* **Corresponding author Habibe Yılmaz:** Department of Pharmaceutical Biotechnology, Faculty of Pharmacy, Trakya University, Edirne, Türkiye; Tel: +90 284 2350180 - 1181; E-mail: habibeyilmaz@trakya.edu.tr

science, it has emerged that the same disease progresses differently in different individuals and the same treatment cannot be applied [1]. Today, the concepts of personalized medicine and precision medicine, in which patients are at the forefront rather than diseases, have entered our lives. According to the National Institutes of Health (NIH), precision medicine is "an emerging approach to disease treatment and prevention that takes into account the individual variability of genes, the environment, and each person's lifestyles." This approach allows to more accurately predict which treatment or measure for a disease will be beneficial for which patient or group of patients. This concept is quite different from the "one size fits all" approach in traditional medicine, which focuses on therapeutic strategies for the average patient [2]. This difference between inter individuals could be caused by genetic factors, age, gender, ethnicity, race, habits, environmental factors reasons may originate. Therefore, despite the differences in patients, the understanding of disease-oriented treatment has led to the waste of drugs, increased costs, and poor patient and physician satisfaction [3].

Throughout the chapter, before establishing the connection between precision medicine and nanotechnology, information about precision medicine and its tools will be given, and then its relationship with nanomedicine will be discussed.

PRECISION MEDICINE HISTORY

Although the concept of precision medicine is not a new concept, it appears even in the hypotheses of Hippocrates, who is considered the father of traditional medicine, centuries ago. According to Sir William Osler, it is more important to know which patient has the disease than to know what type of disease the patient has. Advances in genomics and medicine have accelerated the development of precision medicine. Examples of these are the discovery of the double helix structure of DNA in 1953, the development of Sanger sequencing in 1977, and most importantly, the launch of the Human Genome Project in 1990. The dates and events important to the development of precision medicine are shown in detail in Fig. (1). The concept of precision medicine was first used and accepted by the US National Research Council (NRC) in 2011 after the publication of a report titled *"Toward Precision Medicine: Building a Knowledge Network for Biomedical Research and a New Taxonomy of Disease"* [4]. In this report, it was recommended to classify diseases according to genetic or genomic basis, not symptoms. Then, with the speech of US President Barack Obama in January 2015, the concept of precision medicine was heard by the masses for the first time, and after this speech, there was a great increase in interest and research in precision medicine techniques. While there were approximately 4000 articles in the National Library of Medicine before 2007, it is seen that this number has reached 91,947 in 2022 [5, 6].

Timeline of Precision Medicine

Years and important dates

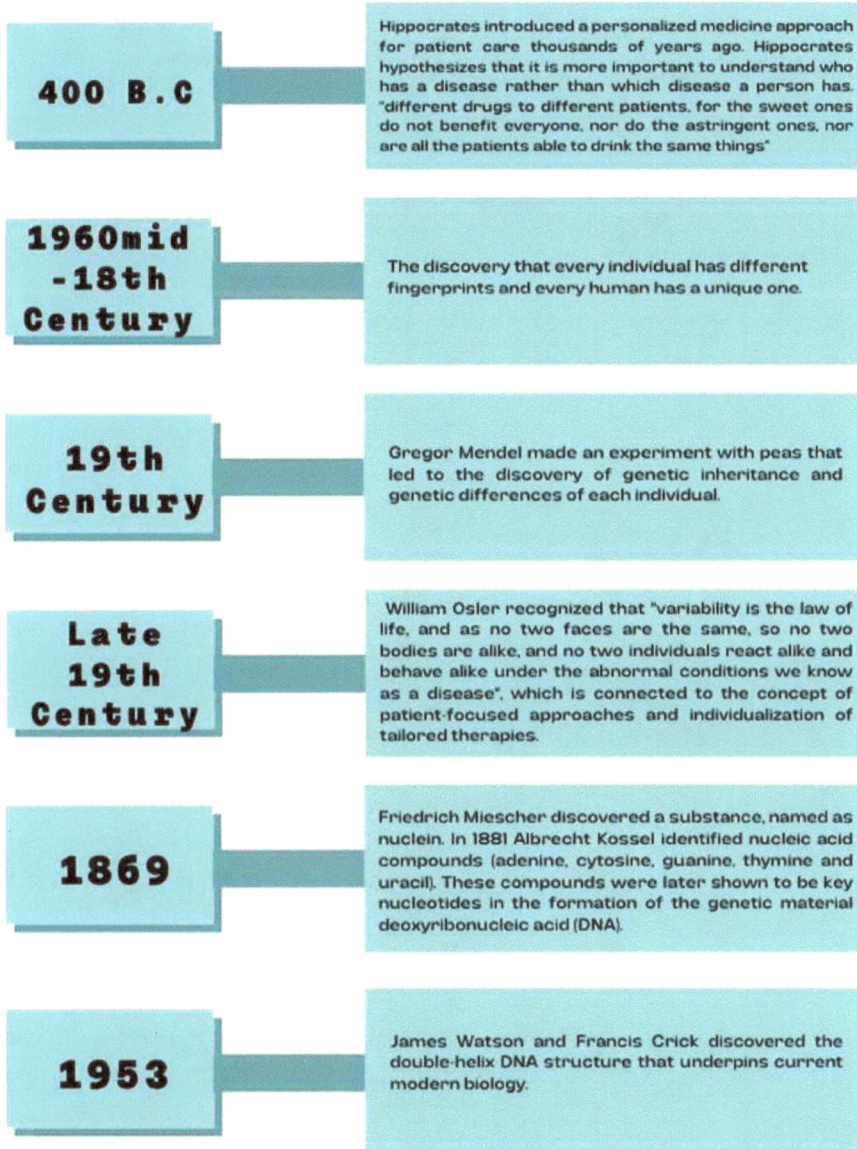

400 B.C	Hippocrates introduced a personalized medicine approach for patient care thousands of years ago. Hippocrates hypothesizes that it is more important to understand who has a disease rather than which disease a person has. "different drugs to different patients, for the sweet ones do not benefit everyone, nor do the astringent ones, nor are all the patients able to drink the same things"
1960mid -18th Century	The discovery that every individual has different fingerprints and every human has a unique one.
19th Century	Gregor Mendel made an experiment with peas that led to the discovery of genetic inheritance and genetic differences of each individual.
Late 19th Century	William Osler recognized that "variability is the law of life, and as no two faces are the same, so no two bodies are alike, and no two individuals react alike and behave alike under the abnormal conditions we know as a disease", which is connected to the concept of patient-focused approaches and individualization of tailored therapies.
1869	Friedrich Miescher discovered a substance, named as nuclein. In 1881 Albrecht Kossel identified nucleic acid compounds (adenine, cytosine, guanine, thymine and uracil). These compounds were later shown to be key nucleotides in the formation of the genetic material deoxyribonucleic acid (DNA).
1953	James Watson and Francis Crick discovered the double-helix DNA structure that underpins current modern biology.

(Fig. 1) contd.....

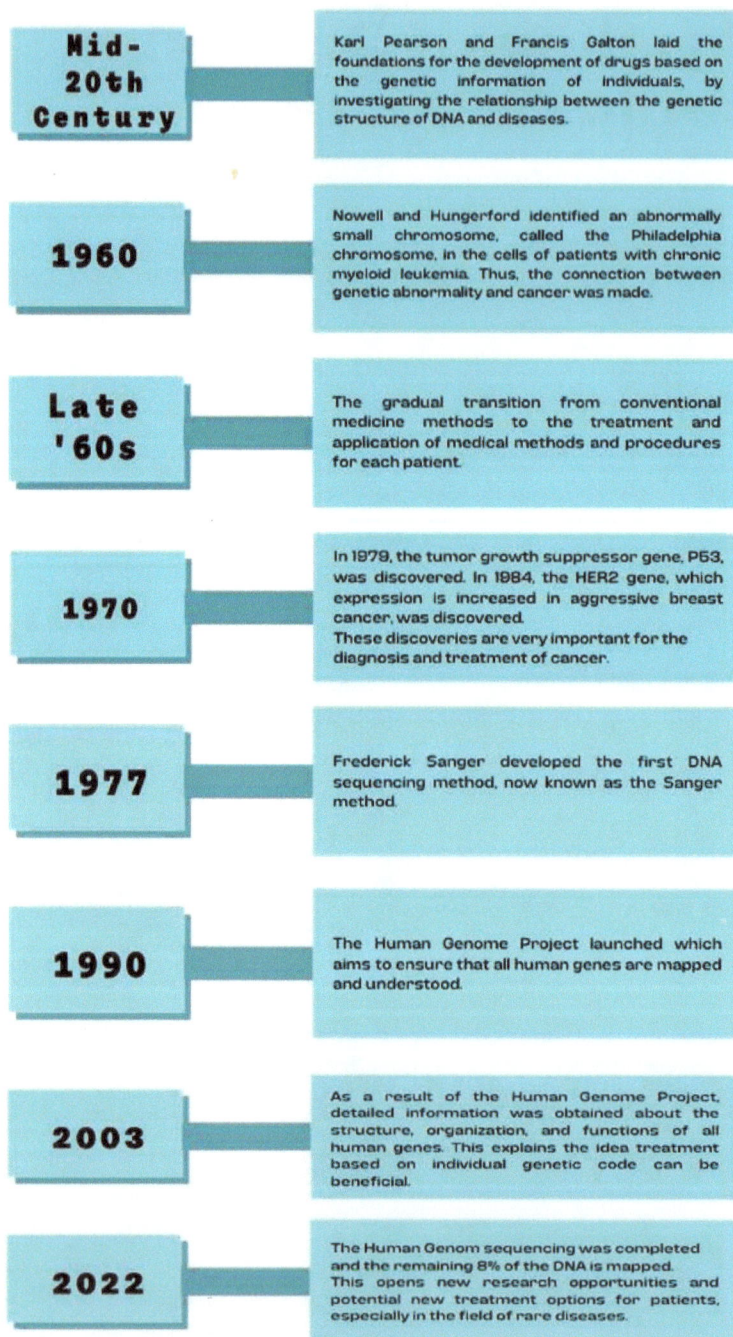

Fig. (1). Timeline of precision medicine [9].

One of the important steps in the field of precision medicine, which came to the fore with the support of US President Barack Obama, was the establishment of the "Precision Medicine Initiative Working Group of the Advisory Committee" by the National Institute of Health (NIH) in 2015. Then the All of Us project, supported by the NIH, was launched. The project in question aims to obtain pharmacogenomic information about diseases by collecting biological and electronic data from 1 million patients. Since 2018, electronic medical records, wearable devices, and of course, genetic results have been obtained from more than 500 thousand patients. Another goal is to design appropriate medication based on these data. The collection of such data requires both digital archives and biobanks. When examined from this aspect, it will be seen that the UK biobank has genetic and electronic health records of 500 thousand patients [7].

It is seen that the developments related to precision medicine in Turkey, which is located in the Eurasian region, are mostly under the responsibility of Health Institutes of Türkiye (TÜSEB). TUSEB made many references to personalized medicine in its 2040 vision report prepared in 2022. According to the report, it has officially announced its decisions such as the development of artificial intelligence, wearable technologies, the establishment of centers where genomic/proteomic/metabolomic tests can be carried out, the inclusion of precision medicine in rare diseases and oncology, and the support of related projects [8]. On 24 January 2020, the Applied Project Collaboration Program in the Field of Individual and Transformational Medicine was opened with the aim of creating a biological sample (tissue, blood, saliva, urine, and stool samples from sick and healthy individuals) bank that will consist of a total of 20.000 individuals, and the ethics committee approved the project on 17 March 2020.

PRECISION MEDICINE TOOLS

In precision medicine, to determine the susceptibility to the disease, understand the clinical course of the disease, and recommend the optimum treatment based on the personal profile, it is aimed to identify and overcome differences caused by genetic, exposure, and lifestyle factors. In order to achieve this goal, it is necessary to have a multi-level approach to patients [10].

The precision medicine approach is not only used to determine which drug and which dose is suitable for the patient during the disease, but also contributes to the provision of preventive health services by determining which diseases individual are prone to.

The advancement of precision medicine depends on the use of a multitude of resources from omics, pharmaco-omics, big data, artificial intelligence, machine learning (ML), environmental, social, and behavioral factors, and integration with preventive and public health. Molecular-level omics approaches such as transcriptomics, metabolomics, genomics, proteomics, and epigenomics provide a deeper understanding of patient conditions, from causes to consequences of diseases [11].

It should not be forgotten that not only state-of-the-art technologies but also proper biobanks are among the precision medicine tools. Biological materials are the samples from which all the genomic, proteomic, transcriptomic, metabolomics-omic technology information we need is obtained. In order to obtain accurate data, it is essential to use an accurate and suitable sample for analysis.

Processes covering the collection, processing, preservation, distribution, and storage of biological samples as well as access to users and ethical issues are within the scope of the activities of biobanks. A living quality management process and standardization are needed for the correct management of processes. The qualifications that biobanks should have are included in the ISO 20387 standard [12]. In ISO 20387, the required standard qualities of the whole process

are expressed under the headings of general, structural, resource, process and quality management system requirements [13].

Omics Approaches

The addition of "omics" to a molecular term refers to a comprehensive or general assessment of a set of molecules. Omics technologies provide the tools needed to study the differences found in DNA, RNA, protein, and cellular molecules within a species or between different species. Omics experiments can yield enormous data on functional and structural changes within the cell. There is an abundance of omics terms derived from different types of biomolecules in cells of the human body, such as genomics, proteomics, metabolomics, transcriptomics, epigenomics *etc* [14].

Genomics, Transcriptomics and Epigenomics

The precision medicine approach is often equated with a genomics approach. This is most likely due to, the fact that genomic approaches, technologies, and data were the first common omics data used in precision medicine. Since the discovery of DNA structure, great progress has been made in understanding the human genome. Sequencing of the entire human genome was obtained with the Human Genome Project. Genomic approaches have been widely adopted in biomedical approaches, and genes and genetic loci involved in human diseases have been successfully identified. These studies provided insight into the complexity of biological systems. Early work in precision medicine involved the use of genomic sequence information to diagnose patients, predict individual disease risk, and measure the suitability and success of certain treatments in individual patients [3, 14]. The development of techniques that decrease the time and cost of genetic analyzes during the human genome project led to the establishment of the first close relationship between precision medicine and genomics.

As a result of all these efforts, when the underlying genetic cause of the existing pathologic condition is determined, the selection or even development of the synthetic or biological molecule needed for the treatment can be in question. We have many drug options available to combat diseases. However, genetic variations determine which treatment individual will respond to. This information brings us the term pharmacogenomics [15].

People within and between populations are thought to be 99.9% genetically similar. It is this difference of 0.1% that changes our genetic variations and thus, the responses to treatments, corresponding to approximately 3 million base pairs. The most common genetic variation is single nucleotide polymorphisms (SNPs) and their distribution in the human genome is heterogeneous. A database of SNPs has also been created in recent years due to its association with diseases and treatment responses [16].

However, a single SNP alone is not always associated with disease or response to treatment. For this reason, the approach of detecting mutations, copy number changes and genome signatures across the entire genomic by whole genome sequencing has adopted [17].

The detection of diseases requires a comparison of genomic data from healthy and diseased individuals, and the Human Genome Project provided us with information about the healthy genome in 2003. Techniques and tools of molecular biology are frequently used in genomic and pharmacogenomic research. As the development of PCR accelerated the Human Genome Project, the Sanger sequencing technique provided information on the correct sequence of base pairs of human genomes. In 2004, the pyrosequencing technique, which provides parallel sequencing, was developed, reducing both time and cost. The discovery of pyrosequencing resulted in the publication of the first next-generation-sequencing (NGS) study in 2005 [18].

Sanger sequencing is a technique developed by F. Sanger in 1977. In the Sanger sequencing method, primer and template are polymerized with DNA polymerase I in the presence of deoxynucleotides (dNTP) and dideoxynucleotides (ddNTPs) which used as inhibitors. As a result of the inability of DNA polymerase I to prolong the reaction in the presence of ddNTP, fragments of different lengths are formed and exhibit a certain pattern when separated by electrophoresis. It is possible to determine the sequence from these obtained patterns [19]. Although it was introduced to the scientific community as a fast method when it was discovered, it requires a lot of labor, time and resources today. On the other hand, with the effect of capillary electrophoresis and pyrosequencing, the next-generation-sequencing method offered much faster analysis. NGS analysis consists of 3 main elements: library preparation, sequencing and data analysis. In the preparation of the library, the DNA template is randomly fragmented, then ligated to platform-specific adapters and then amplified by PCR. There are two different ways of sequencing, short-read and long-read sequencing. Short-read sequencing has a read length of 100-600 bp and long-read sequencing has a read length of 900 kb. Although short-read sequencing is commonly used today, long-read sequencing is more appropriate for more complex or heterozygous sequences. Data analysis, which is another step, has 4 stages: base calling, read alignment, variant identification (SNVs, indels, CNAs, SVs), and variant annotation. For example, in an aggressive and complex pathology such as cancer, it is very important to determine the origin, especially in the case of metastasis. NGS is very useful in the identification of drivers, primary origin, and mutations that cause drug resistance as well as predictive biomarkers [18, 20]. NGS technology is a technique used not only in genomic research, but also in transcriptomic and epigenomic research.

The transcriptome is the sum of all the transcripts of a cell under certain conditions. By obtaining information about all transcripts, including mRNAs, non-coding RNAs and small RNAs, it is possible to detect splicing patterns and other post-transcriptional modifications and to examine changes in expression levels [21]. Either RNA-sequencing or microarray analysis is performed in transcriptomic analyses. Sample preparation steps of RNA sequencing are total RNA isolation, target RNA enrichment, and reverse transcription of RNA into complementary DNA (cDNA) [22]. In microarray analysis, transcripts are hybridized with fluorescently labeled short oligonucleotide sequences, called as probes, attached to solid substrates. Transcript amount can be determined by fluorescence intensity. However, in order to select the appropriate probe in the microarray technique, unlike NGS, some prior knowledge of the relevant organism is required. Transcriptome analysis is used to detect SNP, gene fusions, allele-specific variants related to the diagnosis of the disease [22 - 24].

In the development of diseases, there are environmental effects that cause modifications in DNA as well as variations that occur directly in DNA. Among the modifications affecting DNA, analysis of DNA methylation, histone modification, changes in chromatin structure and non-coding RNAs form the basis of epigenetic studies. Among the analysis techniques used are PCR,

Southern blotting, sequencing and microarray technologies, and mass spectrometry [25, 26].

Metabolomics

Metabolites are biological molecules that arise as a result of biochemical reactions in the organism and trigger epigenetic changes, alter protein activity and facilitate signal transduction *via* triggering signal transduction. In addition, the metabolites themselves are biomarkers. The metabolome is the set of all metabolites in the organism. Metabolites and their levels can change with changing environmental conditions. Nuclear magnetic resonance (NMR) and mass spectrometry (MS) technologies are mainly used in metabolomics research, which provides important data on how the environment and changing conditions affect diseases. Many biological samples, such as tissue, body fluids or cells, can be used in metabolomic analysis. With the developing NMR technologies, structural information can be obtained as well as the amount of metabolites in the samples. In MS analysis, depending on the type of metabolite of interest, different extraction conditions can be used, *e.g.*, isopropyl alcohol extraction if lipid, methanol/acetonitrile extraction for amines and cationic metabolites. Then, following the appropriate chromatographic separation technique, each fraction is subjected to qualitative and quantitative analysis by mass analysis. This information can then be integrated with other -omic data to obtain more comprehensive information about the disease. By integrating this information with other -omics data, more detailed information about the disease can be obtained through metabolite genome-wide association studies (mGWAS) [27 - 29].

Proteomics

The proteome refers to the entire set of proteins of an organism under certain conditions. Proteins, like metabolites, are dynamic molecules that can vary under changing conditions and from organism to organism which also can be used as biomarkers. According to the central dogma theory, genes provide expression of transcripts and transcripts provide expression of proteins. However, just as RNA undergoes transcriptional modifications after synthesis from DNA, proteins also undergo modification after translation. This causes a much greater variety of proteins to be synthesized despite a fixed amount of genetic information. Considering that proteins carry out functional processes in the cell either alone or together as a module, it is obvious that they are a valuable source of information necessary for precision medicine. Therefore, joint evaluation of genomic, transcriptomic and proteomic data is important to identify the correct causes and targets so that diseases can be fully elucidated. This approach, which expresses the evaluation of genotype and phenotype together, is called proteogenomics [30 - 32].

The most widely used technology in proteomics research is mass spectrometry. Time of Flight (TOF), Ion-Trap, Fourier-transform ion cyclotron resonance (FTICR) and quadrupole (Q) analyzers are the most preferred mass analyzers for proteomic research [30]. In proteomics research, biological samples are subjected to the homogenization step, especially if they contain tissue or cells, which is called the bottom-up approach. Following protein extraction, fragmentation with proteolytic enzymes and, if preferred, fractions of peptides should be subjected to mass analysis after collection. Ion exchange (IEX), hydrophilic interaction chromatography (HILIC) and reversed phase (RP) high pressure liquid chromatography (HPLC) are used both in the analysis and fractionation of the peptides formed [31]. However, the top-down approach, in which the intact protein is analyzed, is also becoming widespread with the developing technology [30].

Although MS technology is frequently used, high-throughput techniques such as affinity techniques are also frequently used in proteome research. Affinity-based methods are based on affinity molecules such as antibodies. Proteins of interest with the depletion strategy, which can be particularly useful in targeted-proteomics research, can be depleted from the complex biological environment by means of antibodies. Similarly, with the affinity capture method, it is possible to isolate the targeted protein from the biological environment through the desired anti-antibodies. Protein microarray systems, on the other hand, are systems that enable the recognition of multiple proteins from small samples simultaneously and are especially useful as a disease screening test. Aptamer technology, on the other hand, allows multi-protein recognition with synthetically synthesized oligonucleotides (aptamer) that have a certain affinity for proteins. Molecular imprinting technology, which is among the most popular topics of nanotechnology, also finds its place in affinity-based analysis. Protein-imprinted MIPs enable targeted proteomic studies by isolating the protein of interest from biological fluid in a targeted manner. Nanoparticles can also be used for the detection of biomarkers needed in precision medicine. By utilizing the knowledge that protein corona is formed with the interaction of nanoparticles with biological fluids, it is possible to isolate proteins from patients' blood using nanoparticles, to conduct targeted proteomics studies and to detect biomarkers [30].

DIAGNOSTICS IN PRECISION MEDICINE

In order to combat any disease, primarily, an accurate diagnosis and an appropriate treatment regimen in line with this diagnosis should be applied. For this reason, the necessity of developing new diagnostic systems with applicability in precision medicine has arisen. In particular, research continues on biosensor-based approaches, especially those that can be implanted in the patient and monitored remotely. Therapeutic drug monitoring (TDM) is a widely accepted method in recent years. The data obtained by the successful implementation of TDM is also the subject of precision medicine. For this, biosensors based on electrochemical measurement have been preferred in recent years [33].

In an exemplary study, a paper-based electrochemical lab-on-a-chip device for Alzheimer's disease was designed. They developed a rapid and cost-effective method to determine the appropriate dose for the patient. An easy-to-use design was chosen because the application dose of the choline esterase inhibitors used in the treatment depends on the choline esterase activity and the activity of this enzyme varies from individual to individual. The tool was prepared on the office printer using a sheet of paper. First of all, carbon black/Prussian blue nanoparticles were synthesized. Then, hydrophobic regions were created by adding wax on the paper. A pseudo-reference electrode with silver/chloride ink and a working electrode with graphite ink were created on paper. On top of that, carbon black/Prussian blue nanoparticles were applied to form a film layer. The device was prepared by adding a layer containing cholinesterase enzyme on this layer and a layer separating plasma from whole blood on it. It has been stated that the measurement precision of the device and its usage potential is high [34]. Other researchers, have developed a biosensor that can simultaneously detect CEA, CA153 and CA125 biomarkers with high sensitivity for use in breast cancer precision medicine. The multiplex sensor was developed using graphene/methylene blue–chitosan/antibody and bovine serum albumin on indium tin oxide glass electrode [35]. Another issue as important as the development of point-of-care (POC) biosensors is the use of wearable sensor systems in precision medicine.

Today, implantable/wearable continuous glucose monitoring systems, in which diabetic patients change their sensors at regular intervals such as 6 months, are the best examples of how continuous monitoring improves the quality of life of patients and precision diagnosis. With the integration of this system into the insulin pump, a result such as pumping insulin in case of an increase in blood glucose level, or pumping insulin in accordance with the glucose level, or not pumping at all can be achieved [36]. This is the most successful lifetime example of the success of personalized diagnosis and treatment. In order to achieve this success, integration of electrochemical biosensors with microneedle technology has been utilized. Microneedles emerge as systems that can be mounted on transdermal patches, can reach the stratum corneum and epidermis, and can measure the compound of interest in body fluids when combined with an electrochemical biosensor [33].

As the applications and researches of nanotechnology in the medical field increase, there has been an increased interest in the development of theranostic platforms where nanomaterials can be used not only for treatment or diagnosis alone, but also simultaneously. Biomaterials are used in the custom design of theranostic systems. It is expected that theranostic structures will selectively go to the pathological site and selectively leave the active substance on the target. For this, three different designs are proposed: stable or activatable ligand-based, activated by the pathological microenvironment or activated by externally applied stimuli. Theranostic designs are mostly performed by targeting receptors derived from genetic information, which are highly expressed in the tumor such as folate, ferritin, or only expressed in the tumor such as CD68 and CD163. However, such a design poses a risk, especially for non-pathological tissues with non-tumor expressed receptors. Therefore, the current approach has focused on the creation of unnatural carbohydrate residues on the tumor cell surface in a way that does not interfere with biological processes and then targeting these unnatural receptors [37].

Ultrasonic nanotheranostics is one of the good examples of these applications. In recent years, new nanomaterials that can be used as ultrasound contrast agents have been developed that increase the contrast power of ultrasound for better diagnosis. Approved products on the market are Definity, Optison, Sonazoid and SonoVue. These products are microbubbles containing an inert gas. With recent studies, the drug is also encapsulated into these microbubbles and modified with a ligand that can bind to the target desired to be monitored, and thus the theranostic feature is gained. Another material that can be used as an ultrasound contrast agent is gas vesicles obtained from planktonic microorganisms. These vesicles have the potential to be used for lysosomal function monitoring, especially since they are degraded by lysosomes. In addition to these, liposomes, micelles and metal nanoparticles such as gold and iron oxide are also being investigated as ultrasound contrast agents [38].

THERAPEUTICS IN PRECISION MEDICINE

The important point in the treatment arrangement in precision medicine is to arrange the treatment scheme suitable for the genotypic and phenotypic characteristics of the individual. With this information about the individual, it can be determined which drug should be given in which dose. But beyond that, it is also possible to design new drugs based on genotypic and phenotypic information. New drugs can be obtained by biotechnological means, as well as by synthetic designs or nanotechnological approaches. Both approaches will be explained with examples in the following headings.

Nanomedicines

As knowledge about precision medicine and bio-inspired nanomaterials increases, researchers are able to synthesize nano drug delivery systems with more efficient designs. One of the examples in this regard is the transformation of the static design of nano drug delivery systems into dynamic designs as in nature. As an example, when evaluated on cancer pathology, it is seen that nanomedicine, until recently, is based on passive targeting associated with increased permeability and retention (EPR) effect. However, clinical findings revealed that nanodrugs were taken by transcytosis, not by the EPR effect. Moreover, it is seen that such static systems are not able to self-adjust the dose according to the pathological condition, at best leaving the entire dose on target. For this reason, designing nano-drugs that keep the dose controlled by showing high sensitivity to environmental variability, as hemoglobin exhibits cooperation depending on environmental conditions, makes it possible to adjust the dose according to the pathological condition of the individual required by precision medicine [39].

Another example where precision medicine is used is the nano RNA carrier system that the researchers designed using the findings they obtained as a result of transcriptome analysis. Researchers demonstrated that CD22ΔE12, which is a genetic defect that MAPK, PI3-K/m-TOR and WNT pathway activation, is the driver and possible target of childhood relapsed B-line acute lymphoblastic leukemia. Nanoparticles were obtained by complexing CD22ΔE12 RNA trans-splicing molecule (RTM) with PVBLG-8, which is a helical cationic polypeptide to repair this genetic defect. It has been shown in studies that this prepared nano drug carrier causes regression in the tumor by reducing the genetic defect *in vivo* [40].

Although treatments performed with approaches such as the inhibition of the metabolic pathway or overexpressed receptors for cancer treatment provide a good response for a certain period of time, drug resistance develops in many cancer types and individuals later on. Therefore, for an effective treatment, it is necessary to apply treatment in such a way that the drug resistance mechanism never develops, to overcome the drug resistance or to adopt a new treatment regimen. In this context, a group of researchers found that the Asporin (ASPN) responsible for oxaliplatin resistance was overexpressed in their genomic screening in colon cancer patients with drug resistance, and when its overexpression suppressed, oxaliplatin resistance was reversed. Silencing RNA targeting ASPN and oxaliplatin encapsulated into PEG-G5-PAMAM dendrimers. This systemically applied nanostructure has been shown to provide remarkable tumor therapy, both *in vitro* and *in vivo* [41].

Chimeric antigen receptor (CAR) T cells are a very promising technology for cancer immunotherapy. Although it is quite effective, its cost and complexity of production limits its reach to all patients. Compared to autologous practice, the idea of generating CAR-T cells directly in patients may produce the most clinically effective results, but implementation has its challenges. For such an application, research has been carried out with lentivirus, adenovirus, synthetic polymeric nanocarriers and liposomes that can target directly without infiltrating healthy tissues [42].

Small Molecule Pharmaceutics and Biopharmaceutics

Once the genetic origin of diseases such as cancer is understood, targeted and specific treatments can be designed. Approved by the FDA for use in the treatment of breast cancer in 1998, Trastuzumab was the first biopharmaceutical product designed and approved using sensitive medical information. In 2001, the synthetic molecule imatinib with targeting capability to cancer became the first small molecule drug approved [18]. This was followed by the knowledge, derived with precision medicine tools, that the administration of gefitinib or erlotinib is effective in non-small cell lung cancer patients of Asian population [43].

Developments in precision medicine also facilitate the discovery of new drug molecules, either small molecules or biological molecules, and enable the developed molecules to achieve higher success in clinical research. However, using precision medicine data is not only about developing new molecules, but also the right selection of conventionally used drugs and the use of the right drug combinations.

Temozolomide is currently the gold standard in the treatment of glioblastoma. On the other hand, there are studies showing that everolimus, a mammalian target of rapamycin specific (mTOR) inhibitor, can be used in the treatment of glioblastoma since it is found in 80% of glioblastoma cases derived from precision medicine data. Researchers investigated the synergistic effect of paclitaxel in combination with temozolomide and everolimus, both *in vitro* and *in vivo*. It has been determined that the combination of paclitaxel and everolimus works synergistically and at appropriate doses *in vitro*. They then synthesized a scaffold acetalated dextran for this drug combination that would be implanted directly into the brain. This hydrogel scaffold, developed by the researchers, is designed to be tunable to allow the release of drugs at appropriate precise doses. It has been shown by an *in vivo* study that this developed combination therapy significantly reduces growth of tumors and prevents progression, which lasted approximately 1.5-2 months [44].

Another example of the benefits of choosing the right drug using precision medicine tools is the treatment of pancreatic cancer. Considering the mutations detected in pancreatic cancer with omics technologies, it was possible to increase the survival from 1.5 to 2.5 years with the treatment regimens selected. For example, the Food and Drug Agency (FDA) approved the use of Olaparib as a maintenance treatment in patients with pancreatic cancer with the BRCA1/2 mutation in 2020. In 2018 and 2019, respectively, larotrectinib and entrectinib received rapid approval for pancreatic cancer patients with NTRK fusions. Pembrolizumab, nivolumab, and dostarlimab, developed as programmed cell death protein 1 (PD-1) antibodies in patients with metastatic MMR-D/MSI-H pancreatic cancer, have recently approved [45].

Using the knowledge obtained with precision medicine, a more specific derivative of acetazolamide used in the treatment of glaucoma and epilepsy has been synthesized. Researchers have developed a unique protein-small molecule (PriSM) hybrids platform. A cysteine was added to the molecule in the fibronectin protein library expressed in yeast and acetazolamide was conjugated *via* the maleimide-poly(ethylene glycol) linker. When inhibition rate and specificity were investigated by isolating the species that specifically bind to the carbonic anhydrase enzyme isoforms, it was found to be 9 times more potent and 80 times more specific [46].

CLINICAL TRIALS IN PRECISION MEDICINE

The precision medicine approach has required the updating of clinical research protocols. With the expansion of molecular knowledge of diseases and the recognition that the one-size-fits-all approach is inappropriate, it has led to the development of not only new treatment or imaging tools, but also new clinical designs. The master fixed protocol and intervention approach is now being abandoned. In addition to assigning the appropriate arm after biomarker screening, it is now possible to flexibly change arms throughout the study in admission of patients to the clinical trial. Master protocol, basket and umbrella designs are now gaining acceptance, which make treatments more targeted and increase clinical trial success. Adaptive randomization, dropping or adding arms, seamles trial, reassessment of volunteer numbers, and adaptive enrichment are all possible with these new designs. In the basket design, the study design is made on the molecular marker, not on the pathology. This means, for example, enrolling patients with different cancers with the same molecular marker. In the Umbrella design, volunteers with the same cancer type or histology are included in the study, even if they express different biomarkers, a different targeted therapy study is performed for each individual [47 - 49]. A search of the clinicaltrials.gov website with the keywords "precision medicine, personalized medicine, drug and medication" revealed 785 studies. This data demonstrates that the outlook for clinical research is moving at an accelerating pace towards precision medicine [50].

CONCLUDING REMARKS

As a result, breakthroughs in precision medicine have accelerated, especially after Barack Obama's initiation of precision medicine initiatives in the USA. Individual studies, which were previously aimed at elucidating disease mechanisms, have progressed systematically and with a more holistic perspective. In precision medicine, the detection and classification of genotypic and phenotypic changes for the disease and the application of the treatment regimens determined for these classes are carried out in the patient. In this process, data collected from -omics technologies, biosensors, wearable technologies and information obtained from the patient's anamnesis are used. In light of this information, synthetic, biotechnological drugs or nanomedicine suitable for patients are now used at appropriate doses in many countries. Making adaptations in the treatment model according to the genotypic and phenotypic changes of the patient during the treatment process is another factor that increases the success of the treatment. However, it has not been possible to examine the clinical studies of new drug candidates obtained with the developments in precision medicine with the ongoing protocol designs. This has spurred the development and improvement of clinical research protocols. Although all these developments ensure that the treatment of patients is carried out successfully, factors such as the expensiveness of precision medicine instruments, the lack of accessibility, and the necessity of highly qualified personnel appear as disadvantages. For this reason, there is a need to develop new technologies that will enable the widespread use of these tools, not only for diseases, through precision medicine.

REFERENCES

[1]   A. Blasimme, and E. Vayena, "Legitimation of precision medicine " Tailored-to-You", *Perspect. Biol. Med.,* vol. 59, no. 2, pp. 172-188, 2018.
[http://dx.doi.org/10.1353/pbm.2017.0002]

[2]   *The precision medicine initiative cohort program – building a research foundation for 21st century medicine: Precision medicine initiative (PMI) working group report to the advisory committee to the director.* NIH, 2015, pp. 1-107.

[3]   N. Naithani, S. Sinha, P. Misra, B. Vasudevan, and R. Sahu, "Precision medicine: Concept and tools", *Med. J. Armed Forces India,* vol. 77, no. 3, pp. 249-257, 2021.
[http://dx.doi.org/10.1016/j.mjafi.2021.06.021] [PMID: 34305276]

[4]   U.S.N.R. Council, "Toward precision medicine: Building a knowledge network for biomedical research and a new taxonomy of disease", *Towar. Precis. Med.Build. a Knowl. Netw. Biomed. Res. New. Taxon. Dis.,* pp. 1-128, 2012.

[5]   Precision Medicine. PubMed® National Library of Medicine. No Title. PubMed®.

[6]   M.R. Kosorok, and E.B. Laber, *Precision Medicine.*, vol. 6, pp. 263-286, 2019.
        [http://dx.doi.org/101146/annurev-statistics-030718-105251]

[7]   M.N. Pelter, and R.S. Druz, "Precision medicine: Hype or hope?", *Trends Cardiovasc. Med.*, p.
        S1050-1738(22)00139-6, 2022.
        [PMID: 36375778]

[8]   U.S.N.R. Council, "Toward precision medicine: Building a knowledge network for biomedical
        research and a new taxonomy of disease", *Towar. Precis. Med. Build. Knowl. Netw. Biomed. Res.
        New. Taxon. Dis.*, pp. 1-128, 2012.

[9]   U.S.N.R. Council, "Toward precision medicine: Building a knowledge network for biomedical
        research and a new taxonomy of disease", *Towar Precis Med Build a Knowl Netw Biomed Res a New
        Taxon Dis.*, pp. 1-128, 2012.

[10]   J.A. Leopold, and J. Loscalzo, "Emerging role of precision medicine in cardiovascular disease", *Circ.
        Res.*, vol. 122, no. 9, pp. 1302-1315, 2018.
        [http://dx.doi.org/10.1161/CIRCRESAHA.117.310782] [PMID: 29700074]

[11]   O Strianese, F Rizzo, M Ciccarelli, G Galasso, Y D'agostino, and A Salvati, "Precision and
        personalized medicine: How genomic approach improves the management of cardiovascular and
        neurodegenerative disease", *Genes*, vol. 11, no. 7, p. 747, 2020.

[12]   H. Müller, G. Dagher, M. Loibner, C. Stumptner, P. Kungl, and K. Zatloukal, "Biobanks for life
        sciences and personalized medicine: Importance of standardization, biosafety, biosecurity, and data
        management", *Curr. Opin. Biotechnol.*, vol. 65, pp. 45-51, 2020.
        [http://dx.doi.org/10.1016/j.copbio.2019.12.004] [PMID: 31896493]

[13]   H. Müller, G. Dagher, M. Loibner, C. Stumptner, P. Kungl, and K. Zatloukal, "Biobanks for life
        sciences and personalized medicine: Importance of standardization, biosafety, biosecurity, and data
        management", *Curr. Opin. Biotechnol.*, vol. 65, pp. 45-51, 2020.
        [http://dx.doi.org/10.1016/j.copbio.2019.12.004] [PMID: 31896493]

[14]   M. Olivier, R. Asmis, G.A. Hawkins, T.D. Howard, and L.A. Cox, "The need for multi-omics
        biomarker signatures in precision medicine", *Int. J. Mol. Sci.*, vol. 20, no. 19, p. 4781, 2019.
        [http://dx.doi.org/10.3390/ijms20194781] [PMID: 31561483]

[15]   L. Zhang, R. Parvin, Q. Fan, and F. Ye, "Emerging digital PCR technology in precision medicine",
        Biosens. Bioelectron., vol. 211, p. 114344, 2022.
        [http://dx.doi.org/10.1016/j.bios.2022.114344] [PMID: 35598553]

[16]   W. Mu, and W. Zhang, "Molecular approaches, models, and techniques in pharmacogenomic research
        and development", *Pharmacogenomics*, pp. 273-294, 2013.

[17]   E. Pleasance, A. Bohm, L.M. Williamson, J.M.T. Nelson, Y. Shen, M. Bonakdar, E. Titmuss, V.
        Csizmok, K. Wee, S. Hosseinzadeh, C.J. Grisdale, C. Reisle, G.A. Taylor, E. Lewis, M.R. Jones, D.
        Bleile, S. Sadeghi, W. Zhang, A. Davies, B. Pellegrini, T. Wong, R. Bowlby, S.K. Chan, K.L.
        Mungall, E. Chuah, A.J. Mungall, R.A. Moore, Y. Zhao, B. Deol, A. Fisic, A. Fok, D.A. Regier, D.
        Weymann, D.F. Schaeffer, S. Young, S. Yip, K. Schrader, N. Levasseur, S.K. Taylor, X. Feng, A.
        Tinker, K.J. Savage, S. Chia, K. Gelmon, S. Sun, H. Lim, D.J. Renouf, S.J.M. Jones, M.A. Marra, and
        J. Laskin, "Whole-genome and transcriptome analysis enhances precision cancer treatment options",
        Ann. Oncol., vol. 33, no. 9, pp. 939-949, 2022.
        [http://dx.doi.org/10.1016/j.annonc.2022.05.522] [PMID: 35691590]

[18]   S. Morganti, P. Tarantino, E. Ferraro, P. D'Amico, G. Viale, D. Trapani, B.A. Duso, and G.
        Curigliano, "Complexity of genome sequencing and reporting: Next generation sequencing (NGS)
        technologies and implementation of precision medicine in real life", *Crit. Rev. Oncol. Hematol.*, vol.
        133, pp. 171-182, 2019.
        [http://dx.doi.org/10.1016/j.critrevonc.2018.11.008] [PMID: 30661654]

[19]   F. Sanger, S. Nicklen, and A.R. Coulson, "DNA sequencing with chain-terminating inhibitors", *Proc. Natl. Acad. Sci.,* vol. 74, no. 12, pp. 5463-5467, 1977.
[http://dx.doi.org/10.1073/pnas.74.12.5463] [PMID: 271968]

[20]   B.M. Hussen, S.T. Abdullah, A. Salihi, D.K. Sabir, K.R. Sidiq, M.F. Rasul, H.J. Hidayat, S. Ghafouri-Fard, M. Taheri, and E. Jamali, "The emerging roles of NGS in clinical oncology and personalized medicine", *Pathol. Res. Pract.,* vol. 230, p. 153760, 2022.
[http://dx.doi.org/10.1016/j.prp.2022.153760] [PMID: 35033746]

[21]   Z Wang, M Gerstein, and M. Snyder, *Nrg2484-1. Nat Rev | Genet.,* vol. 10, pp. 57-63, 2009.

[22]   T. Hu, N. Chitnis, D. Monos, and A. Dinh, "Next-generation sequencing technologies: An overview", *Hum. Immunol.,* vol. 82, no. 11, pp. 801-811, 2021.
[http://dx.doi.org/10.1016/j.humimm.2021.02.012] [PMID: 33745759]

[23]   R Lowe, N Shirley, M Bleackley, S Dolan, and T. Shafee, "Transcriptomics technologies", *PLoS. Comput. Biol.,* vol. 13, no. 5, p. 1005457, 2017.
[http://dx.doi.org/10.1371/journal.pcbi.1005457]

[24]   K.H. Liang, "Transcriptomics", *Bioinforma. Biomed. Sci. Clin. Appl.,* pp. 49-82, 2013.
[http://dx.doi.org/10.1533/9781908818232.49]

[25]   T.A. Turunen, M.A. Väänänen, and S. Ylä-Herttuala, "Epigenomics", *Encycl. Cardiovasc. Res. Med.,* pp. 258-265, 2018.
[http://dx.doi.org/10.1016/B978-0-12-809657-4.99575-9]

[26]   K.C. Wang, and H.Y. Chang, "Epigenomics", *Circ. Res.,* vol. 122, no. 9, pp. 1191-1199, 2018.
[http://dx.doi.org/10.1161/CIRCRESAHA.118.310998] [PMID: 29700067]

[27]   C.H. Johnson, J. Ivanisevic, and G. Siuzdak, "Metabolomics: Beyond biomarkers and towards mechanisms", *Nat. Rev. Mol. Cell. Biol.,* vol. 17, no. 7, pp. 451-459, 2016.
[http://dx.doi.org/10.1038/nrm.2016.25] [PMID: 26979502]

[28]   CB Clish, "Metabolomics: An emerging but powerful tool for precision medicine", *Cold. Spring. Harb. Mol. Case. Stud.,* vol. 1, no. 1, p. 000588, 2015.
[http://dx.doi.org/10.1101/mcs.a000588]

[29]   A. Alonso, S. Marsal, A. JuliÃ, A. Julià, and A. James Carroll, "Analytical methods in untargeted metabolomics: state of the art in 2015", *Front. Bioeng. Biotechnol.,* vol. 3, p. 23, 2015.
[http://dx.doi.org/10.3389/fbioe.2015.00023] [PMID: 25798438]

[30]   E.C. Nice, "The status of proteomics as we enter the 2020s: Towards personalised/precision medicine", *Anal. Biochem.,* vol. 644, p. 113840, 2022.
[http://dx.doi.org/10.1016/j.ab.2020.113840] [PMID: 32745541]

[31]   M.Y. Ang, T.Y. Low, P.Y. Lee, W.F. Wan Mohamad Nazarie, V. Guryev, and R. Jamal, "Proteogenomics: From next-generation sequencing (NGS) and mass spectrometry-based proteomics to precision medicine", *Clin. Chim. Acta,* vol. 498, pp. 38-46, 2019.
[http://dx.doi.org/10.1016/j.cca.2019.08.010] [PMID: 31421119]

[32]   A.C. Uzozie, and R. Aebersold, "Advancing translational research and precision medicine with targeted proteomics", *J. Proteomics,* vol. 189, pp. 1-10, 2018.
[http://dx.doi.org/10.1016/j.jprot.2018.02.021] [PMID: 29476807]

[33]   T.D. Pollard, J.J. Ong, A. Goyanes, M. Orlu, S. Gaisford, M. Elbadawi, and A.W. Basit, "Electrochemical biosensors: A nexus for precision medicine", *Drug. Discov. Today.,* vol. 26, no. 1, pp. 69-79, 2021.
[http://dx.doi.org/10.1016/j.drudis.2020.10.021] [PMID: 33137482]

[34]   V. Caratelli, A. Ciampaglia, J. Guiducci, G. Sancesario, D. Moscone, and F. Arduini, "Precision medicine in Alzheimer's disease: An origami paper-based electrochemical device for cholinesterase inhibitors", *Biosens. Bioelectron.,* vol. 165, p. 112411, 2020.

[http://dx.doi.org/10.1016/j.bios.2020.112411] [PMID: 32729530]

[35]   S. Cotchim, P. Thavarungkul, P. Kanatharana, and W. Limbut, "Multiplexed label-free electrochemical immunosensor for breast cancer precision medicine", *Anal. Chim. Acta,* vol. 1130, pp. 60-71, 2020.
[http://dx.doi.org/10.1016/j.aca.2020.07.021] [PMID: 32892939]

[36]   M. Gray, J. Meehan, C. Ward, S.P. Langdon, I.H. Kunkler, A. Murray, and D. Argyle, "Implantable biosensors and their contribution to the future of precision medicine", *Vet. J.,* vol. 239, pp. 21-29, 2018.
[http://dx.doi.org/10.1016/j.tvjl.2018.07.011] [PMID: 30197105]

[37]   H. Kim, G. Kwak, K. Kim, H.Y. Yoon, and I.C. Kwon, "Theranostic designs of biomaterials for precision medicine in cancer therapy", *Biomaterials,* vol. 213, p. 119207, 2019.
[http://dx.doi.org/10.1016/j.biomaterials.2019.05.018] [PMID: 31136910]

[38]   Y. Qin, X. Geng, Y. Sun, Y. Zhao, W. Chai, X. Wang, and P. Wang, "Ultrasound nanotheranostics: Toward precision medicine", *J. Control. Release,* vol. 353, pp. 105-124, 2023.
[http://dx.doi.org/10.1016/j.jconrel.2022.11.021] [PMID: 36400289]

[39]   J. Wilhelm, Z. Wang, B.D. Sumer, and J. Gao, "Exploiting nanoscale cooperativity for precision medicine", *Adv. Drug Deliv. Rev.,* vol. 158, pp. 63-72, 2020.
[http://dx.doi.org/10.1016/j.addr.2020.08.012] [PMID: 32882321]

[40]   F.M. Uckun, L.G. Mitchell, S. Qazi, Y. Liu, N. Zheng, D.E. Myers, Z. Song, H. Ma, and J. Cheng, "Development of polypeptide-based nanoparticles for non-viral delivery of CD22 RNA trans-splicing molecule as a new precision medicine candidate against b-lineage all", *EBioMedicine,* vol. 2, no. 7, pp. 649-659, 2015.
[http://dx.doi.org/10.1016/j.ebiom.2015.04.016] [PMID: 26288837]

[41]   C.Z. Huang, Y. Zhou, Q.S. Tong, Q.J. Duan, Q. Zhang, J.Z. Du, and X.Q. Yao, "Precision medicine-guided co-delivery of ASPN siRNA and oxaliplatin by nanoparticles to overcome chemoresistance of colorectal cancer", *Biomaterials,* vol. 290, p. 121827, 2022.
[http://dx.doi.org/10.1016/j.biomaterials.2022.121827] [PMID: 36228517]

[42]   A. Michels, N. Ho, and C.J. Buchholz, "Precision medicine: *in vivo* CAR therapy as a showcase for receptor-targeted vector platforms", *Mol. Ther.,* vol. 30, no. 7, pp. 2401-2415, 2022.
[http://dx.doi.org/10.1016/j.ymthe.2022.05.018] [PMID: 35598048]

[43]   A. Mazumder, C. Cerella, and M. Diederich, "Natural scaffolds in anticancer therapy and precision medicine", *Biotechnol. Adv.,* vol. 36, no. 6, pp. 1563-1585, 2018.
[http://dx.doi.org/10.1016/j.biotechadv.2018.04.009] [PMID: 29729870]

[44]   E.G. Graham-Gurysh, A.B. Murthy, K.M. Moore, S.D. Hingtgen, E.M. Bachelder, and K.M. Ainslie, "Synergistic drug combinations for a precision medicine approach to interstitial glioblastoma therapy", *J. Control. Release,* vol. 323, pp. 282-292, 2020.
[http://dx.doi.org/10.1016/j.jconrel.2020.04.028] [PMID: 32335153]

[45]   R. Ayasun, T. Saridogan, O. Gaber, and I.H. Sahin, "Systemic therapy for patients with pancreatic cancer: Current approaches and opportunities for novel avenues toward precision medicine", *Clin. Colorectal Cancer,* vol. 22, no. 1, pp. 2-11, 2022.
[PMID: 36418197]

[46]   A.K. Lewis, A. Harthorn, S.M. Johnson, R.R. Lobb, and B.J. Hackel, "Engineered protein-small molecule conjugates empower selective enzyme inhibition", *Cell Chem. Biol.,* vol. 29, no. 2, pp. 328-338.e4, 2022.
[http://dx.doi.org/10.1016/j.chembiol.2021.07.013] [PMID: 34363759]

[47]   M Nikanjam, and R. Kurzrock, "New rationales and designs for clinical trials in the era of precision medicine", *Encycl. Cancer.,* pp. 30-43, 2019.

[48] P. Janiaud, S. Serghiou, and J.P.A. Ioannidis, "New clinical trial designs in the era of precision medicine: An overview of definitions, strengths, weaknesses, and current use in oncology", *Cancer. Treat. Rev.,* vol. 73, pp. 20-30, 2019.
[http://dx.doi.org/10.1016/j.ctrv.2018.12.003] [PMID: 30572165]

[49] D. Dickson, J. Johnson, R. Bergan, R. Owens, V. Subbiah, and R. Kurzrock, "Snapshot: Trial types in precision medicine", *Cell,* vol. 181, no. 1, pp. 208-208.e1, 2020.
[http://dx.doi.org/10.1016/j.cell.2020.02.032] [PMID: 32243791]

[50] "ClinicalTrials.gov", Available From: https://clinicaltrials.gov/ct2/results?cond=&term=precision+medicine&cntry=&state=&city=&dist= (Accesed on 29.12.2022).

CHAPTER 13

In Vitro Applications of Drug-carrying Nanoparticle Systems in Cell Culture Studies

Nur Selvi Günel[1,*], **Tuğba Karakayalı**[2], **Buket Özel**[1] and **Sezgi Kıpçak**[1]

[1] *Ege University, Faculty of Medicine, Department of Medical Biology, Izmir, Turkey*

[2] *Ege University, Faculty of Science, Department of Biochemistry, Izmir, Turkey*

Abstract: The safety and efficacy of each drug candidate, including nanomedicine considered for pharmaceutical use, primarily must be determined *in vitro*. In this context, the most widely used method is cytotoxicity tests, which include cell culture studies. It examines the parameters of membrane integrity, metabolite incorporation, structural alteration, survival and growth in tissue culture, enzyme assays, and the capacity for transplantation within the scope of viability tests. Within the scope of cell culture studies, tests related to apoptosis, which are effective in proper cell cycle, immune system and embryonic development, are also included. Another way to detect cell viability is to detect the biomolecules it expresses. Determination of protein expression is one of the preferred methods in this sense. Within the scope of this chapter, there is information about cell culture-based methods under these main subjects, which are applied to nanomedicines.

Keywords: Annexin V, Apoptosis assay, Capture antibody, Cell culture, Colorimetric assay, Comet assay, Cytotoxicity, DNA Fragmentation, Enzyme-linked Immunosorbent Assay, Flow cytometry, Fluorescence-activated cell sorting, Fluorometric assays, Luminometric assay, Microtiter plate, Multiplex ELISA Polyvinylidene difluoride, Propidium iodide, Real-time viability assay, SDS-PAGE Electrophoresis, Western Blot.

INTRODUCTION

Nanoparticles have currently been used in the analysis and treatment of many illnesses. This means that nanoparticles possess great physical and chemical properties, but need to be biocompatible [1, 2] in order to be safely introduced into the body and have a positive effect on tissues and cells.

* **Corresponding author Nur Selvi Günel:** Ege University, Faculty of Medicine, Department of Medical Biology, Izmir, Turkey; Tel: 902323902262; Fax:+902323420542; E-mail: nur.selvi.gunel@ege.edu.tr

Habibe Yılmaz (Ed.)

In this chapter, we discussed about the methods that are frequently used in cell culture studies. These methods used to determine the cytotoxic and apoptotic effects of nanoparticles on the cell lines of the targeted disease are described. In addition, the protein expression analysis methods used to determine the molecular changes caused by the nanoparticles used are discussed in detail.

CYTOTOXICITY ASSAYS

The cytotoxicity assays are types of biological assessment and concealing test that looks at how nanoparticles influence cell growth, reproduction, and morphology by using tissue cells *in vitro*. There is no better predictor of a nanoparticle's toxicity than its cytotoxicity. Additionally, it is easy to use, fast, sensitive and can prevent toxicity in animals [3, 4]. Definitions of cell viability often center on the concept of survival. Viability assays are also commonly employed to study cell proliferation over time within a population. To determine cell viability, researchers might use either a population-based or a single-cell approach. Population analysis is more efficient than single-cell viability assays, but the resulting data is less specific [5].

Assessment of cell viability following potentially damaging treatments is a common requirement in biological research (such as radiation, heat, chemicals, *etc.*). The major parameters in viability research are membrane integrity, metabolite incorporation, structural alteration, survival and growth in tissue culture, enzyme assays, and the capacity for transplantation. High correlations can be expected between the many indices of viability that can be obtained from these various criteria. However, the criterion for cell viability must be adjusted for the specific aims of each study. For instance, if the membranes of the cells are still in good shape, a viability test that relies on this property will consider them to be alive even though they can't divide any longer [6].

Tests for cytotoxicity and cell viability are put into different groups.

• Trypan blue, eosin, and erythrosin B assay cannot be used with these dyes.

• Colorimetric assays include the MTT, MTS, XTT, WST-1, WST-8, Lactate dehyrogenase, Sulforhodamine B, Neutral red uptake, and Crystal violet assays.

• Fluorometric assays include the alamarBLue and 5-Carboxyfluorescein Diacetate, Acetoxymethyl Ester (CFDA-AM) assays.

• ATP assays and Real-time viability assays are both luminometric assays.

Dye Exclusions

Trypan Blue Assay

The trypan blue assay was one of the oldest techniques to determine whether or not a cell is viable, and it is still frequently employed nowadays. It is predicated on the notion that live cells have an entire cell membrane, allowing the trypan blue dye to pass through. As a result of their membrane's inability to regulate the movement of macromolecules, dead cells absorb trypan blue and appear blue. For the experiment, the cells must be in a single cell suspension, and they are then counted under a microscope using a defined volume haemocytometer or recently developed automated counting equipment. Using these counts, it is reasonably easy to calculate the total number of cells and the proportion of viable cells within a population [5].

It is problematic that cell membrane integrity is used to indirectly measure viability. Even though a cell's membrane integrity has been preserved, its viability may have been impaired. Alternatively, a cell's membrane integrity may be compromised, but the cell may still be able to repair and recover normal viability [7].

Because the amount of dye absorption is evaluated subjectively, there is also the possibility that even minute levels of dye uptake that suggest cell damage will not be identified. This is yet another potential problem. In this context, more non-viable cells with dye uptake are found in examinations with a fluorescence microscope and a fluorescent dye than in examinations with a transmission microscope and trypan blue [7].

Eosin Assay

Measurements of an organism's potential for growth should not be the sole basis for determining whether or not a unicellular entity is alive or dead; rather, the distinction should be made using criteria that are more fundamental and applicable. It has been demonstrated that the ion (eosin) exclusion principle is a straightforward and efficient method for achieving this aim. In cell and tissue cultures, the conditions necessary for making accurate observations have been specified in terms of the concentrations of eosin, serum, and electrolytes. These simplified methods have replaced culturing procedures in studies of the effects of exposure of sensitive cells to pancreatin, desiccation, and tuberculin; storage of cell suspensions without renewal of the medium; and use of such methods to study nutritional or metabolic needs. These methods have also been used to find out what nutritional or metabolic requirements are needed [8].

Erythosin B Assay

It is investigated whether or not erythrosin B can function as an essential exclusion stain for mammalian cells when grown in monolayer culture. Erythrosin B is utilized in the process of determining which cells have membrane damage. fluorescent dyes are what erythrosin B is. Trypan blue is only successful in optimally staining about sixty percent of monolayer cells. It has been found that erythrosin B labeling a cell in two different ways is incompatible with one another. This discovery demonstrates that the viability indicators provided by erythrosin B overlap, lending credence to the reliability of the viability stain despite the fact that the two utilize distinct methodologies for determining membrane permeability. Erythrosin B is a fluorescent vital exclusion dye that is useful, safe, and practical for use with three mammalian cell lines when grown in a monolayer culture. In contrast, trypan blue is typically not recommended for this application [6]. Erythrosin B has formerly been used on yeast cells with great effectiveness in the capacity of a fluorescent vital exclusion dye [9].

Dye Exclusion Assay in Flow Cytometry

A more precise measurement of cell viability can be obtained through the use of flow cytometry if dye exclusion is considered. This is made possible because typan blue is a protein that binds to other proteins and then emits a fluorescence signal that can be detected using a flow cytometer. This allows for the accomplishment of the desired goal. Measurement of exclusion can also be accomplished with the use of propidium iodide or other light-emitting dyes as an alternative. Trypan blue exclusion estimates of living against dead cells reviewed manually as stated above and electronically analyzed by flow cytometry reveal that, in the hands of qualified researchers, the two approaches give results that are remarkably similar to one another. While the flow cytometric methodology is less suitable for determining cell viability during the execution of difficult and time-consuming cell purification processes, the described dye exclusion method is more likely to produce error because of operator subjectivity. This is due to the dye exclusion method's reliance on fluorescently labeled dyes. As a result, a flow cytometric approach should only be used to determine exact counts of the dead cells present in a cell combination [7].

Recent research has resulted in the publication of an "automated fluorescent microscopy viability test." This test determines whether or not cells are viable by employing propidium iodide and a device that counts both living and dead cells in a portable microscope cell counter that contains a microchip. This counter has a limitation which is reported to be able to count cells more correctly and reliably while measuring cell viability more quickly. This gadget might be useful in situations when several cell viability tests need to be gathered fast, as they are required [7, 10].

It is important to note that in addition to amine reactive dyes, agents that bind to phosphatidylserine (Annexin V), and dyes that attach to DNA (ethidium monoazide), other substances can also be employed to determine whether or not a cell is viable. Flow cytometry is often required to make use of these alternative agents, hence their application is restricted to particular scenarios only [7].

COLORIMETRIC ASSAYS

MTT Assay

The 3-(4,5-dimethyl-2-thiazolyl)- 2,5-diphenyl-2H-tetrazolium bromide (methyl thiazolyl tetrazolium; MTT) assay is a colorimetric assay for determining cell metabolism or function that provides a rapid assessment of cell proliferation and cytotoxicity [11]. It is also known as the mitochondrial dehydrogenase performance measurement. The primary principle is as described below: Mitochondrial dehydrogenase in living cells has the ability to break the tetrazole ring, which results in the formation of a purple crystalline formazan rather than the yellow, water-soluble MTT. This chemical does not dissolve in water, however it does dissolve in organic solvents like dimethyl sulfoxide and others. The colorimetric value of absorbance, also known as optical density, is a reflection of the number of cells that have survived and the level of metabolic activity, and there is a positive correlation between the amount of crystals formed and the number of cells and their level of activity [1].

The MTT assay is now the method that is used most frequently for assessing the toxicity of a culture as well as the pace of cell development. The MTT test has a variety of application challenges, despite the fact that it is sensitive to the expansion of nanoparticles [1].

In addition to this, picking the right moment to do the risk assessment is of the utmost importance. Although the MTT assay can detect dose-related toxicity in a short amount of time, it takes a significant amount of time to assess how much time influences nanoparticles. In conclusion, the MTT assay is not only labor-intensive but also less precise than other types of detection tests. The MTT experiment is laborious, fraught with repetition, and yields findings that are only slightly below average. Traditional cytotoxicity assays rely on artificial procedures, such as measuring the number of surviving cells by monitoring platelets. Because humans and the environment both have the potential to impact the results of the test, this can lead to inaccuracies and a more drawn-out testing process.

MTS Assay

The MTS assay (5-(3-carboxymethoxyphenyl)-2-(4,5-dimethyl-thiazoly)-3-(4-sulfophenyl) tetrazolium) is a colorimetric assay. For the purpose of this test, the mitochondrial activity of living cells is used to determine whether or not a tetrazolium salt has been converted into a colored formazan. The amount of formazan formed is proportional to the number of viable cells present in the culture and can be determined using a spectrophotometer at 492 nm [12].

According to findings from earlier studies, the MTS *In Vitro* cytotoxicity test possesses all of the qualities that make for an effective measurement instrument, including simplicity of operation, accuracy, and a speedy capacity to identify hazardous effects [13, 14]. The MTS assay is a specific, fast, and cost-effective *In Vitro* cytotoxicity test. It also has a high level of sensitivity. In comparison to the performance of other toxicological assays, this assay's results are quite acceptable. Because of its ease of use, swiftness, dependability, and low cost, this assay is an excellent choice for cytotoxicity measurement. As a consequence of this, it is suitable for use in on-location toxicological studies [12, 13, 15].

The degree of absorbance that is determined at 492 nm is influenced by the duration of incubation, the type of cells utilized, and the total number of cells. The proportion of MTS detection reagents to the grown cells has an impact on the degree of absorbance that is measured. The incubation period, however, can last up to 5 hours [15 - 17].

XTT Assay

Paull *et al*. have that XTT may be manufactured [18]. When XTT is bioreduced, the resulting product is a brightly colored formazan that contains only live cells. The formazan dye, unlike other tetrazolium salts, which are insoluble in aqueous solutions, can be easily quantified using a scanning multiplate spectrophotometer. This makes it possible to achieve a high degree of accuracy, makes it possible for computers to process data online (including data collecting, calculation, and the generation of reports), and makes it possible to manage a large number of samples in a quick and easy manner. In a tissue culture plate with 96 wells, cells are allowed to incubate in a solution of XTT that is yellow in color. An increase in the total activity of mitochondrial dehydrogenases in a sample is caused by an increase in the number of living cells present in the sample. During this incubation period, an orange formazan solution is formed, the concentration of which is determined spectrophotometrically. The efficacy of the XTT assay is determined by the reductive capacity of live cells that contain active mitochondrial dehydrogenase. As a consequence of this, alterations in viable cell reductive capacity brought on by enzymatic regulation, pH, cellular ion concentration (for example, sodium, calcium, and potassium), cell cycle variation, or other environmental factors may all have the potential to influence the final absorbance reading [19]. This increase has been shown to have a strong correlation with the amount of orange formazan that has formed, as determined by the absorbance [19, 20].

WST-1 Assay

WST-1 was developed by Ishiyama *et al*. [21]. It is a tetrazolium salt that, when combined with mitochondrial dehydrogenase enzymes and an intermediate electron acceptor, leads to the formation of a readily water-soluble formazan. Specifically, the formation of this formazan takes place in the mitochondria. The amount of mitochondrial dehydrogenase present in a cell culture has a direct correlation with the quantity of formazan that is generated. Therefore, the assay determines the level of metabolic activity within the cells. Despite being less hazardous than XTT, WST-1 has a sensitivity that is comparable to that of XTT. Additionally, WST1 does not require an additional step to dissolve the formazan, which is an advantage for conducting large-scale drug screening.

In addition to being straightforward, risk-free, and reliable, this method is highly utilized in the field of cytotoxicity and cell viability testing. In addition, the presence of phenol red indicators does not impede the dye response when the medium is designed for cell culture. On the other hand, the normal incubation time for WST-1 is two hours. It is not yet known if a single addition of WST-1 can accurately reflect the influence of the testing agents at various time points on the general trend of relative cell viability [12].

WST-8 Assay

Tominaga *et al*. synthesized WST-8, tetrazolium salt (1999) [14]. It acts as a chromogenic indicator of cell vitality. Reduction of the slightly yellow WST-8 by viable cells results in an orange formazan product that is directly proportional to the number of viable cells in the range of 200-25,000 cells/well for many cell lines, including non-adherent cells. This product is directly proportional to the number of viable cells. Because the dye has a net negative charge, it is largely unable to penetrate cells. As a viability indicator, WST-8 requires the use of an intermediate electron acceptor for extracellular reduction, such as mPMS. This is necessary to achieve the reduction. By measuring the absorbance at 450 nm, it is possible to determine how much reduced WST tetrazolium is present in the media used for growth. In this way, it is possible to perform tests in real time [22].

WST-8 was found to be more sensitive than other tetrazolium salts for measuring cell viability. In addition to WST-1, WST-8 forms water-soluble formazan after cellular reduction, eliminating the need for an additional step to dissolve the formazan, which is another advantage for the assay. This is further evidence of the superiority of WST-8 over other methods.

Lactate Dehydrogenase Assay

During the 1980s, Decker and his colleagues developed the lactate dehydrogenase (LDH) assay. It is a method that can measure cytotoxicity in immune cells in a short amount of time and with high accuracy [23]. LDH is a stable cytoplasmic enzyme that is released into the cell culture medium upon loss of membrane integrity [24] This loss of membrane integrity is a commonly observed feature that occurs in cells as a result of cellular damage. By oxidizing the reduced nicotinamide adenine dinucleotide (NADH) and catalyzing the production of pyruvate from lactate, LDH can generate NAD +. In this assay, NADH is responsible for the conversion of a yellow tetrazolium salt. It known as INT (iodonitrotetrazolium or 2-(4-iodophenyl)-3-(4-nitrophenyl)-5-phenyl-2H-tetrazolium) into a water-soluble red formazan dye.

The total amount of LDH activity in the culture is represented at 490 nm, and the amount of damaged cells is proportional to these values. The amount of formazan is estimated from these values [20, 25].

Sulforhodamine B Assay

The Sulforhodamine B (SRB) assay was initially created in the year 1990 with the purpose of determining the cytotoxicity of cancer treatments [26]. SRB is an aminoxanthene dye that is brilliant pink in color and has two sulfonic groups. These groups bind to amino acid residues in settings that are mildly acidic but dissociate in situations that are basic. The ability of SRB to attach to cellular proteins serves as the foundation for this treatment. This ability is maintained by the addition of trichloroacetic acid. After that, the dye-bound protein is dissolved in a solution of tris-base, are utilized in the calculation of the number of live cells at 510 nm using spectrophotometer [20, 27].

The SRB assay is a colorimetric method for assessing drug-induced cytotoxicity in adherent and suspended cell cultures. The assay can be performed on both types of cell cultures. This method for measuring cytotoxicity is both quick and sensitive. Because this technique does not rely on detecting metabolic activity, the procedures necessary in customizing the protocol for a particular cell line are substantially reduced. This is one of the advantages of using this technique [28].

Neutral Red Uptake Assay

The neutral red (3-amino-7-dimethylamino-2-methylphenazine hydrochloride) uptake assay, which Borenfreund and Puerner developed, has been designed to assess quantity of live cells there are in monolayer preparations [29]. This assay is based on the ability of living cells to bind to the supravital dye neutral red in lysosomes. An acidified ethanol solution is then used to extract the bound dye from the viable cells, and a spectrophotometer is used to determine the absorbance of the bound dye [20, 30].

In order for cells to take up neutral red, they need to have the ability to keep pH gradients stable through the synthesis of ATP. When the dye is at a pH that is physiologically relevant, it has no net charge. Because of this charge, the dye is able to travel through the cell membrane. Lysosomes have a proton gradient, which allows them to maintain a pH that is significantly lower than that of the cytoplasm. As a direct consequence of this, the dye acquires a positive charge and stays put inside the lysosomes. When the cell dies or when the pH gradient is lowered, the dye cannot be preserved in any amount. In addition, alterations in the cell surface or the membranes of lysosomes can have an effect on the uptake of neutral red by live cells. As a direct consequence of this, it is now possible to distinguish between living, injured, and dead cells [29].

The ability of lysosomes to take up a neutral red dye is a highly sensitive measure of cell health. This assay can be used to evaluate cell viability as well as measure cell proliferation, cytostatic or cytotoxic effects, depending on the seeding density used. Absorbance at a wavelength of 540 nm is evaluated using a multiwell plate-reading spectrophotometer [31].

Crystal Violet Assay

The crystal violet (CVS) assay was created at 1989 for the purpose of determining the cytotoxicity of substances. It is an indirect method for calculating the mortality of cells. The methodology is based on assessing cell adhesion by staining associated cells with CVS, which binds to proteins and DNA. This is done in order to see how well the cells are sticking together. The quantity of CVS in the cell culture diminishes as a result of the inability of dead cells to attach to the surface of the culture dish. Methanol is utilized in the process of extracting the bound dye from live cells, and a spectrophotometer is used to measure the absorbance of the dye once it has been solubilized [20, 32].

This assay outlines a screening procedure that is both quick and reliable, and it may be used to evaluate the impact that chemotherapeutics and other substances have on the ability of cells to survive and proliferate. However, in order to determine the reason for the diminished crystal violet staining, further approaches will need to be utilized, which are discussed in further detail elsewhere [32].

FLUOROMETRIC ASSAYS

Alamar Blue Assay

The nonspecific, enzymatic, and irreversible reduction of the chemical by live cells is the fundamental principle of the Alamar blue fluorometric assay. As a result of an enzyme activity that takes place inside the cells, alamar blue or resazurin is converted into resorufin, which is a pink compound, and then the living cells extract this compound from their medium. The extracted molecule is responsible for the change in color of the medium, which can be monitored in a linear range from 50 to 50,000 cells by employing excitation fluorescence filters in the range of 530-570 nm and emission fluorescent filters in the range of 580-620 nm. The alamar blue assay is a sensitive, straightforward, and risk-free method for monitoring cell viability and proliferation [33– 35]. In addition, this method outperforms the other conventional tests in terms of safety, simplicity, homogeneity, and sensitivity. These tests are used to determine whether or not a cell is viable and whether or not it is dividing (47,48). There are many alamar blue cell viability and proliferation kits that are available for purchase on the market today [20].

CFDA-AM (5-Carboxyfluorescein diacetat) Assay

Another chemical that is utilized in fluorometric analyses of cell viability is 5-CFDA-AM. 5-CFDA-AM, whose structure is similar to that of alamar blue, is a target for the intracellular nonspecific esterase enzymes found in living cells. Carboxyfluorescein, a fluorescent chemical, is produced by the nonspecific enzymatic activity of esterases, which transforms 5-CFDA-AM. Because it is polar, carboxyfluorescein cannot pass through living cells' cellular membranes [36]. Coming up next are a few instances of 5-CFDA-AM tests that are financially accessible: Examples include the 5-CFDA-AM from Invitrogen, the 5-CFDA-AM from Synchem, the 5-CFDA-AM from Biotium, and so on [20].

LUMINOMETRIC ASSAYS

ATP ASSAY

The most reliable approach for determining the amount of ATP in a sample is based on the capacity of firefly luciferase to produce a luminous signal. Although this strategy offered a sensitive manual method of detecting the number of viable eukaryotic cells in culture, its utility for measuring large numbers of samples was limited due to the need for sample manipulation and the fact that the signal was only present for a short period of time.

The development of a stable form and recombinant of luciferase were significant breakthroughs that enabled widespread acceptance of using the ATP assay as a marker for viable cells in the process of drug discovery. ATP assays are used to determine whether or not a cell is able to produce and use energy. Additionally, recombinant luciferase made it possible to produce mass quantities of enzyme in a manufacturing environment that was under strict control, which led to reagents that were more consistent from one batch to the next. It is convenient and flexible to capture data from large batches of multi-well plates since the luminescent signal glows with a half-life of more than five hours after mixing and a quick equilibration period of ten minutes.

As a result of its numerous benefits, the luminous ATP detection assay has emerged as the method of choice in HTS laboratories for determining the level of cell viability. The most significant benefits of this method include the simplicity of the add-mix-measure homogeneous method, the speed with which the assay can be performed, and the low interference from test compounds. In addition, this microplate assay is the most sensitive one available for determining whether or not cells in culture are viable. The ATP assay is often only able to detect a maximum of 10 cells in each well. The reagent may be used on an automated robotic platform because it is stable enough at ambient temperature to do so. Additionally, the assay is immediately scalable from 96 well forms all the way up to 1536 well formats. The fact that the ATP assay lyses cells as soon as the reagent is administered is another key advantage of using it. This advantage allows for the extraction of information on the level of ATP at the point in time when the reagent is added.

Because this method of assaying kills the cells, the sample that has been treated with the ATP detection reagent cannot be utilized for anything else once it has been processed through it. This is one of the disadvantages of the procedure. The use of a sequential format is needed, despite the fact that there are various examples of different assay methods that can be multiplexed. The fact that the ATP assay is sensitive to temperature fluctuations is yet another one of its many drawbacks. Because temperature has an effect on the enzymatic rate of luciferase and, consequently, the luminous signal, it is essential to keep the temperature of the environment consistent in order to achieve the highest level of reproducibility. A potential drawback of all assays based on luciferase is the possibility that test chemicals will inhibit luciferase. This will lead to a drop in signal but will not have any effect on the survival of the cells or the amount of ATP they contain.

Real-Time Viability Assay

The luciferase-based approach is modernized by the real-time viability assay. It is the only cell viability method that allows cells being tested for viability to be monitored in real time. A marine shrimp-derived luciferase and a pro-substrate derived from luciferase are used in this novel approach. A cell-permeable pro-substrate and the luciferase are added to the culture medium using this method; however, the intracellular ATP is not released from the cells themselves. However, the intracellular ATP is not released from the cells themselves. Viable cells, on the other hand, are in charge of converting the pro-substrate into the "substrate" that diffuses into the culture medium. Then, at that point, the luciferase catalyst utilizes the diffused substrate to produce a light sign. Both endpoint assays and applications requiring continuous measurement can benefit from this strategy [5].

APOPTOSIS ASSAYS

Apoptosis is the cell death process which is energy-dependent manner. It is vital mechanism for proper cell cycle, immune system and embryonic development. Improper apoptosis mechanism is associated with many diseases including cancer. Its ability to modulate a cell's life is recognized for its enormous therapeutic potential [37]. In multicellular organisms, cells numbers are tightly controlled not only by the cell division but also by the cell death. If the cell is unnecessary, an intracellular death program is activated that is called apoptosis. So far two main apoptotic ways defined: extrinsic pathway (*via* death receptor) or intrinsic pathway (mitochondrial pathway) [38]. Many apoptosis methods have been developed to detect the apoptotic effect of drugs and nanoparticles upon to cells after its application. These methods depend on functional and structural changes during apoptosis, fragmentation of DNA, cell proliferation and cytotoxicity, mitochondrial damage, immunological detection, and mechanism-based assays. Some of them are given as a scheme below (Fig. **1**) [39].

COLORIMETRIC ASSAYS

Annexin V

Live cells are defined by the asymmetric distribution of different phospholipids between the inner and outer sides of the cell membrane. The choline-containing phospholipids, phosphatidylcholine and sphingomyelin found in the outer side of living cells, and the aminophospholipids phosphatidylethanolamine and phosphatidylserine (PS) found on the cytoplasmic face. During apoptosis, cells expose PS to the outer site of the cell membrane. PS functions like a label for identification by macrophages and for phagocytosis of the dying cell. Annexin V is a method to detect the apoptosis and recognize this PS tranferred to the outer of the membrane. In live cells, Annexin V can not bind, as the molecule can't penetrate the phospholipid bilayer. However, in dead cells, because of lost integrity of cell membrane, Annexin V is able to bind inner side of membrane. For discriminate between live and dead cells PI dye can be added as a membrane-impermeable dye. In this way, live, apoptotic and dead cells can be differentiated by double-labeling for PI and annexin V that can be analyzed by fluorescence microscopy or flow cytometry [40, 41].

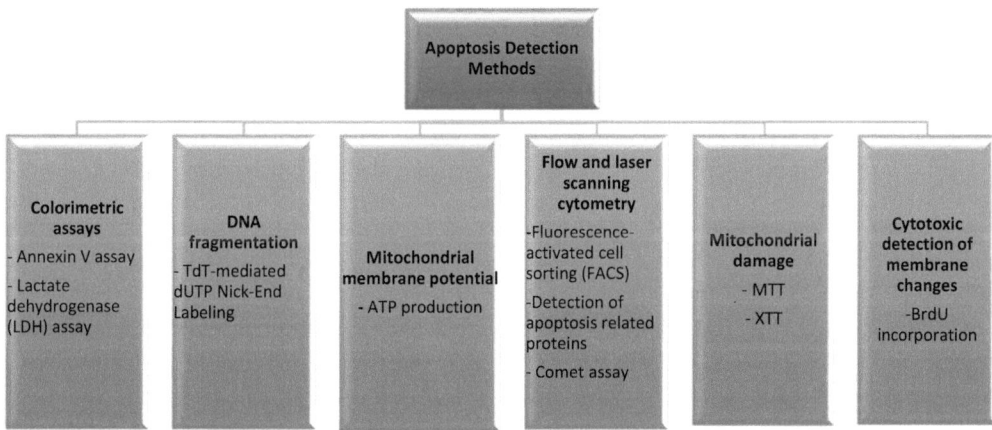

Fig. (1). Types of apoptosis detection method.

LACTATE DEHYDROGENASE ASSAY

Lactate dehydrogenase (LDH) is a steady enzyme found in cytoplasm in all cells. LDH is an enzyme released into the cell culture medium as a result of cell damage during apoptosis and necrosis. LDH activity can be easily measured by reducing a second compound to a product whose properties can be easily measured in the coupled reaction using NADH produced during the transformation of lactate to pyruvate. LDH catalyzes the reduction of NAD+ to NADH and H^+ *via* the oxidation of lactate to pyruvate. Then, diaphorase uses newly formed NADH and H^+ to catalyze the reduction of tetrazolium salt (INT) to formazan, a colored component that can be measured at 492 nm. The increase in the number of dead or plasma membrane damaged cells are correlated to increased LDH activity [42].

DNA FRAGMENTATION

TdT-mediated dUTP Nick-End Labeling (TUNEL)

One of the signs of apoptosis is the nucleosomal cleavage of genomic DNA into small fragments. In late 1992, It was announced the development of the TdT-mediated dUTP-biotin nick end labeling (TUNEL) assay that uses this biochemical distinguishing feature and enables *In Situ* recognition of DNA breaks. The method is based on the synthesis of the polydeoxynucleotide polymer of Tdt (deoxynucleotidyl transferase) followed by the specific binding of the DNA to the 3-OH ends. TUNEL staining is not used for the detection of only apoptotic cells but also non apoptotic cells that undergo DNA repair, necrotic degenerating cells, and mechanical forces cells that have been damaged by or undergone active gene transcription. TUNEL staining is nonspecific in the sense that the assay labels all free 3'OH termini independently of the molecular mechanisms that lead to the development of these termini. For this reason, TUNEL assay should be considered the DNA damage detection method besides its apoptosis detection [43].

Mitochondrial Membrane Potential

Mitochondrial dysfunction is related to apoptosis induction and a major way to the apoptotic pathway. Mitochondrial permeability transition pore opening has been shown to induce depolarization of the transmembrane potential ($\Delta\psi$m). It has been determined that disruption in mitochondrial transmembrane potential and release of some mitochondrial proteins such as cytochrome c and apoptosis-inducing factor into the cytoplasm initiate apoptotic pathways [44]. Briefly, the inner and outer membranes of mitochondria contain a group of proteins belonging to the Bcl-2 superfamily which coordinate caspase activation.

JC-1, is the cationic fluorescent probe, being used for detection changes in $\Delta\psi$m in order to measure apoptosis. JC-1 generates J-aggregates or monomers with two dyes that can be detectable at 585 nm or 530 nm by flow cytometry and fluorescence microscopy. In normal cells, high $\Delta\psi$m filled with JC-1 let for the formation of red fluorescent J-aggregates whereas in low of $\Delta\psi$m, J-aggregates turns into monomers leading to color change from red to green fluorescence [44, 45].

FLOW AND LASER SCANNING CYTOMETRY

Fluorescence-Activated Cell Sorting (FACS)

Fluorescence-activated cell sorting (FACS) is a laser-based method that is used to detect apoptosis. It is well-known that Annexin V and PI are the most used dyes for apoptosis analysis. 4',6-diamidino-2-phenylindole (DAPI), is a fluorescent dye that binds strongly to A-T-rich regions in DNA, can also be performed instead of PI. DAPI can be used combined with Annexin V FITC and fluorescent dye tetramethylrhodamine methyl ester (TMRM) dye so that it measures mitochondrial permeability transition and mitochondrial membrane depolarization. TMRM is a cell-permeable fluorescent probe that indicates active mitochondria and tags living cells. Combining DAPI-Annexin V FITC- TMRM has benefits to distinguish living, dying or dead cells in all cell population. FACS is a rapid and more precise method to measure the number of dying cells in an entire cell population [46].

DETECTION OF APOPTOSIS RELATED PROTEINS

Proteins, related to apoptosis, can be determined by a few methods such as ELISA, quantitative reverse transcription PCR (RT-qPCR), and western blotting. For instance, apoptosis-related proteins such as caspases, Bid (BH3 interacting-domain death agonist), Fas protein with death domain, Bax, NFκB, PARP and, Bid can be detected with western blot. When western blotting is inefficient in detection, RT-qPCR is used for specific and quantitative measurement. Caspases 3 and caspase 7, activated initiator caspases 2, caspase 8, caspase 9, Bid, Bax, Bak, APAF and PARP are suitable to measure apoptosis rate by RT-qPCR. Besides the RT-qPCR and western blotting, ELISA method can be performed to measure the apoptosi-related proteins [39].

COMET ASSAY

For a long time, comet assay (also known as the single-cell gel electrophoresis method) has been used to determine DNA damage and repair in the cells. During repair, the enzymes increase the DNA lesions that can be determined by this method so it can also be used to measure the DNA repair activity. In the cells, DNA strand breaks can be measured by comet assay.

Cells embedded in agarose are lysed with high salt and detergent to form nucleotides. The structures resembling comets, can be analyzed by fluorescence microscopy as a result of high pH electrophoresis. The density of the comet tail relative to the head states the number of DNA breaks. Comet assay is beneficial for testing genotoxicity, environmental contamination, molecular epidemiology, DNA damage and repair [47, 48].

CYTOTOXIC DETECTION OF MEMBRANE CHANGES

BrdU Incorporation

5-Bromo-2 '-deoxyuridine (bromodeoxyuridine (BrdU)) is a thymidine analog readily incorporated into DNA during DNA synthesis. It ensures a precise process of detecting proliferation and apoptosis. BrdU as a thymidine analog incorporates into nuclear DNA so that can be used as a label. BrdU enters into rivalry with thymidine for integration into nuclear DNA throughout the S phase of the cell cycle, DNA repair and, cell degeneration. When BrdU integrates into nuclear DNA, it can be determined with immunohistochemical methods.

Following BrdU is associated with nuclear DNA, samples are incubated with anti-BrdU monoclonal antibodies and nucleases. Then, the samples are incubated with a secondary antibody which has an enzyme that can cause a colorimetric reaction to be observed by bright field microscopy [49].

PROTEIN EXPRESSION ASSAYS

Enzyme-Linked Immunosorbent Assay

Enzyme-linked immunosorbent assay (ELISA) is a method commonly used in research and clinical areas for the specific detection of biomolecules. The basis of ELISA method is set on antigen-antibody reactions. Leveraging this reaction, ELISA is used for ultra sensitive and selective quantitative/qualitative experiments of specific metabolic, including proteins, peptides, nucleic acids, hormones, and secondary metabolites.

Antigens or antibodies labeled with alkaline phosphatase (ALP), horseradish peroxidase (HRP) and β-galactosidase are often used to quantify and detect these molecules [50, 51].

The antigens liquid form is immobilized on a solid phase. Materials like, polyvinyl chloride and rigid polystyrene are generally used for immobilization. The antigen then reacts with a specific antibody identified by the enzyme-labeled secondary antibody. The chromogenic substrate is used for determining the presence of the reaction and color change indicates the presence of antigen. It is appropriate to use fluorogenic substrates for β-galactosidase, while chemilu-minescent substrates are used for ALP and HRP. Enzyme-substrate reactions are completed in 30-60 minutes and the reaction is stopped by adding solutions such as sulfuric acid, sodium hydroxide, sodium carbonate, hydrochloric acid, and sodium azide. Lastly, colored or fluorescent products are determined *via* a microtiter plate reader [52].

Types of ELISA

Direct ELISA

In this method, first the antigen or antibody of interest is immobilized on the microtiter surface through passive absorption. Then, the surface is blocked with proteins such as albumin, gelatin, and casein to prevent non-specific binding and saturate the unbound sites, followed by the labeled conjugated detection antibody (monoclonal or policlonal) or antigen binds to the target protein. After the binding process, the substrate is added, and the signal is generated according to the amount of analyte in the sample (Fig. **2**). Direct ELISA method is rapid and simple however the specificity is low due to the use of a single antibody [53].

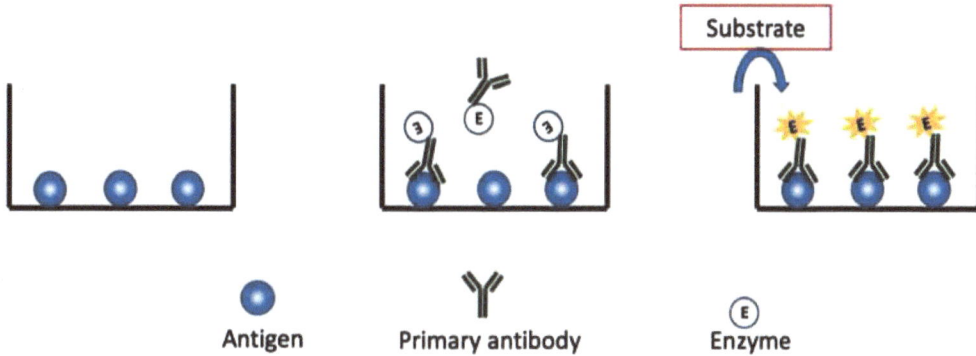

Fig. (2). Schematic representation of direct ELISA.

Indirect ELISA

In the indirect ELISA method, the antigen to be investigated is fixed on polystyrene microtiter plates by passive absorption as in direct ELISA. A blocking buffer is then adjoined to saturate the unbound sites on the surface. After the blocking buffer, the microtiter plate incubates with a primary antibody, which is peculiar for the target antigen. In Indirect ELISA primary antibody is not labeled with any enzyme. After the primary antibody incubation is complete, a secondary antibody conjugated with a specific enzyme is added to the primary antibody, usually HRP is used as the conjugation enzyme. A further widespread method is to use a biotin-linked primary antibody and an enzyme-conjugated streptavidin as a secondary antibody (Fig. **3**) [54]. In both methods, the quantitation of the target molecule is provided according to the color change that occurs by using the substrate. Indirect ELISA is utilizable for protein quantification, antibody screening and epitope mapping. Primary and secondary antibody usage provide stong signal therefore it is more sensitive than direct elisa [55].

Competitive ELISA

In competitive ELISA, targets (antigen or antibody) in the sample and enzyme-labeled competitive targets compete against for linked to the immobilized antibody or antigen. After completion of binding, signal formation is measured by adding a substrate as in other ELISA types (Fig. **4**). If the amount of antigen in the sample is high, less enzyme-labeled competitive antigens will bind to the immobilized antibody, so samples with high concentration of targets give low signal. That is, the amount of target molecule in the sample is inversely proportional to the signal. This method is often used to determine the concentration of antigens in complex sample mixtures with molecules too small to sandwich ELISA method [56].

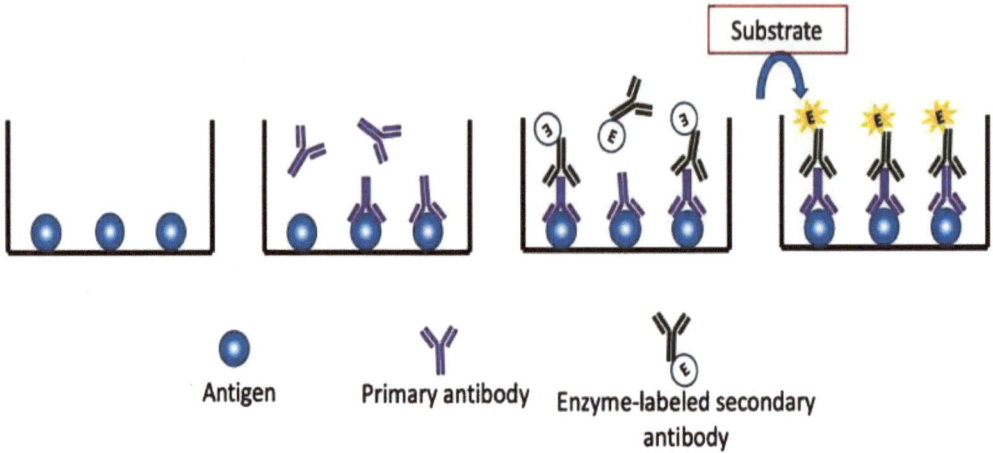

Fig. (3). Schematic representation of indirect ELISA.

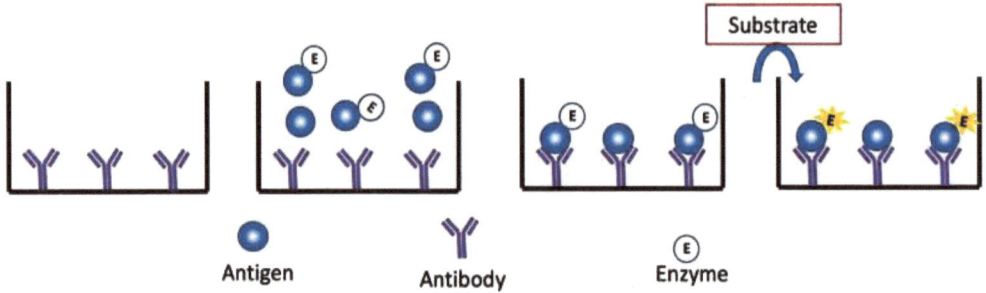

Fig. (4). Schematic representation of competitive ELISA.

Competitive Indirect ELISA

This method is a combination of indirect ELISA and competitive ELISA. The antigen to be investigated is immobilized to the microtiter plate. Then incubated with the free antigen (sample) and the primary antibody, there is competition against the free antibody occurs between the immobilized antigen and the sample antigen. After bindings, adding washing solution for removal to the free antigen and primary antibody bindings. After that, the enzyme-labeled secondary antibody is added to the medium and this secondary antibody binds to the primary antibody bound to the immobilized antigen (Fig. 5). As with competitive ELISA, this method produces a lower signal if there is a high amount of antigen in the sample, as there will be less primary secondary antibody binding. Indirect methods are more sensitive and versatile due to the usage of enzyme-labeled secondary antibody. However, should be attention the cross-reactivity risk [57].

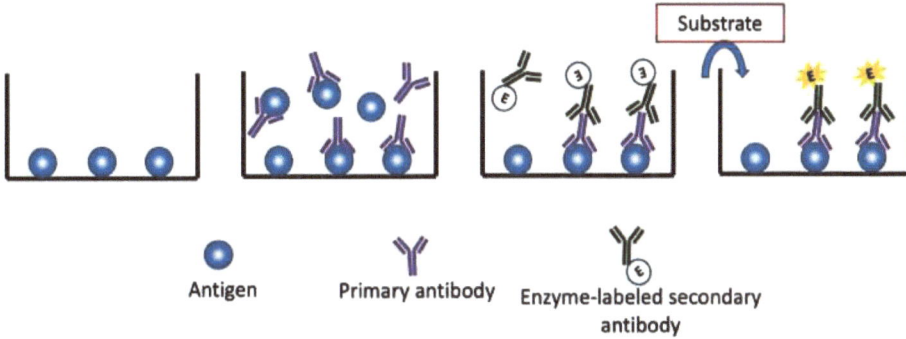

Fig. (5). Schematic representation of indirect competitive ELISA.

Sandwich ELISA

Sandwich ELISA is the most widely used ELISA method, in this system the target antigen in the sample creates a sandwich model by establishing a link between 2 antigens that recognize different epitopes.

First, the capture antibody is immobilized on the microtiter plate and blocked to prevent non-specific binding, then the sample is added to the medium and the target antigen is ready to reaction with the capture antibody. Then, enzyme-labeled antibody, which recognizes the antigen from its different epitope, is added for the color change to occur, and a sandwich model is created (Fig. 6). Signal formation is provided by adding substrate to the medium, the higher the amount of antigen, gives higher signal. Since two antibodies with different epitopes against the target antigen are required, it is costly and is a highly specific method often used for measuring macromolecules [58].

Cell Based ELISA

Cell-ELISA was created on the principle of enzyme immunohistochemistry and ELISA. It is used to detect and quantitate antigens located in the intracytoplasmic, transmembrane or endoplasmic reticulum. Flow cytometry (FACS) and radioimmunobinding tests (RIA), which perform the same function, they are disadvantageous due to their cost, use of radioactive material, and the need for multiple assays. Therefore, cell-ELISA is a simpleton, rapid, low-cost and sensitive, despite this procedure is not a proper method for analyzing mixed cell populations [59].

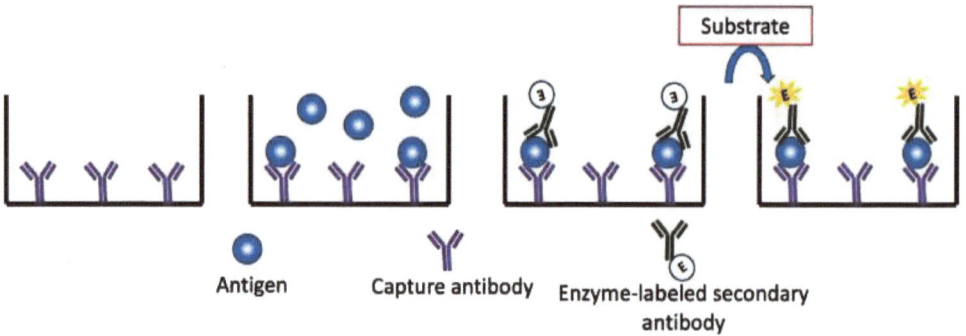

Fig. (6). Schematic representation of sandwich ELISA.

Multiplex ELISA

Multiplex ELISA is a method that allows simultaneous examination of more than one analyte using less sampling amount in a shorter time than standard immunoassay methods. Alike with ELISA, the antibody sandwich method is used; merely the methods of binding to the substrate and detection are different.

Instead of a flat surface, Multiplex ELISA uses sets of distinctively detectable beads that capture analytes in solution. These bead sets classify the analytes, and detecting antibodies are utilized to quantify the analyte (Fig. 7). The utilization of differentially detectable beads permits simultaneous identification and quantification of many analytes in the same sample [60].

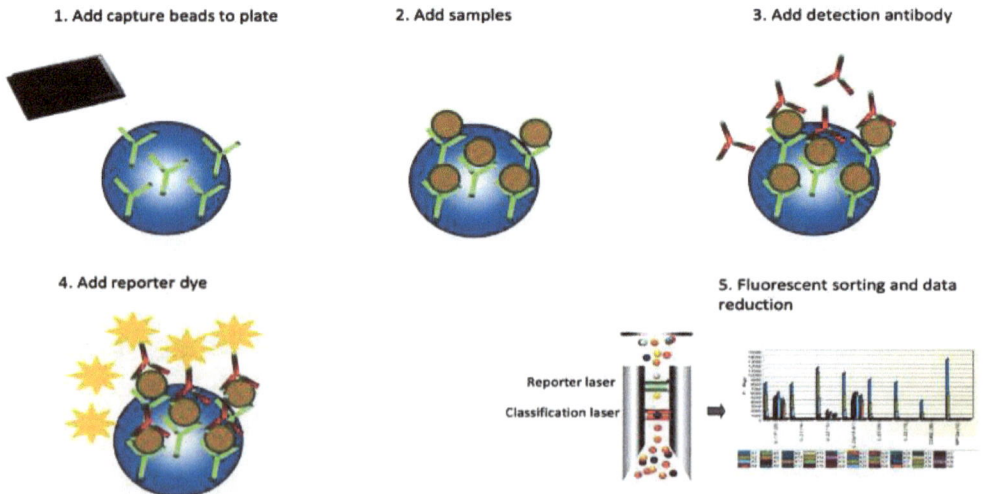

Fig. (7). Workflow of multiplex ELISA, this figure adapted from reference [60].

Western Blot Technique

The western Blot (WB) technique is a molecular biology method utilized to especially determine a desired protein or protein profile from any sample. The presence, size, concentration, and concentration changes of a protein in a tissue, and the comparison of concentrations between different groups can be investigated by means of this technique. The western blot technique also named the protein immunoblot or Western blotting is a high precision measurement and a semi-quantitative molecular technique [61].

The western blot is a method used in detecting the responses of proteins to stimulation and assessing protein expression as either absent or increased-decreased. It is a method also frequently used in the detection of specific protein isoforms, and in the determination of proteins responsible for the emergence of many diseases. In the WB technique, proteins should be extracted from cells effectively in the first step. Cell lysis and protein extraction are performed with suitable buffers. Then, the samples need to be separated by SDS-PAGE gel based on their molecular weights. Afterwards, the transfer of proteins to a solid-phase membrane is realized and the final step is to mark the target protein using the appropriate primary and secondary antibodies to make it visible.

Secondary antibody proteins can be made visible using different methods such as staining, immunofluorescence, and radioactivity (Fig. **8**). Western blot can be applied in a shorter time by using the new developing technologies, thus allowing the analysis of many proteins at the same time [62].

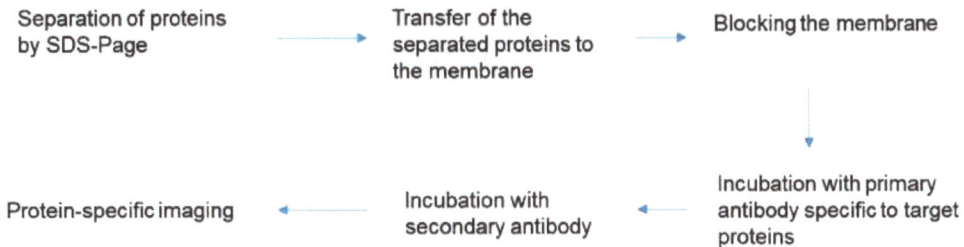

Fig. (8). The Steps of the Western Blot Method.

PREPARATION OF SAMPLES TO BE ABLE TO MAKE ANALYSIS BY WESTERN BLOTTING TECHNIQUE AND SDS-PAGE ELECTRO-PHORESIS

Protein samples can be obtained from tissue or cell culture. An ultrasonic cell disrupter/homogenizer can be used to disrupt cells mechanically. For obtaining the protein sample in cell culture studies, the appropriate number of cells should be seeded in culture flasks. If the cells to be studied are adherent cells, unlike the cells in suspension, it is necessary to wait for 24 hours to permit them to adhere to the flask. Then, the predetermined IC50 concentrations of the active substances will be applied, and the cells to which no substance has been applied will be used as the control group.

Adherent cells will be collected by trypsinization after the required active ingredients are applied within the specified time and proteins will be obtained using the appropriate protein isolation solution.

Various detergents, salts, and buffer solutions can be used to ensure cell dissolution and make proteins soluble. Protease inhibitors are frequently used to prevent the breakdown of proteins, which will be obtained. These steps should be performed inside ice to prevent protein denaturation at this stage.

Total protein concentrations in the resulting cell lysates will be able to determine using the Bradford or Lowry method. The Lowry method is based on the reaction of Follin reagent and copper, while Bradford is a method made by using "Coomassie Brilliant Blue" dye [63]. Protein concentrations are measured spectrophotometrically and when examined by both methods, it was observed that the color intensity changes depending on the amount of protein [64].

Sodium dodecyl sulfate (SDS) is an anionic detergent and its oligomeric proteins are able to divide into the subdivisions. PAGE is a medium in which the electrical force of attraction is applied to separate proteins according to their size [65]. The preparation of the sample to be run on the gel should be determined by the molecular weight of the protein to be analyzed. One of the most important steps is to adjust the pore size through which the proteins can pass.

The pore size can be controlled by using polyacrylamide gels, formed from the polymerization of acrylamide, and bisacrylamide, which have a certain percentage. The pore size is reduced by increasing the acrylamide percentage, and the gel ensures the separation of proteins having lower molecular weight. An equal amount of protein from the quantified protein sample will be loaded on the gels and pre-loading proteins will be denatured, thus the pre-loading proteins are converted to their primary structure [66].

Protein samples to be analyzed are placed in an electrophoresis chamber filled with a buffer and applied a constant current connected to a power source. The buffers used show a different feature according to each gel and that also depends on the conditions, pH, and ionic environment [67].

Care should be taken to transport the samples until the lowest part of the gel for proper separation and detection of protein samples. Proteins with a large molecular weight will be able to transport the higher part of the gel, while proteins with low molecular weight will be visualized at the lower part of the gel.

BLOTTING AND BLOCKING PROCESSES OF PROTEIN SAMPLES

Transfer of the execution-completed proteins to the membrane is done using a wet or semi-wet transfer system, Polyvinylidene difluoride (PVDF) membrane has a more robust structure than that of nitrocellulose membrane and the binding ratio of proteins is higher. In this step, the transfer is realized from the gel to the polyvinylidene difluoride (PVDF) membrane and after this stage, all the processes are realized upon the membrane. The wet transfer is called "the electrophoretic transfer".

At this stage, a Sponge-Filter Paper-Gel-Membrane-Filter Paper-Sponge Sandwich model (the gel and membrane are sandwiched) is formed. Placing proteins in the transfer apparatus, the passage of protons to the membrane in a cold environment is ensured. Protein transfer is provided in a shorter time by using the iBlot systems in the semi-wet model [68, 69].

The chromogenic immunodetection kit is one of the methods used to identify proteins in the post-transfer processes. Firstly, the membranes will be incubated in a shaker with the blocking solution at room temperature. This step is done for the fixation of proteins on membranes. BSA-Bovine Serum Albumin or skimmed milk powder solutions are the most used solutions in the blocking phase. This step is applied to prohibit nonspecific binding between the membrane and the antibody.

PROTEIN DETECTION AND VISUALIZATION

At this stage, to identify and visualize proteins, the membranes should be incubated with primary antibodies, the dilutions of which were prepared according to the manufacturer's instructions, at room temperature. The procedure must primarily be based on confirming the identity of a target protein since we are able to detect it by a specifically directed antibody.

At this stage, it should be taken care of to ensure that the protein we want to study is monoclonal and suitable for the Western blot method. In order to do quantitative analysis, the housekeeping control protein sample should also be studied simultaneously at this step.

It should be taken care of to wash in the required number and time in order to eliminate artifacts from the environment other than protein. The incubation with the secondary antibody will be made in the next step. Finally, the chromogenic substrate will be added to the membranes and the formation of purple-colored protein bands will be observed as a result of incubation for a certain period of time. After the membranes have dried, the bands will be monitored by the imaging system. The displayed bands can be analyzed quantitatively using the ImageJ v2 program [70].

In conclusion, the Western blot technique is a molecular method used frequently in proteomics studies. In addition, it benefits *In Vitro* applications of nanoparticle drug delivery systems, and in detecting the presence, and size of a specific protein.

CONCLUDING REMARKS

In vitro studies are one of the leading methods used to determine the effectiveness of developed drug delivery systems. *In vitro* studies are an important step in comparing drug delivery systems with free-drugs, determining their intracellular uptake status, and evaluating their therapeutic potential. However, *in vitro* studies have some limitations; since the studies are performed at the cellular level, the biological responses that will take place in the living organism may not be determined.

In addition, studying drug-carrying nanoparticle system studies in *in vitro* systems before starting to work on *in vivo* systems provides an advantage in the early detection of possible design errors of the drug-carrying system. Another concept is that well-planned *in vitro* studies provide advantages that can prevent unnecessary animal use in *in vivo* experiments. This concept is very important in terms of ethical principles.

REFERENCES

[1]   W. Li, and J. Zhou, "Study of the *In Vitro* cytotoxicity testing of medical devices (Review)", *Biomed. Rep.,* pp. 617-620, 2015.

[2]   L. H. Reddy, J. L. Arias, and J. Nicolas, "Magnetic nanoparticles: Design and characterization, toxicity and biocompatibility, pharmaceutical and biomedical applications", *Chem. Rev.,* pp. 5818-5878, 2012.

[3]   Q.He ve, and Shi. J, "Mesoporous silica nanoparticle based nano drug delivery systems: synthesis, controlled drug release and delivery, pharmacokinetics and biocompatibility", *Biochim. Biophys. Acta,* vol. 21, no. 16, pp. 5845-5855, 2011.

[4]   A. Kunzmann, B. Andersson, T. Thurnherr, H. Krug, A. Scheynius, and B. Fadeel, "Toxicology of engineered nanomaterials: Focus on biocompatibility, biodistribution and biodegradation", *Biochim. Biophys. Acta,* vol. 1810, no. 3, pp. 361-373, 2011.

[5]   M.J. Stoddart, Cell Viability Assays: Introduction.*içinde Mammalian Cell Viability: Methods and Protocols,,* M.J. Stoddart, N.J. Totowa, Eds., Humana Press, 2011, pp. 1-6.

[http://dx.doi.org/10.1007/978-1-61779-108-6_1]

[6]   A. W. Krause, and W. W. Carley, "Fluorescent erythrosin B is preferable to trypan blue as a vital exclusion dye for mammalian cells in monolayer culture", *J. Histochem. Cytochem.*, vol. 32, no. 10, pp. 1084-1090, 1984.

[7]   W. Strober, "Trypan blue exclusion test of cell viability", *Curr. Protoc. Immunol.*, vol. 111, no. 3, pp. 1-3, 2015.
        [http://dx.doi.org/10.1002/0471142735.ima03bs111]

[8]   J.H. Hanks ve, J.H. Wallace, and W. Strober, "Determination of cell viability", *Proc. Soc. Exp. Biol. Med.*, vol. 98, no. 1, pp. 188-192, 1958.

[9]   K.J. Hutter, and H.E.Y. Eipel, "Short communications advances in determination of cell viability", *Microbiology*, vol. 107, no. 1, pp. 165-167, 1978.

[10]   J. S. Kim, "Comparison of the automated fluorescence microscopic viability test with the conventional and flow cytometry methods", *J. Clin. Lab. Anal.*, vol. 25, no. 2, pp. 90-94, 2011.

[11]   G. Fotakis ve, and J. A. Timbrell, "*In Vitro* cytotoxicity assays: Comparison of LDH, neutral red, MTT and protein assay in hepatoma cell lines following exposure to cadmium chloride", *Toxicol. Lett.*, vol. 160, no. 2, pp. 171-177, 2006.

[12]   M. Larramendy ve, and S. Soloneski, *Genotoxicity: A predictable risk to our actual world.* IntechOpen, 2018, pp. 1-122.

[13]   K. Berg, L. Zhai, M. Chen, and A. Kharazmi, "The use of a water-soluble formazan complex to quantitate the cell number and mitochondrial function ofLeishmania major promastigotes", *Parasitol. Res.*, vol. 80, no. 3, pp. 235-239, 1994.

[14]   H. Tominaga, "A water-soluble tetrazolium salt useful for colorimetric cell viability assay", *Anal. Commun.*, vol. 36, no. 2, pp. 47-50, 1999.

[15]   T. M. Buttke, and J. A. McCubrey, "Use of an aqueous soluble tetrazolium/formazan assay to measure viability and proliferation of lymphokine-dependent cell lines", *J. Immunol. Methods*, vol. 157, no. 1, pp. 233-240, 1993.

[16]   R. Tl, "Comparison of MTT, XTT and a novel tetrazolium compound MTS for *in-vitro* proliferation and chemosensitivity assays", *Mol. Biol. Cell.*, vol. 3, no. 1, pp. 184-190, 1992.

[17]   B. A. Rotter, B. K. Thompson, and S. Clarkin, "Rapid colorimetric bioassay for screening of fusarium mycotoxins", *Nat. Toxins,*, vol. 1, no. 5, pp. 303-307, 1993.

[18]   K. D. Paull, "The synthesis of XTT: A new tetrazolium reagent that is bioreducible to a water-soluble formazan", *J. Heterocycl. Chem.*, vol. 25, no. 3, pp. 911-914, 1988.

[19]   D. A. Scudiero, "Evaluation of a soluble tetrazolium/formazan assay for cell growth and drug sensitivity in culture using human and other tumor cell lines", *Cancer. Res.*, vol. 48, no. 17, pp. 4827-4833, 1988.

[20]   S. Kamiloglu, G. Sari, T. Ozdal, and ve E. Capanoglu, "Guidelines for cell viability assays", *Food. Front.*, vol. 1, no. 3, pp. 332-349, 2020.

[21]   M. Ishiyama, M. Shiga, K. Sasamoto, and M. Mizoguchi, "A new sulfonated tetrazolium salt that produces a highly water-soluble formazan dye", *Chem. Pharm. Bull.*, vol. 41, no. 6, pp. 1118-1122, 1993.

[22]   K. Präbst, H. Engelhardt, S. Ringgeler, ve H. Hübner, "Basic Colorimetric Proliferation Assays: MTT, WST, and Resazurin", içinde *Cell Viability Assays*, c. 1601, D. F. Gilbert ve O. Friedrich, Ed. New York, NY: Springer New York, 2017, pp. 1-17.

[23]   "A quick and simple method for the quantitation of lactate dehydrogenase release in measurements of cellular cytotoxicity and tumor necrosis factor (TNF) activity", *J. Immunol. Methods*, vol. 115, no. 1,

pp. 61-69, 1988.

[24]   F. K.-M. Chan, K. Moriwaki, ve M. J. De Rosa, "Detection of Necrosis by Release of Lactate Dehydrogenase Activity", içinde *Immune Homeostasis: Methods and Protocols*, A. L. Snow ve M. J. Lenardo, Ed. Totowa, NJ: Humana Press, 2013, pp. 65-70.

[25]   P. Kumar, and A. Nagarajan, "Analysis of Cell Viability by the Lactate Dehydrogenase Assay", *Cold Spring Harb. Protoc.,* 2018.

[26]   P. Skehan, "New colorimetric cytotoxicity assay for anticancer-drug screening", *JNCI J. Natl. Cancer Inst,* vol. 82, no. 13, pp. 1107-1112, . 1990.

[27]   V. Vichai, and K. Kirtikara, "Sulforhodamine B colorimetric assay for cytotoxicity screening", *Nat. Protoc.,* vol. 1, no. 3, p. 3, 2006.

[28]   E. A. Orellana, and A.L Kasinski, "Sulforhodamine B (SRB) assay in cell culture to investigate cell proliferation", *Bio.Protoc.,* vol. 6, no. 21, pp. 1984-1984, 2016.

[29]   E. Borenfreund, and J.A. Puerner, "A simple quantitative procedure using monolayer cultures for cytotoxicity assays (HTD/NR-90)", *J. Tissue Cult. Methods.,* vol. 9, no. 1, pp. 7-9.

[30]   G. Repetto, and A. del Peso, "Neutral red uptake assay for the estimation of cell viability/cytotoxicity", *Nat. Protoc.,* vol. 3, no. 7, p. 7, 2008.

[31]   J. Weyermann, D. Lochmann, and ve A. Zimmer, "A practical note on the use of cytotoxicity assays", *Int. J. Pharm.,* vol. 288, no. 2, p. 369p. 376, 2005.

[32]   M. Feoktistova, P. Geserick, and ve M. Leverkus, "Crystal violet assay for determining viability of cultured cells", *Cold Spring Harb. Protoc.,* vol. 2016, no. 4, p. 087379, 2016.

[33]   K. B. Jonsson, A. Frost, R. Larsson, S. Ljunghall, and ve O. Ljunggren, "A new fluorometric assay for determination of osteoblastic proliferation: Effects of glucocorticoids and insulin-like growth factor-I", *Calcif. Tissue Int.,* vol. 60, no. 1, pp. 30-36, 1997.

[34]   E. M. Larson, D. Doughman, D. Gregerson, and ve W. F. Obritsch, "A new, simple, nonradioactive, nontoxic *In Vitro* assay to monitor corneal endothelial cell viability", *Invest. Ophthalmol. Vis. Sci.,* pp. 1929-1933, 1997.

[35]   E.M. Czekanska, "Assessment of cell proliferation with resazurin-based fluorescent dye", In: *içinde Mammalian Cell Viability: Methods and Protocols,* M. J. Stoddart, Stoddart, E.M. Czekanska, NJ. Totowa, Eds., Humana Press, 2011, pp. 27-32.
[http://dx.doi.org/10.1007/978-1-61779-108-6_5]

[36]   R.C Ganassin, "Growth of rainbow trout hemopoietic cells in methylcellulose and methods of monitoring their proliferative response in this matrix", *Meth. Cell. Sci.,* vol. 22, no. 2, pp. 147-152, 2000.

[37]   S. Elmore, "Apoptosis: A review of programmed cell death", *Toxicol. Pathol.,* vol. 35, no. 4, pp. 495-516, 2007.
[http://dx.doi.org/10.1080/01926230701320337] [PMID: 17562483]

[38]   X. Xu, Y. Lai, and Z.C. Hua, "Apoptosis and apoptotic body: Disease message and therapeutic target potentials", *Biosci. Rep.,* vol. 39, no. 1, p. BSR20180992, 2019.
[http://dx.doi.org/10.1042/BSR20180992] [PMID: 30530866]

[39]   G. Banfalvi, "Methods to detect apoptotic cell death", *Apoptosis,* vol. 22, no. 2, pp. 306-323, 2017.
[http://dx.doi.org/10.1007/s10495-016-1333-3] [PMID: 28035493]

[40]   M. Engeland, L.J.W. Nieland, F.C.S. Ramaekers, B. Schutte, and C.P.M. Reutelingsperger, "Annexin V-affinity assay: A review on an apoptosis detection system based on phosphatidylserine exposure", *Cytometry,* vol. 31, pp. 1-9, 1998.
[http://dx.doi.org/10.1002/(SICI)1097-0320(19980101)31:1<1::AID-CYTO1>3.0.CO;2-R] [PMID: 9450519]

[41]   G. Koopman, C.P.M. Reutelingsperger, G.A.M. Kuijten, R.M.J. Keehnen, S.T. Pals, and M.H.J. van Oers, "Annexin V for flow cytometric detection of phosphatidylserine expression on B cells undergoing apoptosis", *Blood.,* vol. 84, no. 5, pp. 1415-1420, 1994.
[http://dx.doi.org/10.1002/(SICI)1097-0320(19980101)31:1<1::AID-CYTO1>3.0.CO;2-R]   [PMID: 9450519]

[42]   P. Kumar, A. Nagarajan, and P.D. Uchil, "Analysis of cell viability by the lactate dehydrogenase assay", *Cold. Spring. Harb. Protoc.,* no. 6, 2018.
[http://dx.doi.org/10.1101/pdb.prot095497]

[43]   D.T. Loo, ''*In Situ* Detection of Apoptosis by the TUNEL Assay: An Overview of Techniques. DNA Damage Detection *In Situ, Ex Vivo*, and *In Vivo*'', (2010). 3–13. doi:10.1007/978-1-60327-409-8_1.
[http://dx.doi.org/10.1007/978-1-60327-409-8_1]

[44]   J.D. Ly, D.R. Grubb, and A. Lawen, "The mitochondrial membrane potential (deltapsi(m)) in apoptosis; an update", *Apoptosis,* vol. 8, no. 2, pp. 115-128, 2003.
[http://dx.doi.org/10.1023/A:1022945107762] [PMID: 12766472]

[45]   S. Desagher, and J.C. Martinou, "Mitochondria as the central control point of apoptosis", *Trends. Cell. Biol.,* vol. 10, no. 9, pp. 369-377, 2000.
[http://dx.doi.org/10.1016/S0962-8924(00)01803-1] [PMID: 10932094]

[46]   F. Wallberg, T. Tenev, and P. Meier, "Analysis of apoptosis and necroptosis by fluorescence-activated cell sorting", *Cold. Spring. Harb. Protoc.,* no. 4, p. prot087387, 2016.
[http://dx.doi.org/10.1101/pdb.prot087387]

[47]   P. Møller, "The comet assay: Ready for 30 more years", *Mutagenesis,* vol. 33, no. 1, pp. 1-7, 2018.
[http://dx.doi.org/10.1093/mutage/gex046] [PMID: 29325088]

[48]   A.R. Collins, "The comet assay for DNA damage and repair: Principles, applications, and limitations", *Mol. Biotechnol.,* vol. 26, no. 3, pp. 249-261, 2004.
[http://dx.doi.org/10.1385/MB:26:3:249] [PMID: 15004294]

[49]   A.M. Crane, and S.K. Bhattacharya, "The use of bromodeoxyuridine incorporation assays to assess corneal stem cell proliferation", *Methods. Mol. Biol.,* vol. 1014, pp. 65-70, 2013.
[http://dx.doi.org/10.1007/978-1-62703-432-6_4]

[50]   S. Comoglio, and F. Celada, "An immuno-enzymatic assay of cortisol using E. coli β-galactosidase as label", *J. Immunol. Methods,* vol. 10, no. 2-3, pp. 161-170, 1976.
[http://dx.doi.org/10.1016/0022-1759(76)90167-8] [PMID: 778271]

[51]   P.K. Nakane, and A. Kawaoi, "Peroxidase-labeled antibody. A new method of conjugation", *J. Histochem. Cytochem.,* vol. 22, no. 12, pp. 1084-1091, 1974.
[http://dx.doi.org/10.1177/22.12.1084] [PMID: 4443551]

[52]   J. C.-T. E. Guidebook and undefined 2009, *Chapter 3 Stages in ELISA*, vol. 516. 1995.

[53]   P. Hornbeck, "Enzyme-linked immunosorbent assays", *Curr. Protoc. Immunol.,* 2001.
[http://dx.doi.org/10.1002/0471142735.im0201s01]

[54]   A. Fan, Z. Cao, H. Li, M. Kai, and J. Lu, "Chemiluminescence platforms in immunoassay and DNA analyses", *Anal. Sci.,* vol. 25, no. 5, pp. 587-597, 2009.
[http://dx.doi.org/10.2116/analsci.25.587] [PMID: 19430138]

[55]   S.D. Gan, and K.R. Patel, "Enzyme immunoassay and enzyme-linked immunosorbent assay", *J. Invest. Dermatol.,* vol. 133, no. 9, pp. 1-3, 2013.
[http://dx.doi.org/10.1038/jid.2013.287] [PMID: 23949770]

[56]   T.O. Kohl, and C.A. Ascoli, "Direct competitive enzyme-linked immunosorbent assay (ELISA)", *Cold Spring Harb. Protoc.,* vol. 2017, no. 7, p. 093740, 2017.
[http://dx.doi.org/10.1101/pdb.prot093740] [PMID: 28679705]

[57]   T.O. Kohl, and C.A. Ascoli, "Indirect competitive enzyme-linked immunosorbent assay (ELISA)", *Cold Spring Harb. Protoc.,* vol. 2017, no. 7, p. 093757, 2017.
[http://dx.doi.org/10.1101/pdb.prot093757] [PMID: 28679706]

[58]   S. Sakamoto, W. Putalun, S. Vimolmangkang, W. Phoolcharoen, Y. Shoyama, H. Tanaka, and S. Morimoto, "Enzyme-linked immunosorbent assay for the quantitative/qualitative analysis of plant secondary metabolites", *J. Nat. Med.,* vol. 72, no. 1, pp. 32-42, 2018.
[http://dx.doi.org/10.1007/s11418-017-1144-z] [PMID: 29164507]

[59]   E.V. Lourenço, and M.C. Roque-Barreira, "Immunoenzymatic quantitative analysis of antigens expressed on the cell surface (cell-ELISA)", *Methods Mol. Biol.,* vol. 588, pp. 301-309, 2010.
[http://dx.doi.org/10.1007/978-1-59745-324-0_29] [PMID: 20012840]

[60]   B. Houser, "Bio-Rad's Bio-Plex® suspension array system, xMAP technology overview", *Arch. Physiol. Biochem.,* vol. 118, no. 4, pp. 192-196, 2012.
[http://dx.doi.org/10.3109/13813455.2012.705301] [PMID: 22852821]

[61]   B. Kurien, and R. Scofield, "Western blotting", *Methods,* vol. 38, no. 4, pp. 283-293, 2006.
[http://dx.doi.org/10.1016/j.ymeth.2005.11.007]

[62]   H. Towbin, T. Staehelin, and J. Gordon, "Electrophoretic transfer of proteins from polyacrylamide gels to nitrocellulose sheets: Procedure and some applications", *Proc. Natl. Acad. Sci.,* vol. 76, no. 9, pp. 4350-4354, 1979.
[http://dx.doi.org/10.1073/pnas.76.9.4350]

[63]   Karina Martínez-Flores, Ángel Tonatiuh Salazar-Anzures, Javier Fernández-Torres, Carlos Pineda, Alberto Aguilar-González Carlos, and Alberto López-Reyes, "Western blot: A tool in the biomedical field", *mediagraphic.,* vol. 6, no. 3, pp. 128-137, 2017.

[64]   H.P. Aslim, and O. Bulut, "Western blot", *J. Vet. Sci.Tech.,* vol. 6, no. 1, pp. 45-56, 2021.
[http://dx.doi.org/10.31797/vetbio.799660]

[65]   H. Schägger, "Tricine–sds-page", *Nat. Protoc.,* vol. 1, no. 1, pp. 16-22, 2006.
[http://dx.doi.org/10.1038/nprot.2006.4] [PMID: 17406207]

[66]   P-C. Yang, and T. Mahmood, "Western blot: Technique, theory, and trouble shooting", *N. Am. J. Med. Sci.,* vol. 4, no. 9, pp. 429-434, 2012.
[http://dx.doi.org/10.4103/1947-2714.100998] [PMID: 23050259]

[67]   M.W. Bolt, and P.A. Mahoney, "High-efficiency blotting of proteins of diverse sizes following sodium dodecyl sulfate-polyacrylamide gel electrophoresis", *Anal. Biochem.,* vol. 247, no. 2, pp. 185-192, 1997.
[http://dx.doi.org/10.1006/abio.1997.2061] [PMID: 9177676]

[68]   B.T. Kurien, and R.H. Scofield, Introduction to protein blotting.*Protein Blotting and Detection: Methods and Protocols.* Humana Press, 2009, pp. 9-22.
[http://dx.doi.org/10.1007/978-1-59745-542-8_3]

[69]   B.T. Kurien, and R.H. Scofield, "Protein blotting: A review", *J. Immunol. Methods,* vol. 274, no. 1-2, pp. 1-15, 2003.
[http://dx.doi.org/10.1016/S0022-1759(02)00523-9] [PMID: 12609528]

[70]   C.T. Rueden, J. Schindelin, M.C. Hiner, B.E. DeZonia, A.E. Walter, E.T. Arena, and K.W. Eliceiri, "ImageJ2: ImageJ for the next generation of scientific image data", *BMC Bioinformatics,* vol. 18, no. 1, p. 529, 2017.
[http://dx.doi.org/10.1186/s12859-017-1934-z] [PMID: 29187165]

CHAPTER 14

An Overview of *In Vivo* Imaging Techniques

Aysa Ostovaneh[1] and **Yeliz Yildirim**[2,3,*]

[1] *Department of Biotechnology, Ege University, Graduate School of Natural and Applied Sciences, 35100, İzmir, Türkiye*

[2] *Department of Chemistry, Faculty of Science, Ege University, İzmir, Türkiye*

[3] *Center of Drug, R&D, and Pharmacokinetic Aplications (ARGEFAR), Ege University, İzmir, Türkiye*

Abstract: Imaging is developing very quickly in various study bases. Nowadays, due to the desire for the technology coming to imaging, it is widely used to detect molecular and structural targets in *in vivo* studies. The aim of developing new non-invasive imaging methods is to provide affordable, high-resolution images with minimal known side effects for studying the biological processes of living organisms. For this purpose, X-ray-based computed tomography (CT), magnetic resonance imaging (MRI), ultrasound (UI), Nuclear imaging methods (positron emission tomography (PET), single-photon emission computed tomography (SPECT)), and optical imaging, are techniques widely used in imaging. Each of these has unique advantages and drawbacks. The background of imaging techniques and their developments have been shown in this chapter and we discuss in detail the use of optical imaging through bioluminescence, fluorescence, and Cerenkov luminescence techniques in various diseases for preclinical applications, early clinical diagnosis, treatment, and clinical studies.

Keywords: Bioluminescence, Biological processes, Computed tomography, Clinical diagnosis, Cerenkov luminescence, Fluorescence, High-resolution images, *in vivo* imaging, Imaging technique, Living organisms, Molecular imaging, Magnetic resonance imaging, Nuclear method, Non-invasive method, Optical imaging, Positron emission tomography, Single-photon emission computed tomography, Technology, Ultrasound and X-ray.

INTRODUCTION

In vivo imaging is an indispensable imaging technique that could be used in research and clinical trials. The visual display capacity and possibilities have attracted great attention and have been developed over the years. This allows researchers and clinicians to delve deep into living systems to display their anatomical features, search for specific biomolecules, and illuminate the varieties

[*] **Corresponding author Yeliz Yildirim:** Department of Chemistry, Faculty of Science, Ege University, İzmir, Türkiye and Center of Drug, R&D, and Pharmacokinetic Aplications (ARGEFAR), Ege University, İzmir, Türkiye;
E-mail: yeliz.yildirim@ege.edu.tr

Habibe Yılmaz (Ed.)

of a complex array of components and ingredients that underlie a wide variety of repositories [1]. Imaging uses signals with different mechanisms and sources, making it possible to visualize critical cellular and molecular processes [2]. Various techniques are developed for *in vivo* imaging: from computerized tomography (CT) [3] to photon-based *in vivo* fluorescence imaging (FLI), *in vivo* bioluminescence imaging (BLI) [4] and Cerenkov luminescence imaging (CLI), magnetic resonance imaging (MRI) [5] and sound wave-based ultrasound imaging (UI) [6], as well as radionuclide-based single photon emission computed tomography (SPECT) [7] and positron emission tomography (PET) [8]. In this chapter, we will first discuss the background, advantages, and drawbacks of all the above imaging techniques. Then we will delve into the detail of the optical imaging technique details and the utilization of these techniques in the diagnosis and treatment of diseases in preclinical *in vivo* studies.

Computerized Tomography (CT) imaging, also called "CAT" (Computed Axial Tomography). The root of tomography is generated from the Greek word "tomos" meaning "slice" and "graphia" which means "describing". British engineer Godfrey Hounsfield and physicist Allan Cormack concocted CT in 1971. It was publicly announced in 1972 [9, 10]. In 1979, Nobel Peace Prize was awarded to Hounsfield and Cormack in two different fields of medicine and science [11]. A CT scan is based on an X-ray. This imaging technique takes a series of X-ray slices from different angles of various parts of the body (bones and blood vessels, *etc.*) and is collected all data from the detector and is transferred to the computer for analysis. Whenever the machine starts turning around, computerized information is obtained comparing conventional X-rays with CT scans. CT images provide more detailed information about a particular area in a cross-section, eliminating image overlap [12].

Magnetic Resonance Imaging (MRI) is a type of non-invasive/non-ionizing scan that utilizes powerful magnetic fields and radio waves that force the proton in the body to arrange with that field to generate images of any part of the body. In this imaging technique, the hydrogen nucleus (a single proton) is used due to its abundance in body fluids and fat [13]. Bloch and Purcell described nuclear magnetic resonance (NMR) in 1946. They won the Nobel Prize for Physics in 1952. Magnetic resonance images were introduced clinically in Nottingham and Aberdeen in 1980 for the first time [14, 15]. MR imaging can assist in the diagnosis of disorders and treatments as well. It is increasingly being used and demanded as several new indications have been created in the last few years [16]. By bypassing the radiofrequency flow through the body, the protons are stimulated, and it causes unsteady protons, which str*etc*h against the pull of the magnetic field. MRI sensors detect the energy released from realigned protons when the magnetic field is switched off. The chemical structure of the molecules can affect the amount of energy that is released from protons. Contrast agents (*e.g.*: element Gadolinium, Manganese, Europium) may be given to a patient intravenously, orally, or intra-articularly before or during the MRI to increase the speed at which protons realign with the magnetic field for more specific types of imaging. Although all metallic properties, whether inside or outside the body, are better not to be used in MRI because of being in a magnetic field, MRI has no known biological hazards. MRI uses radiation which does not cause any significant damage to the tissue as it passes through.

Ultrasound (UI) is a non-invasive technique, used widely in medicine as both a diagnostic and therapeutic tool to image the inside of the body. UI was discovered approximately 10 years before the X-ray (1883), but it found application long after it started to be used in medicine. The first use of this technique refers to the investigation of submarines in the First World War. After that, it was used in the medical field for the first time in 1950. The sound waves are generated by transducers (Ultrasound probes) which work at frequencies much higher than the human hearing range (in the megahertz (MHz) range) [17]. Its mechanism is based on the detection and conversion of ultrasound waves, which are reflected from different tissues of different natures [18]. Because of using non-ionizing radiation, UI is a safe technique even for pregnant. Ultrasound is not only among the most accurate imaging technologies, but they are also the most cost-effective and accessible imaging technique. Its is said that one of the important drawbacks of UI is being dependent on the technician's experts.

Types of Nuclear Imaging Techniques: SPECT and PET determine by accessing the basis of nuclear technology for today's usages in the 1990s. The definition of uranium radiation was described by Henri Becquerel in 1896 and Marie and Pierre Curie in 1898 won the Nobel Prize in Physics for the introduction of "radioactivity" terms. Nowadays, there are various important techniques in nuclear medicine imaging, such as Positron Emission Tomography (PET) imaging or Single-Photon Emission Computed Tomography (SPECT). The first operation of SPECT and PET started simultaneously in the middle of the 20th century. PET imaging was described and conceptualized first by Brownell and Sweet in 1950 [19].

Single-Photon Emission Computed Tomography (SPECT) is defined as a nuclear medicine, a tomographic imaging technique which uses gamma rays for detection. In SPECT, single photon emissions from an introduced radioactive tracer (also known as a probe) are detected to produce a computer-generated image of the local radioactivity distribution in tissues [20]. These probes generally are equipped with a detectable radioactive isotope gamma camera, providing images depending on the type of scan being carried out. A Gamma Camera rotates around the body. Detection of the radioactive particles in the body by a gamma camera collects all data and displays it on the screen as cross-sectional slices from different parts of the body [21]. The important ability of SPECT is to give physiological information, especially tissues. Technetium-99m (^{99m}Tc), Iodine-123 (^{123}I), isotopes are most commonly used, while some other isotopes like Thallium-201 (^{201}Tl) are used to a lesser extent. Because of the expenses of the isotopes, patient health, and parts of the body that need to be scanned, choosing isotopes is crucial [22].

A Positron Emission Tomography (PET) scan is an imaging type that utilizes radiotracers for measuring various metabolic processes in the body. It is commonly used in diagnosing brain or heart disorders as well as cancer. PET scanners work by detecting the radiation given off by a radioactive agent (radiotracer) injected, swallowed, or inhaled, depending upon the site of the different parts of the body being examined [23]. The basis of the PET technique is that of connecting radioactive atoms with the positive beta decay and collecting in a special tissue during the metabolic process of the body. For instance, in PET scans of the brain, depending on the consumption of glucose, a radioactive form of glucose which is known as fluorodeoxyglucose (^{18}FDG), is selected [24]. Other radioactive forms of isotopes like Oxygen-15 (^{15}O), Carbon-11 (^{11}C), Nitrogen-13 (^{13}N), or Gallium-68 (^{68}Ga) are used for different physiological investigations (body fluids, perfusion, *etc.*) [25]. Positrons emitted from the decay of the radionuclide are collected by gamma-sensitive cameras that are spaced 180 degrees apart. After that, the computer processed the images by measuring the volume and concentration of the probe *via* gamma rays to display images on the screen [26]. Using limited quantities of imaging agents for detecting biological changes with high sensitivity is the advantage of PET and SPECT. The information of images taken by SPECT is based on the amount of gamma-emitting radionucleotides which is sensitive but not more than PET [27].

Hybrid systems have been used and preferred recently to combine the two systems for gathering high-resolution images. SPECT and PET imaging provides critical insight into various diseases, but they have limitations. The greatest limitation is the lack of direct anatomic correlation. The utilization of hybrid imaging methods such as SPECT/CT, PET/CT, or PET/MRI provides the benefit of molecular sensitivity and spatial resolution fused with anatomical specificity [28].

OPTICAL IMAGING

Optical imaging is a non-invasive imaging method that relief on the light of photons and that plays an important and powerful role in monitoring cellular and molecular components (organs, tissues, and cells). It is mostly used in immune system reaction gene studies [29, 30]. It is a fast, reasonable, reliable, non-invasive, powerful, and safe technique because of using non-ionizing radiation to evaluate the progression of a disease or the results of treatment. Due to different soft tissue absorption and scattering light, optical imaging can measure the abnormality in the tissue with metabolic changes that act as markers [31].

Bioluminescence, Fluorescence, and Cerenkov luminescence imaging provide unique characteristics for small animal imaging. Bioluminescence imaging (BLI) uses luciferase enzyme to produce images from animal tissues with high resolution for evaluating biological disorders such as abnormality and cell behavior of biological targets and the progression of abnormalities of tumor cells *via* transfection or transduction into the animals. Transfection involves the delivery of a gene into the cell by chemical or physical tools, while transduction injects this gene-driven luminescent probe through the viral agent [32]. Cerenkov luminescence imaging (CLI) is a novel, reasonable optical imaging tool to study charged particles (a β-particle produced by radioactive decay) of adequate energy through the Cerenkov light. It is used to detect unexpected disorders which are not detectable by other methods [33, 34].

In vivo Screening *via* Fluorescence (FLI)

George Gabriel Stokes introduced fluorescence and fluorescent dyes and the invention of the fluorescent microscope in a scientific publication in the middle of the 19th century [35]. Fluorescence is the emission of light due to the absorption of an excitation photon beam (with high energy, shorter wavelength) that returns from the excited level to the basic level [36]. It is used for detecting biological targets, mechanisms, and functional processes in animal models. Fluorescent imaging needs a molecular probe to enter a high-excitation state and a pre-excitation light source that is exogenous to the animal. The probe is surrounded by electrons, which will release longer wavelength photons with lower energy (probe's light) after their excitation states relax. Detecting this emitted light from the probe is carried out by a charge-coupled device (CCD) camera and gives detailed information about any part of the animal body that needs to be detected. In fluorescence imaging, endogenous molecules (such as collagen or hemoglobin), fluorescent proteins (green fluorescent protein [GFP] and related molecules), or fluorescent molecules can be used as optical contrast. All of these optical contrasts are utilized to acquire cellular and molecular images in small animals, but the sources of optical contrasts have the potential to extend to preclinical patient care *in vivo* imaging [37].

In vivo Screening *via* Bioluminescence (BLI)

Bioluminescence is a light-producing phenomenon that emerges because of enzymatic reactions that occur spontaneously in the structure of many living creatures that occur naturally in nature. There are numerous species of creatures in nature that are called luminous creatures, such as bacteria, fungi, and marine and terrestrial creatures like fireflies [38]. In 1555, Conrad Gesner, for the first time, published bioluminescence and in 1956, the two main ingredients of bioluminescence (BLI), luciferin and luciferase, and their reaction were recognized [39]. It cannot be denied that BLI is a sensitive technique that gives detailed information about the biological *in vivo* pathways of translational values in small animals, but one of the defects of BLI imaging is the inability to use this method in humans [40]. Bioluminescence is produced through a chemical reaction taking its source from the live organism, which provides illumination. In the BLI technique, the most common type of luciferase used is Firefly (PhotinusPyralis) luciferase. Luciferase enzyme causes oxidation of the substrate, luciferin, in the presence of cofactors, for instance, O_2 and adenosine triphosphate (ATP), $Mg^{2+,}$ and Ca^{2+}. As a result of this reaction, the light emission raises to the highest level at 600nm (red and infrared spectrum), approximately $\approx$10–12 min after the reaction is finished and it starts to diminish gradually over 60 minutes [41]. One of the distinct features of luciferin is the ability to pass through barriers including the brain, blood, and placenta, and distributed to the tissue after injection [42].

In vivo Screening *via* Cerenkov Luminescence (CLI)

The usage of Cerenkov radiation has been increasing over the last 15 years, making it one of the fascinating optical imaging techniques. Marie Curie and Pierre Curie, for the first time, observed a blue glow from radioactive substances(radium). This phenomenon takes its name from its discoverer, Pavel Cherenkov, and the radiation caused by light transmission. Pavel Cherenkov, Frank, and Tamm won the Nobel Prize in Physics for their finding and explanation of the Cherenkov effect in 1958. CL imaging (CLI) was first described in 2009. The Cerenkov luminescence (CL) is generated by either positively (β^+) or negatively (β^-) charged β particles transferring in the dielectric medium, which is much higher than the phase velocity of light in that medium [43]. In CL imaging (CLI), when the decay processes of numerous isotopes like ^{18}F, ^{64}Cu, ^{131}I, ^{68}Ga, and ^{225}Ac are started, Cerenkov photons are detected with a charge-coupled device (CCD) detected Cerenkov photons and provided images. In fact, CLI connects optical imaging to radionuclide imaging. It can be used in radiation distribution and radiotracers. Even more, CLI is able to utilize in several applications, such as medicinal evaluation, endoscopy, surgery, and dosimetry. Having lower imaging time, being economical, and giving higher resolution results make this technique have priority of usage in comparison with PET and SPECT methods [44, 45]. Despite these advantages, CLI has some limitations, such as the low penetration depth of some Cerenkov photons in tissue, which hampers the utility of the technique. Although CLI needs further development and optimization of its value, it will be used as a modern technique for either preclinical or clinical imaging [46].

General Comparison Between Bioluminescence (BLI), Fluorescence (FLI) Imaging, and Cerenkov (CLI) Imaging

Table **1** summarizes the general comparison of the different methods used in optical imaging. All of them provide two-dimensional (2D) and three-dimensional (3D) images to be obtained. Internal chemical energy is utilized in the bioluminescence (BLI) technique for signal generation, and external light energy and charged particles are utilized for fluorescence (FLI) imaging and Cerenkov (CLI) imaging, respectively. While the mechanism of BLI is based on enzyme-substrate oxidative reaction, FLI is based on the retention or activation of fluorescent agents. The principle of CLI is based on the charged particles generated by radionuclides in the dielectric medium moving with a velocity higher than the phase velocity of light in the same medium. Each imaging technique has different biological applications.

Table 1. General comparison between BLI, FLI and CLI.

–	BLI	FLI	CLI
Energy source	Internal chemical energy	External light energy	Charged particles
Mechanism	Luciferin-luciferase oxidative reaction	Selective retention/activation of fluorescent agents	The charged particles generated by radionuclides in the dielectric medium move with a velocity higher than the phase velocity of light in the same medium
Explanation	Highly specific and sensitive, but more limited	Highly versatile, but more complex	Fast imaging, strong sensitivity
Biological application	1. Providing information of gene expression 2. Providing information of protein–protein interaction 3. GProviding information of cell-mediated immunity and infectious disease, neuroscience, and cancer 4. Providing information of stem cells [47]	1. Information of perfusion and vasculature 2. Fluorescent dyes can act as biomarkers through bonding to a variety of targets 3. Fluorescently suppressed agents by activating specific enzymatic cleavage	1. Functional imaging of β-emitting radioisotopes surface dosimetry in external beam radiotherapy 2. Proton therapy for quality assurance and *in vivo* dosimetry.

The Use Of The Optical Imaging Technique In Biological Studies

In this chapter, we provide examples of *in vivo* studies that used optical imaging techniques for the investigation of diagnosis, progression of disease and treatment as well as pharmacokinetic and biodistribution studies in small animals. For example, Komesli and coworkers designed a self-microemulsifying drug delivery system of the hydrophobic antihypertensive drug Olmesartan Medoxomil (OM-SMEDDS) [48] and labeled with the fluorescent dyes VivoTag® 680 XL and Xenolight® DiR and control dye administrated to mice, for evaluating of the oral biodistribution. *In vivo* imaging showed Vivotag 680® XL and Xenolight® DiR labeled OM-SMEDDS emitted 2 to 24 times stronger emissions than the control dye administered group. For more detailed analyzing, organs were removed for *ex vivo* imaging. Results demonstrated VivoTag® 680XL and Xenolight® DiR were emitted 4 and 1.7 times higher by organs than the control dye group, respectively [49]. In the other study, researchers utilized fluorescence dye to show the biodistribution of the designed nanostructure on non-small cell lung cancer (NSCLC). Lung cancer is one of the fatal cancers in men and women worldwide, and NSCLC is one type of lung cancer with high mortality. This study was conducted by Hamarat and colleagues, two drugs were investigated for the development of nanobubble systems: Pemetrexed is administrated for the treatment of some types of NSCLC and pazopanib generally is administrated for the treatment of soft tissue sarcoma. They designed Pemetrexed and pazopanib loaded nanobubble system with magnetic and ultrasound sensitivity properties (NBs-400) for NSCLC therapy. For the investigation of nanobubble biodistribution, they designed *in vivo* studies and used *in vivo* imaging system. After NSCLC-induced mice reached to the desired tumor size, mice were injected with FITC (Fluorescein isothiocyanate)-loaded NBs-400. As a result of this study, they reported the magnet properties provided, the accumulation of carrier the system, and ultrasound producing explosion of nanobubbles, the biodistribution result of NBs-400 was performed by IVIS using FITC fluorescence dye. Fig. (**1**) demonstrates the effect of magnet on 80.22% of the nanobubble accumulation in the tumor area [50].

Birgül *et al*. evaluated the anticancer activity of (S)-naproxen derivatives against prostate cancer (PCa). PCa is the fatal cancer type in men worldwide, after lung cancer. In this study, several (S)-Naproxen derivatives were synthesized and according to the results of the analyzes, IRDye800-labeled SGK-636 (**5n**) compound was selected and administrated to cancer-induced nude mice for *in vivo* study. Prostate cancer was induced by using LNCaP cells (Fig. **2**). Tumor size and its progression were evaluated by using the IVIS instrument and the Xenolight RediJect 2-DG750 glucose Probe. Due to the abundance of glucose in cancerous tissues than normal tissues, it is used in imaging cancerous tissues. After 15 days, the ROI (region of interest) cancerous area decreased by 55-70 percent when (**5n**) compound was applied. In the non-treated group, tumor activity increased five-fold [51].

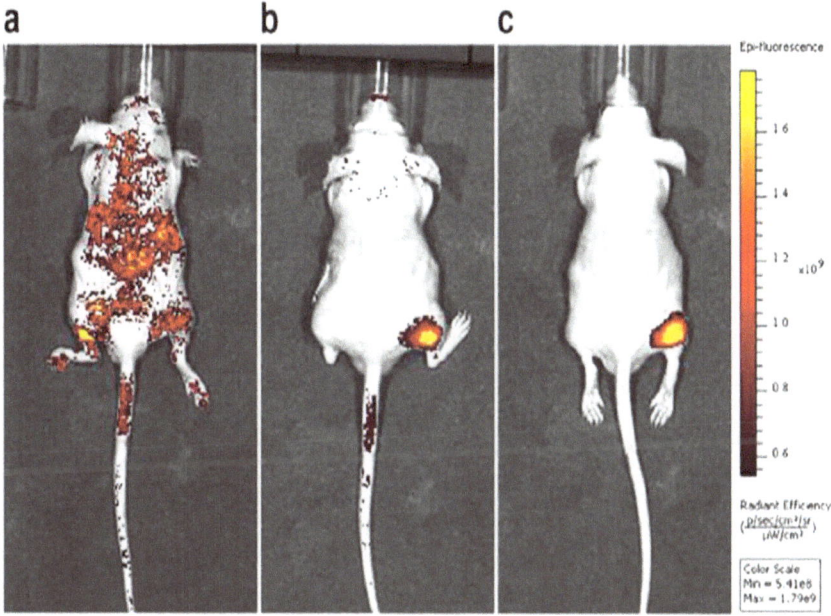

Fig. (1). IVIS images of FITC-loaded-Nbs-400 administrated mice: picture (**a**) shows 30 minutes after injection without magnetic properties; picture (**b**) shows 30 minutes after administration with magnetic properties, and picture (**c**) shows 30 minutes after administration with magnetic properties and ultrasound (Hamarat *et al.*).

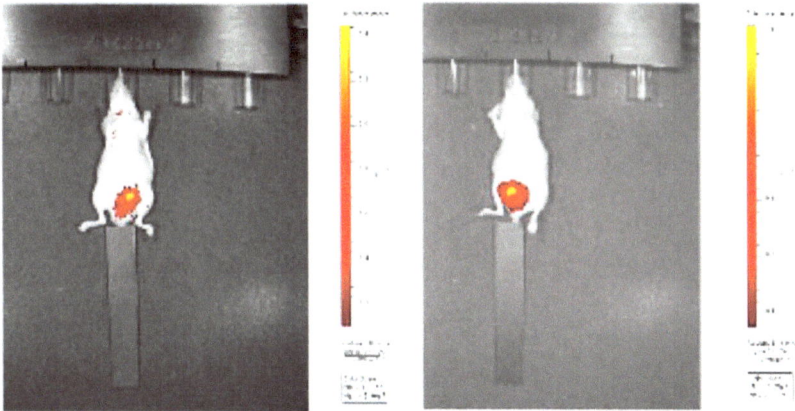

Fig. (2). IVIS spectrum images of prostate cancer induced in nude mice by using LNCaP cells (Birgül *et. al.*).

Alzheimer's disease (AD) is one of the most common age-dependent neurodegenerative diseases, and its increasing due to the aging population [52]. AD is considered as one of the top five causes of death for people over 65 years old. [53]. *In vivo* detection of amyloid plaques is vital to diagnose Alzheimer's disease [54]. Formulation and clinical usages of AD drugs encounter with many challenges due to poor bioavailability of properties, such as an imbalance between fat and water solubility [55]. The neuroinflammation induced by amyloid ß (Aß) in Alzheimer's disease has an important role in the onset and progression of AD.

Zhao *et. al.*, developed a LRGT-loaded nanostructure. They decorated the nanostructure with angiopep-2 peptide and assembled with polyethylene glycol (PEG) (pALRGT) as a glucagon-like peptide 1 receptor (GLP-1R) agonist that was capable of crossing the blood-brain barrier (BBB) and suppressing inflammation associated with AD-related factors. For *in vivo* imaging study, they used Cy5.5 labelled LRGT and pALRGT in nude mice with AD. They demonstrated the effect of pALRGT nanostructures on the reduction of inflammatory microglia and astrocytes in the hippocampal region of AD-induced mice. Moreover, it was observed total repair of cognitive impairment [56]. Ince and coworkers developed a tool with enzymatic Thymoquinone (TQ) release from glucuronide (G), Conjugated Magnetic Nanoparticles (TQGMNP) [57, 58] bimodal for the imaging and treatment of lung cancer. For bioluminescence imaging (BLI), they implanted Luc-A549 cells for tumor induction in nude mice. After the treatment of mice with TQGMNP, it was observed the reduction of xenograft tumor size after 10 and 15 days, respectively. Non-treated animals showed no considerable reduction in tumor volume, as expected [59]. Anaplastic thyroid cancer (ATC) is one of the deadliest cancer types because of occurring during lung metastasis. Natural killer (NK) cells have an important and warrior role against tumor cells in the immune system. Zhu and coworkers evaluated NK effects on cell-based immunotherapy on pulmonary metastasis of ATC. They monitored tumor growth through *in vivo* BL imaging. Cancer was induced through the injection of CAL-62/F into nude mice. For investigation of NK-92MI cells, they were injected into cancerous-nude mice and BLI signals were detected in the lung region after injection. Subsequently, results showed the effect of suppression on the growth of the metastasis in mice that received the injection of and NK cells. The BLI signal intensity in the control group was 2.8-fold higher than that in the NK treatment group [60]. Kamkaew and coworkers used Cerenkov radiation (CR) in photodynamic therapy (PDT) because of the limited tissue penetration of light in PDT. In this study, zirconium-89 (^{89}Zr) was utilized as a CR source to stimulate chlorin e6 (Ce6) to produce reactive oxygen species (ROS). They designed hollow mesoporous silica nanoparticles (HMSNs) as a carrier to encapsulate Ce6 molecules and ^{89}Zr isotope, simultaneously. They demonstrated the [^{89}Zr] HMSN-Ce6 nanostructure that could convey its energy to

Ce6 to produce sufficient ROS to induce the PDT effect in tumor cells and tissues in both *in vitro* and *in vivo* studies. For *in vivo* study, they injected subcutaneously [^{89}Zr] HMSN-Ce6 without using any external light source in a murine breast tumor model. They achieved excellent therapeutic efficacy [61].

CONCLUDING REMARKS

In summary, *in vivo* imaging is developing very fast for providing details about molecular and cellular targets that need to be detected. New multimodal technologies are moving toward the application of imaging to determine physiological disorders, pathways and molecular targets in *in vivo studies*. There are a number of modalities of imaging developing for *in vivo* studies. Due to the advantages and drawbacks of each imaging technique, today using hybrid imaging methods such as SPECT/CT, PET/CT, and PET/MRI, can reduce these defects; optimize the quality and resolution of imaging. Each technique has advantages and limitation for example, optical imaging has potential to give images with high resolution and sensitivity, while, poor tissue penetration is one of the challenges of this technique. It cannot be denied this technique has been recently validated in animal models, but it has the potential to extend to preclinical, disease pathogenesis, diagnosis, patient treatment processes and drug development.

REFERENCES

[1]   J.E. Burdette, *In vivo* imaging of molecular targets and their function in endocrinology.*J. Mol. Endocrinol.,* vol. 40, no. 6, pp. 253-261, 2008.
       [http://dx.doi.org/10.1677/JME-07-0170] [PMID: 18502818]

[2]   F. Bray, J. Ferlay, I. Soerjomataram, R.L. Siegel, L.A. Torre, and A. Jemal, "Global cancer statistics 2018: GLOBOCAN estimates of incidence and mortality worldwide for 36 cancers in 185 countries", *CA Cancer J. Clin.,* vol. 68, no. 6, pp. 394-424, 2018.
       [http://dx.doi.org/10.3322/caac.21492] [PMID: 30207593]

[3]   J. Hsieh, and T. Flohr, "Computed tomography recent history and future perspectives", *J. Med. Imaging,* vol. 8, no. 5, p. 052109, 2021.
       [http://dx.doi.org/10.1117/1.JMI.8.5.052109] [PMID: 34395720]

[4]   JE Lloyd, and EC Gentry, *Bioluminescence. inencyclopedia of insects.,* Academic Press., pp. 101-105, 2009.

[5]   V.P.B. Grover, J.M. Tognarelli, M.M.E. Crossey, I.J. Cox, S.D. Taylor-Robinson, and M.J.W. McPhail, "Magnetic resonance imaging: Principles and techniques: Lessons for clinicians", *J. Clin. Exp. Hepatol.,* vol. 5, no. 3, pp. 246-255, 2015.
       [http://dx.doi.org/10.1016/j.jceh.2015.08.001] [PMID: 26628842]

[6]   A. Carovac, F. Smajlovic, and D. Junuzovic, "Application of ultrasound in medicine", *Acta. Inform. Med.,* vol. 19, no. 3, pp. 168-171, 2011.
       [http://dx.doi.org/10.5455/aim.2011.19.168-171] [PMID: 23408755]

[7]   KM Davis, JL Ryan, VD Aaron, and JB Sims, "PET and SPECT imaging of the brain: History, technical considerations, applications, and radiotracers", In: *InSeminars in Ultrasound, CT and MRI* vol. 41. WB Saunders., 2020, no. 6, pp. 521-529.

[http://dx.doi.org/10.1053/j.sult.2020.08.006]

[8]   EL Ergün, S Saygi, D Yalnizoglu, KK Oguz, and B Erbas, "SPECT-PET in epilepsy and clinical approach in evaluation", In: *InSeminars in Nuclear Medicine* vol. 46. WB Saunders., 2016, no. 4, pp. 294-307.
[http://dx.doi.org/10.1053/j.semnuclmed.2016.01.003]

[9]   C. Richmond, "Sir godfrey hounsfield", *BMJ,* vol. 329, no. 7467, p. 687, 2004.
[http://dx.doi.org/10.1136/bmj.329.7467.687]

[10]   E.C. Beckmann, "CT scanning the early days", *Br. J. Radiol.,* vol. 79, no. 937, pp. 5-8, 2006.
[http://dx.doi.org/10.1259/bjr/29444122] [PMID: 16421398]

[11]   TD DenOtter, and J Schubert, "Hounsfield Unit", 2019.

[12]   J. Sevick, *Evaluating neuroimaging sensitivities to alterations in structural connectivity following mild traumatic brain injury.* (Doctoral dissertation, University of British Columbia).

[13]   A. Berger, "How does it work?: Magnetic resonance imaging", *BMJ,* vol. 324, no. 7328, p. 35, 2002.
[http://dx.doi.org/10.1136/bmj.324.7328.35] [PMID: 11777806]

[14]   ME Packard, P ackard, and E M artin, "Nuclear induction at stanford and the transition to varian", *eMagRes,* 2007.

[15]   I. Rabi, *The Discovery of NMR..* Available From: https://mriquestions.com/who-discovered-nmr.html

[16]   R.C. Hawkes, G.N. Holland, W.S. Moore, and B.S. Worthington, "Nuclear magnetic resonance (NMR) tomography of the brain: A preliminary clinical assessment with demonstration of pathology", *J. Comput. Assist. Tomogr.,* vol. 4, no. 5, pp. 577-586, 1980.
[http://dx.doi.org/10.1097/00004728-198010000-00001] [PMID: 6967878]

[17]   N. Bhandari, B. Thakur, S. Shrestha, and A. Kharel, "Ultrasound of lung", *Nepal. J. Cancer.,* vol. 6, no. 2, pp. 53-65, 2022.
[http://dx.doi.org/10.3126/njc.v6i2.48770]

[18]   A. Carovac, F. Smajlovic, and D. Junuzovic, "Application of ultrasound in medicine", *Acta. Inform. Med.,* vol. 19, no. 3, pp. 168-171, 2011.
[http://dx.doi.org/10.5455/aim.2011.19.168-171] [PMID: 23408755]

[19]   KM Davis, JL Ryan, VD Aaron, and JB Sims, "PET and SPECT imaging of the brain: History, technical considerations, applications, and radiotracers", *Semin. Ultrasound. CT. MR.,* vol. 41, no. 6, pp. 521-529, 2020.
[http://dx.doi.org/10.1053/j.sult.2020.08.006]

[20]   S Yandrapalli, and Y. Puckett, "SPECT Imaging", In: *StatPearls* StatPearls Publishing: Treasure Island, 2023.

[21]   O. Israel, O. Pellet, L. Biassoni, D. De Palma, E. Estrada-Lobato, G. Gnanasegaran, T. Kuwert, C. la Fougère, G. Mariani, S. Massalha, D. Paez, and F. Giammarile, "Two decades of SPECT/CT : The coming of age of a technology: An updated review of literature evidence", *Eur. J. Nucl. Med. Mol. Imaging,* vol. 46, no. 10, pp. 1990-2012, 2019.
[http://dx.doi.org/10.1007/s00259-019-04404-6] [PMID: 31273437]

[22]   B.F. Hutton, "The origins of SPECT and SPECT/CT", *Eur. J. Nucl. Med. Mol. Imaging,* vol. 41, no. S1, suppl. 1, pp. 3-16, 2014.
[http://dx.doi.org/10.1007/s00259-013-2606-5] [PMID: 24218098]

[23]   M Kapoor, and A. Kasi, *PET Scanning.* StatPearls, 2022.

[24]   K. Anand, and M. Sabbagh, "Amyloid imaging: Poised for integration into medical practice", *Neurotherapeutics,* vol. 14, no. 1, pp. 54-61, 2017.
[http://dx.doi.org/10.1007/s13311-016-0474-y] [PMID: 27571940]

[25]   NR Carlson, and NR Carlson, *Physiology of Behavior*, 2000.

[26]   R. Blasberg, "PET imaging of gene expression", *Eur. J. Cancer,* vol. 38, no. 16, pp. 2137-2146, 2002.
[http://dx.doi.org/10.1016/S0959-8049(02)00390-8] [PMID: 12387839]

[27]   H. Herschman, "PET reporter genes for noninvasive imaging of gene therapy, cell tracking and transgenic analysis", *Crit. Rev. Oncol. Hematol.,* vol. 51, no. 3, pp. 191-204, 2004.
[http://dx.doi.org/10.1016/j.critrevonc.2004.04.006] [PMID: 15331078]

[28]   KM Davis, JL Ryan, VD Aaron, and JB Sims, *PET and SPECT imaging of the brain: History, technical considerations, applications, and radiotracers.,* vol. 41, no. 6, pp. 521-529, 2020.*Semin. Ultrasound. CT. MRI.,* vol. 41, no. 6, pp. 521-529, 2020.
[http://dx.doi.org/10.1053/j.sult.2020.08.006]

[29]   K. Licha, and C. Olbrich, "Optical imaging in drug discovery and diagnostic applications", *Adv. Drug Deliv. Rev.,* vol. 57, no. 8, pp. 1087-1108, 2005.
[http://dx.doi.org/10.1016/j.addr.2005.01.021] [PMID: 15908041]

[30]   G. Pirovano, S. Roberts, S. Kossatz, and T. Reiner, "Optical imaging modalities: Principles and applications in preclinical research and clinical settings", *J. Nucl. Med.,* vol. 61, no. 10, pp. 1419-1427, 2020.
[http://dx.doi.org/10.2967/jnumed.119.238279] [PMID: 32764124]

[31]   K. Hansson Mild, R. Lundström, and J. Wilén, "Non-ionizing radiation in swedish health care—exposure and safety aspects", *Int. J. Environ. Res. Public. Health.,* vol. 16, no. 7, p. 1186, 2019.
[http://dx.doi.org/10.3390/ijerph16071186] [PMID: 30987016]

[32]   G.D. Luker, and K.E. Luker, "Optical imaging: Current applications and future directions", *J. Nucl. Med.,* vol. 49, no. 1, pp. 1-4, 2008.
[http://dx.doi.org/10.2967/jnumed.107.045799] [PMID: 18077528]

[33]   J. Zhong, C. Qin, X. Yang, S. Zhu, X. Zhang, and J. Tian, "Cerenkov luminescence tomography for *in vivo* radiopharmaceutical imaging", *Int. J. Biomed. Imaging,* vol. 2011, pp. 1-6, 2011.
[http://dx.doi.org/10.1155/2011/641618] [PMID: 21747821]

[34]   A.E. Spinelli, and F. Boschi, "Novel biomedical applications of cerenkov radiation and radioluminescence imaging", *Phys. Med.,* vol. 31, no. 2, pp. 120-129, 2015.
[http://dx.doi.org/10.1016/j.ejmp.2014.12.003] [PMID: 25555905]

[35]   Y. Koide, Y. Urano, K. Hanaoka, W. Piao, M. Kusakabe, N. Saito, T. Terai, T. Okabe, and T. Nagano, "Development of NIR fluorescent dyes based on Si-rhodamine for *in vivo* imaging", *J. Am. Chem. Soc.,* vol. 134, no. 11, pp. 5029-5031, 2012.
[http://dx.doi.org/10.1021/ja210375e] [PMID: 22390359]

[36]   A.U. Acuña, F. Amat-Guerri, P. Morcillo, M. Liras, and B. Rodríguez, "Structure and formation of the fluorescent compound of Lignum nephriticum", *Org. Lett.,* vol. 11, no. 14, pp. 3020-3023, 2009.
[http://dx.doi.org/10.1021/ol901022g] [PMID: 19586062]

[37]   V. Ntziachristos, E.A. Schellenberger, J. Ripoll, D. Yessayan, E. Graves, A. Bogdanov Jr, L. Josephson, and R. Weissleder, "Visualization of antitumor treatment by means of fluorescence molecular tomography with an annexin V–Cy5.5 conjugate", *Proc. Natl. Acad. Sci.,* vol. 101, no. 33, pp. 12294-12299, 2004.
[http://dx.doi.org/10.1073/pnas.0401137101] [PMID: 15304657]

[38]   G. Pirovano, S. Roberts, S. Kossatz, and T. Reiner, "Optical imaging modalities: Principles and applications in preclinical research and clinical settings", *J. Nucl. Med.,* vol. 61, no. 10, pp. 1419-1427, 2020.
[http://dx.doi.org/10.2967/jnumed.119.238279] [PMID: 32764124]

[39]   C.E. Badr, *Bioluminescence imaging: Basics and practical limitations.* Bioluminescent Imaging, 2014, pp. 1-8.
[http://dx.doi.org/10.1007/978-1-62703-718-1]

[40]   T. Xu, D. Close, W. Handagama, E. Marr, G. Sayler, and S. Ripp, "The expanding toolbox of *in vivo* bioluminescent imaging", *Front. Oncol.,* vol. 6, p. 150, 2016.
[http://dx.doi.org/10.3389/fonc.2016.00150] [PMID: 27446798]

[41]   S.T. Adams Jr, and S.C. Miller, "Beyond D-luciferin: Expanding the scope of bioluminescence imaging *in vivo*", *Curr. Opin. Chem. Biol.,* vol. 21, pp. 112-120, 2014.
[http://dx.doi.org/10.1016/j.cbpa.2014.07.003] [PMID: 25078002]

[42]   Z. Paroo, R.A. Bollinger, D.A. Braasch, E. Richer, D.R. Corey, P.P. Antich, and R.P. Mason, "Validating bioluminescence imaging as a high-throughput, quantitative modality for assessing tumor burden", *Mol. Imaging,* vol. 3, no. 2, pp. 117-124, 2004.
[http://dx.doi.org/10.1162/1535350041464865] [PMID: 15296676]

[43]   R. Robertson, M.S. Germanos, C. Li, G.S. Mitchell, S.R. Cherry, and M.D. Silva, "Optical imaging of Cerenkov light generation from positron-emitting radiotracers", *Phys. Med. Biol.,* vol. 54, no. 16, pp. N355-N365, 2009.
[http://dx.doi.org/10.1088/0031-9155/54/16/N01] [PMID: 19636082]

[44]   Y. Xu, H. Liu, and Z. Cheng, "Harnessing the power of radionuclides for optical imaging: Cerenkov luminescence imaging", *J. Nucl. Med.,* vol. 52, no. 12, pp. 2009-2018, 2011.
[http://dx.doi.org/10.2967/jnumed.111.092965] [PMID: 22080446]

[45]   G. Lucignani, "Čerenkov radioactive optical imaging: A promising new strategy", *Eur. J. Nucl. Med. Mol. Imaging,* vol. 38, no. 3, pp. 592-595, 2011.
[http://dx.doi.org/10.1007/s00259-010-1708-6] [PMID: 21174087]

[46]   N. Kotagiri, G.P. Sudlow, W.J. Akers, and S. Achilefu, "Breaking the depth dependency of phototherapy with Cerenkov radiation and low-radiance-responsive nanophotosensitizers", *Nat. Nanotechnol.,* vol. 10, no. 4, pp. 370-379, 2015.
[http://dx.doi.org/10.1038/nnano.2015.17] [PMID: 25751304]

[47]   C.E. Badr, and B.A. Tannous, "Bioluminescence imaging: Progress and applications", *Trends. Biotechnol.,* vol. 29, no. 12, pp. 624-633, 2011.
[http://dx.doi.org/10.1016/j.tibtech.2011.06.010] [PMID: 21788092]

[48]   J. Patel, A. Dhingani, J. Tilala, M. Raval, and N. Sheth, "Formulation and development of self-nanoemulsifying granules of olmesartan medoxomil for bioavailability enhancement", *Particul. Sci. Technol.,* vol. 32, no. 3, pp. 274-290, 2014.
[http://dx.doi.org/10.1080/02726351.2013.855686]

[49]   Y. Komesli, Y. Yildirim, and E. Karasulu, "Visualisation of real-time oral biodistribution of fluorescent labeled self-microemulsifying drug delivery system of olmesartan medoxomil using optical imaging method", *Drug. Metab. Pharmacokinet.,* vol. 36, p. 100365, 2021.
[http://dx.doi.org/10.1016/j.dmpk.2020.10.004] [PMID: 33191089]

[50]   Ş. Hamarat Şanlier, G. Ak, H. Yilmaz, A. Ünal, Ü.F. Bozkaya, G. Taniyan, Y. Yildirim, and G. Yildiz Türkyilmaz, "Development of ultrasound-triggered and magnetic-targeted nanobubble system for dual-drug delivery", *J. Pharm. Sci.,* vol. 108, no. 3, pp. 1272-1283, 2019.
[http://dx.doi.org/10.1016/j.xphs.2018.10.030] [PMID: 30773203]

[51]   K. Birgül, Y. Yildirim, H.Y. Karasulu, E. Karasulu, A.I. Uba, K. Yelekçi, H. Bekçi, A. Cumaoğlu, L. Kabasakal, Ö. Yilmaz, and Ş.G. Küçükgüzel, "Synthesis, molecular modeling, *in vivo* study and anticancer activity against prostate cancer of (+) (S)-naproxen derivatives", *Eur. J. Med. Chem.,* vol. 208, p. 112841, 2020.
[http://dx.doi.org/10.1016/j.ejmech.2020.112841] [PMID: 32998089]

[52]   G. Livingston, A. Sommerlad, V. Orgeta, S.G. Costafreda, J. Huntley, D. Ames, C. Ballard, S. Banerjee, A. Burns, J. Cohen-Mansfield, C. Cooper, N. Fox, L.N. Gitlin, R. Howard, H.C. Kales, E.B. Larson, K. Ritchie, K. Rockwood, E.L. Sampson, Q. Samus, L.S. Schneider, G. Selbæk, L. Teri, and N. Mukadam, "Dementia prevention, intervention, and care", *Lancet,* vol. 390, no. 10113, pp. 2673-2734, 2017.

[http://dx.doi.org/10.1016/S0140-6736(17)31363-6] [PMID: 28735855]

[53] Q. Ouyang, Y. Meng, W. Zhou, J. Tong, Z. Cheng, and Q. Zhu, "New advances in brain-targeting nano-drug delivery systems for Alzheimer's disease", *J. Drug Target.,* vol. 30, no. 1, pp. 61-81, 2022.
[http://dx.doi.org/10.1080/1061186X.2021.1927055] [PMID: 33983096]

[54] H. Rai, S. Gupta, S. Kumar, J. Yang, S.K. Singh, C. Ran, and G. Modi, "Near-infrared fluorescent probes as imaging and theranostic modalities for amyloid-beta and tau aggregates in alzheimer's disease", *J. Med. Chem.,* vol. 65, no. 13, pp. 8550-8595, 2022.
[http://dx.doi.org/10.1021/acs.jmedchem.1c01619] [PMID: 35759679]

[55] A. Lopalco, and N. Denora, Nanoformulations for drug delivery: Safety, toxicity, and efficacy.*InCompu. Toxicol.* Humana Press: New York, NY, 2018, pp. 347-365.

[56] Y. Zhao, S. Tian, J. Zhang, X. Cheng, W. Huang, G. Cao, Y.Z. Chang, H. Wang, G. Nie, and W. Qiu, "Regulation of neuroinflammation with GLP-1 receptor targeting nanostructures to alleviate Alzheimer's symptoms in the disease models", *Nano. Today.,* vol. 44, p. 101457, 2022.
[http://dx.doi.org/10.1016/j.nantod.2022.101457]

[57] İ İnce, ZB Müftüler, Eİ Medine, ÖK Güldü, V Tekin, S Aktar, E Göker, and P Ünak, "Synthesis of radioiodinated thymoquinone glucuronide conjugated magnetic nanoparticle (125 I-TQG-Fe 3 O 4) and its cytotoxicity and *in vitro* affinity", *ACTA Pharmace. Sci.,* vol. 56, no. 2, .

[58] M. Ediz, U. Avcibaşi, P. Unak, F.Z.B. Muftuler, E.I. Medine, A.Y. Kilcar, H. Demiroglu, F.G. Gumuser, and S. Sakarya, "Investigation of therapeutic efficiency of bleomycin (blm) and bleomycin-glucuronide (blmg) labeled with radioactive (131)i on the cancer cell lines", *Cancer. Biother. Radiopharm.,* vol. 28, no. 4, pp. 310-319, 2013.
[http://dx.doi.org/10.1089/cbr.2012.1316] [PMID: 23350895]

[59] İ. İnce, Z.B. Müftüler, E.İ. Medine, Ö.K. Güldü, G. Takan, A. Ergönül, Y. Parlak, Y. Yildirim, B. Çakar, E.S. Bilgin, Ö. Aras, E. Göker, and P. Ünak, "Thymoquinone glucuronide conjugated magnetic nanoparticle for bimodal imaging and treatment of cancer as a novel theranostic platform", *Curr. Radiopharm.,* vol. 14, no. 1, pp. 23-36, 2021.
[http://dx.doi.org/10.2174/2211556009666200413085800] [PMID: 32282311]

[60] L. Zhu, X.J. Li, S. Kalimuthu, P. Gangadaran, H.W. Lee, J.M. Oh, S.H. Baek, S.Y. Jeong, S.W. Lee, J. Lee, and B.C. Ahn, "Natural killer cell (NK-92MI)-based therapy for pulmonary metastasis of anaplastic thyroid cancer in a nude mouse model", *Front. Immunol.,* vol. 8, p. 816, 2017.
[http://dx.doi.org/10.3389/fimmu.2017.00816] [PMID: 28785259]

[61] A. Kamkaew, L. Cheng, S. Goel, H.F. Valdovinos, T.E. Barnhart, Z. Liu, and W. Cai, "Cerenkov radiation induced photodynamic therapy using chlorin e6-loaded hollow mesoporous silica nanoparticles", *ACS Appl. Mater. Interfaces,* vol. 8, no. 40, pp. 26630-26637, 2016.
[http://dx.doi.org/10.1021/acsami.6b10255] [PMID: 27657487]

SUBJECT INDEX

A

Absorption, photon 86
Acetylcholinesterase 327
Acid 28, 41, 44, 65, 97, 133, 229, 252, 278, 281, 338, 390,
  arachidonic 28
  caproic 44
  hyaluronic 252
  hydrochloric 390
  hydrofluoric 338
  lactic 133, 281
  lactic-coglycolic 229
  polylactic 97, 278
  rosmarinic 41
  tannic 65
Activity 15, 135, 190, 193, 253
  anti-tumorigenic 253
  glycosyltransferase 190
  hydrolase 15
  signaling pathway 135
  transglycosylation 193
Acute myeloid leukemia (AML) 255
Adhesion 127, 192, 199, 257
  bacterial 199
Agents 33, 42, 68, 74, 204, 230, 249, 252, 275, 280, 379, 381
  antifungal 42
  antimicrobial 74
  antineoplastic 204
Alkaline phosphatase 390
Amperometric 326, 328, 329, 330
  dehydrogenase 326
  enzyme biosensor 329
  glucose biosensor 328, 330
  Immunosensor 326
  phenolic compounds biosensor 328
  polyphenol biosensor 329
  sarcosine biosensor 330
  xanthine biosensor 329
Amyloid precursor protein (APP) 129
Analysis 141, 336, 362, 363, 367

metabolomic 363
serum exosome 141
transcriptome 362, 367
transcriptomic 362
voltammetric 336
Anaplastic thyroid cancer (ATC) 412
Antibodies 68, 69, 97, 98, 170, 171, 198, 203, 226, 227, 228, 229, 248, 321, 364, 390, 391, 392, 393
  antiglycan 203
  enzyme-labeled 393
  immobilized 391, 392
  therapeutic 171
Antibody 202, 320, 321
  based biosensors 320, 321
  drug conjugates (ADCs) 202
Anticoagulant nanodevice 4
Antigens 5, 34, 47, 127, 175, 177, 392
  glycolipid 175
  immobilized 392
  producing cells (APCs) 5, 127, 177
  tumor-associated 34, 47
Antitumor activities 72
Apoptosis 5, 89, 90, 91, 92, 127, 193, 195, 246, 247, 248, 254, 255, 386, 387, 388, 389
  detection method 387
  drug-induced 247
  hypoxia-induced 248
  inducing factor (AIF) 90
Apoptotic pathways 70, 388
Applications 60, 61, 66, 74, 83, 84, 251, 296
  antibacterial resistance 74
  biomedical 60, 61, 66, 74
  biosensing 61
  oncological 296
  photocatalysis 251
  photodynamic 84
  phototherapeutic 83
Arthritis, rheumatoid 194
Assay methods, immunosorbent 334
Atherosclerosis 179, 180, 192

G

Gene(s) 69, 124, 133, 143, 144, 202, 225, 244, 246, 253, 259, 356, 360, 363, 407, 409
  efflux transporter 259
  expression 124, 225, 409
  tumor-promoting 144
  tumor supressor 202
Generic 205, 328
    sandwich-type biosensor 328
    disorders 205
Genomic sequence information 361
Genotoxicity 95, 389
  cisplatin-induced 95
Glucose 128, 174, 248, 255
    monomycolate 174
    transporters 128, 248, 255
Glycans glycobiology 167
Glycoconjugates 180, 199
  fucosylation 199
  glycosylated 180
Glycolysis 225, 248
  aerobic 248
  anaerobic 225
Glycoproteomics 206
Glycosaminoglycans 168, 175, 178, 180
Glycosidase inhibitors 204
Glycosidases 166, 183, 187, 191, 193, 205
Glycosides, cyanogenic 194
Glycosphingolipids 123, 168, 173, 175
Glycosylation 197, 204, 205
    inhibitors 204, 205
    pathways 197

H

Healing, diabetic ulcer 46
Heat shock proteins (HSPs) 122, 128, 129, 130
Heavy metal pollution 319
Hedgehog 246, 253
    pathways 246
    signaling inhibitor 253
Helicobacter pylori infections 199
Hematological neoplasms 204
Hematopoietic stem cells (HSCs) 241, 244
Hepatitis 10, 12, 42, 142
  chronic 142
High-density lipoprotein (HDL) 3

High intensity focussed ultrasound (HIFU) 232
High-pressure homogenization 37, 38, 40, 44
  methods 37, 40, 44
  technique 38
High pressure liquid chromatography (HPLC) 32, 33, 41, 364
Hollow mesoporous silica nanoparticles (HMSNs) 412
Human 42, 174, 198, 199
    immunodeficiency virus (HIV) 42, 174, 199
    influenza viruses 198
Huntington's disease 41
Hydrolysis 66, 193
Hydrophilic 35, 87, 88, 98, 120, 173, 252, 255, 281, 333
    photosensitizers 87, 98
    polymer 255

I

Imaging 34, 72, 232, 233, 234, 249, 403, 404, 405
    magnetic resonance 34, 72, 232, 233, 234, 249, 403, 404
    nuclear medicine 405
    photoacoustic 72
Immune 15, 96, 301
    activation 301
    system activation 15, 96
Immunosensors 316, 321, 326
Immunostimulatory adjuvants 34
Immunotherapeutic strategy 259
Immunotherapy 96, 251, 259, 260, 276
    dendritic cell 96
Inflammation, triggering 132
Influenza virus infection 199
Inhibitors 145, 192, 205, 230, 244, 246, 248, 252, 253, 361, 368
    glucose hydrolase 205
    glycolipid 205
    neuraminidase 205
    neurominidase 205
    proteasome 253
    tyrosine kinase 244

W

Western 395, 397
   blot method 395, 397
   blotting technique 395

Z

Zinc finger nucleases (ZFNs) 259